Mushrooms, Humans and Nature in a Changing World

Jesús Pérez-Moreno
Alexis Guerin-Laguette
Roberto Flores Arzú • Fu-Qiang Yu
Editors

Mushrooms, Humans and Nature in a Changing World

Perspectives from Ecological, Agricultural and Social Sciences

Springer

Editors
Jesús Pérez-Moreno
Microbiología, Edafología
Campus Montecillo
Colegio de Postgraduados Montecillo
Texcoco, Estado de México, México

Roberto Flores Arzú
Facultad de Ciencias Químicas y Farmacia
Universidad de San Carlos de Guatemala
Ciudad de Guatemala, Guatemala

Alexis Guerin-Laguette
Microbial Systems for Plant Protection
The New Zealand Institute for Plant & Food
Research Limited, Lincoln, New Zealand

Fu-Qiang Yu
Kunming Institute of Botany
Chinese Academy of Sciences
Kunming, China

ISBN 978-3-030-37377-1 ISBN 978-3-030-37378-8 (eBook)
https://doi.org/10.1007/978-3-030-37378-8

This Springer imprint is published by the registered company Springer Nature Switzerland AG
The registered company address is: Gewerbestrasse 11, 6330 Cham, Switzerland

Prologue

Japan is a country of forest: 68% of the land is forested, a proportion that has been almost constant for the past 200 years (Forestry Agency 2018). It is unusually high among developed countries, for example, 9 % in the UK, 24–30 % in France, Italy, and Spain, and 34 % in the USA (Fenton 2005). The Japanese forest area, ranging from boreal forests in Hokkaido Island to subtropical forests in Bonin and Okinawa Islands, holds a large biodiversity under various climatic (temperature, precipitation) and geological (continental plates, volcanoes) conditions. Local people especially in rural areas have spent their lives tightly connected with the forest for their natural resources. They utilize non-timber products, such as mushrooms, wild vegetables, edible nuts, charcoal, and Japanese lacquer for personal consumption and as a source of income, which contributed to develop rural communities, while accounting for approximately 50% of gross forestry output. Mushrooms occupied more than 80% of the value of non-timber products.

A mushroom is called "ki-no-ko" in Japanese, meaning "child of the tree": "ki" = tree, "no" = of, and "ko" = child because we have the perception that mushrooms grow under trees in forests. Conversely, mushrooms (fungi) are indispensable to trees because ectomycorrhizal species associate with tree roots and support the host nutrition from the soil. The mycorrhizal associations are especially important for the recovery of vegetation after natural disasters such as volcanic eruption, landslide, and tsunami. In addition, saprotrophic species that are not the main topic of this book, clean up the forest floor by decomposing litter and logs, and release nutrients (nitrogen, minerals) from them for tree growth. In this way, mushrooms extensively stimulate the forest productivity.

Many mushroom species have been harvested in the wild and used as food. Of these, matsutake (*Tricholoma matsutake*) was given a special position in Japan for over a millennium. Matsutake firstly appeared in the poem Man-yo-shu, approximately 700 AD, in which amazing matsutake appeared at the mountain ridge in the vicinity of Nara (capital city of ancient Japan) like umbrellas occupying a narrow ridge. This geographic area (central Japan) was originally dominated by broad-leaved Fagaceae trees. However, trees had been harvested repeatedly from the forests to construct buildings of the capital, and people went into forests and collected litter and understory bush for fuel wood. As a result, forest soil became poor but suitable for the growth of Japanese red pine (*Pinus densiflora*), a primary host tree of *T. matsutake*. In the early 1940s, about 12,000 tons of *T. matsutake* were harvested annually. But the harvest has since drastically decreased to less than 100 tons recently. Possible causes are that people changed lifestyle and moved to urban area; therefore, pine forest was no more managed adequately. Another explanation is that pine wilt disease also damaged host trees and reduced the production of *T. matsutake*. It seems that matsutake production is tightly related with lifestyle in Japan. Besides, many wild mushrooms (hiratake, shiitake, shoro, shimeji) have been collected even in the Nara area in ancient times.

In Japan, we have developed techniques for the cultivation of many edible mushrooms because, under natural environmental conditions, mushrooms do not occur constantly due to variable weather conditions and substrate distribution. A saprotrophic species, *Lentinula edodes* (shiitake), has traditionally been cultivated on logs of oaks and other broad-leaved trees, but is now often cultivated on bags containing a mixture of sawdust, grain powder, and other nutrients. The latter cultivation technique has been applied to other saprotrophic species such as *Pleurotus ostreatus* (oyster mushroom), *Pl. eryngii*, *Flammulina velutipes* (enoki-take), *Pholiota microspora* (nameko), *Hypsizygus marmoreus* (buna-shimeji), and *Grifola frondosa* (mai-take). Besides, some ectomycorrhizal species (*Lyophyllum shimeji*, *Boletus* sp., and *Hebeloma* spp.) successfully form fruiting bodies saprotrophically under controlled conditions (Ohta 1994, 1998; Ohta and Fujiwara 2003). Fruiting bodies of *Lactarius akahatsu*, *Rhizopogon roseolus*, *Cantharellus anzutake*, *Lyophyllum shimeji*, and *Tricholoma* spp. have been successfully produced in co-cultures with tree seedlings in laboratory conditions (Kawai 1997; Yamada et al. 2001, 2007, unpublished data). Additionally, *R. roseolus* and *Tr. bakamatsutake* formed fruiting bodies in field experiments (Shimomura and Ariyoshi 2012; Kawai et al. 2018). Development of mushroom cultivation techniques will provide new markets in local areas that differ in vegetation and climate. Ectomycorrhizal mushrooms such as *Tr. matsutake* in Japan and *Tuber magnatum* in Europe are difficult to cultivate. The conservation and management of natural populations of high value edible mycorrhizal fungal species, which are important forest resources as food or medicine for local markets, have therefore become urgent issues.

This book mainly describes recent topics on both basic and applied aspects of edible ectomycorrhizal fungi such as *Tuber* (truffles), *Lactarius* (milk-caps), *Tricholoma matsutake* (Matsutake), and many other species. Taxonomy, ecology, and cultivation but also cultural and social aspects are included from several countries all around the world. Researchers in this field have shared their results and passion at the International Workshop on Edible Mycorrhizal Mushrooms (IWEMM) held every 2–3 years since 1998. In some way, this book is the result of the efforts of all the participants in these meetings carried out during the last two decades, where they have shared scientific experiences, analyzed dynamic regional and national scenarios and inspired novel ideas related with multi-focus approaches of the mycological resources around the globe. The seminal idea for this book was developed at the IWEMM9 in Texcoco, Mexico, in July 2017. As the organizers of the IWEMM10 in Suwa, Nagano, Japan, from 21 to 25 October 2019, it is our great pleasure to present this book which we hope will further inspire young generations of researchers interested in the knowledge, protection, and sustainable use of wild edible fungi.

References

Fenton RT (2005) Japanese forestry and its implications. Marshall Cavendish Academic, Singapore
Forestry Agency (2018) Annual report on forest and forestry in Japan, Fiscal year 2017.
Kawai M (1997) Artificial ectomycorrhiza formation on roots of air-layered Pinus densiflora saplings by inoculation with Lyophyllum shimeji. Mycologia 89:228–232
Kawai M et al (2018) Fruit body production in Tricholoma bakamatsutake after field inoculation. Nara Prefectural Research Center
Ohta A (1994) Production of fruit-bodies of a mycorrhizal fungus, Lyophyllum shimeji, in pure culture. Mycoscience 35:147–151
Ohta A (1998) Fruit-body production of two ectomycorrhizal fungi in the genus Hebeloma in pure culture. Mycoscience 39:15–19
Ohta A, Fujiwara N (2003) Fruit-body production of an ectomycorrhizal fungus in genus Boletus in pure culture. Mycoscience 44:295–300
Shimomura N, Ariyoshi K (2012) Production technique of fruiting bodies by inoculation of the mycelium of Rhizopogon rubesens. Japan Patent P2012-80811A
Yamada A, Ogura T, Ohmasa M (2001) Cultivation of mushrooms of edible ectomycorrhizal fungi associated with Pinus densiflora by in vitro mycorrhizal synthesis. I. Primordium and basidiocarp formation in open-pot culture. Mycorrhiza 11:59–66
Yamada A, Kobayashi H, Ogura T, Fukada M (2007) Sustainable fruit-body formation of edible mycorrhizal Tricholoma species for 3 years in open pot culture with pine seedling hosts Mycoscience 48:104–108

Takashi Yamanaka
yamanaka@affrc.go.jp
Forest Research and Management Organization
Tsukuba, Ibaraki, Japan

Akiyoshi Yamada
akiyosh@shinshu-u.ac.jp
Shinshu University, Minamiminowa, Nagano, Japan

Hitoshi Furukawa
furukawa-hitoshi@pref.nagano.lg.jp
Nagano Prefecture Forestry Research Center
Shiojiri, Nagano, Japan

Jesús Pérez-Moreno
jperezm@colpos.mx
Colegio de Postgraduados, Campus Montecillo, Texcoco
Estado de México, Mexico

Alexis Guerin-Laguette
alexis.guerin@plantandfood.co.nz
The New Zealand Institute for Plant and Food
Research Limited, Lincoln, New Zealand

Contents

About the Authors

Jesús Pérez-Moreno is a Professor at Colegio de Postgraduados, Mexico, with a PhD from the University of Sheffield, England. He has delivered lectures in Canada, China, Colombia, Ecuador, England, France, Guatemala, Japan, Mexico, Morocco, New Zealand, Spain, Thailand, and Sri Lanka; his articles have been cited 2900 times by researchers from 55 countries. Author of 15 books and 23 book chapters; supervisor of 50 MSc and PhDs and invited Professor from 2011 to 2012 at the University of Sheffield and from 2018 to 2019 at the Kunming Institute of Botany, Chinese Academy of Sciences (CAS). During 30 years, he has studied the relationship between mushrooms and humans focusing on sustainable development, poverty alleviation, food security, and mitigation of climate change.

Alexis Guerin-Laguette is a Senior Scientist at The New Zealand Institute for Plant & Food Research Ltd. focusing on edible mycorrhizal fungi (EMF). He has worked in France, Japan, and Canada. His passion for EMF stems from his PhD on *Lactarius deliciosus* and postdoc on *Tricholoma matsutake*. In New Zealand, he extended his activities to truffles. He is the Technical Representative for the NZ Truffle Association, and since 2015 he has collaborated with the Kunming Institute of Botany, CAS. He has authored over 20 peer-reviewed articles and book chapters. He is the Secretary of the International Workshop on Edible Mycorrhizal Mushrooms.

Roberto Flores Arzú is a Professor at the Universidad de San Carlos, Guatemala, with a PhD from the University of Murcia, Spain. He has been a visiting scholar at IRTA in Barcelona; Universities of Tlaxcala, Mexico; Bologna, Pavia, and Cagliari in Italy. He has published papers on the macrofungi of Guatemala and the description and synthesis of ectomycorrhizae of neotropical trees. His current research, in collaboration with scientists from China, Italy, Mexico, and Japan, focuses on macrofungal taxonomy, ectomycorrhizae, and ethnomycology.

Fu-Qiang Yu is an Associate Professor and Director of the General Office at the Kunming Institute of Botany, CAS. He was a visiting scholar at the Department of Botany, Chicago. His research focuses on the diversity, management, and cultivation of hypogeous fungi and highly prized ectomycorrhizal fungi, and he leads a group hosting scholars from New Zealand, Japan, the United States, Mexico, and Spain. He has published over 20 papers and authored 4 patents on the cultivation and management of high-prized mushrooms.

Part I
Introduction

Chapter 1
Setting the Scene

Jesús Pérez-Moreno, Alexis Guerin-Laguette, Roberto Flores Arzú, Fu-Qiang Yu, and Annemieke Verbeken

1.1 Believe It or Not

For many people, the word 'mushroom' is immediately associated with the cultivated button mushroom, known technically as *Agaricus bisporus*. However, mushrooms are of course much more diverse than this single species. Just to give an example, if we type the word 'mushroom' in the Google search bar, 289 million entries will appear. According to the Encyclopaedia Britannica (2019) mushrooms are defined as 'the conspicuous umbrella-shaped fruiting body (sporophore) of certain fungi, typically of the order Agaricales in the phylum Basidiomycota but also of some other groups'. Mushrooms are indeed the spore forming structures of a very large group of living organisms which form a kingdom of their own, the fungi. They constitute an important structural and functional component of the earth ecosystems. The world would not be able to exist as we know it, without fungi. Saprotrophic fungi are responsible for breaking down dead plants, animals, and microbes and recycling the nutrients contained in them, hence bringing them back in the cycle of life. We can say that fungi are the greatest disassemblers of complex compounds in

J. Pérez-Moreno (✉)
Microbiología, Edafología, Colegio de Postgraduados, Campus Montecillo, Texcoco, Mexico
e-mail: jperezm@colpos.mx

A. Guerin-Laguette
The New Zealand Institute for Plant and Food Research Limited, Christchurch, New Zealand

R. Flores Arzú
Departamento de Microbiología, Facultad de CCQQ y Farmacia, Universidad de San Carlos de Guatemala, Guatemala, Guatemala

F.-Q. Yu
Kunming Institute of Botany, Chinese Academy of Sciences, Kunming, China

A. Verbeken
Department of Biology, Research Group Mycology, Ghent University, Ghent, Belgium

J. Pérez-Moreno et al. (eds.), *Mushrooms, Humans and Nature in a Changing World*, https://doi.org/10.1007/978-3-030-37378-8_1

nature. Unlike plants or animals, the body of an individual fungus is hidden underground or in the substrate and is not composed of cells but of networks of filaments (called hyphae individually and mycelium as a whole).

Since their appearance in evolution history, fungi have played a key role in the structure and functioning of natural ecosystems. Fungi not only have adapted themselves to colonize all of the earth's environments but also have contributed to shape them as we know them currently. Let's present just two iconic examples of how these dynamic processes involving fungi have been at play. First, there was a time in the geological history of the earth when life was at risk due to the accumulation of plant debris made up of lignocellulose polymeric compounds and when swamp forests locked up the global atmospheric CO_2. The accumulation of such substrates was enormous and there was no natural process to break them down efficiently. A genetic mutation in a group of fungi (Basidiomycota) empowered them to produce the proper group of enzymes, known as ligninolytic peroxidases, allowing the degradation of polymeric compounds and solving once for good this fundamental problem for life. Since then, a particular group of saprotrophic fungi, particularly belonging to Basidiomycota and mostly mushroom-forming, are responsible for nutrient cycling by degrading very complex polymers such as lignin and celluloses. Those that break down lignin first are the so-called white rotters; those that start degrading the wood by breaking down the celluloses are the brown rotters. In order to elucidate which group of Basidiomycota was involved and to determine the geological period during which this mutation happened, a number of studies have been carried out. Mushroom genomes have been of paramount importance to study the enzyme groups that have the ability to carry out white and brown rots. Indeed, genomes provided an understanding of the mushrooms evolutionary timeline and more accurate timestamps for when Basidiomycota groups might have developed the ability to break down all woody components in plant cell walls. First, it was discovered that the evolution of white rot fungi around 300 million years ago coincided with the end of the coal forming Carboniferous period, impacting the global carbon cycle (Floudas et al. 2012). Later on, it was demonstrated that additional to peroxidases, other enzymes, including those that attack crystalline celluloses, also contributed to the decomposition of polymeric compounds of woody plant cell walls, widening the understanding of the origins of lignin-degrading fungal enzymes (Nagy et al. 2016). More recently, it has been shown that a powerful evolutionary convergence process also existed, and that peroxidases of Polyporales (an important group of wood-decaying Basidiomycota often forming tough and striking mushrooms—also called bracket fungi) acquired the ability to degrade non-phenolic lignin using a tryptophanyl radical interacting with the bulky polymer at the surface of the enzyme (Ayuso-Fernández et al. 2018). It is currently accepted that Basidiomycota mushrooms were the pioneer lignocellulosic decomposers through different evolutionary mechanisms during the Paleozoic era. These mechanisms were keystone evolutionary processes, which modified the whole earth ecosystem structure and function, contributing to shape the world as we currently know it.

A second iconic example is related to the influence of fungi as rain-makers, a fascinating illustration of their influence on life on earth. We know that rain stimulates mushroom growth, however, it has also been discovered that on the other way around, mushroom spores (including species of *Lactarius*, *Suillus*, *Russula* and *Lycoperdon*) act as condensation nuclei, which means that they have surfaces on which water vapour condenses in order to form big water droplets which eventually will produce rain (Hassett et al. 2015). The mechanism starts when the spores secrete mannitol and other water-loving sugars that trigger the formation of a tiny ball of water around individual spores. Interestingly, these carbohydrates are present during the spore discharge but they evaporate once the spore is airborne. Thus, the liquid water modifies the fluid motion and results in a rapid displacement of the spores' centre of mass, imparting momentum and launching it skyward at up to 6.5 km/h (Stolze-Rybczynski et al. 2009). Due to the large number of spores released by mushrooms, this process has a global influence. A single mushroom can release around 30,000 spores every second, and then billions can be produced per day (Money 2011). It has been estimated that globally 50 million tons of mushroom spores are dispersed into the atmosphere each year (Elbert et al. 2007). The global influence of this phenomenon is simply impressive because it implies that those tiny spores produced by mushrooms, influence rain formation and therefore have driven climate patterns around the globe, from rainforests to boreal forests since long ago.

Besides the saprotrophs, another important trophic group of fungi, is composed by those who damage plants, animals, or other microorganisms and live as parasites. It would be surprising to know that in terms of size, the largest organism on earth is a mycelial mat (or body) of the opportunistic parasite edible mushroom *Armillaria ostoyae*, known as Humongous fungus, which covers 965 ha, with a maximum length of 3.8 km and an estimated age of 8500 years. This giant genet (the equivalent designation of individuals for fungi) is located in the Blue Mountains of Oregon, the United States. This fungal genet has a huge mycelium that permeates below the forest floor and parasitizes hundreds of trees. To confirm the size of this organism, molecular analyses called restriction fragment length polymorphism (RFLP) and rapid amplified polymorphic DNA (RAPD) were used. The mushrooms formed by this mycelium appear above ground about once a year, around the base of infected or newly killed trees. Theoretically, this organism may continue growing indefinitely (Ferguson et al. 2003). Paradoxically, this organism grew undetected during millennia until the beginning of the twenty-first century. In other species, it is not as much the mycelium, but rather the mushrooms that can be very big and heavy. The largest and heaviest mushroom is a perennial giant polypore, *Phellinus ellipsoideus*, found in Hainan Island in southern China, where it lives as a saprotrophic species. When it was discovered in 2012, it was 20 years old with an estimated volume of 500,000 cm^3, a length of 10 m, a weight of half a ton and it had around 452 million pores, which might produce 1 trillion spores per day (Dai and Cui 2011). Interestingly, anti-tumor bioactive compounds against liver cancer have been isolated from this mushroom (Zan et al. 2012).

The third trophic group of fungi is probably the most fascinating from both a structural and functional perspectives in natural ecosystems. This group also forms the mushrooms that are the starring players in this book: edible mycorrhizal fungi. The mycorrhizal symbiosis is established between the roots of 90% of plants and thousands of fungal species. This symbiosis played a key role in the colonization of land by plants. The oldest fossil records have been found during the Ordovician and Devonian periods, 460 and 407 million years ago, respectively (Remy et al. 1994; Redecker et al. 2000). The colonization of the earth by plants, and the subsequent development of life on the terrestrial surface, was driven by this symbiosis (Pirozynski and Dalpé 1989; Taylor and Osborn 1996; Brundrett 2002). In the mycorrhizal symbiosis, plants and fungi are in close contact in order to exchange nutrients. Around 10–20% of the carbohydrates photosynthesized by the plants are transferred to the mycorrhizal symbionts, and in return, the fungi transfer macro- and micronutrients and water to their associated host plants, conferring them resistance to stress originated by pathogens, potentially toxic elements, and drought (Smith and Read 2008). Particularly one type of mycorrhiza, called ectomycorrhiza, whose origin dates back to the Cretaceous period is currently established between around 8500 gymnosperm and angiosperm plant species (Brundrett and Tedersoo 2018) and more than 20,000 mushroom species (Comandini et al. 2012). It is possible to state that without this symbiosis, there would be no forests, particularly in boreal, temperate, and some subtropical and tropical areas. Very recently, by analysing over 1.1 million forest plots globally distributed, Steidinger et al. (2019) estimated that ectomycorrhizal trees constitute approximately 60% of tree stems on earth. This symbiosis dominates forests where seasonally cold and dry climates inhibit decomposition, located at high latitudes and altitudes. The authors of this research named this effect 'the Read's Rule' after the British pioneer scientist working on symbiosis, Sir David Read from the University of Sheffield, who first predicted this distribution almost three decades ago (Read 1991). It has been demonstrated that ectomycorrhizal fungi tend to increase the amount of C stored in soil (Averill and Hawkes 2016) and that these types of mushrooms have the ability to modify their local environment to further reduce decomposition rates by mineralizing forest organic compounds (Read and Pérez-Moreno 2003). Through this global model, Steidinger et al. (2019) have predicted massive changes in the symbiotic state of the world's forests that will be linked to strong modifications of the global climate if C emissions continue being unabated by 2070. Another fascinating discovery is that through the connection of different plants and mycelia, very complex networks connect trees in nature (Pérez-Moreno and Read 2004). Through these networks, nutrients, water, and signal compounds are transferred between trees. For this reason, these mycelial networks have been called the 'wood wide web'.

The fruiting bodies of some ectomycorrhizal fungi are edible, and this specific sub-group is the main subject of this book: edible ectomycorrhizal mushrooms, or EEMs. The international commerce of EEMs is worth billions of US dollars annually (Yun and Hall 2004). Additionally, hundreds of bioactive compounds with analgesic, anti-allergic, anti-carcinogenic, anti-bacterial, anti-coagulant, anti-fungal, anti-hypertensive, anti-inflammatory, anti-nociceptive, anti-oxidant, anti-pyretic, anti-venom, anti-viral (including anti-HIV), cholesterol-lowering, hepatoprotective, and with immune enhancement properties have been isolated from more than 100 species of EEMs (Pérez-Moreno and Martínez-Reyes 2014). However, the study of this important group of mushrooms is still in its infancy. Vast mycological diversities of ectomycorrhizal mushrooms around the world have been poorly studied or not studied at all. The inspiration of this book is to encourage studying this important group of mushrooms from the ecological, social, cultural, and economic points of view.

1.2 Early Interactions

In addition to their ecophysiological relevance, mushrooms have played an important role in the development of human civilizations as we know them currently. In different cultures around the world, they have been ancestrally used as food, flavouring, medicine, and sacred elements. Deadly or poisonous species often represent death or disease. More recently, it has been shown that they are the source of bioactive compounds with applications in industrial processes related with the enzyme production of stains, cosmetics, biopesticides, and bioactive compounds with a wide range of uses. The most ancient evidence of mushroom consumption by humans dates back to the Stone Age. An international team led by Robert Power of the Max Planck Institute for Evolutionary Anthropology in Leipzig, Germany, discovered that in the Upper Palaeolithic, humans already ate a variety of plants and mushrooms. They identified spores of Agaricales and Boletales (specifically boletes) in a tooth plaque of an old woman dated 18,700 years old by radiocarbon techniques, found at El Miron cave, Cantabria, in northern Spain (Power et al. 2015). Additionally, the ancestral uses of *Amanita muscaria* (commonly known as the fly agaric) in Siberia, and of *Psilocybe* spp. (known as *teonanácatl* in Aztec language meaning the God's flesh) in sacred ceremonies in Mexico have been subject to a vast body of literature (e.g. Wasson and Wasson 1957; Wasson 1980; Riedlinger 1990). *A. muscaria* was widely used as an entheogen by many of the indigenous peoples of Siberia (Saar 1991b). Ostyak and Vogul tribes in western Siberia and Kamchadal, Koryak, Chukchi tribes in eastern Siberia still use *A. muscaria* for shamanistic rites. The isoxazoles, especially muscimol, pass through the human renal system almost intact, and thus it is possible to ingest them by drinking human urine; this tradition has been reported several times from northern Siberia. Ibotenic acid and muscimol are the active components contained by this mushroom conferring its psychotropic effects. The biochemistry and toxicology associated with the consumption of *A. muscaria* have been extensively reviewed elsewhere (e.g. Michelot and Melendez-Howell 2003). Since ancient times, the entheogenic species of *Psilocybe* have been widely used among native cultures from Mesoamerica. Before the arrival of the Spaniards in the sixteenth century, its use was recorded in various codices and colonial writings (Hernández-Santiago et al. 2017). However, there were no scientific records related to their usage, identity, and biochemistry until the middle of the twentieth century. In 1955, Robert Gordon Wasson and Valentina Wasson became the first Westerners to consume these sacred mushrooms in a ritual ceremony under the guidance of the Mazatec shaman Maria Sabina (Heim and Wasson 1958; Wasson and Wasson 1957; Wasson et al. 1974). They published their discovery in a historic article in the *American Journal Life* in 1957 (Wasson 1957). Later on, the French mycologist Roger Heim was the first to identify the species used in the rituals as *Psilocybe,* including *P. mexicana*, *P. caerulescens*, and *P. zapotecorum*, and the Swiss chemist Albert Hofmann subsequently isolated an indol that was named psilocybin (Hofmann et al. 1958). Currently, these mushrooms are still used in sacred ceremonies by Chatin, Matlatzinc, Mazatec, Nahuatl, Totonac, and Zapotec people in southwestern Mexico (Guzmán 2008).

Mushrooms consumption has been confirmed also during the European Chalcolithic. Ötzi the Iceman, who lived 5300 years ago and was discovered in 1991 at the border of Italy and Austria, carried two types of mushrooms. The first one has been identified as *Fomitopsis betulina* which has strong medicinal properties including anti-bacterial, anti-parasitic, anti-viral, anti-inflammatory, anticancer, neuroprotective, and immuno-modulating activities (Pleszczyńska et al. 2017). The second mushroom was identified as *Fomes fomentarius* which is a very well-known polypore used as a primary tinder for making fire, during millennia because basidiomata bearing traces of human handling have often been found in archaeological sites as ancient as 11,555 year old (Peintner and Pöder 2000). More than 2000 years ago in the book XVI, Chap. 77 entitled *Methods of obtaining fire from wood*, of his *Naturalia History* Pliny mentioned 'Territur ergo lignum liqno, ignem quc concipit attritu, excipiente materia aridi fomitis, fungi, vel foliorum' that can be translated as 'one piece of wood is rubbed against another, and the friction sets them on fire, which is augmented by dry tinder (aridi fomitis), especially by that of fungi and leaves' (Bostock and Riley 1855). Buller (1914) considered that the main tinder referred to by Pliny is the *amadou*, which is made from fruiting bodies of *Fomes fomentarius*. Additionally, there is evidence that this fungus has been used as medicine, due to its haemostatic and anaesthetic properties (Saar 1991a) and antitumour effects on human lung carcinoma cells (Kim et al. 2015). Anecdotally, Saar (1991a) described the ceremonial use of this mushroom among Khanty people from West Siberia, who burnt the basidiomata to produce smoke, when a person passed away until the corpse had been taken out of the house. Both uses, setting fire and healing, were fundamental in the human survival during Palaeolithic times. A clear notion of the association between oaks and edible wild mushrooms appears as early as two millennia ago in the book XVI, at the end of Chap. 11, of Pliny in his classic *Naturalia History*. He quoted 'Such is the multiplicity of the products borne by the robur in addition to its acorns; and not only these, but mushrooms as well, of better or worse quality, the most recent stimulants that have been discovered for the appetite; these last are found about its roots. Those of the quercus are the most highly esteemed'. In his interpretation of the English translation, Bostock and Riley (1855) added a footnote associated with the word mushroom by Pliny ' these were the boletus and the suillus, the last of which seem to have been recently introduced at table in the time of Pliny'. Historically, these records show the paramount importance of such fine observations made by Pliny, more than 2000 years ago, when his original publication was published between 77 and 79 A.D. Much more recently, groups of hunter-gatherers have traditionally used mushrooms as food, medicine, and flavourings in other parts of the world, using simple techniques such as dehydration by sun, wind, or heating over a fire, in order to prolong their shelf-life. We refer here to some native North American cultures (Kuhnlein and Turner 1991), who use species of *Agaricus*, *Calvatia*, *Cantharellus*, *Inonotus Lycoperdon*, *Polyporus*, and *Trichoderma* and to other ethnic groups from different parts of the world that are analysed in this book.

1.3 The Global Interwoven Web Between Mushrooms and Humans

The relationships between fungi, humans, and nature at a global level constitute a complex mosaic due to the enormous diversity of environmental, economic, social, and cultural conditions existing in the world. These have been studied with different degrees of depth in the five continents. However, it is important to recognize that there are areas where scientific knowledge of these relationships is practically nil, others in which they have received little attention and, even in those regions where more studies have been conducted, these are far from being complete. These relationships are also extremely dynamic, especially nowadays under the dramatic environmental, technological, socio-cultural, and global changes that we live. In this introductory chapter, we present some contrasting examples from various areas of the world, where scientific studies with a broad-scope have been developed, studying different aspects of wild edible mushrooms, with emphasis on ectomycorrhizal species. It is not our goal to present a complete picture of these relationships, which would be impossible, but rather to highlight and summarize the situations in various regions distributed across the five continents.

In the American continent, the relationships among mushrooms, humans, and nature can be divided into three different scenarios, those in: (1) Canada and the United States; (2) Mesoamerica, including mainly Mexico and Guatemala; and (3) South America. There are contrasting socioeconomical, historical, cultural, and biological conditions in these areas. In the case of Canada and the United States, there has traditionally been a strong scientific knowledge of the mycological diversity, only comparable with that existing in European countries. Numerous studies on the ecology, taxonomy, and use of wild edible mushrooms have historically been produced (including, for example, Molina et al. 1993; Pilz et al. 2003, 2007; Kuo et al. 2012). The commercialization of wild edible mushrooms fresh or processed, dried, powdered, or preserved in different ways, is common, and numerous companies are devoted to the marketing of this non-timber forest resource (Fig. 1.1). In contrast to the enormous mycological diversity, in general, the plant diversity in these areas is relatively small, with most of the trees being ectomycorrhizal. Cultural diversity and the associated traditional knowledge from the native human cultures have basically disappeared, and there are only few remnants of such ancient traditional knowledge. By the opposite in Mesoamerica, there is a great cultural diversity. For example, Mexico has 68 ethnic groups, each with its own culture, language,

Fig. 1.1 Diversity and commercialization of edible mushrooms in the United States of America. (**a**) Store in San Francisco selling fresh, dried, or processed edible mushrooms including ectomycorrhizal species such as chanterelle, violette chanterelle, porcini, and truffles; (**b**) Sale of porcini (*Boletus edulis* s.l.) and other edible mushrooms in San Francisco; (**c**) Wild American matsutake, *Tricholoma magnivelare*, known from eastern United States of America; (**d**) Packed wild chanterelles ready to be sold in San Francisco

world-wide vision, and natural resources management. Additionally, the biological diversity is huge (Pérez-Moreno et al. 2010), for example the country holds some of the largest diversities of ectomycorrhizal plant genera including 72 taxa of pines and 168 species of oaks. As a consequence of these cultural and biological diversities, Mexico holds one of the largest pools of wild edible mushrooms. It has been estimated that more than 450 species are currently consumed in the country (Fig. 1.2). However, this might be an underestimation due to the relatively reduced number of taxonomists who are currently active in the country. Therefore, the number of wild edible mushrooms is expected to be much larger. This huge traditional knowledge faces enormous challenges including high deforestation rates, rapid cultural erosion, emigration of young people to cities, and acculturation processes. In contrast with these realities, particularly in the last decade, there has been an awakening of the revalorization of the mycological resource. In order to preserve the forests, dozens of mushrooms fairs are carried out every year across the country organized by local ethnic groups. Mycotouristic activities have also started in different ethnic groups mainly in Central and Southern Mexico, and there have been numerous ethnomycological studies in groups which were historically obliterated in the past, for example, the Mixtec (Hernández-Santiago et al. 2016, 2017), Chinantec (López-García et al. 2017), and Tzotzil (Ruan-Soto 2018) people. Some of the Mexican groups are highly mycophilic, for example the Tlahuica people, to whom the first author of this chapter and his research team have been studying during the last 7 years, they are constituted by less than 500 persons who are able to recognize and consume more than 160 wild edible mushrooms (Fig. 1.2a). Some species such as chanterelles (Fig. 1.2b), matsutake, Caesar's mushrooms (Fig. 1.2d), morels, and porcini are exported to the United States, Canada, or Europe from Mexico. The country is one of the most important genetic reservoirs of wild edible mushrooms in the world (Fig. 1.2c).

The Central American region is one of the most diverse on the planet, particularly in terms of plants and mushrooms. Many species are related to those from the Northern and Southern hemisphere parts of the continent because this region constitutes an isthmus which connects both regions. Macrofungal diversity from Guatemala (Fig. 1.3a–c) and Costa Rica are one of the most studied in the area, where over 500 species have been recorded (Flores et al. 2012; Mueller et al. 2006).

However, the diversity is much higher, as many areas have remained unexplored, and recent molecular studies have confirmed high local endemism. For example, Del Olmo-Ruiz et al. (2017) found that fungal species richness was very high in Neotropical montane cloud forests in Mesoamerica and concluded that fungi from this region and from the Caribbean and South America are taxonomically different, with a little overlap of species recorded in these regions. It is foreseeable that Honduras, Belize, El Salvador, Nicaragua, and Panama will provide new mushroom species in the future, because large areas in these countries have remained unexplored. It is also necessary to promote studies in these regions with the support of more mycologists and with internationally funded projects that could provide knowledge and technology not available from local governments and universities. Wild edible mushrooms, especially the ectomycorrhizal ones, constitute an

Fig. 1.2 Mexican mycophily. (**a**) Tlahuica people, who are able to distinguish and consume more than 160 wild edible mushrooms; (**b**) *Cantharellus cibarius* s.l.; (**c**) Diversity of edible and sacred wild mushrooms; (**d**) *Amanita jacksonii*, one of the most preferred mushrooms among the 450 edible species known in the country

Fig. 1.3 Wild edible mushrooms from Central and South America. (**a**) Commercialization of edible mycorrhizal mushrooms (*Lactarius*, *Cortinarius*, *Ramaria*, and *Hydnum*) at San Martín Jilotepeque

important element in the diet of many rural populations in Central America. Mayan populations from Chiapas, in southern Mexico, and Guatemala are well known for their ancestral consumption of ectomycorrhizal mushrooms. In Chap. 4, a review of edible mushrooms, their habitats, and interesting ethnomycological aspects in Guatemala is presented. The country, despite its small size compared with bigger countries such as Mexico, surprises with the consumption of around 100 mushroom species by the local ethnic groups. In Honduras, the Lenca, Chortí, and mestizo indigenous populations, particularly those from the occidental zone, make use of at least 22 different species of mushroom including *Amanita jacksonii*, *A. rubescens* s.l., *Boletus pinophilus* s.l., *Cantharellus cibarius* s.l., *Lactarius indigo*, and other species in *L.* sect. *Deliciosi*, and, *Hydnum repandum*; all of these coming from pine-oak forests and locally known as *choros*, *juanillas*, *chequecas*, or *canturinas* (Vega 2018; Sarmiento and Fontecha 2015; Marineros et al. 2015). In El Salvador, despite its scarce native forests, the most known and demanded mushroom is the *tenquique* (*Pseudofistulina radicata*), but the local consumption of species belonging to *Cantharellus*, *Amanita,* and *Lactarius* is also important. The highest fungal diversity of the country can be found in the montane cloud forests from the volcanic chain, where more than 100 wild species have been identified (Aguilera 2016), including new species of *Laccaria* and *Boletales* (Delgado 2010; Toledo 2013). Nicaragua has been scarcely studied but the country should hold unique wild species in pine-oak forests, the stands of *Pinus caribaea* and its extensive savannahs (Stevens et al. 2001). Despite the fact that mainly saprotrophic species have been recorded to date, the potential existence of edible ectomycorrhizal species is almost certain. Costa Rica and Panama are one of the most important megadiverse countries of the planet. There have been at least four books that show the richness of the fungal diversity from Costa Rica, including a surprising number of species associated with local oaks that resemble those from the Northeastern region of the continent (Halling and Mueller 2002). Nevertheless, and despite the high diversity consumption of wild edible mushrooms is low, which might be explained by the small indigenous population compared to those existing in the rest of the region. Panamá, which possesses one of the most important neotropical forests of the continent, also has some well-known edible mycorrhizal species such as *Lactarius indigo* and *Laccaria* spp.; however, local consumption of mushrooms is also low and in general the country has been considered as mycophobic (Vega and de León 2018). Central America is quite a singular territory on the planet that deserves to be studied at a deeper level. The influence of its tectonic-volcanic origin, its location among the oceans, the rich native biodiversity, as well as the different ancestral cultures, including the Maya culture, make this region an almost unopened treasure

Fig. 1.3 (continued) market in Guatemala; (**b**) Mari Flor Gómez Chalí, student of Universidad de San Carlos with fresh *Lactarius* at San Juan Sacatepéquez market in Guatemala; (**c**) Gabriel González collecting the edible saprotrophic mushroom *Pseudofistulina radicata* near Antigua, Guatemala; (**d**) *Suillus luteus* known as *callampas* growing in plantations of *Pinus radiata* Biobío, Chile; (**e**) Trade of *Cyttaria* sp., known as *digüeñe* o *dihueñe* which grows in *Nothofagus* forests in Biobío, Chile; (**f**) *Rhizopogon* sp. known as *papitas* (little potatoes), an ectomycorrhizal mushroom in exotic plantations of *Pinus radiata* in Los Ángeles, Chile that offers knowledge and answers to current discussion regarding conservation and sustainable use of natural resources. Local wild edible mushrooms are not only the result of their inherent capability to adapt and evolve but they can also be a source of food and income. They are an example of appreciation and respect from humans towards nature and an important source of secondary metabolites useful in the green and medicinal industries. The situation in Central America can be summarized in the statement pointed out by Piepenbring et al. (2011) for the mushrooms of tropical Panama 'highly diverse, mostly unknown, and further mycological field work is urgently needed because habitats are being destroyed and fungi specific to them are lost forever'. Meantime, the checklist of fungi in Panama contains more than 2700 species (Piepenbring 2013).

In South America, despite the enormous diversity and the presence of ethnic groups, the diversity of wild edible mushrooms remains poorly explored in most of the countries due to the complex political and social situations. However, some studies are being developed in some south-American countries. For example, in Chile, the consumption of *Suillus* species, called 'callampa' constitutes an important income for the owner of pine plantations, due to their high productivity. These edible ectomycorrhizal mushrooms can be frequently found in domestic stores packed by a number of companies usually dried and sliced (Fig. 1.3d). Also, other ectomycorrhizal mushrooms from *Rhizopogon*, known as *papitas* (little potatoes) are consumed in young stages (Fig. 1.3f). One of the most characteristic species is *Cyttaria* spp. coming from *Nothofagus* forests, which is sold in domestic markets (Fig. 1.3e). Two chapters in this book deal with the mushroom diversity in Colombia and Argentina.

The situation in Europe is highly diverse. There are mycophilic and mycophobic countries; the mycobiotas of a number of countries have historically been studied while others (mainly in eastern Europe) have received less attention; the strongest and oldest evidences of mushrooms used by humans have been recorded in this continent; and the commercialization of a wide variety of edible, medicinal, and nutraceutical mushrooms ranges from fresh to entirely processed products (Fig. 1.4). As several chapters of this book (Chaps. 7–9, 12, and 13) deal with detailed analyses of the situation in this continent, they will not be discussed further here.

Africa is the world's second-largest continent with more than 30 million km^2 covering approximately 20% of the Earth's land. From the mycological point of view, it continues being a mysterious continent full of surprises and largely unexplored. Northern Africa, including the Atlas Mountains with abundant *Cedrus atlantica*, *Quercus*, and *Pinus* forests is extremely rich in wild edible mushrooms including black truffles, chanterelles, morels, black trumpets, and species of *Ramaria*. In arid and subarid regions, proliferates one of the most attractive complexes of fungal species: the desert truffles, included mainly in the genera *Terfezia* and *Tirmania*. These truffles establish ectomycorrhizal associations with shrubs in the family Cistaceae and also with oak and pine trees. They constitute an important source of food for marginal groups including Berbers or Amazighs in Morocco, Algeria, Tunisia, and Libya. The technique used, mainly by women, to gather these truffles is heart-breaking: the women hit the soil with a stick, until they detect, based

Fig. 1.4 Aspects of useful wild mushrooms in Europe. (**a**) Truffle hunting in a Southern Spain forest; (**b**) *Fomitopsis betulina* growing on birch in Bolzano, Northern Italy (this is the mushroom that Otzi the iceman was carrying 5300 years ago); (**c**) *Laetiporus sulphureus*, the chicken of the woods, a highly esteemed edible wild mushroom around the world, growing on beech in a North Yorkshire forest in central England; (**d**) Fresh Porcini mushrooms (*Boletus* spp.) known as 'the king of the mushrooms' in Southern Ukraine; (**e**) Commercialization of processed products mostly including black truffles, porcini, and morels in Cahors, France

on their enormous experience, a different sound emitted by the different texture of the soil where the ascomata of the desert truffles grow. Then, they remove the surface soil to collect the ascomata. Gathering 1 kg of truffles implies to be crouched down for hours daily (Fig. 1.5a) in order to get food or currency to survive (Fig. 1.5b). In the scenario of sub-Saharan Africa, there are some similitudes and differences compared with that of northern Africa. Infinitely rich in ecotypes, biodiversity, natural resources, and traditions, the sub-Saharan Africa is also a treasure box full of

Fig. 1.5 People and edible mushrooms in Africa. (**a**) Berber woman collecting desert truffles, hitting the soil with a stick in a traditional way in Morocco; (**b**) Tajine of dessert truffles *Terfezia* spp. in Morocco; (**c**) Dr. N'golo Kole taking a bite of *Termitomyces shimperi* in Togo; (**d**) One of the rare occasions where local people were collecting boletes in Malawi; (**e**) Children assist in the selling of *Cantharellus platyphyllus*, *C. congolensis*, and *C. rufopunctatus* in Malawi; (**f**) *Cantharellus* harvest of the day in Burundi; (**g**) Mixture of *Lactifluus*, *Cantharellus*, and *Amanita* at a colourful market in Tanzania

edible mushrooms (Fig. 1.5c–g). There, the ethnomycological pathways are as diverse as the vegetation types and ethnical groups. Some populations do not eat wild mushrooms at all, but in many areas, fungi constitute a very important part of the local diet. Especially in countries were ectomycorrhizal trees dominate the vegetation, the traditional knowledge and use of edible mycorrhizal mushrooms is historical and highly significant. Although also present in the tropical rainforest (where they occur among other fungi with local clusters of *Gilbertiodendron* trees), ectomycorrhizal fungi are most abundant and form the largest fruiting bodies in subtropical woodlands: the Sudanian woodlands in West Africa, the Zambezian woodlands (miombo woodlands) in East, Central, and Southern Africa. Miombo woodlands form the most extensive vegetation type (across 2.8 million km^2) in Africa and are seasonally dry deciduous woodlands (White 1983; Frost 1996). These woodlands are dominated by trees of the leguminous genera *Brachystegia*. The name miombo refers to species of the genus *Brachystegia* (Smith and Allen 2004), *Julbernardia*, and *Isoberlinia* (Caesalpinioideae). The most popular edible fungi genera associated with these trees are *Cantharellus*, *Amanita*, *Russula*, *Lactifluus* and to a lesser extent *Lactarius*. The same genera (often represented by different species) are preferred and collected in many West African countries covered in Sudanian woodlands. Boletes are very common and represented by many genera and species but, strangely enough, usually not consumed. Apparently the changing colour by oxidation, a phenomenon common in many boletes, and the soft, spongy texture once they are mature, are characteristics that make them very unappetizing for indigenous people. All but three countries on the list of 25 poorest countries in the world are sub-Saharan African countries. Hence, it should come as no surprise that especially in rural areas of the least developed countries, people are very dependent on non-timber forest products as an important supplement to their crops. The arrival of the rain season, and of the first mushrooms with it, means the end of the dry season and the end of the traditional famine period because the food reserves are exhausted and the mushrooms pop up faster than the newly planted crops. Especially women and children are collecting for their own supplies but also to sell part of the harvest at local markets or along the road. Wild edible mushrooms play an important role in the diet of local people in East Africa and neighbouring countries (Buyck 1994; Morris 1984; Pegler and Piearce 1980; Rammeloo and Walleyn 1993; Härkönen et al. 2003). According to Degreef (1992), yearly consumption can be estimated at 30 kg/inhabitant in rural areas, and 15 kg/inhabitant in the city for the wet (and thus production) season only. Natural miombo forest produces on average of 150 kg edible mushrooms/ha every year (De Kesel et al. 2017). But this huge production is under threat. In some woodlands, the rainy season is restricted and unimodal; in other parts, a longer rainy season is separated by a short dry period from the second rainy season. The length of the dry season is variable and dry season fires commonly occur, but a changing climate also means an unreliable wet season, shifting in time in some countries, or even completely lacking. It goes without saying that this has an important effect on the availability of the edible mushrooms for the local people. Furthermore, the forests are often mismanaged and over-exploited, and turned into cropland and pastures, or replaced by exotic plantations at an alarming rate. Unsustainable utilization leads to impoverishment of the ecosystem's integrity and long-term survival/productivity, which in turn reduces revenues derived from the forest. Moreover, watersheds get degraded and together with erosion lead to desertification. In some countries (e.g. Ivory Coast), where much of the original woodland has been replaced by plantations or secondary forest, people tend to eat more saprotrophic species, especially wood-decayers and litter saprotrophs. Popular genera are *Pleurotus*, *Lentinus*, *Volvariella*, and *Chlorophyllum*. The latter genus has a dubious reputation because of the toxic *C. molybdites*, but a recent molecular study (Ge et al. 2018) showed that four more species are common in Africa, among which are the edible *C. palaeotropicum and C. hortense*. It is also with those groups of saprotrophic species that we see cultivation programs with low-cost technology originating. Even the cosmopolitan *Schizophyllum commune* is cultivated in some African countries (Osemwegie et al. 2014). A popular and delicious genus occurring as well in the woodlands as in gardens and plantations is *Termitomyces*, associated with termite hills. The taste and texture of *Termitomyces* species, from the very small *T. microcarpus* (which you have to collect by the hundreds to fill a plate) to the giant *T. titanicus*, the largest edible mushroom in the world, are excellent. When conducting ethnomycological research in Africa and asking people about their general appreciation of mushrooms as food, it is interesting to notice that they rank mushrooms somewhere between chicken and crocodile (pers. obs. in Zimbabwe) or between chicken and fish (pers. obs. in Malawi, Togo). They consider them meat rather than vegetables, which is exactly the opposite to the general opinion in many temperate countries where people still consider fungi to be plants and mushrooms a kind of vegetable. Up to 20 years ago, an important part of these species harvested and offered for sale only had local names and were new to science. But in this molecular era where the importance of world-wide sampling becomes more and more clear, the interest in African mycology is also increasing. Recently, ectomycorrhizal groups such as chanterelles (Buyck et al. 2013) and milkcaps (Verbeken and Walleyn 2010) were the subject of many studies. Good descriptions, pictures, and identification tools are more common now and stimulate mycologists inside and outside Africa to contribute to the exploration of the diversity. Moreover, some recent publications focus purely on edible mushrooms taking into account the new phylogenetic insights (De Kesel et al. 2017; Härkönen et al. 2015).

In Asia, the highest diversity of edible ectomycorrhizal mushrooms is harboured by China. Due to the diverse habitats and plant species created by the great variation in climate and topography, China has the world's richest diversity of wild edible mushrooms, with 1020 edible species recorded (Wu et al. 2019), around 75% of which are distributed in Southwestern China. Additionally, 692 medicinal species have been recorded in the country, including the world-wide famous Lingzhi (*Ganoderma spp.*) which is a mushroom that has been renowned in China for more than 2,000 years (Cao et al. 2012). Every county in Yunnan has at least one mushroom market trading wild edible mushrooms harvested from surrounding forests (Fig. 1.6a). At the main markets such as in Kunming and Nanhua, hundreds of tons of wild edible mushrooms change hands daily during mushroom season from June

Fig. 1.6 Aspects of wild edible mushrooms in China. (**a**) Mushuihua wild edible mushroom market at Kunming; (**b**) *Thelephora ganbajun* at Ciba wild edible mushroom market, Kunming; (**c**) *Tricholoma matsutake* young fruiting bodies at Shangri-La wild edible mushroom market; (**d**) An experimental plantation for the cultivation of *Tuber indicum*; (**e**) Commercialization of *Tuber indicum* in Yunnan; (**f**) *Lactifluus volemus* sold in southeastern Yunnan

to October. A total of 321 species, belonging to 101 genera, and 47 families have been identified as wild mushrooms traded in the local markets. Of these commercial mushrooms, Boletaceae was the best represented family, with 27 genera and 23.05% (74 spp.) of the total species. *Ramaria* being the most species-rich genus in the Gomphaceae family, with 22 species and 6.83% of the total (Wang and Liu 2002; Wang et al. 2004; Yu and Liu 2005; Dai et al. 2010; Cui et al. 2015; Tang et al. 2015; Wu et al. 2015; Yang 2015). More than 164 species commonly traded and 60 dominant commercial species in the genera *Boletus*, *Cantharellus Lactarius*, *Ramaria*,

Russula, *Termitomyces*, *Thelephora* (Fig. 1.5b), *Tricholoma*, *Tuber*, etc. are found in the Yunnan's local markets. The foreign income produced from wild mushroom exportation is over US$100 million every year in China. Marketing of *Tricholoma matsutake*, and a few additional species, such as *Ophiocordyceps sinensis*, *Tuber indicum*, and *Boletus bainiugan*, has significantly improved the local economy in the last few years (Wang and Yang 2006). Harvesting wild mushrooms is an important livelihood and generates 15–90% of these people' annual income. In the last 10 years, over 1000 tons of fresh fruiting bodies of matsutake have been exported from Southwestern China annually. More than 40 counties in Yunnan are reported to harvest matsutake (Yang et al. 2009). In the Shangri-La region, northwest Yunnan, harvesting matsutake can result in an annual return of over 10,000 Chinese Yuan (about US$1500) for an average family. The natural production of wild edible mushrooms has declined since large-scale commercial harvesting started in the 1990s. A variety of efforts have been deployed to protect wild edible mushrooms. The most important has been the forest ownership reformation which occurred in 2008, giving farmers the right to manage forest products including wild mushrooms. A few regulations have been launched such as prohibiting the harvesting of immature matsutake (Fig. 1.6c) and truffles. Experimental plantations have been set up for truffle cultivation, and the production of *T. indicum* has begun (Fig. 1.6d). Other attempts to cultivate truffles (*T. borchii*, *T. melanosporum*, and *T. sinoaestivum*), milk cap mushrooms (*Lactarius deliciosus*, *L. hatustake*, and *L. vividus*), and to understand the biology, ecology, and cultivation potential of edible mushrooms are being undertaken (Geng et al. 2009; Wang et al. 2019). However, conservation of the precious wild edible mushroom remains a vital and urgent issue (Fig. 1.6e–f).

Australasia constitutes a very unique continent from the biogeographic, historic, and biodiversity perspectives. New Zealand is exceptional in many aspects, and edible mushrooms of Aotearoa are no exception. From a geological point of view, New Zealand has been isolated from all other continents for 80 million years (Dawson and Lucas 2000; Wallis and Trewick 2009), and this situation has generated a high level of endemism, as well as extraordinary features such as the lack of land mammals (except bats) and flightless birds like the iconic kiwis (Murphy et al. 2019). From a human point of view, New Zealand was mass-populated only very recently, first about 700 years ago by Māori people from the Pacific (Walter et al. 2017) followed by European then worldwide settlers since about 350 years. These characteristics created a very original situation for mushrooms. Except Southern beech forests (*Nothofagus* spp.), which are confined to specific areas, most native New Zealand trees (i.e. magnificent Podocarps and other conifers such as Kauri) live symbiotically with arbuscular mycorrhizal fungi and therefore do not produce ectomycorrhizal fruiting bodies (Orlovich and Cairney 2004). Native ectomycorrhizal fungi are associated only with three woody plant genera, *Nothofagus*, *Leptospermum*, and *Kunzea*, and are very different from ectomycorrhizal fungi associated with Fagaceae and Pinaceae in the Northern Hemisphere (Orlovich and Cairney 2004). The short period of human colonization also explains the lack of knowledge concerning the edibility of mushrooms found in beech forests. Several saprotrophic fungi endemic to New Zealand are closely related to commercial

Fig. 1.7 (**a**) *Cyclocybe* (*Agrocybe*) *parasitica* fruiting on poplar, Lincoln Farm of the New Zealand Institute for Plant & Food Research (PFR), Lincoln, New Zealand; (**b**) *Boletus edulis*, Christchurch, New Zealand; (**c**) Cultivated *Lactarius deliciosus*, PFR-Lincoln Farm; (**d**) *Tuber borchii* (bianchetto truffle) plantation (to the right of the picture) on the PFR-Lincoln Farm, note the white tags corresponding to bianchetto truffles marked in the grassy aisle alongside the truffière; (**e**) the dog Mila, a vizsla and wonderful bianchetto expert, PFR-Lincoln plantation; (**f**) Cultivated bianchetto truffles, PFR-Lincoln Farm; (**g**) Tewnion Truffière, Canterbury, New Zealand; (**h**) An amazing 'truffle machine', the spoodle Cassie hidden by her harvest of Périgord black truffles at Tewnion Truffière; (**i**) A Périgord black truffle (*Tuber melanosporum*) grown at Tewnion Truffière, cut and photographed by Chef Vaughan Mabee of Amisfield, Central Otago, New Zealand; (**j**) The first Canterbury Truffle Festival, Riccarton Market, Christchurch, July 2015, has launched a series of winter truffle attractions now popular in Canterbury; (**k**) Participants of the New Zealand Truffle Association' 2019 conference in Christchurch visiting a black truffle plantation during the post-conference field trip; (**l**) Pleasure for the eyes and the mouth made with the Périgord black truffle by Chef Vaughan Mabee of Amisfield, Central Otago, New Zealand

species of the northern hemisphere, e.g. *Lentinula novae-zelandiae* a Shiitake relative (Johnston 2009), *Cyclocybe* (*Agrocybe*) *parasitica* (Fig. 1.7a) a Poplar mushroom relative (*A. cylindrica*) (Cooper 2012). Some New Zealand species are recently the object of cultivation trials (Buchanan pers. comm.) and would greatly complement the currently limited choice of cultivated exotic mushrooms (mostly *Agaricus bisporus* followed by *Pleurotus pulmonarius* and *Lentinula edodes*). A few wild edible saprotrophic species present in New Zealand are exotic, e.g. *Agaricus campestris*, *A. arvensis*, *Marasmius oreades*, *Lepista nuda*. Regarding ectomycorrhizal mushrooms, the arrival of Europeans has considerably changed the situation: early settlers brought with them broadleaf trees and conifers from Europe and, in doing so, imported accidentally several edible ectomycorrhizal mushrooms associated with these trees: Porcini (*Boletus edulis* s.l.) (Wang et al. 1995) (Fig. 1.7b), Slippery jack (*Suillus luteus*), Shoro (*Rhizopogon rubescens*), Birch bolete (*Leccinum scabrum*) while other famous edible species never made it to New Zealand by accident, e.g. economically valuable Truffles (*Tuber* spp.) (Bulman et al. 2010), Chanterelles (*Cantharellus cibarius* s.l.), Milkcaps (*Lactarius* section *Deliciosi*) (Fig. 1.7c), Caesar's mushroom (*Amanita section Caesarea*), etc. Interestingly, nowadays, New Zealand has developed an important agricultural economy based on the cultivation, or farming, of exotic plant and animal species: kiwifruit (native to China), grapevine, cattle and dairy farms, sheep, pine timber, etc. Again, edible mycorrhizal mushrooms are no exception. In the late 1970s, Dr. Ian Hall, an arbuscular mycorrhiza scientist inspired by European researchers working on truffle cultivation, envisioned to grow these delicacies in New Zealand. Free of the cultural bonds that characterized many European truffle experts and growers, he dared raising the pH of agricultural farm land up to levels compatible with the cultivation of the Périgord black truffle (*Tuber melanosporum*) (pH $\approx$ 8), by adding considerable amount of lime. The efforts of his team were rewarded in 1993 when a truffle orchard in Gisborne produced the first ascocarp of *T. melanosporum* ever produced in the southern hemisphere (Guerin-Laguette 2008). This triggered the development of a truffle and edible mycorrhizal fungi industry in New Zealand. After years of patient incubation and technical challenges, New Zealand is now producing over a ton of truffles in commercial truffières, mostly *T. melanosporum* and *T. borchii* (Fig. 1.7d–i), and also *T. aestivum*, *T. brumale*, as well as commercial quantities (in very restricted areas to date) of saffron milk caps (*Lactarius deliciosus*) in pine orchards (Wang et al. 2019; this book, Chap. 5). Like truffles, saffron milk cap were successfully introduced purposely to New Zealand as part of a research programme initiated in 1997. New Zealand has now the opportunity, through further applied and basic research, to grow a profitable and sustainable edible mycorrhizal fungi industry that can fit with its lifestyle and image of gourmet country, further attracting tourists from all over the world (Fig. 1.7j). For almost 30 years, The New Zealand Truffle Association (NZTA) has pioneered and supported the development of an edible mycorrhizal fungi industry in New Zealand. The NZTA organizes a national conference every year (Fig. 1.7k). The Périgord black truffle, black diamond of the French cuisine, is now inspiring very talented New Zealand chefs (Fig. 1.7l). More high value ectomycorrhizal fungi could be cultivated in the future, either native or exotic, providing that biotechnical and biochemical research is carried out to demonstrate their value and to show that the proposed cultivation of selected, new, exotic species (e.g. pine-specific *Lactarius sanguifluus*) in already disturbed environments would increase overall biodiversity without having negative impact on native areas of New Zealand.

Acknowledgements The credit pictures are as follows: Figs. 1.1a–d and 1.3a–c are from Roberto Flores Arzú; Fig. 1.5c–g are from Annemieke Verbeken; Fig. 1.4d is from Alexander Bondarchuk; Fig. 1.6a–d are from Fu-Qiang Yu; Fig. 1.2c is from Lucila Aragón Carrillo; Fig. 1.2d is from Jorge Manuel Mejia Bojorquez; Fig. 1.3c is from José María González; Fig. 1.3d–f are from José Leonardo García Rodríguez; Fig. 1.7, all photos are by The New Zealand Institute for Plant & Food Research except Fig. 1.7b, courtesy Wang Yun; Fig. 1.7e, courtesy Will Clayton; Fig. 1.7g, h, courtesy Wayne Tewnion; Fig. 1.7i, l, courtesy Vaughan Mabee; Figs. 1.2a, b, 1.4a–c, e, 1.5a, b and 1.6e, f are from Jesús Pérez-Moreno. Alexis Guerin-Laguette is grateful to Nicholas Cummings for helpful suggestions on the Australasia section. Jesús Pérez-Moreno acknowledges the financial support from CONACyT 2018-07-01EXTV and COMECyT EST2018-005 to develop a sabbatical stay at the Kunming Institute of Botany, Kunming, China.

Funding from the CONACyT project "Biotechnologies of highly-prized or bioculturally-important edible ectomycorrhizal mushrooms for reforestation with native Mexican trees using frontier scientific methodologies" is also acknowledged.

References

Aguilera E (2016) Hongos de El Salvador. http://elsyaguilera.blogspot.com/2016/06/hongos-de-el-salvador.html

Averill C, Hawkes CV (2016) Ectomycorrhizal fungi slow soil carbon cycling. Ecol Lett 19:937–947

Ayuso-Fernández I, Ruiz-Dueñas FJ, Martínez AT (2018) Evolutionary convergence in lignin-degrading enzymes. Proc Natl Acad Sci U S A 115:6428–6433

Bostock J, Riley HT (1855) The natural history of Pliny. Henry G. Bohn, London

Brundrett MC (2002) Coevolution of roots and mycorrhizas of land plants. New Phytol 154:275–304

Brundrett MC, Tedersoo L (2018) Evolutionary history of mycorrhizal symbioses and global host plant diversity. New Phytol 220:1108–1115

Buller AR (1914) The fungus lore of the Greeks and Romans. Trans Br Mycol Soc 5:21–66

Bulman SR, Visnovski SB, Hall IR, Guerin-Laguette A, Wang Y (2010) Molecular and morphological identification of truffle-producing *Tuber* species in New Zealand. Mycol Progress 9:205–214. https://doi.org/10.1007/s11557-009-0626-0

Buyck B (1994) Ubwoba: les champignons comestibles de l'ouest du Burundi. Administration Générale de la Coopération au Développement, Bruxelles

Buyck B, Kauff F, Cruaud C, Hofstetter V (2013) Molecular evidence for novel *Cantharellus* (Cantharellales, Basidiomycota) from tropical African miombo woodland and a key to all tropical chanterelles. Fungal Divers 58:281–298

Cao Y, Wu SH, Dai YC (2012) Species clarification of the prize medicinal *Ganoderma* mushroom "Lingzhi". Fungal Divers 56:49–62

Comandini O, Rinaldi AC, Kuyper TW (2012) Measuring and estimating ectomycorrhizal fungal diversity: a continuous challenge. In: Pagano M (ed) Mycorrhiza: occurrence in natural and restored environments. Nova Science Publishers, New York, pp 165–200

Cooper J (2012) Mycological Notes 19: *Agrocybe parasitica* and related species. https://www.funnz.org.nz/sites/default/files/MycNotes19-AgrocybeParasitica_2.pdf

Cui YY, Feng B, Wu G et al (2015) Porcini mushrooms (*Boletus* sect. Boletus) from China. Fungal Divers 82:189–212
Dai YC, Cui BK (2011) *Fomitiporia ellipsoidea* has the largest fruiting body among the fungi. J Fungal Biol 115:813–814
Dai YC, Zhou LW, Yang ZL et al (2010) A revised checklist of edible fungi in China. Mycosystema 29:1–21
Dawson J, Lucas R (2000) Nature guide to the New Zealand forest. Random House, Auckland, New Zealand. 312 pp
De Kesel A, Kasongo B, Degreef J (2017) Champignons comestibles du Haut-Katanga (R.D. Congo). ABC Taxa 17:296. (In French)
Degreef J (1992) Les champignons comestibles, leur importance pour les populations africaines. Défis Sud 1992:38–39. (In French)
Del Olmo-Ruiz M, García-Sandoval R et al (2017) Current knowledge of fungi from Neotropical montane cloud forests: distributional patterns and composition. Biodivers Conserv 26:1919. https://doi.org/10.1007/s10531-017-1337-5
Delgado S (2010) Diversidad y abundancia de macromicetos del bosque Las Lajas, del área natural complejo San Marcelino, Santa Ana, Sonsonate, El Salvador. Thesis from Universidad de El Salvador. http://ri.ues.edu.sv/id/eprint/847/1/10136599.pdf
Elbert W, Taylor PE, Andreae MO, Pöschl U (2007) Contribution of fungi to primary biogenic aerosols in the atmosphere: wet and dry discharged spores, carbohydrates, and inorganic ions. Atmos Chem Phys 7:4569–4588
Encyclopaedia Britannica (2019) Mushroom (last updated 1 Aug 2019). www.britannica.com/science/mushroom. Accessed 14 Aug 2019
Ferguson BA, Dreisbach TA, Parks CG, Filip GM, Schmitt CL (2003) Coarse-scale population structure of pathogenic *Armillaria* species in a mixed-conifer forest in the Blue Mountains of northeast Oregon. Can J For Res 33:612–623
Flores R, Comandini O, Rinaldi AC (2012) A preliminary checklist of macrofungi of Guatemala, with notes on edibility and traditional knowledge. Mycosphere 3:1–21
Floudas D et al (2012) The Paleozoic origin of enzymatic lignin decomposition reconstructed from 31 fungal genomes. Science 336:1715–1719
Frost P (1996) The ecology of Miombo woodlands. In: Campbell BM (ed) The Miombo in transition: woodlands and welfare in Africa. Center for International Forestry Research, Bogor, Indonesia, pp 11–55
Ge ZW, Jacobs A, Vellinga EC, Sysouphanthong P, van der Walt R, Lavorato C, An Y-F, Yang ZL (2018) A multi-gene phylogeny of *Chlorophyllum* (Agaricaceae, Basidiomycota): new species, new combination and infrageneric classification. Mycokeys 32:65–90
Geng LY, Wang XH, Yu FQ et al (2009) Mycorrhizal synthesis of *Tuber indicum* with two indigenous hosts, *Castanea mollissima* and *Pinus armandii*. Mycorrhiza 19:461–467
Guerin-Laguette A (2008) La trufficulture en Nouvelle-Zélande: 20 ans déjà. [The truffle cultivation in NZ: 20 years already]. Proceedings of the international conference 'La culture de la truffe dans le monde [The truffle cultivation in the world], In French with English abstract, (ed. INRA and Le Causse Corrézien), Brive-la-Gaillarde, France, 2 February 2007, pp 151–162
Guzmán G (2008) Hallucinogenic mushrooms in Mexico: an overview. Econ Bot 62:404–412
Halling R, Mueller G (2002) Agaric and boletes in neotropical oakwoods. In: Watling R, Farkland J et al (eds) Tropical mycology, Macromycetes, vol 1. CABI Publishing, Wallingford, UK
Härkönen M, Niemelä T, Mwasumbi L (2003) Tanzanian mushrooms. Edible, harmful and other fungi. Norrlinia 10:1–200
Härkönen M, Niemelä T, Mbindo K, Kotiranta H, Piearce G (2015) Zambian mushrooms and mycology. Norrlinia 29:1–208
Hassett MO, Fischer MWF, Money NP (2015) Mushrooms as rainmakers: how spores act as nuclei for raindrops. PLoS One 10:e0140407. https://doi.org/10.1371/journal.pone.0140407
Heim R, Wasson RG (1958) Les champignons hallucinogènes du Mexique - Etudes ethnologiques, taxinomiques, biologiques, physiologiques et chimiques. Archives du Muséum National d'Histoire Naturelle 6:1–445

Hernández-Santiago F, Pérez-Moreno J, Xoconostle-Cázares B, Almaraz-Suárez JJ, Ojeda-Trejo E, Mata G, Díaz-Aguilar I (2016) Traditional knowledge and use of wild mushrooms by Mixtecs or Ñuu savi, the people of the rain, from Southeastern Mexico. J Ethnobiol Ethnomed 12(35):1–22

Hernández-Santiago F, Martínez-Reyes M, Pérez-Moreno J et al (2017) Pictographic representation of the first dawn and its association with entheogenic mushrooms in a 16th century Mixtec Mesoamerican Codex. Sci Fungorum 46:19–28

Hofmann A, Heim R, Brack A et al (1958) Psilocybin, ein psychotroper wirkcstoff aus dem Mexikanishchen Rauchpilz *Psilocybe mexicana* Heim. Experientia 14:107–109

Johnston PR (2009) Causes and consequences of changes to New Zealand's fungal biota. N Z J Ecol 34:175–184

Kim SH, Jakhar R, Kang SC (2015) Apoptotic properties of polysaccharide isolated from fruiting bodies of medicinal mushroom *Fomes fomentarius* in human lung carcinoma cell line. Saudi J Biol Sci 22:484–490

Kuhnlein HV, Turner NJ (1991) Traditional plant foods of Canadian indigenous peoples. Food Nutr Hist Anthropol 8:1–533

Kuo M, Dewsbury DR, O'Donnell K, Carter MC, Rehner SA, Moore JD, Moncalvo JM, Canfield SA, Stephenson SL, Methven AS, Volk TJ (2012) Taxonomic revision of true morels (*Morchella*) in Canada and the United States. Mycologia 104:1159–1177

López-García A, Jiménez-Ruiz M, Pérez-Moreno J (2017) Vocablos relacionados con el recurso micológico en el idioma de la cultura chinanteca de la Sierra Norte del estado de Oaxaca, México. Sci Fungorum 46:9–18. (In Spanish)

Marineros L, Vega H, Bejarano L (2015) Contribución al conocimiento de los hongos comestibles en las regiones Lenca y Chortí del occidente de Honduras. BIOMA 35:35–47. http://www.ethnobiology.net/wp-content/uploads/Contribucio%CC%81n-al-conocimiento-de-los-hongos-comestibles-en-las-regiones-lenca-y-chorti%CC%81-del-occidente-de-Honduras.pdf. Accessed 7 May 2019

Michelot D, Melendez-Howell LM (2003) *Amanita muscaria*: chemistry, biology, toxicology, and ethnomycology. Mycol Res 107:131–146

Molina R, O'Dell T, Luoma D, Amaranthus M, Castellano M, Russell K (1993) Biology, ecology, and social aspects of wild edible mushrooms in the forests of the Pacific Northwest: a preface to managing commercial harvest. Gen. Tech. Rep. PNW-GTR-309. U.S. Department of Agriculture, Forest Service, Pacific Northwest Research Station, Portland, OR

Money NP (2011) Mushroom. Oxford University Press, Oxford, UK

Morris B (1984) Macrofungi of Malawi. Some ethnobotanical notes. Bull Br Mycol Soc 18:48–57

Mueller G, Halling R, Carranza J et al (2006) Saprotrophic and ectomycorrhizal macrofungi of Costa Rican oak forest. In: Kapelle M (ed) Ecology and conservation of neotropical Montane oak forest. Springer, Berlin, pp 55–68

Murphy EC, Russell JC, Broome KG, Ryan GJ, Dowding JE (2019) Conserving New Zealand's native fauna: a review of tools being developed for the Predator Free 2050 programme. J Ornithol 160:883–892. https://doi.org/10.1007/s10336-019-01643-0

Nagy LG, Riley R, Tritt A et al (2016) Comparative genomics of early-diverging mushroom-forming fungi provides insights into the origins of lignocellulose decay capabilities. Mol Biol Evol 33:959–970

Orlovich DA, Cairney JWG (2004) Ectomycorrhizal fungi in New Zealand: current perspectives and future directions. N Z J Bot 42:721–738

Osemwegie OO, Okhuoya AJ, Dani AT (2014) Ethnomycological conspectus of West African mushrooms: an awareness document. Adv Microbiol 4:39–54

Pegler DN, Piearce GD (1980) The edible mushrooms of Zambia. Kew Bull 35:475–491

Peintner U, Pöder R (2000) Ethnomycological remarks on the Iceman's fungi. In: Bortenschlager S, Oeggl K (eds) The Iceman and his natural environment. The man in the ice, vol 4. Springer, Vienna, pp 143–150

Pérez-Moreno J, Martínez-Reyes M (2014) Edible ectomycorrhizal mushrooms: biofactories for sustainable development. In: Guevara-Gonzalez R, Torres-Pacheco I (eds) Biosystems engineering: biofactories for food production in the century XXI. Springer, Cham, pp 151–233

Pérez-Moreno J, Read DJ (2004) Los hongos ectomicorrízicos, lazos vivientes que conectan y nutren a los árboles en la naturaleza. Interciencia 29:239–247

Pérez-Moreno J, Lorenzana Fernández A, Carrasco Hernández V, Yescas-Pérez A (2010) Los hongos comestibles silvestres del Parque Nacional Izta-Popo, Zoquiapan y Anexos. Colegio de Postgraduados, SEMARNAT, CONACYT, Texcoco, Mexico. (In Spanish)

Piepenbring M (2013) References to literature including records of species of fungi in Panama. Compiled by Meike Piepenbring. Smithsonian Tropical Research Institute, Panama. https://biogeodb.stri.si.edu/fungi. Accessed 25 Sept 2019

Piepenbring M, Caballero E, Fournier J et al (2011) Pioneer forays for fungi in the Darién Province in Eastern Panama: quintuplicating the knowledge on fungi in this area by five days of fieldwork. Biodivers Conserv 20:2511. https://doi.org/10.1007/s10531-011-0085-1

Pilz D, Norvell L, Danell E, Molina R (2003) Ecology and management of commercially harvested chanterelle mushrooms. Gen. Tech. Rep. PNW-GTR-576. U.S. Department of Agriculture, Forest Service, Pacific Northwest Research Station, Portland, OR

Pilz D, McLain R, Alexander S, Villarreal-Ruiz L, Shannon B, Wurtz TL, Parks CG, McFarlane E, Baker B, Molina R, Smith JE (2007) Ecology and management of morels harvested from the forests of western North America. Gen. Tech. Rep. PNW-GTR-710. U.S. Department of Agriculture, Forest Service, Pacific Northwest Research Station, Portland, OR

Pirozynski KA, Dalpé Y (1989) Geological history of the Glomaceae with particular reference to mycorrhizal symbiosis. Symbiosis 7:1–36

Pleszczyńska M, Lemieszek MK, Siwulski M et al (2017) *Fomitopsis betulina* (formerly *Piptoporus betulinus*): the Iceman's polypore fungus with modern biotechnological potential. World J Microbiol Biotechnol 33:83. https://doi.org/10.1007/s11274-017-2247-0

Power RC, Salazar-García DC et al (2015) Microremains from El Mirón Cave human dental calculus suggest a mixed plant-animal subsistence economy during the Magdalenian in Northern Iberia. J Archaeol Sci 60:39–46

Rammeloo J, Walleyn R (1993) The edible fungi of Africa south of the Sahara. Scripta Botanica Belgica 5:1–62

Read DJ (1991) Mycorrhizas in ecosystems. Experientia 47:376–391

Read DJ, Pérez-Moreno J (2003) Mycorrhizas and nutrient cycling in ecosystems–a journey towards relevance? New Phytol 157:475–492

Redecker D, Kodner R, Graham LE (2000) Glomalean fungi from the Ordovician. Science 289:1920–1921

Remy W, Taylor TN, Hass H, Kerp H (1994) Four hundred-million-year-old vesicular arbuscular mycorrhizae. Proc Natl Acad Sci U S A 91:11841–11843

Riedlinger TJ (ed) (1990) The sacred mushroom seeker: Essays for R. Gordon Wasson, Historical, ethno- and economic botany series, vol 4. Discorides, Portland

Ruan-Soto F (2018) Sociodemographic differences in the cultural significance of edible and toxic mushrooms among Tsotsil towns in the Highlands of Chiapas, Mexico. J Ethnobiol Ethnomed 14:32. https://doi.org/10.1186/s13002-018-0232-9

Saar M (1991a) Fungi in Khanty folk medicine. J Ethnopharmacol 31:175–179

Saar M (1991b) Ethnomycological data from Siberia and northeast Asia on the effect of *Amanita muscaria*. J Ethnopharmacol 31:157–173

Sarmiento E, Fontecha G (2015) Conocimiento tradicional de los hongos en el occidente de Honduras. Ciencia y Tecnología 13:19–29. https://doi.org/10.5377/rct.v0i13.1710. Accessed 7 May 2019

Smith P, Allen Q (2004) Field guide to the trees and shrubs of the miombo woodlands. Royal Botanic Gardens, Kew

Smith SA, Read DJ (2008) Mycorrhizal symbiosis. Academic, London

Steidinger BS, Crowther TW et al (2019) Climatic controls of decomposition drive the global biogeography of forest-tree symbioses. Nature 569:404–413

Stevens WD et al (eds) (2001) Flora of Nicaragua, vol 85(1). Botanical Garden Press, Missouri. http://www.mobot.org/MOBOT/research/nicaragua/vegetacion.shtml

Stolze-Rybczynski JL, Cui Y et al (2009) Adaptation of the spore discharge mechanism in the Basidiomycota. PLoS One 4:41

Tang LP, Yang ZL, Zeng NK et al (2015) Atlas of higher fungi from Lancang river valley. Yunnan Science and Technology Press, Kunming
Taylor TN, Osborn JM (1996) The importance of fungi in shaping the paleoecosystem. Rev Palaeobot Palynol 90:249–262
Toledo J (2013) Inventario de Macrohongos del Área Natural Protegida Parque del Bicentenario El Espino – Bosque Los Pericos. Salvanatura, El Salvador. http://www.salvanatura.org/wp-content/uploads/2015/08/INVENTARIO-HONGOS-_PDB_MLQ-2012.pdf. Accessed 7 May 2019
Vega H (2018) El mundo de los Hongos. https://www.researchgate.net/publication/323663090_Presentacion_sobre_los_hongos_comestibles_de_la_zona_Lenca_de_Honduras_especifica-mente_en_los_alrededores_del_PNM_Celaque
Vega A, de León J (2018) Hongos Comestibles de Panamá. Silvestres, Cultivados y su potencial de Industrialización. Centro de Investigación en Recursos Naturales, Universidad Autónoma de Chiriquí. http://alaccta.org/wp-content/uploads/2018/04/7-Hongos-Comestibles-de-Panam%C3%A1.-ALACCTA-2018.pdf
Verbeken A, Walleyn R (2010) Monograph of *Lactarius* in tropical Africa. Fungus flora of tropical Africa. National Botanic Garden, Belgium
Wallis GP, Trewick SA (2009) New Zealand phylogeography: evolution on a small continent. Mol Ecol 18:3548–3580
Walter R, Buckley H, Jacomb C, Matisoo-Smith E (2017) Mass migration and the Polynesian settlement of New Zealand. J World Prehist 30:351–376. https://doi.org/10.1007/s10963-017-9110-y
Wang XH, Liu PG (2002) Resources investigation and studies on the wild commercial fungi in Yunnan. Biodivers Sci 10:318–325
Wang L, Yang ZL (2006) Wild edible fungi of the Hengduan Mountains, southwestern China. In: Christoph K, Yang YP, Weyerhäuser H et al (eds) The sustainable harvest of non-timber forest products in China. Sino-German Symposium Proceedings. Sino-German Center, Beijing, pp 58–65
Wang Y, Sinclair L, Hall IR, Cole ALJ (1995) *Boletus edulis sensu lato*: a new record for New Zealand. N Z J Crop Hortic Sci 23:227–231
Wang XH, Liu PG, Yu FQ (2004) Color atlas of wild commercial mushrooms in Yunnan. Yunnan Science and Technology Press, Kunming
Wang R, Guerin-Laguette A, Huang LL, Wang XH, Butler R, Wang Y, Yu FQ (2019) Mycorrhizal syntheses between *Lactarius* spp. section Deliciosi and *Pinus* spp. and effects of grazing insects in Yunnan, China. Can J For Res 49:616–627. https://doi.org/10.1139/cjfr-2018-0198
Wasson RG (1957) Seeking the magic mushroom. Life Magazine, June 10, New York
Wasson RG (1980) The wondrous mushroom: Mycolatry in Mesoamerica. McGraw-Hill, New York
Wasson VP, Wasson RG (1957) Mushrooms, Russia and history. Pantheon Books, New York
Wasson G, Cowan F, Rhodes W (1974) María Sabina and her Mazatec mushroom velada, Ethnomicological studies, vol 3. Harcourt Brace Jovanovich, New York
White JC (1983) The vegetation of Africa. A descriptive memoir to accompany the UNESCO/AETFAT/UNSO vegetation map of Africa. UNESCO, Paris, p 356
Wu G, Li YC, Zhao K et al (2015) Four new genera of the fungal family Boletaceae. Fungal Divers 81:1–24
Wu F, Zhou LW, Yang ZL, Bau T, Li TH, Dai YC (2019) Resource diversity of Chinese macrofungi: edible, medicinal and poisonous species. Fungal Divers 98:1–76
Yang ZL (2015) Atlas of the Chinese species of Amanitaceae. Science Press, Beijing
Yang XF, Wilkes A, Yang YP et al (2009) Common and privatized: conditions for wise management of matsutake mushrooms in northwest Yunnan Province, China. Ecol Soc 14:30
Yu FQ, Liu PG (2005) Species diversity of wild edible mushrooms from *Pinus yunnanensis* forests and conservation strategies. Biodivers Sci 13:58–69
Yun W, Hall IR (2004) Edible ectomycorrhizal mushrooms: challenges and achievements. Can J Bot 82:1063–1073
Zan L, Haiying B, Bau T, Liu H, Cui B (2012) An antitumor component from *Fomitiporia ellipsoidea*. J Microbiol Biotechnol 22:1482–1485

Part II
Biodiversity and Cultivation

Chapter 2
Edible Ectomycorrhizal Fungi and Their Cultivation in China

Yun Wang, Fu-Qiang Yu, Chunxia Zhang, Chengyi Liu, Mei Yang, and Shuhong Li

2.1 Introduction

China is currently the largest producer of edible mushrooms in the world. Production of fresh mushrooms has reached 20 million tons annually, accounting for 70% of total world production; a value just behind those for cereals, cotton, oils, vegetables, and fruit (Chang and Miles 1997). More than 70 fungal species have been cultivated and approximately 50 commercialized (Mao 1998). The main cultivated varieties are *Lentinula edodes*, *Pleurotus ostreatus*, *Auricularia auricula-judae*, *Flammulina velutipes*, *Agaricus bisporus*, *Ganoderma linzhi* and *Auricularia polytricha*. *Volvariella volvacea*, *Tremella fuciformis*, *Pholiota nameko*, *Pleurotus citrinopileatus*, *Hericium erinaceus*, *Dictyophora* species, *Wolfiporia cocos*, and *Cordyceps militaris* are also commonly, commercially cultivated but with less production. Other species such as *Pleurotus eryngii*, *Agrocybe chashingu*, *Pleurotus nebrodensis*, *Tremella aurantialba*, *Agaricus blazei*, *Grifola frondosa*, *Coprinus comatus*,

Y. Wang
Yunnan Institute for Tropical Crop Research, Jing Hong, Sipsongpanna, China

Kunming Institute of Botany, Chinese Academy of Sciences, Kunming, China

Biotechnology and Germplasm Resources Institute, Yunnan Academy of Agricultural Sciences, Kunming, China

F.-Q. Yu
Kunming Institute of Botany, Chinese Academy of Sciences, Kunming, China

C. Zhang
Yunnan Institute for Tropical Crop Research, Jing Hong, Sipsongpanna, China

C. Liu · M. Yang
Panzhihua city Academy of Agricultural and Forestry Sciences, Panzhihua, Sichuan, China

S. Li (✉)
Biotechnology and Germplasm Resources Institute, Yunnan Academy of Agricultural Sciences, Kunming, China

J. Pérez-Moreno et al. (eds.), *Mushrooms, Humans and Nature in a Changing World*, https://doi.org/10.1007/978-3-030-37378-8_2

Stropharia rugosoannulata, *Ganoderma leucocontextum*, *Oudemansiella radicata* (Fig. 2.1), and *Morchella* species (Fig. 2.2, see also Chap. 6 by Yu et al.) are newly commercialized mushroom species, with future market potential.

China has a long history of mushroom cultivation and utilization. Several mushroom species, such as *Auricularia auricula-judae* (estimated cultivation date, 600 AD), *Flammulina velutipes* (800–900 AD), *Lentinula edodes* (1000–1100 AD), and *Volvariella volvacea* (1700 AD), were first successfully cultivated in China (Chang and Miles 1989). However, research into the cultivation of edible mycorrhizal mushrooms is not as advanced as for saprobic species, and also lags behind the Western world, despite China's richness in EMMs. Chinese research into EMMs has made great progress over the last 30 years. China has become the biggest exporter of matsutake (*Tricholoma matsutake*) and is newly emerging as a truffle-producing nation (Wang 1995; Wang et al. 1997, 2008, 2017; Wang and Hall 2006; Wang and Liu 2009). EMMs are gourmet foods and sources of livelihood in China (Wang and Hall 2004; Wang et al. 2008, 2017). Cash earnings from the harvesting of EMMs can make up 20–40% (or even 90%) of an annual family income (Wang and Hall 2004; Wang and Liu 2011b). Some EMMs, such as *Tuber indicum*, *Lactarius hatsudake*, and *Suillus luteus*, have been successfully cultivated. Efforts to protect wild truffles, matsutake, and other EMMs have been made (Wang and Chen 2014). Research into truffles, porcini, matsutake, and others has revealed that China has wide diversity of EMMs (Wang et al. 2006; Zang 2006; Yang 2015; Yu 2007; Tian et al. 2009; Zheng 2010; García-Montero et al. 2010; Wang and Liu 2011a, b; Zhang et al. 2013; Lin 2016). In 2007, the fifth International Workshop on Edible Mycorrhizal Mushrooms (IWEMM5) was successfully held in Yunnan. Truffle festivals, matsutake festivals, and wild edible mushroom festivals have recently been held in Panzhihua, Sichuan, and Diqing, Nanhua, Kunming, Yunnan (Fig. 2.3). We provide here a brief introduction to EMM research in China.

Fig. 2.1 Cultivated *Oudemansiella radicata* in a plastic tunnel house, Yunnan

Fig. 2.2 Cultivated *Morchella* sp., Kunming, Yunnan

Fig. 2.3 Visiting a *Tuber indicum* forest of *Pinus yunnanensis* during the third International Panzhihua Truffle Festival, Sichuan, 2014

2.2 Diversity of Edible Ectomycorrhizal Mushrooms

China has some of the richest biodiversity on earth. There are more than 30,000 seed plants, among which 17,000 are endemic (Wang 1979). The plant flora has ancient origins and was conserved due to it being much less impacted by the continental glacier during the Quaternary ice age than some other parts of Eurasia. China's vast territory has great climatic variation, including boreal, temperate, warm-temperate, subtropical, and tropical zones, giving rise to various types of terrestrial ecosystems including boreal forests, temperate coniferous and broadleaf forests, warm temperate deciduous broadleaf forests, subtropical evergreen broadleaf forests, and tropical monsoon forests. Rainfall reduces from east to west in the north, and forests are replaced by meadow steppe, typical steppe, desert steppe, steppe desert, typical desert, and extremely arid desert. Such rich ecological diversity nourishes great fungal diversity including diversity of EMMs. There are estimated to be around 900 wild edible mushroom species in China, of which about 350 are EMMs (Mao 1998; Lin 2016). The most economically important EMMs, such as truffles, matsutake, porcini, chanterelles, shimeji (*Lyophyllum*), shoro (*Rhizopogon*), and closely related species, all grow in China. In addition, although most species of *Scleroderma* are poisonous or inedible, the unique and interesting *Scleroderma yunnanensis* (Fig. 2.4) is endemic to Yunnan and is a popular EMM locally. In *Thelephora*, the only edible species is *Thelephora ganbajun*; the most expensive delicacy at wild mushroom markets in SW China (Fig. 2.5).

Fig. 2.4 *Scleroderma yunnanensis* at a wild mushroom market, Puwen, Yunnan

Fig. 2.5 *Thelephora ganbajun* (left) and *Tricholoma matsutake* (right) at a wild mushroom market, Nanhua, Yunnan

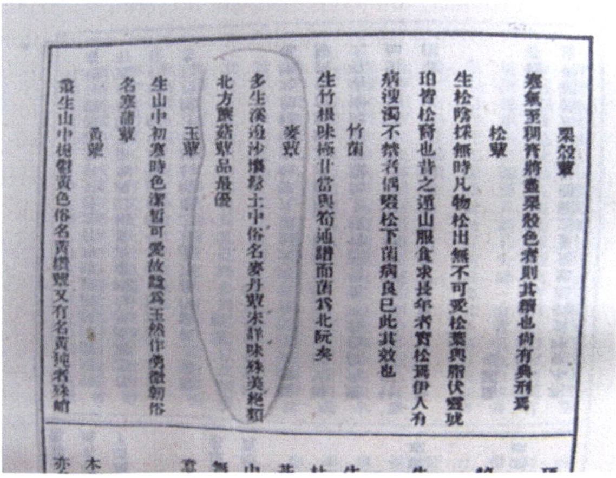

栗殼蕈
寒氣至稠齊將盡栗殼色者則其穉也尚有典刑焉
松蕈
生松陰採無時凡物松出無不可愛松葉與脂伏靈琥
珀皆松裔也昔之遁山服食求長年者實松與伊人有
病溲濁不禁者偶啜松下菌病良已此其效也
竹菌
生竹根味極甘當與笋通譜而菌于北阮矣
麥蕈
多生溪邊沙壤鬆土中俗名麥丹蕈未詳味殊美絕類
北方蘑菇蕈品最優
玉蕈
生山中初寒時色潔皙可愛故殺為玉然作羹微韌俗
名寒蒲蕈
黃蕈
叢生山中槐樹黃色俗名黃槿蕈又有名黃純者殊嫩

Fig. 2.6 *Rhizopogon* sp. recorded as "Wheat mushroom, growing on sandy soils" (circled in pencil)

2.2.1 *Truffles*

China has a long history of mushroom use, and a few of edible and medicinal species are recorded in ancient Chinese texts. Surprisingly, there are no comments about true truffles (*Tuber*). The only hypogeous basidiomycete of a *Rhizopogon* species was described in "Jun-Pu", an ancient mushroom text (Chen 1245) (Fig. 2.6), and known as "mai-xun" (the mushroom that fruits during the wheat harvesting season). This species is also recorded in some ancient Japanese mushroom books (Kobayashi 1983) with the name "song-lu" (dew of pine). There has been a recent trend in many Chinese publications and websites to call truffles "song-lu" (Wang and Liu 2011b).

It was not until 1985 when *Tuber taiyuanense*, the first Chinese truffle species was scientifically reported by B. Liu. Since then, more than 40 truffle species have been reported, with more than 70% of these new to science, and more species will yet be described from China. Quite a few are good edibles and have been commercialized, including *T. indicum*, *T. pseudohimalayense*, *T. sinoaestivum*, and *T. panzhihuanense*. Since the 1990s, increasing quantities of Chinese black truffles have been sold in international and domestic markets, and the harvesting and trading of truffles is becoming a multimillion dollar industry. These developments indicate that China has much richer truffle diversity than previously expected and may be a centre of world truffle biodiversity (Jeandroz et al. 2008; García-Montero et al. 2010; Wang and Liu 2011b; Bonito et al. 2013). China is clearly a newly emerging truffle country (Wang and Liu 2009).

2.2.1.1 Black Truffles

There is a tradition of harvesting truffles for food in Huidong County and Panzhihua, Sichuan Province, although there are no written records. A black truffle species has been collected and traded locally in the Huidong region for a long time (Fig. 2.7).

Local names for truffle species include "wu-niang-tong" (fruit without mother), "songmao-fuling" (*Wolfiporia cocos*-like mushroom beneath pine needles), and "God's fruit". One of these was named as a new species, *T. sinense*, by Chinese mycologists in 1989 based on the collection from Huidong, Sichuan (Tao et al. 1989). Now the *T. sinense* has been considered as one of species in the *T. indicum* complex. *Tuber indicum* was named based on a collection from the small town of Mussoorie on the southwest slope of India's Himalayan foothills in 1892. *Tuber sinense* has been treated as a synonym of *T. indicum* by both Chinese and

Fig. 2.7 A local truffle market in the Huidong region, Sichuan

international mycologists (Zhang et al. 2005; Wang et al. 2006; Chen 2007). However, *T. sinense* slightly differs morphologically from *T. indicum* in spore ornamentation and shape. Unfortunately, no DNA sequence was available from *T. indicum* at the time and then we could not firmly recognize whether *T. sinense* was a co-species of *T. indicum* or not. In 1987, Zhang and Minter re-examined the type collection of *T. indicum* and described a new species, *T. himalayense*, based on a part of this collection which has slightly different spore ornamentation. *Tuber formosanum* was initially reported from Taiwan but was not validly published due to the lack of a designated type. This species has been re-described with a lectotype (Qiao et al. 2013). *Tuber formosanum* is closely related to *T. indicum* but differs morphologically and phylogenetically (Huang et al. 2009; Qiao et al. 2013). Then, there are several names referring to the Chinese black truffles, including *T. indicum*, *T. sinense*, *T. formosanum*, and *T. himalayense*, but their morphological and phylogenetic relationships are not clear. It has been suggested that these should be referred to as "the *T. indicum* complex" (Wang et al. 2006; Chen 2007; Wang and Liu 2009). Species in this complex are mainly found in south-western China: Yunnan, Sichuan, and Tibet. Lately, however, it has been found in Beijing and Hebei Province of north China, Liaoning, and Heilongjiang Provinces of north-eastern China. The complex is also present in Japan and the Korean Peninsula (Fukiharu et al. 2006; Yamanaka et al. 2001). However, the phylogenetic relationships among these different populations from the south-west, north and north-east China need further research.

The *T. indicum* complex is closely related to *T. melanosporum* morphologically and phylogenetically (Song 2005; Zhang et al. 2005; Wang et al. 2006; Huang et al. 2009; Bonito et al. 2013; Qiao et al. 2013; Qiao et al. 2018) and they are comparable sibling species. However, the *T. indicum* complex has much more genetic and ecological diversity than *T. melanosporum*. Recent research into this complex has revealed that it has two pedigrees, ecotypes, subspecies or species: Lineages A and B (Bonito et al. 2013; Qiao et al. 2018). The *T. indicum* in Lineage A are found mainly on the Yunnan Plateau with its relatively cool and moist climate, while those in Lineage B are found in hot dry river valleys (Figs. 2.8 and 2.9). The *T. indicum* complex has mycorrhizal associations with both conifers and broadleaf trees, including those in the Pinaceae (*Pinus yunnanensis*, *P. armandii*, and another conifer *Keteleeria evelyniana*), the Fagaceae (*Quercus* spp., *Lythocarpus* spp., *Castanopsis* spp., *Castanea sativa*, and *Cyclobalanopsis delavayi*), the Corylaceae (*Corylus yunnanensis*), the Juglandaceae (*Platycarya strobilacea*) (Fig. 2.10), and the Salicaceae (*Populus yunnanensis*). This complex can adapt to a wide range of soils, from acid clay to calcareous (pH 6–8), with poor to rich organic matter. Soils that are calcareous or rich in organic matter have better production and better quality truffles (Fig. 2.11).

Since the 1990s, increasing quantities of Chinese black truffles of the *T. indicum* complex have been exported to Europe. One export problem has been the bad impression given by the sale of immature or degenerated Chinese truffles at European markets following a 1–2 month delay in arrival. However, members of the *T. indicum* complex are almost as delicious as *T. melanosporum*. These sister species share almost the same volatile organic compounds dominated by alcohols

Fig. 2.8 A *Tuber indicum* forest of *Keteleeria evelyniana* in a dry-hot valley in Yunnan

(48–57%), aldehydes (4–27%), and aromatics (9–30%). However, *T. indicum* contains a lower quantity of sulfur compounds (Splivallo et al. 2011; Fang et al. 2013). Recently it was discovered that the quantities of volatile organic compounds especially sulfur compounds and ketones, increase, while aldehydes decrease, with increasing maturity. Only mature fruit bodies of the *T. indicum* complex contained 2-methyl-1-propanol (up to 19%), a very important component of truffle flavor.

Fig. 2.9 A *Tuber indicum* forest of *Pinus armandii*, Yunnan Plateau, Yunnan

Fig. 2.10 *Platycarya strobilacea*

Unfortunately, the harvest of Chinese black truffles starts as early as August when they are not mature, with no or very little aroma (Fig. 2.12) (Wang unpublished).

Tuber sinoaestivum is another Chinese black truffle species (Fig. 2.13) which is closely related to the European summer truffle, *T. aestivum* (Zhang et al. 2011). It has been found in Huidong of Sichuan Province and in Gongshan, Baoshan, and

Fig. 2.11 *Tuber indicum* produced in soils with rich organic matter, Gongshan, Yunnan

Fig. 2.12 Immature *Tuber indicum* harvested in September

Dongchuan of Yunnan Province. It grows in habitats moister than the *T. indicum* complex. The *T. indicum* complex can share the same forests as *T. sinoaestivum*; however, the former has a much wider distribution. *Tuber sinoaestivum* is harvested from August to early the following year, but is not ripe until November. *Tuber pseudoexcavatum* is another, but less valuable, edible black truffle species in China (Fig. 2.14). It also often grows in the same habitats as the *T. indicum* complex. A recently published new black truffle species, *Tuber pseudobrumale*, was found closely related to *T. pseudoexcavatum* and *T. brumale* (Li et al. 2014).

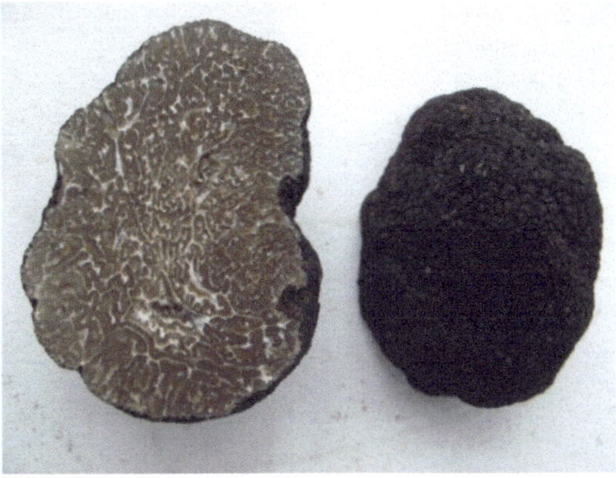

Fig. 2.13 *Tuber sinoaestivum*

Fig. 2.14 *Tuber pseudoexcavatum*

2.2.1.2 White Truffles

Over the last 30 years, a greater number of white truffle species have been reported from China compared to black ones; more than 30 white truffles have been discovered, and over 80% are new to science. All are edible but only a few have commercial value. *Tuber panzhihuanense* is a recently discovered and commercialized Chinese white truffle with large fruit bodies and good aroma (Fig. 2.15). *Tuber liyuanum* (Fig. 2.16) is another newly described white truffle species with commercial potential (Fan and Cao 2012). These species are found under pine forests in Yunnan and Sichuan. White truffles are better adapted to arid climates than black species in China. In the last 5–6 years, the climate has become drier due to much less rainfall in Yunnan and Sichuan, and the yield of the *T. indicum* complex has reduced dramatically, while more white truffles have been discovered.

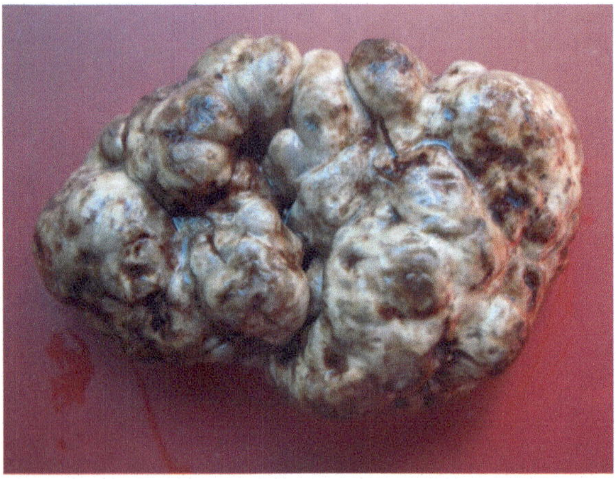

Fig. 2.15 *Tuber panzhihuanense*

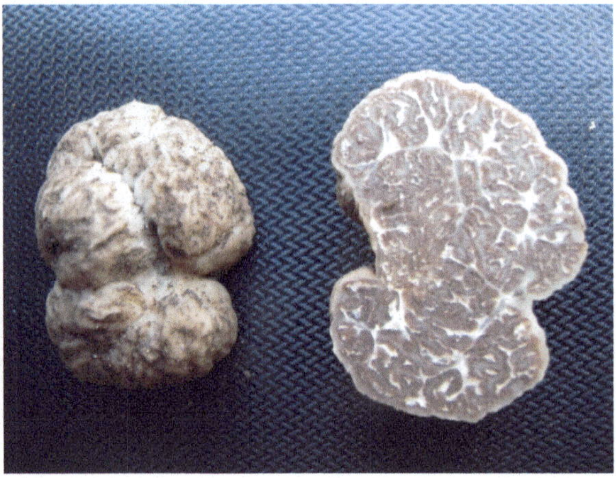

Fig. 2.16 *Tuber liyuanum*

2.2.1.3 Cuisine

With growing awareness of truffles by everyone in China, a whole new truffle cuisine has developed. It ranges from simple stir-fried or sautéed truffles to hot-pots with meats and vegetables, to variations in combinations of both Chinese and Western style cooking (Figs. 2.17 and 2.18). However, it will take time to understand the essence of truffle cooking.

With this new passion for processing truffles, the Chinese truffle industry has developed many different products, including truffle wines, sliced, dried, or frozen truffles, and frozen fresh truffles. The truffle wines, in particular, are considered mysterious beverages and are available in various brands (Fig. 2.19). Improved methods of storage, packaging, and transportation are needed to preserve the quality of harvested truffles.

Fig. 2.17 Crevettes grillées à la truffe [Grilled shrimps with truffle]

Fig. 2.18 Truffle rice

2.2.1.4 Conservation and Cultivation

The price of Chinese black truffles has soared from a few US dollars to US$150/kg as international and domestic demands have increased every year since 1990s. Commercial interests have resulted in large-scale exploitation involving unrestricted plundering of truffles. This has severely damaged truffle habitats and truffle resources (Figs. 2.20 and 2.21). Truffle production has recently declined so much that truffles in early exploited regions have become extinct. Management, restoration, and conservation of truffle resources in China are becoming urgent. The recent use of trained truffle dogs for harvest is a positive strategy. The establishment of natural reserves for truffles has begun. Recently, experimental plots for restoration of *T. indicum* resources have been established at the truffle exploited areas of Luliang, Yunnan. After only 2 years of efforts with re-inoculation and conservation, brûlés are recovering (Figs. 2.22 and 2.23) and the production of *T. indicum* increased dramatically (unpublished data).

Fig. 2.19 One of the truffle wines made with *Tuber indicum*

Fig. 2.20 A severely damaged *Tuber sinoaestivum* woodland of *Pinus armandii* in Huidong, Sichuan

The establishment of plantations of truffle-inoculated mycorrhizal trees has been successful. One plantation with *T. formosanum* in Taiwan and another four with *T. indicum* in Guizhou, Sichuan, and Yunnan have produced truffles (Fig. 2.24). A few more plantations are waiting for production (Fig. 2.25).

Fig. 2.21 *Populus yunnanensis* plantation in Luliang, Yunnan, before its conservation

Fig. 2.22 *Populus yunnanensis* plantation in Luliang, Yunnan, after its conservation

Fig. 2.23 Brûlés recovering after conservation of a *Populus yunnanensis* plantation, Luliang, Yunnan

Fig. 2.24 A *Tuber indicum* plantation with pine, chestnut and oak trees in Shilin, Yunnan

Both native and foreign tree species have been successfully mycorrhized by *T. indicum*, and the mycorrhizal seedlings have been planted out. The Chinese tree species are *Pinus yunnanensis*, *P. armandii*, *Keteleeria evelyniana*, *Cyclobalanopsis glauca*, *Quercus franchetii*, *Castanopsis delavay*, *Castanea molissima*, *Platycarya*

Fig. 2.25 A clear brûlé developed around planted, 3 years old *Castanea mollisima* tree inoculated with *T. indicum*, Panzhihua, Sichuan

strobilacea, *Corylus yunnanensis*, and *Populus yunnanensis*. The introduced trees are *Quercus robur*, *Corylus avellana*, and *Carya illinoensis*. Pine and broadleaf trees mycorrhized by *T. panzhihuanense* and *T. borchii* were also produced and outplanted in plantations. Production from seedlings inoculated with *T. melanosporum* and *T. aestivum* is in progress. Compared with the rapid deterioration of truffle environments, it is clear that these efforts are not enough to protect truffle resources.

2.2.2 *Desert Truffles and Other Hypogeous Fungi*

2.2.2.1 Desert Truffles

Five desert truffle species have been recorded in Chinese literature. Actually, only two species, *Mattirolomyces terfezioides* and *Terfezia parvocarpus*, are present in China (Tao 1988; Zhang 1990). *Mattirolomyces terfezioides* grows under pine trees and on farm soils. *Terfezia parvocarpus* has been found in Mount Wulingshan, Xinglong County, Hebei, growing with *Salix wallichiana* trees. However, the mycorrhizal status of these species needs further research. China has a large area of desert and semi-desert, and these have not been searched for desert truffles. It is quite possible that truffles occur in these areas. More than 100 other hypogeous species in 40 genera and 23 families have been reported in China (Li et al. unpublished data). Shoro and others are often found at mushroom markets.

2.2.2.2 Shoro

The "shoro" (*Rhizopogon roseolus*) was the fourth most commonly consumed mushroom in Japan, around 200 years ago (Okumura 1989), and it is recorded in ancient fungal books (Kobayashi 1983). Chinese shoro is known as "ji-yaozi" (chicken kidney) and is commonly collected and traded in Yunnan and Sichuan (Fig. 2.26). Its previous Latin name was *R. rubescens*, but it was found that Chinese shoro consists of three new species *Rhizopogon jiyaozi*, *R. flavidus*, and *R. sinoalbidus* (Li et al. 2016).

2.2.2.3 *Choiromyces*

Choiromyces aff. *meandriformis* is one of the edible ones, commonly found in Shanxi, Xinjiang, and the Inner Mongolian Region (Fig. 2.27). It was previously reported as *C. vinosus* and *C. alveolatus* (Tao 1988; Zhang 1990). The taxonomy of the Chinese *Choiromyces* species awaits further research. They are found mainly with *Picea* trees such as *P. schrenkiana* and *P. crassifolia*.

Fig. 2.26 *Rhizopogon jiyaozi*

Fig. 2.27 *Choiromyces* aff. *meandriformis*

Fig. 2.28 Yang-yanjing (*Astraeus hygrometricus*) sold at a mushroom market, Shidian, Yunnan

2.2.2.4 Yang-Yanjing (*Astraeus hygrometricus*)

Astraeus hygrometricus is another edible hypogeous species, which is known as "Yang-Yanjing" (Sheep's eye) or "Potato mushroom" and sold at local mushroom markets in Yunnan and Sichuan Provinces (Fig. 2.28). It grows with *Pinus yunnanensis* or *Castanea mollissima*. The outer layer, the exoperidium, is thick and initially inseparable from the inner layer (endoperidium). At maturity, the exoperidium bursts open into several pointed "rays" when moistened, but the rays are closed over the exoperidium when dry. When the exoperidium opens up, the spores are blown away. However, when the rays are closed, the spores are dispersed by insects. The insects can eat the gleba through opening up a hole (Fig. 2.29) from the bottom of fruiting body where there is a peridium gap (Fig. 2.30). There is another similar edible species, *Astraeus asiaticus*, in the north of Thailand (Phosri et al. 2004).

Fig. 2.29 A gleba of *Astraeus hygrometricus* eaten by insects through a hole

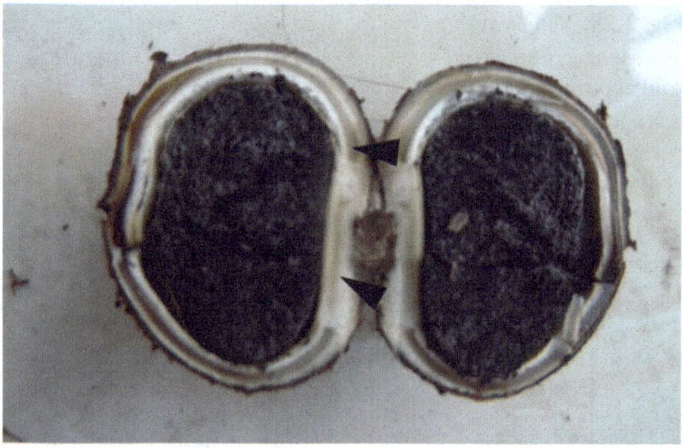

Fig. 2.30 *Astraeus hygrometricus*: Note the peridium gap (arrowheads) at the fruiting body bottom

2.2.3 Porcini and Related Edible Mycorrhizal Mushrooms

China has rich resources of edible boletes with hundreds of species (Li and Song 2001). *Boletus edulis* (porcini) and closely related species are the most important species for export in the form of sliced and dried or brined mushrooms. Annually, 50,000 tons of edible boletes are harvested and sold on domestic and international markets. Wild porcini and related boletes are found all over China, especially in the north-east, south, and south-west.

Boletus aff. *edulis* is one of the most popular edible mycorrhizal mushrooms in China and is called "white bolete" in Yunnan (Fig. 2.31). It grows with pine or oak or in mixed pine-oak forests and fruit from June to August. *Leccinum extremiorientale* is another good edible mycorrhizal fungus. As delicious as *B. edulis*, it gives large yields with great commercial value.

Fig. 2.31 *Boletus* aff. *edulis* sold at a wild mushroom market in Kunming, Yunnan

Fig. 2.32 One of "Jianshouqing" boletes, *Boletus speciosus*

Boletus bicolor, *B. speciosus*, *B. magnificus*, *B. subsplendidus*, and other similar species (Fig. 2.32), known as "jian-shou-qing" (turning blue when bruised or cut), are likeable boletes and are even more popular than *B.* aff. *edulis* in Yunnan. Members of this "jian-shou-qing" group can cause hallucinogenic reactions and even death, but the local people continue to enjoy them. Five poisonous boletes have been detected commonly mixed with commercial edible ones in Yunnan mushroom markets. These are *Tylopilus eximius*, *T. microsporus*, *Pulveroboletus ravenelii*, *Chiua virens*, and *Heimioporus retisporus* (Li et al. 2011).

Research into the taxonomy, ecology, and processing of Chinese boletes is a huge task due to their diversity. Conservation and cultivation has begun, but without any sound progress at the moment.

Suillus luteus and closely related species are very common EMMs all over China. They are commonly dried and kept for winter consumption. *Suillus luteus* has been

Fig. 2.33 *Phlebopus portentosus*

successfully cultivated by mycorrhization of *Pinus massoniana* in the Lishui region, Zhejiang (Han and Ren 2016).

Phlebopus portentosus is an edible mushroom in the Boletinellaceae (Fig. 2.33). It has been considered a facultative mycorrhizal fungus (Kumla et al. 2012). However, based on over 10 years of our research, this fungus shouldn't be considered ectomycorrhizal and can be cultivated using methods for saprophytic mushrooms (Ji et al. 2011; Zhang et al. 2017).

2.2.4 Russulaceae

There are many edible mycorrhizal *Lactarius* species in China. Hundreds and even thousands of kilograms of edible *Lactarius* mushrooms change hands daily at wild mushroom markets. *Lactarius deliciosus*, *L. hatsudake*, *L. volemus*, *L. vividus*, and *L. hygrophoroides* are important edible mycorrhizal species in China. *Lactarius hatsudake* is the most favorable edible mushroom in Hunan Province, and it is known as "Hanjun" (chill or cold season mushroom). Seedlings of *Pinus massoniana* mycorrhized by *L. hatsudake* in Hunan have produced fruit bodies 3–4 years after planting in plantations in 2001 and now give an average yearly production of 670 kg/ha (Tan et al. 2008; Wang et al. 2013). Research into the production of mycorrhizal trees inoculated with *Lactarius* species from the section *Deliciosi* has been a successful and recent focus in Yunnan (Wang et al. 2019, see also Chap. 6 by Yu et al.).

Russula virescens, known as "qingtou-jun" (green head mushroom), is a popular edible species in south-western China. Another favorable and expensive edible *Russula* is *R. griseocarnosa*; it was previously misidentified as *R. virosa* (Wang et al. 2009) and is called "dahong-jun" (bright red mushroom) by the local people (Fig. 2.34). Consumers believe that this mushroom is beneficial to human health,

Fig. 2.34 *Russula griseocarnosa*

especially for pregnant women, and there is scientific evidence to support the claim (Wang et al. 2009). Dried dahong-jun can be sold for over 800 Chinese yuan/kg (around US$130/kg). This species is found under evergreen oaks in tropical and sub-tropical China.

The yield of *L. volemus* and "dahong-jun" has been increased by the use of *L. volemus* spores as inocula in natural habitats in Yunnan (see also Chap. 6 by Yu et al.).

2.2.5 Shimeji

Lyophyllum shimeji (Honshimeji, true-shimeji) is a well-known edible ectomycorrhizal fungus, as famous as matsutake in Japan. *Lyophyllum shimeji* and its closely related species are also considered one of most delicious mushrooms in China, especially in south-western and north-eastern China (Fig. 2.35). It is widely distributed in East Asia and northern Europe. Nineteen *Lyophyllum* species have been reported from China, including *L. shimeji*, which is mainly associated with pines. It grows with *Pinus yunnanenis* and *P. kesiya* var. *langbianensis* at lower elevations, but can be associated with *Quercus aquifolioides* and *Pinus densata* at higher elevations (over 3000 m). Other edible *Lyophyllum* species, such as *L. decastes* and *L. fumosum*, are also present in China. Based on research by Japanese mycologists, it has two genotypes: mycorrhizal and saprophytic and can be cultivated using mycorrhizal or saprophytic methods (Ohta 1994, 1998). Meanwhile, *Lyophyllum shimeji* is not a typical ectomycorrhizal fungus as it produces a very thin and unstructured mantle in the mycorrhizae.

Fig. 2.35 *Lyophylum shimejii*

Fig. 2.36 A species from the *Amanita hemibapha* complex, Diqing, Yunnan

2.2.6 *Amanita*

"Caesar's mushroom" is a yellow–orange *Amanita* species that was once called *A. caesarea* in China, but has now been properly identified as one member of the *Amanita hemibapha* complex (Fig. 2.36). It grows in moist areas of China, especially the south-west, and has large genetic variation. Another good edible *Amanita* species is *A. vaginata.* These edible *Amanita* species are harvested for food and are traded locally. Some poisonous ones are sometimes also collected and sold at local markets. Unfortunately, poisoning incidents caused by *Amanita* species occur every mushroom season in China.

2.2.7 Chanterelles and Craterellus

The *Cantharellus cibarius* complex and related species, and *Craterellus cornucopioides* and its related species, are some of the most popular EMMs in the world and are very common at wild mushroom markets in China (Fig. 2.37). *Cantharellus cinnabarinus* is one of the most common species in Yunnan but is less appreciated. At least 15 species of *Cantharellus* have been recorded from southwestern China; 7 of these are new to science. They grow with many conifers (*Abies*, *Picea*, and *Pinus*) and broadleaf trees (*Quercus*, *Carpinus*, and *Castanea*) in alpine and subalpine regions. In southwestern China, they are the main group of EMMs associated with pine forests, such as *P. yunnanensis*, *P. massoniana*, *P. kesiya* var. *langbianensis*, *P. densata*, and exotic pines. The *Cantharellus cibarius* complex includes some of the most important EMMs exported to Europe from China (Tian et al. 2009). Despite its economic and ecological importance, its cultivation has not been studied in China.

2.2.8 Albatrellus

Nineteen species of *Albatrellus* have been recorded in China with several species new to science. A few have been examined for their biotrophic status in the field and the laboratory, and they have been found to be ectomycorrhizal (Zheng 2010). However, further research is needed to confirm their mycorrhizal status and features. Some, such as *Albatrellus elisii* known as "yellow huzhang-jun" (yellow tiger-paw mushroom), are collected for food and are traded locally (Fig. 2.38). See also Chap. 6 by Yu et al. for more details about this genus.

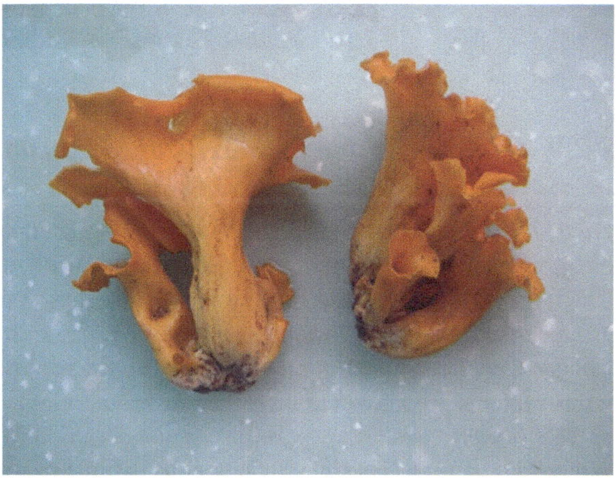

Fig. 2.37 *Cratarellus aureus* at a market in Puer, Yunnan

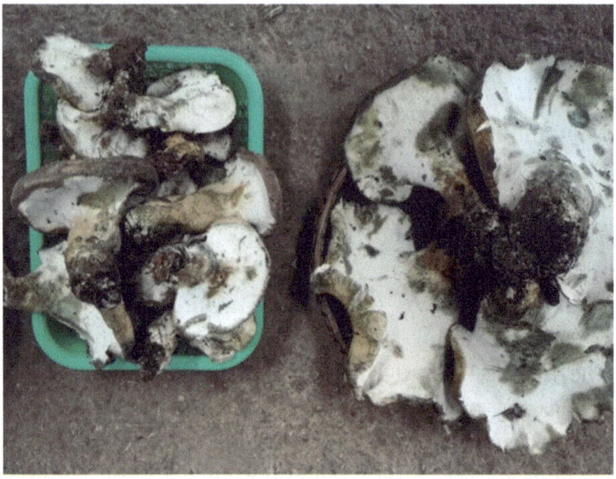

Fig. 2.38 *Albatrellus elisii* sold at a mushroom market, Yunnan

Fig. 2.39 *Thelephora ganbanjun* growing at a forest of *Pinus yunnanensis*, Yunnan

2.2.9 Unique Edible Mycorrhizal Mushrooms

Thelephora ganbajun (Fig. 2.39) and closely related species (2–3 species) are the most expensive and highly favored edible mycorrhizal fungi in Yunnan (see also Chap. 6 by Yu et al.). Research into cultivation has made good progress. Ganbajun produces sclerotia that plays an important role in its life circle (Fig. 2.40). Mycelial isolates have been made (unpublished data). Environment management and habitat improvements are known to increase mushroom yield.

Scleroderma species are widespread and common from tropical to boreal regions and are described as inedible or poisonous in mushroom books and scientific papers. Very interestingly, Yunnan has one edible and sought after new species *S. yunnanense* (Fig. 2.41) (see Chap. 6 by Yu et al. for more details).

Fig. 2.40 Sclerotia of *Thelephora ganbanjun*

Fig. 2.41 *Scleroderma yunnanense* collected from a *Pinus yunnanensis* forest, Yunnan

Acknowledgements This paper has been financed by grants from the National Science Foundation of China (No. 31560009), the Special fund of Sci-Tech Innovation System Construction of the Yunnan Institute for Tropical Crops Research of Yunnan Province (No. RF2015-5), the Special fund for technical development of Yunnan Research Institutions (2015DC014) and the Chinese Academy of Sciences' "the light of the west" personnel training project (No. 326). We would like to thank Toni Atkinson for editing the paper and Shangping Wan and Lin Li for their help with some references.

References

Bonito G, Smith ME, Nowak M et al (2013) Historical biogeography and diversification of truffles in the Tuberaceae and their newly identified southern hemisphere sister lineage. PLoS One 1(8):E52765. https://doi.org/10.1371/journal.pone.0052765

Chang ST, Miles PG (1997) Historical record of the early cultivation of *Lentinus* in China. Mushroom J Tropics 7(3):31–37

Chang ST, Miles PG (1989) Edible Mushrooms and Their Cultivation. Boca Raton, Florida Press.
Chen J (2007) Taxonomy and phylogeny of the genus *Tuber* in China (Pezizales, Ascomycetes, Ascomycotina). PhD Thesis, Graduate School of the Chinese Academy of Sciences, Beijing
Chen RY (1245) Jun-Pu. In: Kobayashi (ed) Japanese and Chinese mushroom history & Kolklore 1983. Guangchuan Press, Tokyo, p 87
Fan L, Cao JZ (2012) Two new species of white truffle from China. Mycotaxon 121:297–304
Fang SP, Pu B, Chen AJ et al (2013) Effects of maturity degree on the aroma compounds of *Tuber indicum*. Mycosystema 32(6):1020–1027
Fukiharu T, Obajima T, Shimadate R et al (2006) Fungal flora in Chiba Pref. Central Japan (IV) *Tuber indicum*, the first record of hypogeous ascomycetes collected in Chiba. J Natl Hist Museum Inst Chiba 9(1):1–6
García-Montero LG, Díaz P, Massimo GD et al (2010) A review of research on Chinese *Tuber* species. Mycol Progress 9:315–335
Han CZ, Ren XM (2016) Research progress on *Suillus luteus*. Northern Hortic 6:195–199
Huang JY, Hu HT, Shen WC (2009) Phylogenetic study of two truffles, *Tuber formosanum* and *Tuber furfuraceum*, identified from Taiwan. FEMS Microbiol Lett 294:157–171
Jeandroz S, Murat C, Wang YJ et al (2008) Molecular phylogeny and historical biogeography of the genus *Tuber*, the 'true truffles'. J Biogeogr 35:815–829
Ji KP, Cao Y, Zhang CX et al (2011) Cultivation of *Phlebopus portentosus* in southern China. Mycol Prog 10:293–300. https://doi.org/10.1007/s11557-010-0700-7
Kobayashi Y (1983) Japanese and Chinese mushroom history and folklore. Hirokawa, Tokyo, p 136
Kumla J, Bussaban B, Suwannarach N et al (2012) Basidiome formation of an edible wild, putatively ectomycorrhizal fungus, *Phlebopus portentosus* without host plant. Mycologia 104:597–603. https://doi.org/10.3852/11-074
Li L, Zhao YC, Zhou DQ et al (2016) Three new species of *Rhizopogon* from Southwest China. Phytotaxa 282(2):151–163
Li SH, Zhao YC, Yu FQ et al (2011) Systematics of the easily confusing poisonous boletes from Yunnan wild mushroom markets. Edible Fungi China 5:9–12
Li SH, Heng LY, Liu CY et al (2014) Two new truffles species, *Tuber alboumbilicum* and *Tuber pseudobrumale* from China. Mycol Prog 13:1157–1163
Li TH, Song B (2001) Species and distributions of Chinese edible boletes. Acta Edulis Fungi 9(2):22–30
Lin XM (2016) Ectomycorrhizal fungi in China (English). Science Press, Beijing
Mao XL (1998) Economic fungi of China. Science Press, Beijing
Ohta A (1994) Production of fruiting-bodies of a mycorrhizal fungus, *Lyophyllum shimeji*, in pure culture. Mycoscience 35:147–151
Ohta A (1998) Culture conditions for commercial production of *Lyophyllum shimeji*. Mycoscience 39:13–20
Okumura A (1989) Healthy mushroom dishes. Hoikusha, Osaka
Phosri C, Watling R, Martín MP et al (2004) The genus *Astraeus* in Thailand. Mycotaxon 89(2):453–463
Qiao P, Liu PG, Hu HT et al (2013) Typification of *Tuber formosanum* (Tuberaceae, Pezizales, Ascomycota) from Taiwan, China. Mycotaxon 123:293–299
Qiao P, Tian W, Liu PG et al (2018) Phylogeography and population genetic analyses reveal the speciation of the *Tuber indicum* complex. Fungal Genet Biol. https://doi.org/10.1016/j.fgb.2018.02.001
Splivallo R, Ottonello S, Mello A (2011) Truffle volatiles: from chemical ecology to aroma biosynthesis. New Phytol 189:688–699
Song MS (2005) Taxonomic and molecular systematic studies on *Tuber* in China. PhD Thesis, Beijing Institute of Microbiology, Chinese Academy of Sciences, Beijing
Tan ZM, Danell E, Shen AR (2008) Successful cultivation of *Lactarius hatsudake*—an evaluation with molecular methods. Acta Edulis Fungi 15(3):85–88

Tao K (1988) Taxonomic study of hypogeous fungi from China. MSc Thesis, Shanxi University, China

Tao K, Liu B, Zhang DC (1989) A new species of the genus *Tuber* from China. J Shanxi Univ 12:215–218

Tian XF, Liu PG, Shao PG (2009) Research status and prospect of the genus *Cantharellus* Adans. ex Fr. (Cantharellaceae, Basidiomycota). Microbiology 36(10):1577–1586

Wang HS (1979) Basic natures of Chinese Flora. Acta Geograph Sin 34(3):224–237

Wang XH, Yang ZL, Li YC et al (2009) *Russula griseocarnosa* sp. nov. (Russulaceae, Russulales), a commercially important edible mushroom in tropical China: mycorrhiza, phylogenetic position, and taxonomy. Nova Hedwigia 88(1–2):269–282

Wang Y (1995) *Tricholoma matsutake*. PhD Thesis, Crop & Food Research, University of Otago, New Zealand

Wang Y, Chen YL (2014) Recent advances of cultivation of mycorrhizal Mushrooms. In: Solaiman ZM et al (eds) Mycorrhizal fungi: use in sustainable agriculture and land restoration, Soil biology, vol 41, pp 395–397. https://doi.org/10.1007/978-3-662-45370-4_23

Wang R, Guerin-Laguette A, Huang LL, Wang XH, Butler R, Wang Y, Yu FQ (2019) Mycorrhizal syntheses between *Lactarius* spp. section *Deliciosi* and *Pinus* spp. and effects of grazing insects in Yunnan, China. Can J For Res 49:616–627. https://doi.org/10.1139/cjfr-2018-0198

Wang Y, Cummings N, Guerin-Laguette A (2013) Cultivation of Basidiomycete Edible Ectomycorrhizal Mushrooms: *Tricholoma*, *Lactarius* and *Rhizopogon*. In: Zamboneli A, Bonito G (eds) Edible Ectomycorrhizal Mushrooms, Soil biology, vol 34. https://doi.org/10.1007/978-3-642-33823-6_16

Wang Y, Evans LA, Hall IR (1997) Ectomycorrhizal fungi with edible fruiting bodies. 1. *Tricholoma matsutake* and related species. Econ Bot 51(3):311–327. https://doi.org/10.1007/978-3-642-33823-6_16

Wang Y, Fuqiang Y, Zhang CX et al (2017) *Tricholoma matsutake*–an edible mycorrhizal mushroom of high socio-economic relevance in China. Scientia Fungorum 46:55–61

Wang Y, Hall IR (2004) Edible mycorrhizal mushrooms, challenges and achievements. Can J Bot 82:1063–1073

Wang Y, Hall IR (2006) Matsutake – a natural biofertilizer ? In: Hand book of Microbial Biofertilizers. Ed. by M. Raid. Haworth Press, Binghampton, N.Y.

Wang Y, Liu PG (2009) Achievements and challenges of research on Chinese truffles. Acta Bot Yunnanica 31(Suppl 16):1–9

Wang Y, Liu PG (2011a) Truffle research and cultivation in China. In: 3o Congresso Internazionale di Spoleto sul Tartufo, Spoleto, 25–28 November, 2009, pp 258–263

Wang Y, Liu PG (2011b) Verification of Chinese names of truffles and their conservation in natural habitats. Plant Divers Resour 33(6):625–642

Wang Y, Liu PG, Chen J, Hu HT (2008) China – a newly emerging truffle-producing nation. In: INRA and Le Causse Corrézien (ed) Proceedings of the International Conference 'La culture de la truffe dans le monde' [Truffle cultivation in the world], Brive-la-Gaillarde, France – 2 Février 2007, pp 35–44

Wang YJ, Tan ZM, Zhang DG et al (2006) Phylogenetic and populational study of the *Tuber indicum* complex. Mycol Res 110:1034–1045

Yamanaka K, Namba K, Tajiri A (2001) Asian Black truffles. In: Courvoisier M, Olivier JM, Chevalier G (eds) Proceedings of the 5th Congrès International Science et Culture de la Truffe, Aix-en-Provence, 4–6 Mars 1999. Fédération Française des Trufficulteurs, Paris, pp 117–120

Yang ZL (2015) Atlas of the Chinese species of Amanitaceae. Science Press, Beijing

Yu FQ (2007) Taxonomy and phylogeny of *Tricholoma matsutake* group (Tricholomataceae, Agaricales) and its allied species. PhD dissertation, Kunming Institute of Botany, Chinese Academy of Sciences, Kunming

Zang M (2006) Flora fungorum sinicorum 22 Boletaceae (I). Science Press, Beijing, pp 1–215

Zhang BC (1990) Study on systematic position of *Tuber* at Tuberales and Chinese *Tuber* species. PhD Thesis, Beijing Institute of Microbiology

Zhang CX, Xu XE, Liu J et al (2013) *Scleroderma yunnanense*, a new species from South China. Mycotaxon 125:193–200
Zhang CX, He MX, Liu J et al (2017) A unique edible bolete – *Phlebopus portentosus* in southern China. J Agric Sci Technol B 7(6):386–394
Zhang JP, Liu PG, Chen J (2011) *Tuber sinoaestivum* sp. nov, a renowned edible truffle in southwestern China and its biogeographical implication. Mycotaxon 122:73–82
Zhang LF, Yang ZL, Song DS (2005) A phylogenetic study of commercial Chinese truffles and their allies: taxonomic implications. FEMS Microbiol Lett 245:85–92
Zheng LD (2010) Taxonomy and phylogeny of Albatrellaceae. PhD Thesis, Graduate School of Chinese Academy of Sciences, Beijing

Chapter 3
Climate Change, Biotechnology, and Mexican Neotropical Edible Ectomycorrhizal Mushrooms

Jesús Pérez-Moreno, Magdalena Martínez-Reyes, Faustino Hernández-Santiago, and Ivette Ortiz-Lopez

As for the future, it is not a question of predicting it, but of making it possible

Antoine de Saint-Exupéry, Citadelle (1948)

3.1 Rationale

Currently, we are living in a geological era called Anthropocene. This term was proposed in 2000 in a meeting in Mexico, by the Dutch chemist Paul Crutzen, winner of the Nobel Prize in Chemistry in 1995, and later on published along with the American biologist Eugene F. Stoermer (Crutzen and Stoermer 2000). The main reasons argued by these authors to propose this geological era were: the enormous increase of human population since the industrial revolution; the excessive use of fossil fuels whose production took hundreds of millions of years; the release of nitrous oxides associated with the combustion of this non-renewable energy sources; the excessive release of SO_2 to the atmosphere by coal and oil burning; the transformation of the land surface by human activities estimated by Lovelock (1992) to be of the order of 70%; the increase of the extinction rates of the species, e.g., from 1000 to 10,000 times in tropical rain forests estimated since more than two decades and a half ago to grow excessively (Wilson 1992); the increase of greenhouse gases into the atmosphere including mainly CO_2 by more than 30% (Vitousek et al. 1997) and CH_4 by even more than 100%; and the release of toxic substances such as chlorofluorocarbons capable of destroying atmospheric ozone. All these changes originated by human activity and the huge population explosion that has taken place from 790 million in 1750 to 7.7 billion people today (Population Matters 2019).

J. Pérez-Moreno (✉) · M. Martínez-Reyes · F. Hernández-Santiago · I. Ortiz-Lopez
Colegio de Postgraduados, Campus Montecillo, Edafología,
Texcoco, estado de México, Mexico
e-mail: jperezm@colpos.mx

J. Pérez-Moreno et al. (eds.), *Mushrooms, Humans and Nature in a Changing World*, https://doi.org/10.1007/978-3-030-37378-8_3

After his proposal, the concept of Anthropocene has been analyzed on stratigraphic grounds, showing that the earth has suffered strong changes that have leaved a global stratigraphic signature distinct from that of Holocene or of previous Pleistocene interglacial phases, encompassing novel biotic, sedimentary, and geochemical changes (Zalasiewicz et al. 2008). Since its re-evaluation by Crutzen (2002), the concept has gained acceptance in the recent scientific literature (Chakraborty 2019; Svensen et al. 2019) and currently there is for example a scientific journal entitled *Anthropocene* that started in 2013, only devoted to research related with this topic (Anthropocene 2019).

The concept of Anthropocene is strongly linked with the biggest environmental problem that humankind faces currently: global change. Land use change, deforestation, destruction of natural ecosystems, and other human activities have contributed to a thousandfold increase in global extinctions in the Anthropocene compared to the presumed prehuman background rate, with negative profound effects on ecosystem functioning and service (Hooper et al. 2012; Rosenberg et al. 2019). Two iconic examples are the following: in a study over 27 years in Germany a midsummer decline of 82% of flying insect biomass was recorded (Hallmann et al. 2017), and it has been estimated that in the United States, the population of monarch butterflies fell by 80% in the last 20 years (Vidal and Rendón-Salinas 2014). These evidences have been the inception that the sixth Earth's mass extinction has already started (Barnosky et al. 2011). Despite the fact that it has been estimated that around 30% of the world's land area is covered by forests; they are being destroyed at an alarming rate. According to the World Bank, in 1990, the forested global area was 4128 million hectares and by 2015 this area had decreased to 3999 million hectares, i.e., between 1990 and 2015, the earth lost 1.3 million square kilometres of forest, which accounted for over 3% of the total forest area (FAO 2015; World Bank 2016). Deforestation is the major cause of loss of biodiversity, and therefore natural forest conservation is vital for stemming this loss (IPCC 2019). The distribution of the forests on earth is not homogeneous, ten countries hold around 90% of the primary forest of the world (FAO 2005). The decisions, actions, and success related with reforestation, afforestation, or establishment of forest plantations in these ten countries will be of paramount importance in the worldwide maintenance of the forest masses in the future. An iconic recent example of the interconnections of global change and deforestation occurs some months ago. One of the most important forested areas, and wildlife reservoirs worldwide is the Amazon, with an estimated 31 of primary forests of the planet (FAO 2005). According to satellite images, Brazil's National Institute for Space Research counted more than 41,000 "fire spots" between 1 January and 24 August last year (Escobar 2019) which might account for the largest deforestation in the last decades in Brazil, with an estimated loss of 6159 km^2, which is four times greater than those registered in 2010 (Mongabay 2019). Additionally, habitat loss are increasing the risks from zoonoses, including Ebola, Zika and COVID19 (Zohdy et al. 2019; Gropp 2020), which have had direct costs of more than US$100 billion, with that figure jumping to several trillion dollars if the outbreaks become human pandemics (UNEP 2016), like is happening now with COVID19.

Globally, Mexico ranks 8th in the ranking of native forests (FAO 2005). However, historically, the deforestation rates of the country have been one of the highest in the planet. For example, in the period from 1990 to 2005, Mexico lost an average of 318,000 hectares of forest annually. This loss represented one of the highest deforestation rates in the world, being exceeded from 1990 to 2010 only by Brazil, Indonesia, Sudan, Myanmar, Nigeria, and Tanzania (GFR 2010). As a consequence, Mexico currently presents serious environmental problems in terms of soil erosion and greenhouse gas emissions, the country occupying the 13th place in the latter category internationally according to Myers-Madeira (2008) and one of the first 12 Carbon dioxide emitting countries according to UNEP (2012) and Olivier et al. (2013). Mainly during the last decades, about 2 billion hectares of land have been degraded due to human activities. This constitutes about 15% of the total surface of the planet, which has affected more than 1200 million people. In the particular case of Mexico, it has been estimated that 71% of the country's soils are degraded, that is, 142 million of the country's 200 million hectares have physical, chemical, or biological degradation problems and 1.42 million hectares of soil are lost every year (Moncada de la Fuente et al. 2013). Reforestation, which emerge a solution to alleviate this problem, represents a complex challenge, which includes among other technical limitations, a low survival rate when trees are transplanted from greenhouse or nurseries to field. This survival is particularly dramatic for some species of enormous forest importance such as pines whose transplant survival rate varies from 0% to 20% in Mexico. This is particularly alarming because the country is one of the main centres of biodiversity of ectomycorrhizal trees with forest importance, including, for example, 176 species of *Quercus* (Valencia and Flores-Franco 2006) and 72 taxa of *Pinus* (Perry 1991). Additionally, the country has a large cultural diversity which includes 68 ethnic groups, each one with its own cosmovision, natural resources management, and traditional knowledge developed during millennia. As a consequence, the country holds one of the largest diversities of edible wild mushrooms worldwide, with more around 450 species consumed by the different ethnic groups and more than 5000 names used to designate them. This genetic diversity and ethnological traditional knowledge constitute one of the most important in the world (Pérez-Moreno et al. 2014) (Fig. 3.1). However, Mexico faces enormous challenges which include the change of land use, a quick acculturation process with the associated loss of traditional knowledge generated during millennia, and migration of young people from small communities to big cities. One of the reasons that explains the low rate of survival in the case of pine trees is the lack of ectomycorrhizal symbionts in the roots of most forest species produced in the greenhouse, whose presence is required when they grow in natural conditions. However, inoculation with ectomycorrhizal fungi has not been traditionally included in the production of pine trees in the country. Therefore, the biotechnological development of inoculation with trees of forest importance with ectomycorrhizal fungi is an urgent necessity in Mexico, in order to have successful reforestation. In this context, a criterion that has gained enormous importance in the selection of ectomycorrhizal symbionts is currently its edibility, due to the enormous social, economic, and environmental importance of ectomycorrhizal edible fungi in Mexico.

Fig. 3.1 Mexico is one the largest genetic and cultural reservoirs of edible ectomycorrhizal mushrooms. (**a**) *Pinus hartwegii* forest at 3800 masl; (**b**) *Turbinellus floccosus* a popular edible Fig. 3.1 (continued) mushroom in *Abies religiosa* forest; (**c**) "Hongueras" or mushrooms gatherers ready to start the recollection of wild edible mushrooms; (**d**) *Boletus edulis* s.l. one of the most appreciated mushrooms worldwide; (**e**) *Amanita basii*, one of the species belonging to the Caesar's mushroom complex in central Mexico

The mycorrhiza is a symbiosis that has played a key role in the structure and functioning of the land ecosystems. Actually, the colonization of the land was only possible with the combination of plants and its associated mycorrhizal fungi, according to fossil evidence from the Ordovicic and Devonic periods, around 460 million years ago (Remy et al. 1994; Redecker et al. 2000). Currently, one type of mycorrhiza called ectomycorrhiza plays a key role in the maintenance of the forest masses mainly in the low latitudes, high altitudes, and areas with low mineralization rates according to the recently coined "Read's rule" (Steidinger et al. 2019). These authors estimated that around 60% of tree stems on earth have their roots colonized by ectomycorrhizal fungi. The ectomycorrhiza is a widely distributed symbiosis around the planet mainly in boreal and temperate forests in the northern Hemisphere and also in some subtropical and tropical areas in South America, Africa, and Eastern Asia (Read and Perez-Moreno 2003). This symbiosis is established between around 8500 plants (Brundrett and Tedersoo 2018) and more than 20,000 fungal species (Comandini et al. 2012). Through this symbiosis, the forest trees are able to have access to mineral and organic nutrients, and water; and in exchange, they provide carbon compounds to their associated mutualistic mycorrhizal fungi (Smith and Read 2008). Recently, Chen et al. (2019) demonstrated for the first time that ectomycorrhizal symbiosis also play a key role in the control of root pathogens under field conditions, and therefore strongly influence the forest structure of subtropical forests in China, by suppressing the attack of pathogens, increasing the seedling survival rates, and therefore driving the species composition of forests ecosystems. Additionally, ectomycorrhizal symbionts increase the tolerance of host trees to drought, through the involvement of leaf membrane lipid metabolism and increased ability to maintain a stable chloroplast membrane functional integrity under water stress (Sebastiana et al. 2019). It has been demonstrated that ectomycorrhizal fungi also provide protection to their host to potentially toxic elements through a number of mechanisms including extracellular chelation by excreting di- and tricarboxylic acids or oxalic acids, cell wall binding mainly in the Hartig net and external mycelium, reduced uptake, and increased efflux of heavy metals and production of intracellular thiol-rich chelators, such as methallothioneins, glutathionines, phytochelatins, and compartmentation within vacuoles (Khullar and Reddy 2016, 2019; Kalsotra et al. 2018). Due to all of these advantages, countries with a forest tradition have usually incorporated mycorrhizal technologies into their production programs of forest trees. The main sources for ectomycorrhization have been either the use of mycelium or spores. In the case of spores, usually Gasteromycetes have been used, mainly included in the genera *Pisolithus*, *Rhizopogon*, and *Scleroderma* due to the huge amounts of spores that they produce (Charya and Garg 2019). However, the use of pilea of Agaricales and hymenium (which constitute the fertile part of the sporomes, containing the spores) of Boletales has received little attention worldwide. In this scenario, two decades ago, we decided to start a research line in order to produce controlled ectomycorrhized plants inoculated with native edible ectomycorrhizal fungi. The challenge was to generate a biotechnological innovation simple, efficient, and cheap. In this work, we present a synthesis of the advances of this biotechnology that has been developed during the last 20 years, describing the beneficial effects assessed in more than 100 combinations of ectomycorrhizal phytobionts and mycobionts and the future challenges in order to scale this biotechnology and produce controlled mycorrhized plants with reforestation, afforestation, forest plantations, and restauration of degraded areas in the country.

3.2 Summarized Analysis

More than two decades ago, we started in Colegio de Postgraduados some researches related to the controlled mycorrhization of trees native to Mexico by using mycelium of ectomycorrhizal fungi as a source of inoculum (Pérez-Moreno 1992). However, this biotechnology of inoculum production has the disadvantages of being expensive, requiring complex infrastructure, having a short shelf life and reduced genetic diversity, given that from the more than 20,000 species of known ectomycorrhizal fungi, only a small number has possibilities to be isolated in culture media. For these reasons, the possibility of generating a different alternative began to be a goal. Knowing that there was a whole body of knowledge related to the use of fungal spores of ectomycorrhizal Gasteromycetes mainly from species belonging to the genera *Pisolithus*, *Scleroderma*, and *Rhizopogon* as a source of inoculum (Lakshangy and Sandeep 2019; Repác 2011; Martin et al. 2003; Turjaman et al. 2005; García-Rodríguez et al. 2006; Castellano et al. 1985; Bruns et al. 2009), an alternative source of spores was searched. Studies were then started during the following years, first obtaining spore prints of Agaricales and Boletales, on sterilized paper, and then preparing solutions by adding sterile distilled water to inoculate trees with 10^6 to 10^8 spores per plant. Species of the genera *Amanita*, *Boletus*, *Cantharellus*, *Hebeloma*, *Helvella*, *Laccaria*, *Ramaria*, *Russula*, *Suillus*, and *Turbinellus* were mainly tested. In all cases, low or no ectomycorrhizal colonization were observed, independently of the evaluated host trees. From the beginning, the idea of generating a biotechnology fulfilling three criteria was raised: methodological simplicity, low cost, and high mycorrhization efficiency. This with the purpose of generating a useful technology for the mosaic of social, financial, and cultural conditions of nurseries in Mexico, where around 80% have only very basic infrastructure. Then we began to evaluate as a source of inoculum pilea of Agaricales and hymenia of Boletales, included in more than 20 genera. One of the criteria for selecting mycobionts was their edibility, due to the fact that Mexico constitutes one of the most important reservoirs of wild edible fungi worldwide, with more than 450 wild species of edible fungi (Fig. 3.1). Additionally, we used as selection criteria: pioneer species of ectomycorrhizal mushrooms, which grew in conditions of low fertility and which were abundant in inoculum banks in Mexican natural ecosystems (Obase et al. 2009; Ishida et al. 2008; Nara 2006; Garibay-Orijel et al. 2013). Since then, we have studied mainly 18 species of edible ectomycorrhizal mushrooms included in the genera *Hebeloma*, *Helvella*, *Laccaria*, and *Suillus* (Table 3.1). In the case of tree species, the investigations have been mainly focused in pine species native to

Table 3.1 Studied combinations of Mexican Neotropical trees inoculated with edible ectomycorrhizal mushrooms and their growth effect in terms of shoot, root, and total biomass under greenhouse conditions

Cn	Host plant	Mycobiont species	Growth effects				Reference
			Shoot	Root	Total	Stem diameter	
1	*Abies religiosa* (Kunth) Schltdl. et. Cham.	*Inocybe splendens* R. Heim*	+++	+	++	++	Sandoval (2010)
2	***A. religiosa***	***Laccaria proxima* (Boud.) Pat.**	+++	++	+++	++	Ramírez (2018)
3	*A. religiosa*	*Suillus brevipes* (Peck) Kuntze	++	+	+	++	Sandoval (2010)
4	***Pinus arizonica* Engelm.**	***L. proxima***	+++	+++	+++	++	Cortés (2019)
5	***P. arizonica***	***L. proxima***	++	++	+	++	Cruz-Ibáñez et al. (2019)
6	***P. ayacahuite* Ehrenb. ex Schltdl.**	***Hebeloma mesophaeum* (Pers.) Quél.**	++	++	++	–	Arteaga-León (2014); Arteaga-León et al. (2018)
7	***P. ayacahuite***	***Helvella* cf. *lacunosa* Afzel.**	++	++	++	–	Arteaga-León (2014); Arteaga-León et al. (2018)
8	***P. cembroides* Zucc.**	***L. proxima***	+	+	+	–	Zuñiga (2018)
9	***P. cembroides***	***L. proxima* (+ *Cohnella* sp.)**	+++	+++	+++	++	Zuñiga (2018)
10	***P. durangensis* Ehren**	***L. laccata* (Scop.) Cooke**	Nd	Nd	+++	+++	Luciano et al. (2018b)
11	***P. durangensis***	***L. proxima***	++	+++	++	++	Cruz-Ibáñez et al. (2019)
12	***P. durangensis***	***L. proxima***	Nd	Nd	++	++	Rosales-Sánchez et al. (2017)
13	***P. greggii* Engelm.**	***Amanita rubescens* Pers.**	+++	+++	+++	+++	Cruz-Ibáñez et al. (2018)
14	***P. greggii***	***Boletus clavipes [= B. edulis* Bull.*]***	++	+	++	+++	Carrera (2004)

(continued)

Table 3.1 (continued)

Cn	Host plant	Mycobiont species	Growth effects				Reference
			Shoot	Root	Total	Stem diameter	
15	***P. greggii***	***H. leucosarx* P.D. Orton**	+++	+++	+++	+++	Rentería-Chávez et al. (2017)
16	***P. greggii***	***H. mesophaeum***	Nd	Nd	+++	+++	Carrasco (2014)
17	***P. greggii***	***H. mesophaeum***	+++	+++	+++	+++	Martínez-Reyes et al. (2012)
18	***P. greggii***	***H. mesophaeum***	++	++	++	Nd	Pérez-Moreno et al. (2009)
19	***P. greggii***	***H. mesophaeum* + *L. bicolor* + *S. pseudobrevipes***	++	++	++	Nd	Pérez-Moreno et al. (2009)
20	***P. greggii***	***H. mesophaeum* s.l.**	+++	++	+	+	Méndez-Neri et al. (2011)
21	***P. greggii***	***H. mesophaeum* s.l. + *L. laccata* s.l. + *S.* cf. *pseudobrevipes***	+++	++	+	+	Méndez-Neri et al. (2011)
22	***P. greggii***	***L. bicolor* (Maire) P.D. Orton**	++	+	++	Nd	Pérez-Moreno et al. (2009)
23	***P. greggii***	***L. bicolor***	+++	++	+++	++	Rentería-Chávez et al. (2017)
24	***P. greggii***	***L. laccata***	++	+	++	+++	Carrera (2004)
25	***P. greggii***	***L. laccata***	+++	++	+++	++	Rentería-Chávez et al. (2017)
26	***P. greggii***	***L. laccata* s.l.**	+++	+	+	+	Méndez-Neri et al. (2011)
27	***P. greggii***	***L. proxima***	Nd	Nd	+++	+++	Carrasco (2014)
28	***P. greggii***	***L. proxima*[a]**	+++	+++	+++	++	Cortés (2019)
29	***P. greggii***	***L. proxima*[b]**	+++	+++	+++	++	Cortés (2019)

30	***P. greggii***	***L. proxima***[c]	+++	+++	+++	++	Cortés (2019)
31	***P. greggii***	***L. proxima***[a]	–	+	–	+	Morales-Rivera (2017)
32	***P. greggii***	***L. proxima***[b]	–	++	++	+	Morales-Rivera (2017)
33	***P. greggii***	***L. proxima***[c]	–	–	–	–	Morales-Rivera (2017)
34	***P. greggii***	***Morchella* sp.**	+++	+++	+++	+++	Cruz-Ibáñez et al. (2018)
35	***P. greggii***	***S. brevipes***	Nd	Nd	+++	+++	Carrasco (2014)
36	***P. greggii***	***S. brevipes + L. proxima + H. mesophaeum***	Nd	Nd	+++	+++	Carrasco (2014)
37	***P. greggii***	***S. pseudobrevipes* A.H. Sm. and Thiers**	++	++	++	+++	Carrera (2004)
38	***P. greggii***	***S. pseudobrevipes***	+	-	+	++	Flores-Armas (2017)
39	***P. greggii***	***S.* cf. *pseudobrevipes***	+++	++	+	+	Méndez-Neri et al. (2011)
40	***P. greggii***	***S. pseudobrevipes***	++	+	++	Nd	Pérez-Moreno et al. (2009)
41	***P. greggii***	***S. pungens* Thiers and A.H. Sm.**	Nd	Nd	+++	+++	Luciano et al. (2018b)
42	***P. greggii***	***S. pungens***[d]	+++	+++	+++	+++	Ortiz-Lopez (2019)
43	***P. greggii***	***S. pungens***[e]	+++	+++	+++	+++	Ortiz-Lopez (2019)
44	***P. greggii***	***S. pungens***	Nd	Nd	+++	+++	Ortiz-Lopez et al. (2018)
45	***P. hartwegii* Lindl.**	***Clavulina* cf. *cinerea* (Bull.) J. Schröt.**	++	++	++	++	Perea-Estrada et al. (2009)
46	***P. hartwegii***	***Hebeloma* sp.**	++	++	++	+	Perea-Estrada et al. (2009)
47	*P. hartwegii*	*I. splendens∗*	–	+	-	+	Nuñez (2013)
48	*P. hartwegii*	*I. splendens∗*	++	+	+	++	Sandoval (2010)
49	***P. hartwegii***	***L. bicolor***	Nd	Nd	+++	+++	Lazo-Montaño et al. (2018)
50	***P. hartwegii***	***L. laccata***	Nd	Nd	+++	+++	Lazo-Montaño et al. (2018)

(continued)

Table 3.1 (continued)

Cn	Host plant	Mycobiont species	Growth effects				Reference
			Shoot	Root	Total	Stem diameter	
51	***P. hartwegii***	***Laccaria* sp.**	++	+++	+++	++	Perea-Estrada et al. (2009)
52	*P. hartwegii*	*S. brevipes*	–	+	–	+	Nuñez (2013)
53	*P. hartwegii*	*S. brevipes*	++	++	++	++	Sandoval (2010)
54	***P. hartwegii***	***S. pungens***	Nd	Nd	+++	++	Lazo-Montaño et al. (2019)
55	***P. maximartinezii* Rzedowski**	***Helvella lacunosa* Afzel.**	+	++	+	+	Almaraz (2019)
56	***P. maximartinezii***	***H. lacunosa* (+ *Azospirillum brasilense*)**	+	+	+	+	Almaraz (2019)
57	***P. maximartinezii***	***L. proxima***	++	++	++	+	Almaraz (2019)
58	***P. maximartinezii***	***L. proxima***	+	+	+	+	Fuentes and Garzón (2017)
59	***P. maximartinezii***	***L. proxima* (+ *A. brasilense*)**	+	++	+	+	Almaraz (2019)
60	***P. maximartinezii***	***S. tomentosus* Singer**	+	+	+	+	Fuentes and Garzón (2017)
61	***P. montezumae* Lamb.**	***H. mesophaeum***	+++	+++	+++	+++	Barragán-Soriano et al. (2018)
62	***P. oaxacana* Mirov.**	***L. trichodermophora* G.M. Muell.**	+++	+++	+++	+++	Hernández (2016)
63	***P. patula* Schl. et Cham**	***Amanita caesarea* (Scop.) Pers.**	+	+	+	+	Carrera (2004)
64	***P. patula***	***A. fulva* Fr.**	+	+	+	++	Carrera (2004)
65	***P. patula***	***B. clavipes* [= *B. edulis*]**	++	+	++	++	Carrera (2004)
66	***P. patula***	***B. clavipes* [= *B. edulis*]**	+++	+++	+++	+++	Carrera (2004)
67	***P. patula***	***C.* cf. *cinerea***	–	+	–	+	Perea-Estrada et al. (2009)
68	***P. patula***	***Hebeloma alpinum* (J. Favre) Bruchet**	+++	+++	+++	++	Carrasco-Hernández et al. (2011)

69	***P. patula***	***H. leucosarx***	+++	+++	+++	++	Carrasco-Hernández et al. (2011)
70	***P. patula***	***H. mesophaeum***	+++	+++	+++	++	Carrasco-Hernández et al. (2011)
71	***P. patula***	***H. mesophaeum + H. alpinum + H. leucosarx***	+++	+++	+++	++	Carrasco-Hernández et al. (2011)
72	***P. patula***	***Hebeloma* sp.**	++	++	++	+	Perea-Estrada et al. (2009)
73	***P. patula***	***L. bicolor***	+++	+++	+++	++	Carrasco-Hernández et al. (2011)
74	***P. patula***	***L. laccata***	+++	+++	+++	++	Carrasco-Hernández et al. (2011)
75	***P. patula***	***L. laccata + L. bicolor + L. proxima***	+++	+++	+++	++	Carrasco-Hernández et al. (2011)
76	***P. patula***	***L. laccata + L. bicolor + L. proxima. + H. mesophaeum + H. alpinum + H. leucosarx***	+++	+++	+++	++	Carrasco-Hernández et al. (2011)
77	***P. patula***	***L. proxima***	+++	+++	+++	++	Carrasco-Hernández et al. (2011)
78	***P. patula***	***Laccaria* sp.**	+++	++	++	+	Perea-Estrada et al. (2009)
79	***P. patula***	***S. granulatus* (L.) Roussel**	+	+	+	+	Carrera (2004)
80	***P. patula***	***S. pseudobrevipes***	+++	++	+++	++	Carrera (2004)
81	***P. pringlei* Shaw**	***H. alpinum***	+++	+++	+++	++	López-Gutiérrez et al. (2018)
82	***P. pringlei***	***L. trichodermophora***	+++	+++	+++	++	López-Gutiérrez et al. (2018)
83	***P. pseudostrobus* Lindl.**	***H. alpinum***	+++	+++	+++	++	Carrasco-Hernández et al. (2011)
84	***P. pseudostrobus***	***H. leucosarx***	+++	+++	+++	++	Carrasco-Hernández et al. (2011)

(continued)

Table 3.1 (continued)

Cn	Host plant	Mycobiont species	Growth effects				Reference
			Shoot	Root	Total	Stem diameter	
85	***P. pseudostrobus***	***H. mesophaeum***	+++	+++	+++	++	Carrasco-Hernández et al. (2011)
86	***P. pseudostrobus***	***H. mesophaeum***	++	++	++	Nd	Pérez-Moreno et al. (2009)
87	***P. pseudostrobus***	***H. mesophaeum* + *H. alpinum* + *H. leucosarx***	+++	+++	+++	++	Carrasco-Hernández et al. (2011)
88	***P. pseudostrobus***	***H. mesophaeum* + *L. bicolor* + *S. pseudobrevipes***	++	++	++	Nd	Pérez-Moreno et al. (2009)
89	***P. pseudostrobus***	***H. mesophaeum* s.l.**	+++	+++	++	++	Méndez (2007)
90	***P. pseudostrobus***	***H. mesophaeum* s.l. + *L. laccata* s.l. + *S.* cf. *pseudobrevipes***	++	+	–	–	Méndez (2007)
91	***P. pseudostrobus***	***L. bicolor***	+++	+++	+++	++	Carrasco-Hernández et al. (2011)
92	***P. pseudostrobus***	***L. bicolor***	+	+	+	Nd	Pérez-Moreno et al. (2009)
93	***P. pseudostrobus***	***L. laccata***	+++	+++	+++	++	Carrasco-Hernández et al. (2011)
94	***P. pseudostrobus***	***L. laccata* + *L. bicolor* + *L. proxima***	+++	+++	+++	++	Carrasco-Hernández et al. (2011)
95	***P. pseudostrobus***	***L. laccata* + *L. bicolor* + *L. proxima* + *H. mesophaeum* + *H. alpinum* + *H. leucosarx***	+++	+++	+++	++	Carrasco-Hernández et al. (2011)
96	***P. pseudostrobus***	***L. laccata* s.l.**	+	–	+	+	Méndez (2007)
97	***P. pseudostrobus***	***L. proxima***	+++	+++	+++	++	Carrasco-Hernández et al. (2011)
98	***P. pseudostrobus***	***L. proxima***	+++	+++	+++	+++	Jiménez (2011)

99	***P. pseudostrobus***	***S.* cf. *pseudobrevipes***	++	++	+	+	Méndez (2007)
100	***P. pseudostrobus***	***S. pseudobrevipes***	++	+	+	+	Flores-Armas (2017)
101	***P. pseudostrobus***	***S. pseudobrevipes***	+	++	++	Nd	Pérez-Moreno et al. (2009)
102	***P. teocote* Schiede ex Schltdl.**	***L. laccata***	Nd	Nd	+++	+++	Luciano et al. (2018b)
103	***P. teocote***	***L. proxima***	Nd	Nd	+++	+++	Luciano et al. (2018b)
104	***Quercus castanea* Née**	***L. trichodermophora***	+++	+++	+++	+++	Hernández (2016)

Cn = combination number; ∗Recorded as edible in India (Sharma and Gautam 2015); – = no increase; + = increase from 0% to 30%; ++ = increase from 31% to 60%; +++ = increase >60%; all of the increases compared with non-inoculated plants. Nd = No data; Inoculation dose of [a]1, [b]2, and [c]3 g. Inoculation with hymenia of mature[d] and young[e] basidiomata. The nomenclature of scientific names for fungi is based on Index Fungorum (2019) and for forest species in USDA database (2019). Scientific names in bold type correspond to experiments carried out in Colegio de Postgraduados under the supervision of the authors of this chapter. Some of these contributions have been published and others are currently in the form of Thesis, and their formal publications are in the process of being completed.

Mexico useful for reforestation, plantation establishment, and restoration of degraded areas. To date, 19 tree species have been evaluated: 16 from *Pinus*, 1 from *Abies*, and 2 from *Quercus* distributed along the north, centre, and southeast of the country. Research has also been conducted on species that are threatened according to the International Union for Conservation of Nature (IUCN) such as *Pinus maximartinenzii* and *P. chiapensis*. Special attention has been focused on *Pinus greggii* which is a fast-growing tree species, adapted to drought, and one of the Mexican pine species with the greatest potential in reforestation programs (e.g., Martínez-Reyes et al. 2012; Méndez-Neri et al. 2011; Ramírez-Herrera et al. 2005). Interestingly, this species is also found in the red list of the IUCN which is considered as vulnerable (Farjon 2013).

After more than two decades of research, a technological package has been generated that meets the requirements set out from the beginning. Ground pilea have been used, as a source of inoculum, or hymenia of mushrooms with short shelf life as *Suillus* species (e.g., Flores-Armas 2017). Subsequently, these pilea are dried and ground before being used. Despite the fact that storage at 4 °C maintain a certain viability for up to 1 or 2 years, the optimal condition is their use as soon as possible, not more than 6–9 months after their preparation. Additionally, slurries have been used successfully, in which case, it is required to have the plants ready to inoculate. For the elaboration of this latter type of inoculum, fresh pilea are cut from their stipes or hymenia are separated from their contexts in the case of fungi with short shelf lives, and subsequently they are ground along with distilled sterile water (Figs. 3.2 and 3.3). The shelf life of these solutions is variable, but in general terms, it is very short, varying from 1 week to 2 months depending on the species involved. The advantages of these techniques in relation to the use of mycelial inoculation are the following: (1) inoculum preparation is cheap and efficient; (2) the genetic diversity involved is large; (3) inoculums prepared with these techniques include microbiomes that are naturally associated with sporomes; and (4) the methodological simplicity for the elaboration of these inoculums allows a wide use in a range of rustic nursery conditions.

In general terms, there has been a beneficial effect both in terms of growth of the shoots, roots, total biomass, and stem diameter in the inoculated plants (Fig. 3.4). Table 3.1 summarizes the results obtained in terms of growth as a result of ectomycorrhizal inoculation using inoculum sources prepared with the described methodologies. In most cases, there have been beneficial effects on inoculated plants, over 60% in growth variables compared to non-inoculated plants, mainly in *Pinus arizonica*, *P. cembroides*, *P. greggii*, *P. montezumae*, *P. oaxacana*, *P. patula*, *P. pringlei*, and *P. pseudostrobus*. In other cases, there have been modest increases in growth, from 30% to 60%, and in very few cases, there has been low or no beneficial effect, e.g. in *Pinus hartwegii*, *P. leiophylla*, *P. montezumae*, *P. pseudostrobus*, *P. rudis*, and *P. teocote* inoculated with some mycobionts. However, even these tree species have shown beneficial effects when they are inoculated with the proper mycobionts, which demonstrate variations in terms of genetic compatibility depending on the different phytobiont–mycobiont combinations. Our research has also included the study of increases in nutritional terms to the shoot, root, and total plants as a result of inoculation (Table 3.2). Differential increases have been demonstrated depending

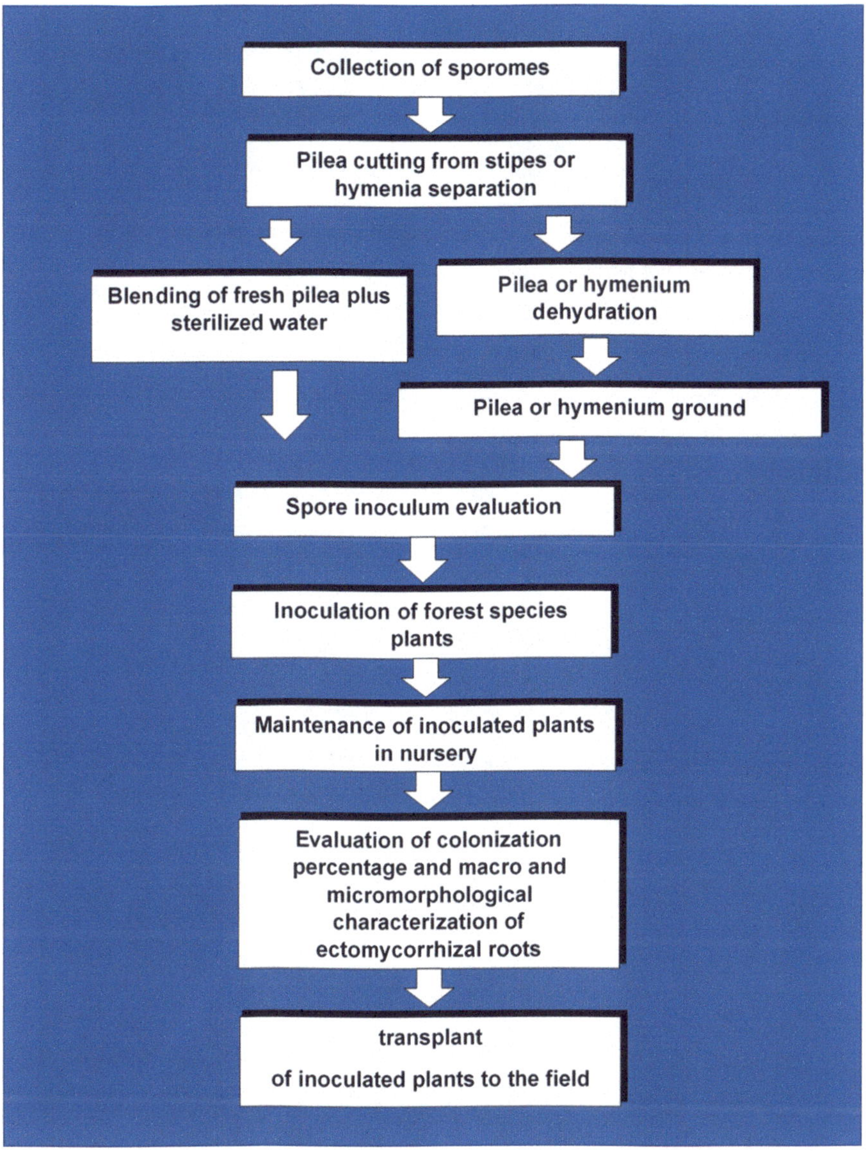

Fig. 3.2 General protocol used to produce ectomycorrhizal inoculum by using fresh or dried pilea or hymenium of edible Agaricales and Boletales

on the combinations for both macronutrients (N, P, K, Ca, and Mg) and micronutrients (Mn, Fe, Zn, Cu, B, and Na). In some cases, these increases have reached values greater than 60% compared to non-inoculated plants, for example, in the case of *Pinus pringlei* with *Hebeloma alpinum*, these increases were recorded in 9 of 10 nutrients evaluated, demonstrating enormous efficiency in terms of nutritional

Fig. 3.3 Production of ectomycorrhizal inoculum. (**a**) Pilea of *Laccaria* spp.; (**b**) Dried and ground pilea of *Laccaria* spp. ready to be used as inoculum; (**c**) Slurries of *Laccaria proxima* (left) and *Suillus pseudobrevipes* (right) ready to be used as inoculum; (**d**) General view of experiments in greenhouse using different combinations of pine and ectomycorrhizal mushroom species

Fig. 3.4 Beneficial effect of inoculation with ectomycorrhizal inoculum. (**a**) *Pinus greggii* inoculated with *Hebeloma mesophaeum* and *Laccaria laccata* compared with non-inoculated plants. (**b**) *Quercus castanea* with (left) or without (right) *Hebeloma mesophaeum*; (**c**) *Pinus greggii* with (left) or without (right) *Amanita rubescens*; (**d**) *Pinus hartwegii* with (left) or without (right) *Suillus pungens*; (**e**) Bioassays with 15 pine species inoculated with different Agaricales and Boletales

transfer in this specific combination (López-Gutiérrez et al. 2018). In general terms worldwide, there is a shortage of studies related to micronutrient transfer as a result of ectomycorrhizal inoculation and this should be a research area that deserved more attention, due to the enormous importance of micronutrients in the physiology and metabolism of trees of forest importance.

Table 3.2 Effect on nutrient contents of ectomycorrhizal combinations of Mexican Neotropical trees, including as a source of inoculum edible wild mushrooms

No	Host plant	Mycobiont	Macronutrients					Micronutrients						Reference
			N	P	K	Ca	Mg	Mn	Fe	Zn	Cu	B	Na	
1	***Abies religiosa* (Kunth) Schltdl. et. Cham.**	***Laccaria proxima* (Boud.) Pat.**	+++	+++	+++	+++	+++	++	++	+++	+++	Nd	+++	Ramírez (2018)
2	***Pinus arizonica* Engelm.**	***L. proxima***	+++	+++	+++	Nd	Nd	Nd	Nd	Nd	Nd	Nd	Nd	Cortés (2019)
3	***P. ayacahuite* Ehrenb. ex Schltdl.**	***Hebeloma mesophaeum* (Pers.) Quél.**	++	++	+	++	++	+	–	Nd	Nd	Nd	Nd	Arteaga-León et al. (2018)
4	***P. ayacahuite***	***Helvella* cf. *lacunosa* Afzel.**	++	++	+	+	+	+	++	Nd	Nd	Nd	Nd	Arteaga-León et al. (2018)
5	***P. cembroides* Zucc**	***L. proxima***	++	++	+		+	+	++	++	+	Nd	Nd	Zuñiga (2018)
6	***P. greggii* Engelm.**	***H. leucosarx* P.D. Orton**	+++	+++	+++	+++	+++	-	+++	+++	+++	+++	+++	Rentería-Chávez et al. (2017)
7	***P. greggii***	***H. mesophaeum***	+++	+++	+++	+++	+++	Nd	Nd	Nd	Nd	Nd	Nd	Martínez-Reyes et al. (2012)
8	***P. greggii***	***H. mesophaeum***	+++	+++	Nd	Nd	Nd	Nd	Nd	Nd	Nd	Nd	Nd	Pérez-Moreno et al. (2009)
9	***P. greggii***	***H. mesophaeum* + *L. bicolor* + *S. pseudobrevipes* A. H. Sm. and Thiers**	+++	++	Nd	Nd	Nd	Nd	Nd	Nd	Nd	Nd	Nd	Pérez-Moreno et al. (2009)
10	***P. greggii***	***L. bicolor* (Maire) P.D. Orton**	++	++	Nd	Nd	Nd	Nd	Nd	Nd	Nd	Nd	Nd	Pérez-Moreno et al. (2009)
11	***P. greggii***	***L. bicolor***	+++	+++	+++	+++	+++	–	+++	+++	+++	+++	++	Rentería-Chávez et al. (2017)
12	***P. greggii***	***L. laccata* (Scop.) Cooke**	+++	+++	+++	+++	+++	–	+++	+++	+++	+++	+++	Rentería-Chávez et al. (2017)
13	***P. greggii***	***L. proxima*[a]**	+++	+++	+++	Nd	Nd	Nd	Nd	Nd	Nd	Nd	Nd	Cortés (2019)
14	***P. greggii***	***L. proxima*[b]**	+++	+++	+++	Nd	Nd	Nd	Nd	Nd	Nd	Nd	Nd	Cortés (2019)
15	***P. greggii***	***L. proxima*[c]**	+++	+++	+++	Nd	Nd	Nd	Nd	Nd	Nd	Nd	Nd	Cortés (2019)
16	***P. greggii***	***Suillus pseudobrevipes***	+++	+	Nd	Nd	Nd	Nd	Nd	Nd	Nd	Nd	Nd	Pérez-Moreno et al. (2009)

17	***P. greggii***	***S. pungens* Thiers and A.H. Sm.**[d]	+++	+++	+++	Nd	Nd	+++	+++	Nd	+++	Nd	Nd	Ortiz-Lopez (2019)
18	***P. greggii***	***S. pungens***[e]	+++	+++	+++	Nd	Nd	+++	+++	Nd	+++	Nd	Nd	Ortiz-Lopez (2019)
19	***P. maximartinezii* Rzedowski**	***Helvella lacunosa***	++	++	+	Nd	Nd	Nd	+	++	Nd	Nd	++	Almaraz (2019)
20	***P. maximartinezii***	***H. lacunosa* (+ *Azospirillum brasilense*)**	+	++	++	Nd	Nd	Nd	+	++	Nd	Nd	+++	Almaraz (2019)
21	***P. maximartinezii***	***L. proxima***	++	++	++	Nd	Nd	Nd	–	++	Nd	Nd	+	Almaraz (2019)
22	***P. maximartinezii***	***L. proxima* (+ *A. brasilense*)**	+	++	++	Nd	Nd	Nd	–	+	Nd	Nd	++	Almaraz (2019)
23	***P. montezumae* Lamb.**	***H. mesophaeum***	+++	+++	+++	Nd	Nd	Nd	Nd	Nd	Nd	Nd	Nd	Barragán-Soriano et al. (2018)
24	***P. oaxacana* Mirov.**	***L. trichodermophora* G.M. Muell.**	+++	+++	+++	+++	+++	+++	+++	++	++	+++	Nd	Hernández (2016)
25	***P. pringlei* Shaw**	***L. trichodermophora***	+++	+++	+++	+++	+++	+++	+++	+++	–	+++	Nd	López-Gutiérrez et al. (2018)
26	***P. pringlei***	***H. alpinum* (J. Favre) Bruchet**	+++	+++	+++	+++	+++	+++	+++	+++	–	+++	Nd	López-Gutiérrez et al. (2018)
27	***P. pseudostrobus* Lindl.**	***H. mesophaeum***	++	++	Nd	Nd	Nd	Nd	Nd	Nd	Nd	Nd	Nd	Pérez-Moreno et al. (2009)
28	***P. pseudostrobus***	***H. mesophaeum + L. bicolor + S. pseudobrevipes***	++	+++	Nd	Nd	Nd	Nd	Nd	Nd	Nd	Nd	Nd	Pérez-Moreno et al. (2009)
29	***P. pseudostrobus***	***L. bicolor***	++	++	Nd	Nd	Nd	Nd	Nd	Nd	Nd	Nd	Nd	Pérez-Moreno et al. (2009)
30	***P. pseudostrobus***	***S. pseudobrevipes***	++	++	Nd	Nd	Nd	Nd	Nd	Nd	Nd	Nd	Nd	Pérez-Moreno et al. (2009)
31	***Quercus castanea* Née**	***L. trichodermophora***	+++	+++	+++	+++	+++	+++	+++	+++	+++	+++	Nd	Hernández (2016)

No = Combination number; – = no increase; + = increase from 0% to 30%; ++ = increase from 31% to 60%; +++ = increases >60%; all of the increases compared with non-inoculated plants. Nd = No data. Inoculation dose of [a]1, [b]2, and [c]3 g. Inoculation with hymenia of mature[d] and young [e] basidiomata. The nomenclature of scientific names for fungi was based on Index Fungorum (2019) and for forest species in USDA database (2019). Experiments of all combinations were carried out in Colegio de Postgraduados under the supervision of the authors of this chapter. Some of these contributions have been published and others are currently in the form of thesis, and their formal publications are in the process of being completed.

In recent years, we have been carrying out studies related to the influence of ectomycorrhizal inoculation on physiological variables including photosynthetic rate, chlorophyll a, b, and total contents, as well as carotenes (Table 3.3). The evaluation of these variables is essential for the understanding of a complex phenomenon that can explain the reasons why there have been increases in growth in the associated hosts. Another area of enormous importance is the coinoculation of ectomycorrhizal fungi with plant growth promoting bacteria. It has been shown that the co-inoculation of these groups of microorganisms has caused synergism in plant growth, nutritional transfer, and physiological quality of host plants, for example, in combinations of *Pinus cembroides* with *Laccaria proxima* plus the N-fixing bacteria *Cohnella* (Zuñiga 2018) and *Pinus montezumae* with *Hebeloma mesophaeum* plus the phosphate-solubilizing bacteria *Azospirillum brasilense* (Barragán-Soriano et al. 2018). In order to verify that the beneficial effects recorded in terms of plant growth, nutritional increases, and improvement in the physiological quality of inoculated plants, detailed studies of colonization percentages have been carried out in all bioassays (Table 3.4). Although there have been variations in these colonizations, in most cases, they have been greater than 60%, reaching up to 100%, which has demonstrated the effectiveness of the biotechnology (Fig. 3.5). In recent years, field studies have begun, transplanting ectomycorrhizal and non-ectomycorrhizal trees (Table 3.5). In most cases, non-ectomycorrhizal trees have had very low survival rates ranging from 0% to 20%, similar to those recorded historically in Mexico. In contrast, in the case of bioassays using controlled ectomycorrhizal trees, higher survival rates have been recorded depending on the combinations evaluated. In some cases, the survival rates of ectomycorrhizal plants have been around 85%, for example, in the cases of *Pinus greggii* inoculated with *Suillus brevipes* (Luciano 2018), with *S. pseudobrevipes* (Luciano et al. 2018c), with *S. pungens* (Cardoso 2017), and *P. hartwegii* with *S. brevipes* (Nuñez 2013). Interestingly, when co-inoculations have been performed, synergistic results have been observed, for example in the case of *P. greggii* co-inoculated with *H. mesophaeum* + *L. bicolor* + *S. pseudobrevipes* (Pérez-Moreno et al. 2009).

The presence of abundant ectomycorrhizal external mycelium, which is the structure responsible for nutritional translocation, has often been observed in the experiments associated with high mycorrhization. Additionally, sporome formation of ectomycorrhizal fungi, mainly species of the *Laccaria* and *Hebeloma* genera, has been frequently recorded (Fig. 3.6). And in the case of field experiments, the first Neotropical ectomycorrhizal fungus *Suillus pseudobrevipes* and *S. pungens* has begun, with the initial production of sporomes, which has started as early as after 2 years of transplantation (Fig. 3.7).

3.3 Conclusions

An innovating biotechnology using the pilea or the hymenium of Agaricales or Boletales as a source of inoculum developed in Mexico, during the last two decades is presented. The pilea and the hymenia of sporomes are a suitable source for the

Table 3.3 Effect on physiological variables of Mexican neotropical trees ectomycorrhizal combinations, under greenhouse conditions

No	Host plant	Mycobiont	Photosynthetic rate	Cholorophyll contents			Carotens	Reference
				a	b	Total		
1	***Pinus arizonica* Engelm.**	***Laccaria proxima* (Boud.) Pat.**	++	+++	+++	+++	+++	Cortés (2019)
2	***P. cembroides* Zucc.**	***L. proxima***	++	++	++	++	+++	Zuñiga (2018)
3	***P. cembroides***	***L. proxima* (+ *Cohnella* sp.)**	+++	++	++	++	+++	Zuñiga (2018)
4	***P. greggii* Engelm.**	***L. proxima*[a]**	+++	+++	+++	+++	+++	Cortés (2019)
5	***P. greggii***	***L. proxima*[b]**	+++	+++	+++	+++	+++	Cortés (2019)
6	***P. greggii***	***L. proxima*[c]**	+++	+++	+++	+++	+++	Cortés (2019)
7	***P. greggii***	***Suillus pungens* Thiers and A.H. Sm[d]**	+++	++	+	++	+++	Ortiz-Lopez (2019)
8	***P. greggii***	***S. pungens*[e]**	+++	++	++	++	++	Ortiz-Lopez (2019)
9	***P. montezumae* Lamb.**	***Hebeloma mesophaeum* (Pers.) Quél.**	+++	+	+	+	+	Barragán-Soriano et al. (2018)
10	***P. montezumae***	***H. mesophaeum* (+ *Azospirillum brasilense*)**	+++	+	+	+	+	Barragán-Soriano et al. (2018)

No = Combination number; + = increase from 0% to 30%; ++ = increase from 31% to 60%; +++ = increases >60%; all of the increases compared with non-inoculated plants. Inoculation dose of [a]1, [b]2, and [c]3 g. Inoculation with hymenia of young[d] and mature basidiomata[e]. The nomenclature of scientific names for fungi was based on the Index Fungorum (2019) and for the forest species in USDA database (2019). Experiments of all combinations were carried out in Colegio de Postgraduados under the supervision of the authors of this chapter. Some of these contributions have been published and others are currently in the form of thesis, and their formal publications are in the process of being completed

Table 3.4 Effect on colonization rates in evaluated Mexican neotropical ectomycorrhizal combinations, under greenhouse conditions, using a source of inoculum either edible ectomycorrhizal mushrooms

No	Host plant	Mycobiont	Colonization rate	Reference
1	*Abies religiosa* (Kunth) Schltdl. et. Cham.	*Inocybe splendens* R. Heim*	+	Sandoval (2010)
2	***A. religiosa***	***Laccaria proxima* (Boud.) Pat.**	+++	Ramírez (2018)
3	***Pinus arizonica* Engelm**	***L. proxima***	++	Cruz-Ibáñez et al. (2019)
4	***P. arizonica***	***L. proxima***	+++	Cortés (2019)
5	***P. ayacahuite* Ehrenb. ex Schltdl.**	***Hebeloma alpinum* (J. Favre)**	+	Carrera-Martínez et al. (2017)
6	***P. ayacahuite***	***Suillus luteus* (L. Fries) Gray**	+	Carrera-Martínez et al. (2017)
7	***P. ayacahuite***	***S. granulatus***	+	Santiago-Hernández et al. (2017)
8	***P. cembroides Zucc.***	***L. proxima***	++	Pérez-Téllez et al. (2017)
9	***P. cembroides***	***Hebeloma leucosarx* P.D. Orton**	++	Pérez-Téllez et al. (2017)
10	***P. cembroides***	***L. proxima***	++	Zuñiga (2018)
11	***P. cembroides***	***L. proxima* (+ *Cohnella* sp.)**	+++	Zuñiga (2018)
12	***P. chiapensis* (Martínez) Andresen**	***L. proxima***	++	Caballero (2019)
13	***P. chiapensis***	***Ramaria* sp.**	+++	Caballero (2019)
14	***P. chiapensis***	***Suillus pseudobrevipes* A.H. Sm. and Thiers**	+++	Caballero (2019)
15	***P. durangensis* Ehren**	***L. laccata* (Scop.) Cooke**	+++	Luciano et al. (2018b)
16	***P. durangensis***	***L. proxima***	++	Cruz-Ibáñez et al. (2019)
17	***P. engelmannii* Carr.**	***L. proxima***	+++	Ríos-García et al. (2017a)
18	*P. engelmannii*	*Russula delica* Fr.	+++	Martínez-Nevárez et al. (2016)
19	***Pinus greggii* Engelm.**	***Amanita rubescens* Pers.**	+++	Cruz-Ibáñez et al. (2018)
20	***P. greggii***	***Boletus clavipes [= B. edulis* Bull.]**	+	Carrera (2004)
21	***P. greggii***	***Hebeloma leucosarx* P.D. Orton**	+++	Rentería-Chávez et al. (2017)
22	***P. greggii***	***H. mesophaeum* (Pers.) Quél.**	+++	Martínez-Reyes et al. (2012)
23	***P. greggii***	***H. mesophaeum* s.l.**	+++	Méndez-Neri et al. (2011)

24	***P. greggii***	***H. mesophaeum***	++	Pérez-Moreno et al. (2009)
25	***P. greggii***	***H. mesophaeum***	+++	Villegas (2014)
26	***P. greggii***	***H. mesophaeum* + *L. bicolor* + *S. pseudobrevipes***	++	Pérez-Moreno et al. (2009)
27	***P. greggii***	***H. mesophaeum* s.l. + *L. laccata* s.l. + *S.* cf. *pseudobrevipes***	++	Méndez-Neri et al. (2011)
28	***P. greggii***	***L. bicolor* (Maire) P.D. Orton**	++	Pérez-Moreno et al. (2009)
29	***P. greggii***	***L. bicolor***	+++	Rentería-Chávez et al. (2017)
30	***P. greggii***	***L. laccata***	++	Carrera (2004)
31	***P. greggii***	***L. laccata* s.l.**	+++	Méndez-Neri et al. (2011)
32	***P. greggii***	***L. laccata***	+++	Rentería-Chávez et al. (2017)
33	***P. greggii***	***L. laccata***	+++	Villegas (2014)
34	***P. greggii***	***L. proxima***	+++	Díaz-Balbuena (2017); Franco-Ramírez et al. (2019)
35	***P. greggii***	***L. proxima*[a]**	+++	Cortés (2019)
36	***P. greggii***	***L. proxima*[b]**	+++	Cortés (2019)
37	***P. greggii***	***L. proxima*[c]**	+++	Cortés (2019)
38	***P. greggii***	***L. proxima*[a]**	+	Morales-Rivera (2017)
39	***P. greggii***	***L. proxima*[b]**	++	Morales-Rivera (2017)
40	***P. greggii***	***L. proxima*[c]**	++	Morales-Rivera (2017)
41	***P. greggii***	***Morchella* sp.**	+++	Cruz-Ibáñez et al. (2018)
42	***P. greggii***	***S. brevipes* (Peck) Kuntze**	+++	Luciano (2018)
43	***P. greggii***	***S.* cf. *pseudobrevipes***	+++	Méndez-Neri et al. (2011)
44	***P. greggii***	***S. pseudobrevipes***	++	Carrera (2004)
45	***P. greggii***	***S. pseudobrevipes***	+++	Díaz-Balbuena (2017)
46	***P. greggii***	***S. pseudobrevipes***	++	Flores-Armas (2017)
47	***P. greggii***	***S. pseudobrevipes***	+++	Luciano et al. (2018c)
48	***P. greggii***	***S. pseudobrevipes***	+++	Pérez-Moreno et al. (2009)

(continued)

Table 3.4 (continued)

No	Host plant	Mycobiont	Colonization rate	Reference
49	***P. greggii***	***S. pungens* Thiers and A.H. Sm.**	+++	Luciano et al. (2018b)
50	***P. greggii***	***S. pungens***[d]	+++	Ortiz-Lopez (2019)
51	***P. greggii***	***S. pungens***[e]	+++	Ortiz-Lopez (2019)
52	***P. greggii***	***S. pungens***	+++	Ortiz-Lopez et al. (2018)
53	***P. greggii***	***T. terrestris***	+++	Luciano et al. (2018b)
54	***P. hartwegii* Lindl.**	***Clavulina* cf. *cinerea* (Bull.) J. Schröt.**	+	Perea-Estrada et al. (2009)
55	***P. hartwegii***	***Hebeloma* sp.**	+	Perea-Estrada et al. (2009)
56	*P. hartwegii*	*I. splendens*∗	+++	Nuñez (2013)
57	***P. hartwegii***	***Laccaria* sp.**	–	Perea-Estrada et al. (2009)
58	***P. hartwegii***	***L. bicolor***	+++	Lazo-Montaño et al. (2018)
59	***P. hartwegii***	***L. laccata***	+++	Lazo-Montaño et al. (2018)
60	*P. hartwegii*	*S. brevipes*	++	Nuñez (2013)
61	*P. hartwegii*	*S. brevipes*	+	Sandoval (2010)
62	***P. hartwegii***	***S. pungens***	+++	Lazo-Montaño et al. (2019)
63	***P. maximartinezii* Rzedowski**	***Hebeloma alpinum***	+	Carrera-Martínez et al. (2017)
64	***P. maximartinezii***	***L. proxima***	+	Carrera-Martínez et al. (2017)
65	***P. maximartinezii***	***S. pseudobrevipes***	+	Carrera-Martínez et al. (2017)
66	***P. maximartinezii***	***Helvella lacunosa* Afzel.**	+++	Almaraz (2019)
67	***P. maximartinezii***	***H. lacunosa* Afzel. (+ *A. brasilense*)**	+++	Almaraz (2019)
68	***P. maximartinezii***	***L. proxima***	+++	Almaraz (2019)
69	***P. maximartinezii***	***L. proxima***	++	Fuentes and Garzón (2017)
70	***P. maximartinezii***	***L. proxima* (Boud.) Pat. (+ *Azospirillum brasilense*)**	+++	Almaraz (2019)
71	***P. maximartinezii***	***S. tomentosus* Singer**	+++	Fuentes and Garzón (2017)
72	***P. montezumae* Lamb.**	***H. mesophaeum***	+++	Barragán-Soriano et al. (2018)

73	***P. montezumae***	***H. mesophaeum***	+++	Villegas (2014)
74	***P. montezumae***	***H. leucosarx***	+++	Villegas-Olivera et al. (2017)
75	***P. montezumae***	***L. laccata***	+++	Villegas (2014)
76	*P. montezumae*	*L. trichodermophora* G.M. Muell.	+++	Galindo-Flores et al. (2015); Rodríguez-Gutiérrez et al. (2017)
77	*P. montezumae*	*L. bicolor*	+++	Rodríguez-Gutiérrez et al. (2017)
78	*P. montezumae*	*S. tomentosus*	+++	Galindo-Flores et al. (2015)
79	***P. oaxacana* Mirov.**	***L. trichodermophora***	+++	Hernández (2016)
80	***P. oocarpa* Schiede ex Schltdl**	***A. rubescens***	+	Almazán (2019)
81	***P. oocarpa***	***H. leucosarx***	+++	Almazán (2019)
82	***P. oocarpa***	***H. leucosarx***	+++	Villegas-Olivera et al. (2018)
83	***P. oocarpa***	***H. lacunosa***	++	Almazán (2019)
84	***P. oocarpa***	***L. laccata***	+++	Almazán (2019)
85	***P. oocarpa***	***L. laccata***	+++	Villegas-Olivera et al. (2018)
86	***P. oocarpa***	***L. proxima***	++	Almazán (2019)
87	***P. oocarpa***	***S. brevipes***	+++	Villegas-Olivera et al. (2018)
88	***P. patula* Schl. et Cham.**	***A. caesarea* (Scop.) Pers.**	+	Carrera (2004)
89	***P. patula***	***A. fulva* Fr.**	+	Carrera (2004)
90	***P. patula***	***Boletus clavipes [= B. edulis* Bull *]***	++	Carrera (2004)
91	***P. patula***	***C.* cf. *cinerea***	+	Perea-Estrada et al. (2009)
92	***P. patula***	***H. alpinum* (J. Favre) Bruchet**	++	Carrasco-Hernández et al. (2011)
93	***P. patula***	***H. leucosarx***	++	Carrasco-Hernández et al. (2011)
94	***P. patula***	***H. mesophaeum***	+	Carrasco-Hernández et al. (2011)
95	***P. patula***	***H. mesophaeum + H. alpinum + H. leucosarx***	+++	Carrasco-Hernández et al. (2011)
96	***P. patula***	***Hebeloma* sp.**	+	Perea-Estrada et al. (2009)
97	***P. patula***	***L. bicolor***	+++	Carrasco-Hernández et al. (2011)
98	***P. patula***	***L. laccata***	+++	Carrasco-Hernández et al. (2011), Portillo-Manzano et al. (2017)

(continued)

Table 3.4 (continued)

No	Host plant	Mycobiont	Colonization rate	Reference
99	***P. patula***	***L. laccata + L. bicolor + L. proxima***	+++	Carrasco-Hernández et al. (2011)
100	***P. patula***	***L. laccata + L. bicolor + L. proxima + H. mesophaeum + H. alpinum + H. leucosarx***	+++	Carrasco-Hernández et al. (2011)
101	***P. patula***	***L. proxima***	+++	Carrasco-Hernández et al. (2011)
102	***P. patula***	***Laccaria* sp.**	+++	Perea-Estrada et al. (2009)
103	***P. patula***	***S. granulatus* (L.) Roussel**	++	Carrera (2004)
104	***P. patula***	***S. pseudobrevipes***	+++	Carrera (2004)
105	***P. pringlei* Shaw**	***H. alpinum***	+++	López-Gutiérrez et al. (2018)
106	***P. pringlei***	***L. trichodermophora***	+++	López-Gutiérrez et al. (2018)
107	***P. pseudostrobus* Lindl.**	***H. alpinum***	++	Carrasco-Hernández et al. (2011)
108	***P. pseudostrobus***	***H. leucosarx***	+++	Carrasco-Hernández et al. (2011)
109	***P. pseudostrobus***	***H. mesophaeum***	++	Carrasco-Hernández et al. (2011)
110	***P. pseudostrobus***	***H. mesophaeum***	+++	Heredia (2011)
111	***P. pseudostrobus***	***H. mesophaeum***	+++	Pérez-Moreno et al. (2009)
112	***P. pseudostrobus***	***H. mesophaeum* s.l.**	+++	Méndez (2007)
113	***P. pseudostrobus***	***H. mesophaeum + H. alpinum + H. leucosarx***	+++	Carrasco-Hernández et al. (2011)
114	***P. pseudostrobus***	***H. mesophaeum + L. bicolor + S. pseudobrevipes***	+++	Pérez-Moreno et al. (2009)
115	***P. pseudostrobus***	***H. mesophaeum* s.l. + *L. laccata* s.l. + *S.* cf. *pseudobrevipes***	+++	Méndez (2007)
116	***P. pseudostrobus***	***L. bicolor***	+++	Pérez-Moreno et al. (2009)
117	***P. pseudostrobus***	***S. pseudobrevipes***	+++	Pérez-Moreno et al. (2009)
118	***P. pseudostrobus***	***Hellvella* sp. + *Laccaria* sp. + *Russula* sp. + *Ramaria* sp. + *Cantharellus* sp. + *Boletus* sp. + *Morchella* sp. + *Lactarius* sp.**	+++	Ramírez (2017)

119	***P. pseudostrobus***	***L. bicolor***	+++	Carrasco-Hernández et al. (2011)
120	***P. pseudostrobus***	***L. laccata***	+++	Carrasco-Hernández et al. (2011)
121	***P. pseudostrobus***	***L. laccata***	+++	Ramírez (2017)
122	***P. pseudostrobus***	***L. laccata* s.l**	+++	Méndez (2007)
123	***P. pseudostrobus***	***L. laccata + L. bicolor + L. proxima***	+++	Carrasco-Hernández et al. (2011)
124	***P. pseudostrobus***	***L. laccata + L. bicolor + L. proxima + H. mesophaeum + H. alpinum + H. leucosarx***	+++	Carrasco-Hernández et al. (2011)
125	***P. pseudostrobus***	***L. proxima***	+++	Carrasco-Hernández et al. (2011)
126	***P. pseudostrobus***	***L. proxima***	+++	Jiménez (2011)
127	***P. pseudostrobus***	***S. pseudobrevipes***	++	Flores-Armas (2017)
128	***P. pseudostrobus***	***S.* cf. *pseudobrevipes***	+++	Méndez (2007)
129	***P. teocote* Schiede ex Schltdl.**	***L. laccata***	+++	Luciano et al. (2018a)
130	***P. teocote***	***L. proxima***	+++	Luciano et al. (2018b)
131	***Quercus castanea* Née**	***L. trichodermophora***	+++	Hernández (2016)
132	***Q. uxoris* McVaugh**	***L. proxima***	+	Ríos-García et al. (2017b)

No = combination number; ∗Recorded as edible in India (Sharma and Gautam 2015); – = no increase; + = increases from 0% to 30%; ++ = increases from 31% to 60%; +++ = increases >60%. Nd = No data; Inoculation dose of [a]1, [b]2, and [c]3 g. Inoculation with hymenia of young[d] and mature basidiomata[e]. The nomenclature of scientific names for fungi was based on Index Fungorum (2019) and for forest species in USDA database (2019). Scientific names in bold type correspond to experiments carried out in Colegio de Postgraduados under the supervision of the authors of this chapter. Some of these contributions have been published and others are currently in the form of thesis, and their formal publications are in the process of being completed.

Fig. 3.5 General views of beneficial effect of ectomycorrhizal inoculation and colonization. (**a**) *Pinus pringlei* with (left) and without (right) *L. trichodermophora*; (**b**) Foliar area of *Arbutus xalapensis* with (left) or without (right) *Laccaria trichodermophora*; (**c**–**f**) Ectomycorrhizas of: (**c**) *Pinus oaxacana* with *Laccaria trichodermophora*; (**d**) *Quercus castanea* with *Laccaria trichodermophora*; (**e**) *Pinus greggii* with *Laccaria bicolor*; (**f**) *Pinus pseudostrobus* with *Hebeloma alpinum*

Table 3.5 Effect on increased survival under field conditions in Mexican neotropical trees ectomycorrhizal combinations

No	Host tree	Mycobiont	Increases in survival rates	Reference
1	*Abies religiosa* (Kunth) Schltdl. et. Cham.	*Inocybe splendens* R. Heim*	+	Sandoval (2010)
2	*A. religiosa*	*Suillus brevipes* (Peck) Kuntze	+	Sandoval (2010)
3	***Pinus greggii* Engelm.**	***H. mesophaeum* (Pers.) Quél.**	+	Carrasco (2014)
4	***P. greggii***	***H. mesophaeum***	+	Pérez-Moreno et al. (2009)
5	***P. greggii***	***H. mesophaeum + L. bicolor + S. pseudobrevipes***	+++	Pérez-Moreno et al. (2009)
6	***P. greggii***	***L. bicolor* (Maire) P.D. Orton**	++	Pérez-Moreno et al. (2009)
7	***P. greggii***	***L. proxima* (Boud.) Pat.**	+	Carrasco (2014)
8	***P. greggii***	***S. brevipes* (Peck) Kuntze**	++	Carrasco (2014)
9	***P. greggii***	***S. brevipes***	+++	Luciano (2018)
10	***P. greggii***	***S. brevipes + L. proxima + H. mesophaeum***	+	Carrasco (2014)
11	***P. greggii***	***S. pseudobrevipes* A.H. Sm. and Thiers**	++	Pérez-Moreno et al. (2009)
12	***P. greggii***	***S. pseudobrevipes***	+++	Luciano et al. (2018c)
13	***P. greggii***	***S. pungens* Thiers and A.H. Sm.**	+++	Cardoso (2017)
14	***P. greggii***	***S. pungens***	++	Luciano et al. (2018b)
15	*P. hartwegii* Lindl.	*I. splendens**	+++	Nuñez (2013)
16	*P. hartwegii*	*I. splendens**	++	Sandoval (2010)
17	*P. hartwegii*	*S. brevipes*	+++	Nuñez (2013)
18	*P. hartwegii*	*S. brevipes*	++	Sandoval (2010)
19	***P. patula* Schl. et Cham**	***S. pungens***	+	Cardoso (2017)
20	***P. pringlei* Shaw**	***H. alpinum* (J. Favre) Bruchet**	++	López-Gutiérrez et al. (2018)
21	***P. pringlei***	***L. trichodermophora* G.M. Muell.**	+	López-Gutiérrez et al. (2018)
22	***P. pseudostrobus* Lindl.**	***H. alpinum***	+	Jiménez (2011)
23	***P. pseudostrobus***	***H. leucosarx* P.D. Orton**	++	Jiménez (2011)
24	***P. pseudostrobus***	***H. mesophaeum***	++	Jiménez (2011)

(continued)

Table 3.5 (continued)

No	Host tree	Mycobiont	Increases in survival rates	Reference
25	***P. pseudostrobus***	***H. mesophaeum + H. alpinum + H. leucosarx***	+	Jiménez (2011)
26	***P. pseudostrobus***	***L. bicolor***	+	Jiménez (2011)
27	***P. pseudostrobus***	***L. laccata* (Scop.) Cooke**	++	Jiménez (2011)
28	***P. pseudostrobus***	***L. laccata + L. bicolor + L. proxima***	+	Jiménez (2011)
29	***P. pseudostrobus***	***L. laccata + L. bicolor + L. proxima + H. mesophaeum + H. alpinum + H. leucosarx***	++	Jiménez (2011)
30	***P. pseudostrobus***	***L. proxima***	+	Jiménez (2011)

No = combination number; ∗Recorded as edible in India (Sharma and Gautam 2015); + = increases from 0% to 30%; ++ = increase from 31% to 60%; +++ = increase >60%; all of the increases compared with non-inoculated plants. The nomenclature of scientific names for fungi was based on Index Fungorum (2019) and for forest species in USDA database (2019). Scientific names in bold type correspond to experiments carried out in Colegio de Postgraduados under the supervision of the authors of this chapter. Some of these contributions have been published and others are currently in the form of thesis, and their formal publications are in the proccess of being completed

Fig. 3.6 Formation of sporomes in *Pinus greggii* inoculated with edible ectomycorrhizal mushrooms. (**a**) *Hebeloma mesophaeum*; (**b**) *Laccaria bicolor*; (**c**) close-up to (**d**) *Hebeloma mesophaeum* and *Laccaria trichodermophora* in the same pot; (**d**) General view of a tree of *Pinus greggii* co-inoculated with *Hebeloma mesophaeum* and *Laccaria trichodermophora* producing sporomes of both species

production of bioinoculants useful in the production of controlled ectomycorrhizal plants. The results of more than 100 combinations of phytobionts and mycobionts are analyzed. In general, inoculation with these techniques enhances the growth, nutrient contents, and physiological quality of the associated neotropical hosts. Before this, the use of sporomes of Gasteromycetes, mainly in the genera *Pisolithus*, *Scleroderma*, and *Rhizopogon* were known to be useful in the production of inoculants. The new biotechnology is simple, efficient, and cheap and then it has a great potential to be used in

Fig. 3.7 Studies under field conditions with *Pinus greggii*. (**a**) Abundant external ectomycorrhizal mycelium of *Hebeloma leucosarx*; (**b**) Mexican *campesino* holding a tree inoculated with *Hebeloma alpinum* showing white external ectomycorrhizal mycelium; (**c**) Field bioassays with different Agaricales and Boletales with individual meshes protecting the trees against herbivory; (**d**) adjacent ditch, filled with mulch, an ancient Mexican technique used to enhance water supply to a recently planted tree; (**e**, **f**) view of a site before and 5 years after plantation of *Pinus greggii* trees inoculated with *Suillus pseudobrevipes*; (**g**) Formation of sporome of *Suillus pseudobrevipes* in the base of inoculated trees 2 years after plantation

greenhouses or nurseries with basic infrastructure in developing countries. In this case, the microbiome associated with the pilea or the hymenia of the sporomes rather to be excluded constitute part of the bioinoculants. The validation of this biotechnology is demonstrated by the high ectomycorrhizal colonization rates recorded in most cases. Studies under field conditions proved to dramatically increase the survival rates, compared with non-inoculated trees. Despite the initial success of this innovation, due to the large number of ectomycorrhizal mycobionts and phytobionts to be tested, is possible to state that the development of this biotechnology is in its infancy.

Acknowledgements The credit pictures are as follows: Fig. 3.1d is from Jazmín Cortes-Sarabia and Fig. 3.6e–g are from Violeta Carrasco-Hernández. All of the other figures are from the authors of this chapter. Jesús Pérez-Moreno acknowledges the financial support from CONACyT 2018-07-01EXTV and COMECyT EST2018-005 to develop a sabbatical stay at the Kunming Institute of Botany, Kunming, China. The authors acknowledge funding from the CONACyT project "Biotechnologies of highly-prized or bioculturally-important edible ectomycorrhizal mushrooms for reforestation with native Mexican trees using frontier scientific methodologies".

References

Almaraz LIA (2019) Diversidad de hongos ectomicorrízicos y su relación con rizobacterias sobre el crecimiento de plantas de *Pinus maximartinezii* Rzed. (in Spanish). Thesis of Master Science, Colegio de Postgraduados, San Luis Potosí, México

Almazán CPJ (2019) Crecimiento de *Pinus oocarpa* Schiede ex Schltdl. inoculado con especies de hongos ectomicorrízicos comestibles (in Spanish). Thesis of Biologist, Instituto Tecnológico de Ciudad Altamirano, Guerrero, México

Anthropocene Journal (2019) Available from www.sciencedirect.com/journal/anthropocene. Accessed 4 Oct 2019

Arteaga-León C (2014) Ectomicorrización con hongos de la Sierra Nevada, México en el crecimiento y movilización nutrimental de *Pinus ayacahuite* Ehrenb. ex Schltdl. (in Spanish). Doctor of Science Thesis, Postgrado en Edafología, Colegio de Postgraduados, Montecillo, Estado de México, México

Arteaga-León C, Pérez-Moreno J, Espinosa-Victoria D et al (2018) Ectomycorrhizal inoculation with edible fungi increases plant growth and nutrient contents of *Pinus ayacahuite*. Rev Mex Biodivers 89:1089–1099. https://doi.org/10.22201/ib.20078706e.2018.4.2235

Barnosky AD, Matzke N, Tomiya S et al (2011) Has the Earth's sixth mass extinction already arrived? Nature 471:51–57

Barragán-Soriano JL, Pérez-Moreno J, Almaraz-Suárez JJ et al (2018) Inoculation with an edible ectomycorrhizal fungus and bacteria increases growth and improves the physiological quality of *Pinus montezumae* Lamb. Revista Chapingo Serie Ciencias Forestales y del Ambiente 24:3–16. https://doi.org/10.5154/r.rchscfa.2017.01.010

Brundrett MC, Tedersoo L (2018) Evolutionary history of mycorrhizal symbioses and global host plant diversity. New Phytol 220:1108–1115

Bruns TD, Peay KG, Boynton PJ et al (2009) Inoculum potential of *Rhizopogon* spores increases with time over the first 4 yr of a 99-yr spore burial experiment. New Phytol 181:463–470

Caballero VMG (2019) Importancia biocultural y potencial biotecnológico de hongos silvestres comestibles en la comunidad zapoteca de Santa Catalina Quierí, Yautepec, Oaxaca (in Spanish). Thesis of Biologist, Instituto Tecnológico del Valle de Oaxaca, Oaxaca, México

Cardoso Del RM (2017) Caracterización metagenómica del microbioma asociado a ectomicorrizas y supervivencia en campo de dos pinos inoculados con *Suillus pungens* y *Thelephora terrestris* (in Spanish). Thesis of Master Science, Postgrado en Edafología, Colegio de Postgraduados, Montecillo, estado de México, México

Carrasco HV (2014) Inoculación de hongos silvestres comestibles en pinos nativos de México (in Spanish). Doctor of Science Thesis, Postgrado en Edafología, Colegio de Postgraduados, Montecillo, estado de México, México

Carrasco-Hernández V, Pérez-Moreno J, Espinosa-Hernández V et al (2011) Contenido de nutrientes e inoculación con hongos ectomicorrízicos comestibles en dos pinos neotropicales. Revista chilena de historia natural 84:83–96. https://doi.org/10.4067/S0716-078X2011000100006

Carrera NA (2004) Inoculación con hongos ectomicorrízicos en *Pinus patula* Schlt. et Cham. y *Pinus greggii* Engelm. en invernadero (in Spanish). Thesis of Master Science, División de Ciencias Forestales, Universidad Autónoma de Chapingo, Texcoco de Mora, estado de México, México

Carrera-Martínez A, Ríos-García U, Martínez-Reyes M et al (2017) First record of synthesis between three ectomycorrhizal edible mushrooms and *Pinus maximartinezii* a Mexican endangered native species. In: Pérez-Moreno J, Guerin-Laguette A (eds) Mushrooms, humans and nature in a changing world, Proceedings of the 9th International Workshop on Edible Mycorrhizal Mushrooms (IWEMM9) Colegio de Postgraduados and CONACYT, Texcoco, Mexico, p 74

Castellano MA, Trappe JM, Molina R (1985) Inoculation of container-grown Douglas-fir seedlings with basidiospores of *Rhizopogon vinicolor* and *R. colossus*: effects of fertility and spore application rate. Can J For Res 15:10–13

Carrera-Martínez A (2017) Identificación morfológica, anatómica y molecular de hongos ectomicorrízicos asociados con *Pinus ayacahuite* Ehrenberg (in Spanish). Thesis of Biologist, Instituto Tecnológico del Valle de Oaxaca, Oaxaca, México

Chakraborty A (2019) Mountains as vulnerable places: a global synthesis of changing mountain systems in the Anthropocene. GeoJournal (in press). https://doi.org/10.1007/s10708-019-10079-1

Chen L, Swenson NG, Ji N et al (2019) Differential soil fungus accumulation and density dependence of tres in a subtropical forest. Science 366:124–128

Comandini O, Rinaldi AC, Kuyper TW (2012) Measuring and estimating ectomycorrhizal fungal diversity: a continuous challenge. In: Pagano M (ed) Mycorrhiza: occurrence in natural and restored environments. Nova Science Publishers, New York, NY, pp 165–200

Cortés GP (2019) Efecto de la inoculación del hongo comestible ectomicorrízico *Laccaria proxima* en el crecimiento, contenido nutrimental y calidad fisiológica de *Pinus arizonica* y *Pinus greggii* (in Spanish). Thesis of Biologist. Centro Universitario de Ciencias Biológicas y Agropecuarias, Universidad de Guadalajara, Jalisco, México

Crutzen PJ (2002) Geology of mankind. Nature 415:23. https://doi.org/10.1038/415023a

Crutzen PJ, Stoermer EF (2000) The anthropocene. Global Change Newsletter 41:17–18

Cruz-Ibáñez EM, Martínez RM, Pérez MJ et al (2018) Inoculación líquida con hongos silvestres comestibles ectomicorrízicos en el crecimiento y desarrollo de *Pinus greggii* en invernadero. In: 1er Congreso Nacional de Investigación en Ciencia e Innovación de Tecnologías Productivas, Altamirano, Guerrero. Mexico, pp 15–25

Cruz-Ibáñez EM, Ortiz LI, Martínez RM (2019) Micorrización de pinos con el hongo silvestre comestible *Laccaria proxima*, sin el uso de fertilizantes. In: 1er Simposio y 2do Seminario Internacional de Educación Ambiental y Desarrollo Sustentable (SIEA 2019). Texcoco, estado de Mexico, pp 40–53

Charya LS, Garg S (2019) Advances in methods and practices of ectomycorrhizal research. In: Meena SN, Naik MM (ed) Advances in Biological Science Research: A Practical Approach. Academic Press, London, UK, pp 303–325

Díaz-Balbuena Y (2017) Biotecnología de inoculación de árboles de importancia forestal en México con hongos comestibles ectomicorrízicos (in Spanish). Thesis of Biologist, Instituto Tecnológico de Ciudad Altamirano, Guerrero, México

Escobar H (2019) Amazon fires clearly linked to deforestation, scientist say. Science 365:853

FAO (2005) Global forest resources assessment. FAO, UN Food and Agriculture Organization, Rome

FAO (2015) Global forest resources assessment. FAO, UN Food and Agriculture Organization, Rome

Farjon A (2013) *Pinus greggii*. The IUCN Red List of Threatened Species 2013: e.T32630A2822678. https://doi.org/10.2305/IUCN.UK.2013-1.RLTS.T32630A2822678.en

Flores-Armas L (2017) Inoculación con el hongo ectomicorrízico comestible *Suillus pseudobrevipes* en *Pinus pseudostrobus* y *Pinus greggii* (in Spanish). Thesis of Biologist, Instituto Tecnológico Superior de Perote, Veracruz, México

Franco-Ramírez A, Pérez-Moreno J, Sánchez-Viveros G et al (2019) Coinoculation of *Pinus greggii* Engelm. arbuscular and ectomycorrhizal fungi. In: International Conference on Mycorrhiza, Merida, Mexico, Mycorrhiza for a sustainable world, Mexico, p 179

Fuentes GO, Garzón TA (2017) Aspectos biotecnológicos y de biodiversidad ectomicorrízica de *Pinus maximartinezii* Rzedowski en la mixteca Oaxaqueña (in Spanish). Thesis of Forestal Engineer. Instituto Tecnológico Superior de San Miguel el Grande, Tlaxiaco, Oaxaca, México

Galindo-Flores G, Castillo-Guevara C, Campos-López A et al (2015) Caracterización de las ectomicorrizas formadas por *Laccaria trichodermophora* y *Suillus tomentosus* en *Pinus montezumae*. Bot Sci 93:855–863

García-Rodríguez JL, Pérez-Moreno J, Aldrete A et al (2006) Caracterización del hongo silvestre ectomicorrízico *Pisolithus tinctorius* (Pers.) Coker et Couch en cultivo y en simbiosis con eucalipto y pino. Agrociencia 40:665–676

Garibay-Orijel G, Morales-Marañon E, Domínguez-Gutiérrez M et al (2013) Caracterización morfológica y genética de las ectomicorrizas formadas entre *Pinus montezumae* y los hongos presentes en los bancos de esporas en la Faja Volcánica Transmexicana. Revista Mexicana de Biodiversidad 84:153–169

GFR (2010) Global forest resources assessment 2010. Main report. FAO Forestry Paper 163. FAO, Rome

Gropp RE (2020) Catastrophes and Disasters: Do We Have a Scientific Infrastructure to Inform Decision-making?. BioScience 70:199–199

Hallmann CA, Sorg M, Jongejans E et al (2017) More than 75 percent decline over 27 years in total flying insect biomass in protected areas. PLoS One 12:e0185809. https://doi.org/10.1371/journal.pone.0185809

Heredia AC (2011) Hongos comestibles silvestres en una zona otomí y dinámica de colonización ectomicorrízica entre *Pinus pseudostrobus* y *Hebeloma mesophaeum* (in Spanish). Thesis of Master Science, Postgrado en Edafología, Colegio de Postgraduados, Montecillo, estado de México, México

Hernández SF (2016) Biotecnología, Etnomicología y Mesofauna asociada con hongos ectomicorrízicos comestibles en la mixteca oaxaqueña (in Spanish). Doctor of Science Thesis, Postgrado en Edafología, Colegio de Postgraduados, Montecillo, estado de México, México

Hooper DU, Adair EC, Cardinale BJ (2012) A global synthesis reveals biodiversity loss as a major driver of ecosystem change. Nature 486:105–108. https://doi.org/10.1038/nature11118

Index fungorum (2019) Index fungorum database with contributions from almost 500 institutions, projects and individuals leaded by Paul Kirk from CABI. Available from http://www.indexfungorum.org/Names/Names.asp (consulted on different dates until May 19, 2019)

IPCC (2019) Intergovermental panel on climate change, Climate Change and Land, An IPCC Special Report on climate change, desertification, land degradation, sustainable land management, food security, and greenhouse gas fluxes in terrestrial ecosystems, Summary for Policymakers. WMO, UNEP. 41p

Ishida TA, Nara K, Tanaka M et al (2008) Germination and infectivity of ectomycorrhizal fungal spores in relation to their ecological traits during primary succession. New Phytol 180:491–500

Jiménez M (2011) Estudio etnomicológico y biotecnológico de hongos silvestres comestibles ectomicorrízicos en *Pinus pseudostrobus* y evaluación de bacterias promotoras de crecimiento vegetal (in Spanish). Thesis of Master Science, Postgrado en Edafología, Colegio de Postgraduados, Montecillo, estado de México, México

Kalsotra T, Khullar S, Agnihotri R et al (2018) Metal induction of two metallothionein genes in the ectomycorrhizal fungus *Suillus hymalayensis* and their role in metaltolerance. Microbiology 164:868–876

Khullar S, Reddy MS (2016) Cadmium induced glutathione bioaccumulation mediated by γ-glutamylcysteine synthetase in ectomycorrhizal fungus *Hebeloma cylindrosporum*. Biometals 32:101–110

Khullar S, Reddy MS (2019) Cadmium and arsenic responses in the ectomycorrhizal fungus *Laccaria bicolor*: glutathione metabolism and its role in metal(loid) homeostasis. Environ Microbiol Rep 11:53–61. https://doi.org/10.1111/1758-2229.12712

Lakshangy CS, Sandeep G (2019) Advances in methods and practices of ectomycorrhizal research. In: Meena SN, Naik MM (eds) Advances in biological science research: a practical approach. Academic Press, New York, NY, pp 303–325. Available from https://www.sciencedirect.com/science/article/pii/B9780128174975000197?via%3Dihub

Lazo-Montaño A, Martínez RM, Pérez MJ et al (2018) Biotecnología de la inoculación de *Pinus hartwegii* con un consorcio de hongos silvestres comestibles ectomicorrízicos del género *Laccaria*. In: 1er Congreso Nacional de Investigación en Ciencia e Innovación de Tecnologías Productivas, Altamirano, Guerrero, Mexico, pp 26–36

Lazo-Montaño A, Ríos GU, Pérez MJ (2019) Bioinoculación de *Pinus hartwegii* (pino de las alturas), como una estrategia ante el cambio climático. In: 1er Simposio y 2do Seminario Internacional de Educación Ambiental y Desarrollo Sustentable (SIEA2019), Texcoco, estado de México, Mexico, pp 54–67

López-Gutiérrez A, Pérez-Moreno J, Hernández-Santiago F et al (2018) Nutrient mobilization, growth and field survival of *Pinus pringlei* inoculated with three ectomycorrhizal mushrooms. Bot Sci 96:286–304

Lovelock J (1992) Gaia una ciencia para curar el planeta. Oasis, Barcelona

Luciano DKI (2018) Biotecnología de los hongos ectomicorrízicos *Suillus brevipes* y *Thelephora terrestris* asociados con el pino neotropical *Pinus greggii* (in Spanish). Thesis of Biologist, Instituto Tecnológico de Ciudad Altamirano, Guerrero, México

Luciano DKI, Cortés SJ, Martínez RM et al (2018a) Síntesis de la ectomicorríza y formación de esporomas de *Laccaria laccata* asociada a *Pinus teocote* Schiede ex Schltdl. un árbol resistente a sequía. In: XII Congreso Nacional de Micología, Xalapa, Veracruz, México, p 370

Luciano DKI, Martínez RM, Cortés SJ et al (2018b) Aspectos biotecnológicos de la inoculación de pinos con hongos ectomicorrizicos en invernadero y campo. In: 1er Congreso Nacional de Investigación en Ciencia e Innovación de Tecnologías Productivas, Altamirano, Guerrero, México, pp 37–47

Luciano DKI, Martínez RM, Pérez MJ et al (2018c) Bioensayos exitosos utilizando pinos ectomicorrizados con hongos comestibles en el Área natural protegida Jardín Botánico Cozcaquauhco, Molino de Flores, Texcoco, estado de México. In: La Red Nacional de Áreas Naturales Protegidas en su II Congreso Internacional de ANP's, Aguascalientes, México, p 47

Martin TP, Harris JR, Eaton GK et al (2003) The efficacy of ectomycorrhizal colonization of pin and scarlet oak in nursery production. J Environ Hort 21:45–50

Martínez-Nevárez LE, Sarmiento-López H, Sígala-Rodríguez JA et al (2016) Respuesta a la inoculación inducida de *Russula delica* Fr. en plantas de *Pinus engelmannii* Carr. en vivero. Revista Mexicana de Ciencias Forestales 7:108–117

Martínez-Reyes M, Pérez-Moreno J, Villarreal-Ruiz L et al (2012) Crecimiento y contenido nutrimental de *Pinus greggii* Engelm. inoculado con el hongo comestible ectomicorrízico *Hebeloma mesophaeum* (Pers.) Quél. Revista Chapingo Serie Ciencias Forestales y del Ambiente 18:183–192 https://doi.org/10.5154/r.rchscfa.2010.11.112

Méndez NM (2007). Inoculación de pinos de importancia forestal útiles en restauración de áreas degradadas con hongos comestibles ectomicorrízicos (in Spanish). Thesis of Master Science, Postgrado en Edafología, Colegio de Postgraduados, Montecillo, estado de México, México

Méndez-Neri M, Pérez-Moreno J, Quintero-Lizaola R et al (2011) Crecimiento y contenido nutrimental de *Pinus greggii* inoculado con tres hongos comestibles ectomicorrízicos. Terra Latinoamericana 29:73–81

Moncada de la Fuente J, Anaya Garduño M, Ortiz-Solorio C et al (2013) Suelo: protejamos al suelo que nos da vida. Colegio de Postgraduados, SAGARPA, Texcoco

Mongabay (2019). Available from https://news.mongabay.com/2019/09/brazils-satellite-agency-resumes-releasing-deforestation-data/. Accessed 6 Oct 2019

Morales-Rivera IL (2017) Porcentaje de colonización ectomicorrízica en *Pinus greggii* originado por la inoculación con diferentes dosis del hongo comestible *Laccaria próxima* (in Spanish). Thesis of Biologist, Instituto Tecnológico Superior de Perote, Veracruz, México

Myers-Madeira EC (2008) Policies to reduce emissions from deforestation and degradation (redd) in developing countries. Resources for the Future, Washington, DC

Nara K (2006) Ectomycorrhizal networks and seedling establishment during early primary succession. New Phytol 169:169–178

Nuñez RE (2013) Inoculación micelial de *Pinus hartwegii* con hongos ectomicorrizógenos provenientes de la cuenca del Rio Magdalena D.F. en invernadero (in Spanish). Thesis of Master Science, Universidad Autónoma de Tlaxcala, Tlaxcala, México

Obase K, Tamai Y, Yajima T et al (2009) Mycorrhizal synthesis of four ectomycorrhizal fungi in potted *Populus maximowiczii* seedlings. Mycoscience 50:143–145

Olivier JGJ, Janssens-Maenhout G, Muntean M et al (2013) Trends in global CO_2 emissions: 2013 report. PBL Netherlands Environmental Agency of the European Commission's Joint Research Center, The Hague

Ortiz-López I (2019) Biotecnología de inoculación de *Pinus greggii* Engelm. con el hongo silvestre comestible ectomicorrízico *Suillus pungens* Thiers & A.H. Sm. (in Spanish). Thesis of Biologist, Universidad de la Sierra Juárez, Oaxaca, México

Ortiz-López I, Hernández AF, Martínez RM et al (2018) Ectomicorrizas de *Pinus greggii* inoculado con himenios jóvenes y maduros del hongo comestible *Suillus pungens*. In: 1er Congreso Nacional de Investigación en Ciencia e Innovación de Tecnologías Productivas, Altamirano, Guerrero, México, pp 48–58

Perea-Estrada VM, Pérez-Moreno J, Villarreal-Ruiz L et al (2009) Humedad edáfica, nitrógeno y hongos ectomicorrízicos comestibles en el crecimiento de pino. Rev Fitotec Mex 32:93–102

Pérez-Moreno J (1992) Aspectos ecológicos, aislamiento de cepas, síntesis de micorriza y pruebas "in vitro" contra patógenos de diversos hongos ectomicorrízicos del Parque Nacional Zoquiapan, Edo. de México. (in Spanish). Thesis of Master Science, Postgrado en Edafología, Colegio de Postgraduados, Montecillo, estado de México, México

Pérez-Moreno J, Martínez-Reyes M, Lorenzana-Fernández A et al (2009) Social and biotechnological of wild edible mushrooms in México, with emphasis in the Izta-Popo and Zoquiapan National Parks. Acta Botanica Yunnanica (Suppl XVI):55–61

Pérez-Moreno J, Lorenzana-Fernández A, Medel-Ortiz R et al (2014) Los hongos ectomicorrizicos de México: una perspectiva global. In: Alvarez J, Rodríguez-Guzmán MP, Alarcón A (eds) Biodiversidad microbiana de México. UNAM y Colegio de Postgraduados, Mexico

Pérez-Téllez J, Martínez-Reyes M, Hernández-Santiago F et al (2017) Inoculation with two ectomycorrhizal edible fungi in *Pinus cembroides* subsp. *orizabensis* a drought-resistant species. In: Pérez-Moreno J, Guerin-Laguette A (eds) Mushrooms, humans and nature in a changing world, Proceedings of the 9th International Workshop on Edible Mycorrhizal Mushrooms (IWEMM9) Colegio de Postgraduados and CONACYT, Texcoco, Mexico, p 125

Perry J (1991) The pines of Mexico and Central America. Timber Press, Portland, OR

Population Matters (2019) Contador de población mundial, en el sitio web Population Matters. Available from http://populationmatters.org/

Portillo-Manzano E, López-Reyes L, Tapia–Hernández A et al (2017) Biotechnological application of plant growth promoting bacteria and *Laccaria laccata* in the growth of *Pinus patula*. In: Pérez-Moreno J, Guerin-Laguette A (eds) Mushrooms, humans and nature in a changing world, Proceedings of the 9th International Workshop on Edible Mycorrhizal Mushrooms (IWEMM9) Colegio de Postgraduados and CONACYT, Texcoco, Mexico, p 126

Ramírez CE (2017) Etnomicología en la zona Tlahuica-Pjiekakjoo del estado de México. (in Spanish). Thesis in Licentiate in Sustainable Development, Universidad Intercultural del Estado de México, estado de México, México

Ramírez RK (2018) Transferencia nutrimental y crecimiento de *Abies religiosa* (Kunth) Schltdl. et. Cham con dos inóculos de hongos ectomicorrízicos, en dos sustratos (in Spanish). Thesis of Master Science, Postgrado en Edafología, Colegio de Postgraduados, Montecillo, estado de México, México

Ramírez-Herrera C, Vargas-Hernández JJ, López-Upton J (2005) Distribución y Conservación de las poblaciones naturales de *Pinus greggii*. Acta Botánica Mexicana 72:1–16

Read DJ, Perez-Moreno J (2003) Mycorrhizas and nutrient cycling in ecosystems - a journey toward relevance? New Phytol 157:475–492

Redecker D, Kodner R, Graham LE (2000) Glomalean fungi from the Ordovician. Science 289:1920–1921

Remy W, Taylor TN, Hass H et al (1994) Four hundred-million-year-old vesicular arbuscular mycorrhizae. Proc Natl Acad Sci 91:11841–11843

Rentería-Chávez MC, Pérez Moreno J, Cetina-Alcala VM et al (2017) Transferencia de nutrientes y crecimiento de *Pinus greggii* Engelm. inoculado con hongos comestibles ectomicorrícicos en dos sustratos. Rev Argent Microbiol 49:93–104

Repác I (2011) Ectomycorrhizal inoculum and inoculation techniques. In: Rai M, Varma A (eds) Diversity and biotechnology of ectomycorrhizae. Soil biology, vol 25. Springer, New York, NY. https://doi.org/10.1007/978-3-642-15196-5_3

Ríos-García U, Carrera MA, Pérez MJ et al (2017a) Potencial biotecnológico del hongo comestible ectomicorrízico *Laccaria proxima* en *Pinus engelmanii*. In: Primer Congreso Nacional de Recursos Naturales, Sistemas de Producción e Innovación Tecnológica, Oaxaca, México, p 35

Ríos-García U, Carrera-Martínez A, Leyva-López JC et al (2017b) Firs record of the ectomycorrhizal synthesis between *Laccaria proxima* and *Quercus uxoris* a vulnerable species of the moist montane cloudy forest of the mazatec región, Mexico. In: Pérez-Moreno J, Guerin-Laguette A (eds) Mushrooms, humans and nature in a changing world, Proceedings of the 9th International Workshop on Edible Mycorrhizal Mushrooms (IWEMM9) Colegio de Postgraduados and CONACYT, Texcoco, Mexico, p 134

Rodríguez-Gutiérrez I, Ramírez-Martínez D, Jacob-Cervantes V et al (2017) Mycorrhization of *Pinus montezumae* with simpatric and allopatric edible ectomycorrhizal species of *Laccaria* spp. In: Pérez-Moreno J, Guerin-Laguette A (eds) Mushrooms, humans and nature in a changing world, Proceedings of the 9th International Workshop on Edible Mycorrhizal Mushrooms (IWEMM9) Colegio de Postgraduados and CONACYT, Texcoco, Mexico, p 137

Rosales-Sánchez JC, Martínez RM, Fuentes GO et al (2017) Síntesis ectomicorrizica de *Laccaria proxima*, un hongo comestible silvestre ectomicorrizico y *Pinus duranguensis*. In: IV Encuentro Internacional sobre Biotecnología en la UATx, Tlaxcala. México, p 35

Rosenberg KV, Dokter AM, Blancher PJ et al (2019) Declive of the North American avifauna. Science 366:120–124

Sandoval GI (2010) Efecto de los hongos ectomicorrizógenos en el crecimiento y supervivencia de plántulas de *Pinus hartwegii* Lindl. y *Abies religiosa* (Kunth. et Cham.): Un enfoque para la restauración de ambientes deteriorados en la cuenca del Río Magdalena D.F (in Spanish). Thesis of Master Science. Universidad Nacional Autónoma de México. México

Saint- Exupéry A de (1948) Citadelle. Gallimard, Paris

Santiago-Hernández I, Peréz-Moreno J, Jiménez-Ruiz M et al (2017) Growth and nutrient content of *Pinus ayacahuite* inoculated with *Suillus granulatus* in the northern mountain range of Oaxaca, Mexico. In: Pérez-Moreno J, Guerin-Laguette A (eds) Mushrooms, humans and nature in a changing world, Proceedings of the 9th International Workshop on Edible Mycorrhizal Mushrooms (IWEMM9) Colegio de Postgraduados and CONACYT, Texcoco, Mexico, p 142

Sharma SK, Gautam N (2015) Chemical, bioactive, and antioxidant potential of twenty wild culinary mushroom species. BioMed Research International 2015:1–12

Sebastiana M, Duarte B, Monteiro F et al (2019) The leaf lipid composition of ectomycorrhizal oak plants shows a drought-tolerance signature. Plant Physiol Biochem 144:157–165. https://doi.org/10.1016/j.plaphy.2019.09.032

Steidinger BS, Crowther TW et al (2019) Climatic controls of decomposition drive the global biogeography of forest-tree symbioses. Nature 569:404–413

Smith SE, Read DJ (2008) Mycorrhizal Symbiosis. Academic Press, London, UK

Svensen HH, Marit RB, Kyrre K (2019) The past as a mirror: deep time climate change exemplarity in the anthropocene. Culture Unbound 11:1–23. Available from http://www.cultureunbound.ep.liu.se

Turjaman M, Yutaka T, Hendrik S et al (2005) Inoculation with the ectomycorrhizal fungi *Pisolithus arhizus* and *Scleroderma* sp. improves early growth of Shorea pinanga nursery seedlings. New Forests 30:67–73

UNEP (2012) The emmissions gap report 2012. United Nations Environment Programme, Nairobi

UNEP (2016) UNEP Frontiers 2016 Report: Emerging issues of environmental concern. United Nations Environment Programme, Nairobi

USDA ARS (2019) National Genetic Resources Program: Germplasm Resources Information Network (GRIN). Available from https://www.ars-grin.gov/

Valencia AS, Flores-Franco G (2006) Catálogo de Autoridad Taxonómica del género *Quercus*, Fagaceae en México. Facultad de Ciencias, UNAM, Base de Datos SNIB-CONABIO proyecto CS008, Mexico

Vidal O, Rendón-Salinas E (2014) Dynamics and trends of overwintering colonies of the monarch butterfly in Mexico. Biol Conserv 180:165–175

Villegas OJA (2014) Factores que influyen en la formación de esporomas y micorrización de dos hongos comestibles ectomicorrízicos asociados con un pino neotropical (in Spanish). Thesis of Master Science, Postgrado en Edafología, Colegio de Postgraduados, Montecillo, estado de México, México

Villegas-Olivera JA, Pérez-Moreno J, Mata G et al (2017) Type of lingt, basidiomata frequency and development of two species of edible ectomycorrhizal mushrooms associated with neotropical pines. In: Pérez-Moreno J, Guerin-Laguette A (eds) Mushrooms, humans and nature in a changing world, Proceedings of the 9th International Workshop on Edible Mycorrhizal Mushrooms (IWEMM9) Colegio de Postgraduados and CONACYT, Texcoco, Mexico, p 149

Villegas-Olivera JA, Pérez MJ, Sánchez VG et al (2018) Inoculación de hongos comestibles ectomicorrízicos en *Pinus oocarpa* (Schiede ex Schltdl.) especie de importancia forestal en México. In: 1er Congreso Nacional de Investigación en Ciencia e Innovación de Tecnologías Productivas, Altamirano, Guerrero, México, pp 59–69

Vitousek PM, Mooney HA, Lubchenco J et al (1997) Human domination of Earth's ecosystems. Science 277:494–499

Wilson EO (1992) The diversity of life. Penguin Books, London

World Bank (2016) World Development Indicators 2016. World Bank, Washington, DC. https://doi.org/10.1596/978-1-4648-0683-4

Zalasiewicz J, Mark WM, Smith AG et al (2008) Are we now living in the Anthropocene? GSA Today 18:4–8. https://doi.org/10.1130/GSAT01802A.1

Zohdy S, Schwartz TS, Oaks JR (2019) The coevolution effect as a driver of spillover. Trends in Parasitology 35:399–408

Zuñiga CAJ (2018) Inoculación de *Pinus cembroides* Zucc. con un hongo ectomicorrízico comestible y una bacteria auxiliadora de la micorrización en dos sustratos (in Spanish). Thesis of Master Science, Postgrado en Edafología, Colegio de Postgraduados, Montecillo, estado de México, México

Chapter 4
Diversity and Importance of Edible Ectomycorrhizal Fungi in Guatemala

Roberto Flores Arzú

4.1 Introduction

Guatemala is located on the northern side of the Central American isthmus. It is a small country that has been included in the list of megadiverse countries on earth and is a biodiversity hotspot (Myers et al. 2000), where Nearctic and Neotropical flora and fauna converge (see Fig. 4.1). There is few published information about mushroom diversity and, in particular, edible mycorrhizal ones that exist in the territory and their association with their symbionts. Guatemala has a wide variety of endemic species, and it seems to be the limit of the natural distribution of some species that are found in the northern hemisphere such as *Boletus edulis*, *Gomphus clavatus*, *Lactarius salmonicolor*, and *Catathelasma ventricosa*, among others (Boa 2004). The purpose of this chapter is to provide scientific information about the wide diversity of mycorrhizal mushrooms that exist in Guatemala and are associated with different local Neotropical species of the genera *Abies*, *Alnus*, *Pinus*, and *Quercus* (Flores et al. 2012). This research includes also relevant information about the distribution of wild edible mushrooms, endemism, local presence of species complexes, habitat, threats, and ethnomycology.

4.2 Natural History and Local Vegetation in Guatemala

Guatemala is one of the megadiverse countries in the world in terms of biological wealth and is considered a hotspot of biodiversity (Myers et al. 2000). Guatemala's flora has 538 species with local endemism, of which 285 are endemic to Mesoamerica

R. Flores Arzú (✉)
Unidad de Biodiversidad, Tecnología y Aprovechamiento de Hongos—UBIOTAH—,
Departamento de Microbiología, Facultad de Ciencias Químicas y Farmacia,
Universidad de San Carlos de Guatemala, Ciudad de Guatemala, Guatemala

J. Pérez-Moreno et al. (eds.), *Mushrooms, Humans and Nature in a Changing World*, https://doi.org/10.1007/978-3-030-37378-8_4

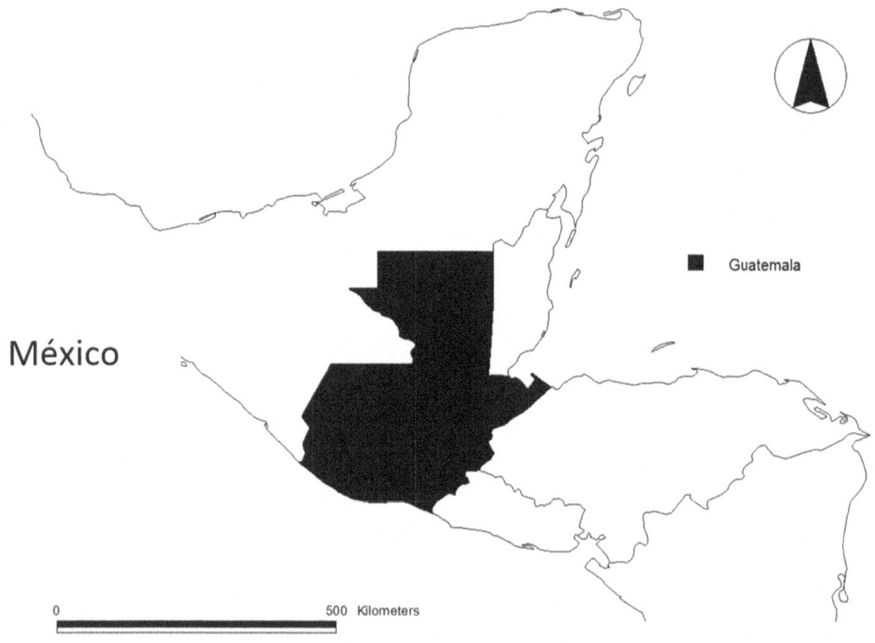

Fig. 4.1 Map location of Guatemala. The country itself comprises the majority of Mesoamerica

(Véliz et al. 2013). Local fungal diversity is a reservoir for species whose origins date back to the natural migration of plants and fungi from the old continent of North America (Van der Hammen and Hooghiemstra 2001; Halling and Mueller 2002; Halling, Osmudson and Neves 2007), prior to the formation of the Central American isthmus at the end of the Cretaceous period.

The geological history of the isthmus, which was characterized by displacement, floods, and submarine soil emersions, as a result of the collision of tectonic plates and strong volcanic activity, led to the formation of an extensive territorial bridge that facilitated the biological migration from North to South and vice versa, at the end of the Miocene and during the Pliocene. In this long period of time, Nearctic and Neotropical plants established in Central America (Gentry 1982).

For millions of years, part of what is now Guatemala (known as Nuclear Central America) belonged to the North American continent, receiving migration of Nearctic flora and fauna, from both the western and eastern zones, that over time, with the influence of climate change (global warming and ice ages) and the continuous changes in the isthmus, produced isolation of species, particularly in the elevated areas. For this reason, natural populations of *Abies* between 2000 and 3000 m a.s.l. (Véliz et al. 2014), *Juniperus* in altitudes of 2700–3400 m (with the exception of *J. comitana* that is located at 1200–2200 m, in two specific areas of the country), *Alnus*, *Arbutus*, and *Comarostaphylis* between 2000 and 3000 m (Véliz et al. 2014), *Cupressus* (1200–3300 m), *Pinus* (400–3600 m), *Quercus* (400–3000 m),

Liquidambar (800–2200 m), *Taxodium* (300–2500 m), and *Taxus* (1000–3000 m) occur in Guatemala. Some of these genera developed local species as *Abies guatemalensis* Rehd., *Juglans guatemalensis* (= *J. olanchana*), *J. steyermarkii*, *Pinus tecunumanii*, *P. caribaea* var. *hondurensis*, and *Litsea guatemalensis* (URL 2004; Véliz et al. 2014).

Tropical high forests, pine-oak forests, and those of conifers tend to be located in the more humid life zones described by Holdridge et al. (1971) and constitute the best climate and vegetation reference to understand the natural diversity of the country. Guatemala has 10 physiographical regions, 7 biomes, 14 ecoregions, 66 ecosystems (41 natural and 25 with anthropogenic intervention), and 14 life zones, in accordance with the Holdridge system (Castañeda 2008).

The humid zones correspond to the so-called humid, very humid, pluvial, and montane subtropical forests. Low and middle forests and savanna are located in the area known as dry forest. In the semiarid areas, the genus *Prosopis* (Fabaceae), better known as mesquite (Gutiérrez-Espeleta and Frenkel, 2005), predominates. There are also special ecosystems such as wetlands and mangroves in the Pacific and Atlantic coasts.

The country naturally possesses seven genera of conifers (*Abies*, *Cupressus*, *Juniperus*, *Pinus*, *Podocarpus*, *Taxodium*, and *Taxus*). That is why Veblen (1977) went on to write that no other region of so low latitude (approximately 14–16° N) is characterized by such a large variety of taxa of conifers.

Podocarpus, the only conifer from South America in the country, has three species including one endemic, *P. guatemalensis* Standl., in the mountainous and wet zone in the central highlands at altitudes between 1500 and 3200 m.

The genera *Abies* and *Acer* represent organisms of great ecological value and very reduced populations in Guatemala that find their most southern location at the continental level. *Abies guatemalensis* Rehder is located in the higher altitudes of the western highlands and as relictual manner in the two highest mountain systems of Eastern Guatemala (Sierra de Las Minas and Cerro Miramundo). *Acer saccharum* subsp. *skutchii* (Rehder) A.E. Murray is found only in Sierra de Las Minas, which is considered one of the greatest wealth zones of biodiversity in Mesoamerica (Azurdia, 2004; Fundación Defensores de la Naturaleza, 2010). Some of these places are pointed in a map in Fig. 4.2.

Pines and oaks in Guatemala constitute the dominant vegetation in the Guatemalan highlands, from 1200 to 2600 m a.s.l. (Islebe and Véliz 2001), and depending on the altitude and geographical position can be found isolated or mixed with other tree genera, particularly *Cupressus*, *Alnus*, *Arbutus*, *Buddleja*, *Persea*, *Fraxinus*, *Prunus*, *Ostrya*, *Carpinus*, *Alfaroa*, *Brunellia*, *Myrica*, and *Clethra*, all Laurasic genera (Gentry, 2001), or can be found even with Neotropical genera such as *Montanoa*, *Olmediella*, *Oreopanax*, *Nectandra*, *Clusia*, *Ficus*, and *Annona*. At lower altitude areas, these can be found with Neotropical latifoliate species such as *Birsomina*, *Bursera*, *Cedrela*, *Curatella*, and *Platymiscium* and leguminous plants such as *Gliricidia*, *Erythrina*, *Leucaena*, *Acacia*, *Caesalpinia*, *Inga*, *Calliandra*, and *Dalbergia*, but to a lesser amount. The lower altitude areas, close to the shores of the Pacific, do not have populations of pine or oak in a natural way.

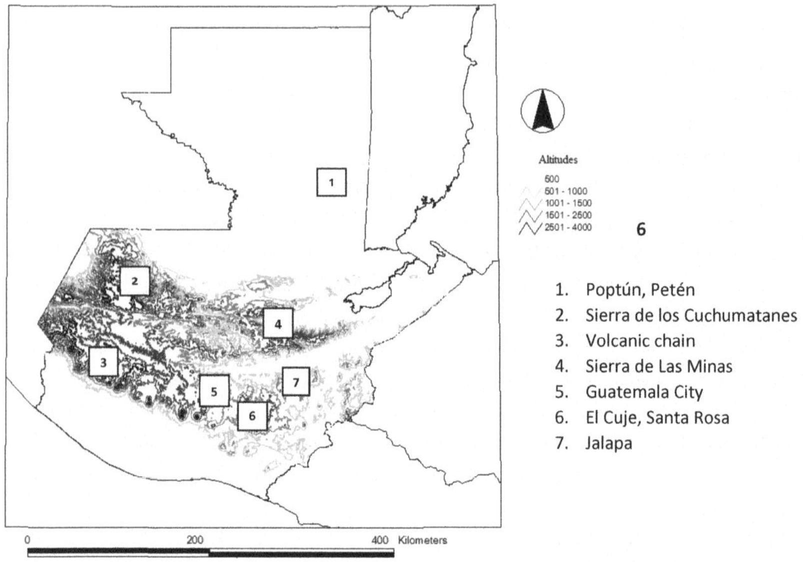

Fig. 4.2 Altitudinal zone map of Guatemala. The numbers indicate the main places where fungi have been collected. The Sierra de Los Cuchumatanes and the volcanic chain show the highest altitudes, around 4000 m. The rest of the highlands in the country have altitudes ranging from 800 to 2900 m

Several species of pine and oak trees that grow in Guatemala are distributed all over Mesoamerica, but most are located in sites with specific conditions of temperature, humidity, and soil. The genus *Pinus* has at least ten recognized species in the country: *P. ayacahuite*, *P. strobus* var. *chiapensis*, *P. caribaea*, *P. maximinoi*, *P. montezumae*, *P. tecunumanii*, *P. devoniana*, *P. oocarpa*, *P. pseudostrobus*, and *P. hartwegii*. The species with the widest distribution is *P. oocarpa*, which is also the most widespread in the tropical region, from the center of Sonora (Mexico) to the north of Nicaragua (Styles 1993). *Pinus caribaea* is the only species found naturally in the lowlands of Petén and Belize, toward the Atlantic Coast, at altitudes up to 400 m and tropical temperature conditions. *Pinus hartwegii*, in contrast, is the only species that grows in altitudes up to 4000 m according to URL (2004), in different mountainous systems and in the volcanic chain. *Pinus ayacahuite*, usually in the western area, produces the largest and robust pine trees in the country, with up to 3 m in diameter and up to 50 m in height. *Pinus devoniana* is the unique species in the Pino-Encino Ecoregion in Mexico and Guatemala (Farjon 2013; UVG 2009). *Pinus tecunumanii*, a species with large, straight shafts, makes reference to Tecún Umán, a Maya-K'iche' hero during the period of the Spanish conquest (INAB 1999; URL 2004).

On the other hand, due to the fact that pine trees are open-pollinated plants, high hybridization occurs between species, such as between *P. caribaea* and *P. oocarpa*

(Styles, Stead and Rolph 1982), a situation that enriches the local diversity, and constitutes a source of research for their ecological benefits and importance for forestry. In addition, it has been observed that pine forests replace wet montane forests that disappear due to their own fragility or immoderate logging (Islebe 1993).

With regard to oaks' diversity, Valencia (2004) reported 20 oak species that are common to México and Guatemala: *Q. peduncularis* Née, *Q. tristis* Liebm., *Q. sapotiifolia* Liebm., *Q. skinneri* Benth., *Q. oleoides* Schltdl. and Cham., *Q. segoviensis* Liebm., *Q. benthamii* A. DC., *Q. candicans* Née (= *Q. acuminata* M. Martens and Galeotti), *Q. castanea* Née, *Q. crassifolia* Humb. and Bonpl. (= *Q. brachystachys* Benth.), *Q. crispipilis* Trel., *Q. elliptica* Née, *Q. corrugata* Hook and Arn., *Q. insignis* M. Martens and Galeotti, *Q. lancifolia* Cham. and Schltdl., *Q. oocarpa* Liebm., *Q. pilarius* Trel. (= *Q. insignis* var. *pilarius* (Trel.) E. Murray), *Q. polymorpha* Schltdl. and Cham., and *Q. vicentesis* Trel. and 25 species, according to the UVG (2009). Recently, Quezada et al. (2018) made an accurate collection in the country and determined 29 oak species in Guatemala.

Standley and Steyermark (1952), who studied in detail the complex flora of Guatemala, came initially to consider about 40 species of oaks, although it is likely that some resulted from the high hybridization and morphological variation at the species level, a fact that complicates the identification of the local species. From a global point of view, Guatemala and Mexico are the two world centers of diversification of the genus *Quercus*, with more than 160 species (Valencia 2004). In comparison with Mexico, the number of oak species in Guatemala is quite high in relation to its small territorial space.

It is important to note that pine-oak forests in Guatemala constitute an important eco-region in Mesoamerica, known as "Central American Pine-Oak Forests" (URL 2004), but depending on the different geographical regions and life zones that exist in Guatemala, the oaks can be found in associations, particularly striking as the *Quercus–Podocarpus* in Sierra de Las Minas (URL 2004), *Quercus–Alnus* in the highlands of central-western region, or *Quercus–Clethra–Montanoa* in the Chicabal volcano, Quetzaltenango (Palacios 2013). *Clethra* is a high tree species easily confused with wide-leaf oaks that frequently occurs in the volcanic chain, Sierra de Los Cuchumatanes, and Sierra de las Minas (Véliz et al. 2014). The dominance of pine-oak forest is located between 800 and 2200 m a.s.l., with a temperature range between 14 and 25 °C and a precipitation of 900–3700 mm/year (Pérez et al. 2008).

Up to now, pines and oaks remain the primary source of energy for many homes in the rural area of the highlands of Guatemala, oaks (*Quercus* spp.) being the preferred choice because of the quality of its fuelwood and charcoal, a situation frequent in the rest of Central American countries (Brown and Kappelle 2001). Although its wood is in high demand, and despite its large ecological and economic importance, there is no reforestation program with oaks and only natural regeneration occurs. Nevertheless, *Q. peduncularis*, the species with the highest distribution in the country, has good regeneration rates and seed production and can be found in different ecosystems and altitudes. Some oak local species have a good potential for forestry. *Q. skinneri*, located in very reduced areas of the midland (bocacosta) region of the Pacific, and as remnants at the center of the country, produces large trees and big accorns, up to 5 cm diameter, that can could be used for reforestation. *Q. oleoides*, in Petén, forms scattered stands alone or in association with *Pinus caribaea*, and shows resistances to drought and high temperatures; and *Q. peduncularis*, with its large distribution in the country and host for many edible mycorrhizal mushrooms (Morales et al. 2002; Bran et al. 2003a), is an interesting species for uses in reforestation.

A number of initiatives, such as the Alliance for the Conservation of Pine-Oak Forests (UVG 2009), are being developed in Guatemala to maintain this important diversity, particularly in the biomes: mountain forest and humid subtropical and tropical broadleaf forest, and the five ecoregions (Central American Montane Forests, Central American Pine-Oak Forests, Humid Forests of the Sierra Madre of Chiapas, Montane Forest of Chiapas, and Humid Forests of Petén-Veracruz), which include the volcanic chain, Boca Costa, highlands, and limestone highland. This large number of ecosystems makes this small geographical area one of the richest in biological diversity on the planet (URL 2004).

4.3 Fungi and Forests in Guatemala

For a better understanding of the diversity of ectomycorrhizal fungi in Guatemala, it is important to bear in mind that the humid mountainous forests in the country have much more phytogeographic affinity with Mexico than with Costa Rica and South America (Islebe and Kappelle 1994) and that the Mesoamerican forests constitute one of the richest centers of biodiversity and endemism, with more than 9000 species of endemic plants (Myers 1998).

The biodiversity values of these forests are very high, even when compared with those from the tropical rain forest (Brown and Kappelle 2001). For example, Islebe and Véliz (2001) reported 831 plant species in the humid mountainous forest of Acatenango volcano. Although the mushrooms from the volcanoes in Guatemala have not been studied, the diversity and endemism of these could be high as a result of the isolation of each volcano. Despite their biological richness, entire communities in mountainous forests are at the highest risk of extinction because of the disappearance of vegetation patches that could act as a refugee for species during the dynamic orographic and climatic history of the region (Van der Hammen and Hooghiemstra 2001). Guatemala has 50,000 ha of montane humid forest—one of the smallest areas in Central America—of which fortunately 20,000 ha is protected (Brown and Kappelle 2001). Pictures of the different conifer-hardwood forests in Guatemala are shown in Fig. 4.3.

Most studies on fungal diversity of Guatemala have been carried out in mixed forests in the highlands while only a few in tropical forests. Much of this information can be found in undergraduate theses, research reports, memories of international congresses, and indexed publications (Flores and Simonini 2000; Bran et al. 2003a; Morales et al. 2010; Flores et al. 2012). The diversity found and identified to date primarily includes genera and species of ectomycorrhizal fungi similar to those

Fig. 4.3 Main forest types in Guatemala. (**a**) *Abies guatemalensis* forest in Sierra de los Cuchumatanes, Huehuetenango, 3400 m a.s.l.; (**b**) conifer-broadleaf forest in Quetzaltenango, ca. 2300 m; (**c**) pine-oak forest near Guatemala City, ca. 1800 m; (**d**) pine forest in San Juan Sacatepéquez, ca. 1500 m; (**e**) oak forest in Parque Florencia, Sacatepéquez, ca. 1700 m; (**f**) mixed broadleaf forest remnants in Jutiapa, ca. 800 m

of the northern hemisphere; however, endemism appears to be an important feature of the mycobiota of Guatemala.

The distribution of mycorrhizal mushrooms in Guatemala appears to be conditioned by the presence of the plant symbiont (*Abies*, *Alnus*, *Pinus*, *Quercus*, and others). Areas with other symbionts such as *Arbutus*, *Acer*, *Juglans*, and *Liquidambar*, or tropical environments with broadleaf forests (particularly Leguminosaceae), have not been studied yet, and little is known about their mycorrhizal mushrooms. The findings of *Cantharellus*, *Clavulina*, *Inocybe*, *Tylopilus*, *Boletellus*, and *Sarcodon*, associated to *Dycimbe* (Caesalpinioideae) and *Pakaraimaea* (Dipterocarpaceae) in Guyana (Fulgenzi et al. 2008; Henkel et al. 2011, 2012, 2014; Smith et al. 2013; Grupe et al. 2015), are an incentive for the research in tropical areas of Guatemala, to go deeper into the distribution, adaptation capability, origin of these mushrooms, and their application in the conservation of ecosystems. In addition, tropical plants such as *Coccoloba* spp. (Polygonaceae), *Neea* spp. (Nyctaginaceae), *Guapira* spp. (Nyctaginaceae), and *Pisonia* are present in Guatemala, and they have been reported to form ectomycorrhizae (Haug et al. 2005; Moyersoen 2012). It should also be mentioned that, the presence of ectomycorrhizal boletales in a tropical dry forest in the north of Yucatan, Mexico, was discovered (Hasselquist et al. 2011), possibly associated with these plants. In 2019, fruit bodies of *Phlebopus* sp. were also found with tropical local trees in the Pacific coast as well as *Pisolithus* and *Scleroderma* (personal observations).

The studies carried out to date in the country show that most macrofungi belong to Basidiomycetes and Ascomycetes and that they are typical of the northern hemisphere except some species showing affinity with a southern hemisphere origin. The most actualized listing of macrofungi collected in Guatemala (Flores et al. 2012) has been carried out in recent years with the sampling and collections of the Research Unit in Biodiversity, Technology and Use of Mushrooms (Unidad de Biodiversidad, Tecnología y Aprovechamiento de Hongos (UBIOTAH)) of the Faculty of Chemical Sciences and Pharmacy at the University of San Carlos de Guatemala (USAC). This unit maintains the main local mycological herbarium, Micoteca Rubén Mayorga Peralta, internationally recognized with the code MICG by the Index Herbarium of the New York Botanical Garden. The herbarium has a collection of more than 6000 mushroom samples, preserved and available for study and exchange of information (http://micoteca.usac.edu.gt).

Because the research in Guatemala about macrofungi, particularly edible ectomycorrhizal mushrooms, is relatively recent and because the results obtained to date come from projects funded primarily by the USAC, the current information is scarce and does not exhaust the enormous fungal diversity of the country.

The first studies in Guatemala focusing on ectomycorrhizal fungi began with Sommerkamp (1984), who studied the mushrooms of the Biotope for the Conservation of the Quetzal, Mario Dary Rivera, in Baja Verapaz, and then with another study on the edible mushrooms for sale in the markets of the departments of the country (Sommerkamp 1990). In 1997, the first study on ectomycorrhizal fungi associated with *Abies* and *Pinus* in Sierra de Los Cuchumatanes, at altitudes between 3000 and 3600 m a.s.l., which was extended to similar ecological zones in the volcanic chain during the drought caused by the El Niño phenomenon in 1998, took place. This study improved our understanding of similarities and differences between mushroom populations in the sampling sites and allowed to make inferences about the influence of vegetation, soil, altitude, and isolation by geographical barriers on mushroom distribution (Flores et al. 2002). The first study on local forest mycorrhizal plant production was developed with this project, achieving a 75%

survival of *P. rudis* seedlings planted in the field within the area of Tuicoyg, after 2 years of monitoring and without additional watering.

In parallel, another study was developed on the diversity of ectomycorrhizal fungi in native populations of *Pinus caribaea* in Poptún, Petén (around 400 m a.s.l.), on limestone soils. Surprisingly, many species typical from temperate coniferous forests of the country, such as *Lactarius indigo*, *Tylopilus plumbeoviolaceus*, and abundant red *Russula*, were found (Flores and Bran 2018). However, the major number of specimens belonged to *Pisolithus* and *Scleroderma*, which are common in the central-eastern highland forests with warmer climate and sandy soil. *Pisolithus* was even found producing fruiting bodies between the burned bark of some pine trees, as a way of protection against forest fires.

Both studies made it possible to carry out successful trials of pine seedling production (*P. caribaea*, *P. oocarpa*, *P. pseudostrobus*, *P. maximinoi*, *P. hartwegii*, and *P. ayacahuite*) mycorrhized by native fungal strains (*Laccaria bicolor*, *Suillus brevipes*, *Alpova* sp., *Pisolithus arhizus*, *Rhizopogon evadens*, and *Scleroderma geaster*). A bank of strains of ectomycorrhizal (*Boletus*, *Laccaria*, *Lactarius*, *Suillus*, *Rhizopogon*, *Tylopilus*) and saprophytic (*Pleurotus*, *Agrocybe*, *Neolentinus*, *Lepista*) fungi of edible traditional use was also created in the same faculty (Flores et al. 2002).

In addition to the above, the synthesis of mycorrhizae of *L. indigo* with *P. oocarpa*, *P. pseudostrobus*, and *P. hartwegii* was later achieved in a container, using a peat-vermiculite substrate (Flores et al. 2005). To investigate the capacity of the species to form mycorrhizae with other pines, the synthesis was tested with *P. nigra*, *P. pinaster*, and *P. halepensis*, obtaining equally positive results (Díaz et al. 2007). These have been the first studies on mycorrhizal synthesis with *L. indigo*, one of the edible species with the highest demand in Mesoamerica.

Years later, a field research was conducted to collect ectomycorrhizae of *Quercus* with *L. indigo*, to corroborate its symbiosis, due to the fact that *L. indigo* is found either in oak forests, pine forests, or mixed pine-oak forests. Two pure stands were selected, one in Cerro Alux (department of Guatemala) and the other in Pueblo Nuevo Viñas (department of Santa Rosa), both without any pine present. It was found that the ectomycorrhizae of *Q. peduncularis–L. indigo* are long, thin, simple to branched, and as it happens with pines, they have a pale-orange color that changes to green-to-blue when damaged, with rhizomorphs of equal color. The exuded latex is orange and changes to dark green with time. This change of color was also observed under in vitro conditions, using semi-culture medium BAF, where the strains initially produced yellowish-orange hyphae that become greenish, and rhizomorphs with blue-green color. Also, it was possible to observe that aerial compact mycelium was able to produce the characteristic blue-indigo coloration.

One goal is the local production of mycorrhized oak seedlings with *L. indigo* to promote reforestation with oaks, not only for its high demand as fuelwood, particularly in rural areas, but also to achieve the production of edible mushrooms as an add-on product. The species with the greatest interest could be *Q. peduncularis*, *Q. tristis*, *Q. acatenanguensis*, *Q. skinneri*, and *Q. oleoides*.

On the other hand, considering the high consumption of *Lactarius deliciosus* sensu lato in Guatemala, in 2001, a research to produce mycorrhized local pines (*P. oocarpa*, *P. pseudostrobus*, and *P. hartwegii*) with three strains of *Lactarius* aff. *deliciosus* isolated in the country was conducted. Unfortunately, only a few ectomycorrhizas were produced, which later disappeared because of contaminants (unpublished data). When inoculated with strains of *L. deliciosus* Fr., harvested in Spain, the results were positive with abundant mycorrhizae (Flores et al. 2005; Díaz et al. 2007). To learn more about the features of the Guatemalan strains, in vitro cultures were made, finding differences with respect to the Spanish strains (Flores et al. 2008b). New researches are currently conducted in the USAC to produce pine seedlings mycorrhized by local edible *Lactarius miniatosporus*, *L.* aff. *subpurpureus*, and *L. deliciosus s.l.*, as well as *Quercus* spp. seedlings mycorrhized by *L. indigo*.

Another studied species, *L. rimosellus*, associated with *Quercus* in Guatemala and Mexico, produces dark-orange mycorrhizae with abundant hirsute hyphae, similar to those produced at the base of the fruiting bodies. Genetic sequences from the ITS region were obtained, being deposited in GenBank (Comandini et al. 2012). This particular species, which resembles *Laccaria*, has been reported as edible only in Jacaltenango, a small rural town close to the border between Guatemala and Mexico (Bran et al. 2003b).

As far as *Laccaria* genus is concerned, mycorrhizae of a strain of *L. bicolor* were produced with *Abies guatemalensis* seedlings. The strain was collected at 3600 m a.s.l. in a young stand of *P. hartwegii* in Sierra De Los Cuchumatanes. The strain not only produced mycorrhizae but also small fruiting bodies in the culture containers (Flores et al. 2008a). The use of *L. bicolor* and *Inocybe* spp. has been recommended to farmers who wish to cultivate fir in communal nurseries in high mountains of Guatemala, since we observed fruit bodies of both genera growing with seedlings of *Abies* grown in polyethylene bags in local soil. Those seedlings appeared to be larger and healthier than those not producing fruit bodies.

The interinstitutional collaboration allowed to perform mycorrhization assays with native strains of *Boletus edulis* and local pines, one in greenhouse under controlled conditions (University of Murcia, Spain) and the other one in vitro (Instituto de Recerca i Tecnologia Agroalimentaria-IRTA, Barcelona, Spain). The results were negative, despite achieving abundant production of mycelium as inoculum and the visible absence of pollutants (Díaz et al. 2009). This is in accordance with Ortega-Martínez et al. (2011), who concluded that *B. edulis* is a climax species in the successional stage of mycorrhizal fungi that requires mature trees for its development and fruiting. Recently, Endo et al. (2014) observed the formation of small mycorrhizae in seedlings of *Pinus densiflora* after inoculation with a Japanese strain of *B. edulis*, but these disappeared when changed from laboratory to greenhouse conditions. To date, mycorrhizal syntheses of *B. edulis* have also been achieved with hazelnut plants (*Corylus avellana*) and more recently with *Cistus ladanifer* in Europe (Águeda et al. 2006, 2008; Mediavilla et al. 2016).

4.4 Fungi and *Abies* in Guatemala

The studies with *Abies* have yielded interesting information about the symbiotic relationship of this genus in latitudes as far as South Guatemala. Most of the macrofungi present in pure stands or pine-cypress-fir mixed forests of Sierra de Los Cuchumatanes, Totonicapán—the best coniferous forests of the country and under the zealous care of its Maya-K'iche' settlers—and in the mountainous volcanic region of San Marcos show many similarities with reports from North America, including genera and species found in the west coast of the United States, which is a novelty for the few publications in Central America.

Halling (1996, 2001) and Halling et al. (2007) have explained how plants and fungi could have been established in the Central American isthmus as a product of the co-migration from North America, under the influence of the climatic changes of the different geological periods and considering the shift of host plants, from Coniferae to Fagaceae. Halling and Mueller (2002, 2005) found that many genera of ectomycorrhizal fungi associated with *Quercus* in Costa Rica have a clear relationship with the mycobiota from the East coast of North America, but obviously, the mandatory route toward the South was the complex volcanic-mountainous territory of Guatemala. For this reason, the diversity of ectomycorrhizal fungi in Guatemala must be greater than that from Costa Rica but remains poorly known for the following reasons: the greater diversity of plant symbionts, the large temperate subtropical areas, and the existence of high-elevated areas since the end of the Cretaceous, including volcanic cones, which are all biological reservoirs.

Abies guatemalensis is a climax species with timely distribution in Guatemala and Chiapas, Mexico. The associated mycobiota, only studied at the level of fruiting bodies, have been easy to identify to the genus level. However, local species show some differences in comparison with those from North America and Europe. Common genera are *Amanita*, *Cantharellus*, *Cortinarius*, *Hydnum*, *Hygrophorus*, *Helvella*, *Inocybe*, *Lactarius*, *Morchella*, *Ramaria*, *Russula*, *Tricholoma*, and *Thelephora*. In Guatemala, the most striking species in these forests are *Amanita* aff. *rubescens*, *Cantharellus* aff. *cibarius*, *Cortinarius* aff. *odorifer*, *Gomphus* aff. *kauffmanii*, and *G. floccosus*, *Helvella lacunosa*, *Hydnum umbilicatum*, *Hygrophorus* aff. *pudorinus*, *Lactarius* aff. *salmonicolor*, *Lactarius* aff. *uvidus*, *L. mexicanus*, *Russula olivacea*, and *Morchella* aff. *esculenta*. The mixed forests of *Abies* in the eastern part of the country, in Cerro Miramundo and Sierra de Las Minas, possess a complex and unique diversity. In Miramundo, fruit bodies of *Gomphus* aff. *kauffmanii* measure up to 40 cm in length. A particular feature of the young and mature specimens of this species is a strong bitter flavor and thinner carpophores with less thick scales in comparison with those from Mexico.

Due to the fact that *A. guatemalensis* is in high demand as Christmas tree in Guatemala, a situation that led to include it in the list of threatened species of CITES, its plantation (cropping) was encouraged in different areas of the highlands. This was done through the use of seeds and chaff from local fir forests and also from mixed pine forests. Therefore, one can find in these plantations different mushrooms to those seen in natural forests. An example has been the discovery of abundant sporomes of *Scleroderma* sp. (prevalent in pine-oak forests at lower altitudes) in one of the best plantations in San Marcos, close to 3000 m, as well as carpophores of *Boletus edulis* and *Suillus* in a fir plantation in Tecpán, where the original pine forest was once cut down (personal observation).

Currently, a study on *L.* aff. *salmonicolor* is being carried out, a remarkable species of the section *Deliciosi* associated with *Abies*, with a much more intense orange color and in high demand by local populations, particularly of the ethnicities Mam, Chuj, and K'iche'. Preliminary phylogenetic analyses indicate that this is a local species, very close to *L. aestivus* in Oregon and Washington, USA, associated with *Abies* and recently described by Nuytinck and Ammirati (2014).

In 2016, the study on the diversity of *Lactarius* section *Uvidus* associated with *Abies* was retaken, as there are different species in the two areas with the greatest abundance of fir trees in the west of the country. Porras and Flores (2016) confirmed by microscopy the presence of *L. montanus* and other species, likely to be cryptic.

In Guatemala, pure or mixed forests with *Abies* are rich in mushroom diversity but do not exceed the diversity seen in pine-oak forests, contrary to what Burrola-Aguilar et al. (2013) reported for temperate zones in the Neotropical region of Mexico.

4.5 Fungi and *Pinus* in Guatemala

The pine forests in Guatemala show differences in their mushroom composition according to their geographical distribution and ecosystems.

The original forests of *P. caribaea* in Petén (northern Guatemala) and Belize are found on limestone soil with humid tropical environments, mixed with broadleaves and sometimes forming almost-pure stands. The most representative mycorrhizal species found in these forests are *Boletus guatemalensis*, *Tylopilus leucomycelinus*, *T. plumbeoviolaceus*, *Pulveroboletus ravenelii*, *Pisolithus arhizus*, *Scleroderma texense*, and three species of *Lactarius* section Deliciosi: *L. salmoneus* (the only one with white-tomentose pileus), *L. indigo*, and *Lactarius* aff. *paradoxus*. Two very useful references to understand the diversity of mycorrhizal mushrooms in the *P. caribaea* forests in Belize, and therefore in Petén, are Kropp (2001) and Ortiz-Santana et al. (2007). Indeed, the pine-oak forests in Petén and Belize have the same origin and are only separated by a political border.

One of the most interesting findings in the stands of *P. caribaea* was *Amanita muscaria* (*A. muscaria* var. *persicina*) in Poptún, Petén, in 1997 (Flores 2003; Rinaldi et al. 2005). This variety was also found by Orson Miller and Tim Baroni in pine forests in Belize (personal communication). Actually, this variety has been recognized as a new species: *A. persicina* (Dav. T. Jenkins) Tulloss and Geml, according to Index Fungorum. In Guatemala, *A. muscaria* var. *flavivolvata* seems to be restricted to temperate-cold pine forests between 2200 and 3000 m in the western highlands of Guatemala; however, Cáceres (2011) reported the species in a pine-oak forest in Comalapa (central highlands). The discovery of the species in warm, tropical climate is a novelty. Ethnomycology inferences on *A. muscaria* in the Maya area are discussed below.

In contrast to the temperate pine forests in Guatemala, the forests of *P. caribaea* rarely host the complexes *Amanita caesarea*, *Cantharellus cibarius*, and *Lactarius deliciosus*. Kropp (2001) only mentions the finding of *C. cibarius* in his report from Belize where fruiting bodies of *C. cibarius* are similar to those found in Petén and southeast of Guatemala. Recently, a new species similar to *Craterellus ignicolor* has been found, in Poptún, but there is no hint about its local edibility (personal observation).

The white amanitas and red russulas are present in these warm forests but have not been studied carefully.

The pine forests in the South and East regions (toward El Salvador and Honduras), with volcanic soils, are composed mainly of *P. oocarpa*, *P. pseudostrobus*, *P. maximinoi*, and probably hybrids thereof. The climate of these areas is considerably drier and warmer than at higher altitude. The mycobiota associated with these forests resembles that of the Caribbean region, despite the existence of complex geographical barriers: the high Sierra de Las Minas to the North and the extensive river Motagua, surrounded by a dry thorny forest. Many species of Boletales and *Lactarius* from Petén are also present in the Southeast, though with greater abundance and diversity. However, some species seem to be present only in the South such as *Ramaria zippelli*, *Gyroporus umbrinosquamosus*, *G. cyanescens*, *Retiboletus griseus*, *R. ornatipes*, *Butyriboletus* sp., *Sutorius* sp., *Cantharellus minor*, and *C. cinnabarinus*. *Veloporphyrellus conicus*, an infrequent species, is also present there and reported from Belize by Yan-Chun et al. (2014). More recently, between 2017 and 2018, several field trips were done to Sierra de Las Minas (El Progreso), Jalapa, Jutiapa, and Santa Rosa in order to collect specimens, and the results were outstanding: many species and genera found were not yet recorded in Guatemala and Mesoamerica. Endemism in these places seems to be higher than in the western highland of the country. *Boletus guatemalensis*, with large fruit bodies and their typical red pores, was also found in Sierra de Las Minas and, surprisingly, further South in Santa Rosa in association with oaks; this particular species is currently being studied at molecular and phylogenetic levels, in order to determine its position in the current taxonomy. Some of the boletes present in the eastern zone are shown in Fig. 4.4.

To date, the temperate pine forests in the central and western highlands show greater mushroom diversity than those from the eastern regions. The major genera of ectomycorrhizal basidiomycetes are *Amanita*, *Albatrellus*, *Alpova*, *Boletus*, *Cantharellus*, *Catathelasma*, *Chalciporus*, *Coltricia*, *Craterellus*, *Cortinarius*, *Hysterangium*, *Hydnum*, *Hygrocybe*, *Hygrophorus*, *Inocybe*, *Laccaria*, *Lactarius*, *Melanogaster*, *Octavianina*, *Russula*, *Pisolithus*, *Phellodon*, *Porphyrellus*, *Pulveroboletus*, *Ramaria*, *Rhizopogon*, *Sarcodon*, *Scleroderma*, *Suillus*, *Tricholoma*, *Thelephora*, and *Tylopilus*. Also, ascomycete genera such as *Helvella*, *Gyromitra*, *Morchella*, and *Peziza* are present (Flores et al. 2002).

Fig. 4.4 Wild boletes recently found in the eastern highlands of Guatemala: (**a**) *Aureoboletus* sp. nov.; (**b**) *Fistulinella* sp. nov.; (**c**) *Caloboletus* sp. nov.; (**d**) *Suillus* aff. *spraguei*; (**e**) *Boletus guatemalensis*; (**f**) red pores of *B. guatemalensis*; (**g**) *Butyriboletus* sp. nov.; (**h**) *Phylloporopsis* aff. *boletinoides*

With regard to hypogeous species, it is important to note that these do not fructify abundantly as in other latitudes, as personally observed. The finding of *Alpova*, *Hysterangium*, *Melanogaster*, *Octavianina*, and *Rhizopogon*, all belonging to the order Boletales, has often been accidental with only a few specimens collected (Bran et al. 1997; Flores et al. 2002). To date, no *Tuber* species have been found in Guatemala, the closest find being *T. regimontanum* in northern Mexico, in symbiosis with *Q. polymorpha* (Guevara et al. 2008).

The fungal diversity changes depending on the forest and its age (Martínez-Peña et al. 2012). Although in Guatemala no study has analyzed yet mushroom production in function of the age of the trees and the surrounding vegetation, it has been noted that red-pine forests (subgenus *Pinus*) are more productive and diverse in mushrooms than white-pine forests (subgenus *Strobus*), which are often mixed with *Cupressus lusitanica*. In a recent visit to pine forests in the western highlands, it was possible to confirm that the diversity and quantity of mushrooms in stands of *P. ayacahuite* were lower than those seen in stands of *P. tecunumanii*, *P. rudis*, and *P. montezumae*.

The most representative species in the temperate pine forests from Guatemala are *Amanita muscaria*, *A. rubescens*, *Boletus edulis*, *B. neo-pinophilus* sp. nov., *Cantharellus cibarius* complex, *Hydnum repandum*, *H. umbilicatum*, *Lactarius indigo*, *L. deliciosus*, *L. thyinos*, *Russula brevipes*, *R. nigricans*, *Sarcodon* sp., *Suillus luteus*, *S. decipiens*, *S. brevipes*, *Scleroderma geaster*, *Tylopilus porphyrosporus*, and *T. chromapes*.

The Boletales occupy an important place in the mushroom diversity of Guatemala, and almost all the genera described to date have been found mainly associated with pine-oak forests.

Two species of great ecological and evolutive interest are *B. edulis* and *B. pinophilus*. The first one can be found in the forests at high altitude of the country, associated with *P. hartwegii*, *P. tecunumanii*, and even with *A. guatemalensis*, spanning a natural distribution from North America to Sierra de las Minas in Guatemala. The second one is found in mixed pine-oak-madrone forests and is distinguished from the sensu stricto European species by its smaller size, reddish stipe, notorious garlic flavor, and differences at the molecular level (Flores 2003). This species extends to Honduras, in pine forests (Elia Sarmiento and Osberth Morales, pers. comm.). Pictures of edible *Boletus* (porcini group) species are shown in Fig. 4.5.

4.6 Fungi and *Quercus* in Guatemala

In the central highlands, the pine-oak forests predominate, while in the west, the coniferous forests predominate. These highlands are home to the majority of the indigenous population of Guatemala, original ethnic groups who also own the vast traditional knowledge about edible, poisonous, and toxic fungi, a knowledge that goes back to pre-Columbian times. A further description of ethnomycology in Guatemala is given below.

Fig. 4.5 Some edible *Boletus* (porcini group) in Guatemala: (**a**) *Boletus* sp. nov.; (**b**) *Boletus* sp. nov.; (**c**) *Boletus edulis* of San Marcos (volcanic chain); (**d**) *Boletus neo-pinophilus* sp. nov. nom. prov., from Quetzaltenango; (**e**) *Boletus* aff. *variipes* (sp. nov.) from Sierra de las Minas; (**f**) *Boletus variipes* var. *fagicola* from Sierra de las Minas

The diversity of mushrooms in oak or pine-oak forests is even greater than in the coniferous forests. Although many similitudes have been found between pine-oak forests, it is feasible to find species with characteristics that suggest endemism and species with disjunct distribution. Halling and Mueller (2002) pointed out that in mountainous forests of Central America, particularly in oak forests, there are degrees of endemism for mushrooms, with species having a local distribution and others a very restricted distribution. They also noted that there was a high level of similarity between genera, but not at the species level. This is the case of *Boletus* in Guatemala, and it probably extends to other species complexes such as *Cantharellus cibarius*, *Amanita caesarea*, *Lactarius deliciosus*, and *Morchella esculenta*.

Many edible species of mushrooms sold in rural and urban markets come from surrounding oak and pine-oak forests. The mycorrhizal genera more frequent in *Quercus* forests in Guatemala are *Albatrellus*, *Amanita*, *Astraeus*, *Aureoboletus*, *Austroboletus*, *Boletus*, *Boletopsis*, *Boletellus*, *Cantharellus*, *Craterellus*, *Chalciporus*, *Clavariadelphus*, *Chroogomphus*, *Craterellus*, *Coltricia*, *Coriolus*, *Cortinarius*, *Dermocybe*, *Exsudoporus*, *Gyroporus*, *Hebeloma*, *Heimiella*, *Heimioporus*, *Hydnum*, *Hygrophorus*, *Hebeloma*, *Heimioporus*, *Hydnellum*, *Hygrophorus*, *Hygrocybe*, *Gomphus*, *Inocybe*, *Laccaria*, *Lactarius*, *Leccinum*, *Paxillus*, *Pisolithus*, *Phylloporus*, *Pulveroboletus*, *Retiboletus*, *Ramaria*, *Rhizopogon*, *Russula*, *Sarcodon*, *Scleroderma*, *Sistotrema*, *Strobilomyces*, *Sutorius*, *Thelephora*, *Tricholoma*, *Tylopilus*, *Xerocomellus*, and *Xerocomus*. Among the Ascomycetes, 10 species of *Helvella* and various *Pezizales* have been found, which are just about to be published. Some of these species are shown in Fig. 4.6.

Several species present in oak forests in Guatemala have a wide distribution, from southern Mexico to Colombia (Halling and Mueller 2002). It is possible to find sub-genera, new species, and/or varieties in each country. Some local *Cantharellus* and *Lactarius* species are shown in Fig. 4.7.

Some of the most representative mushroom species in oak forests in Guatemala are *Amanita jacksonii*, the *A. caesarea* complex, *A. costaricensis*, *A. flavoconia*, *A. conara*, *Austroboletus subvirens*, *A. gentilis*, the species complex similar to *Boletus variipes* and *B. atkinsonii*, *Boletellus russelli*, *B. ananas*, *Boletus betula* (= *Heimioporus betula*), *B. frostii* (= *Exsudoporus frostii*), *B. floridanus* (= *E. floridanus*), *Boletopsis grisea*, *Cantharellus lateritius*, the *C. cibarius* complex, *C. atrolilacinus*, *Craterellus ignicolor*, *C. calyculus*, *Gomphus clavatus*, *Lactarius indigo*, *L. psammicola*, *L. argillaceifolius*, *L. rimosellus*, *L. volemus*, *Phylloporus centroamericanus*, *Russula vesicatoria*, *R. virescens*, *R. delicata*, *Ramaria araiospora*, *R. botrytis*, *R.* aff. *violaceibrunneus*, *Sarcodon imbricatus*, *Scleroderma verrucosum*, *Tricholoma caligatum*, and *T. flavovirens*, among others.

The more studied oak forests to date have been those close to Guatemala City, at an altitude of 1500–2000 m a.s.l., with predominance of *Q. peduncularis*. Within these, *Boletus luteoloincrustatus* (Flores and Simonini 2000), an endemic species, was found as well as the majority of the species mentioned above. Although the diversity of mushrooms in these forests is high, there are species only found in certain areas such as *A. costaricensis*, *B. variipes* var. *fagicola*, *Boletinellus rompelii*, *G. clavatus*, and several species that appear to be endemic, particularly of the genera *Amanita*, *Boletus*, *Cortinarius*, *Gomphus*, *Helvella*, *Inocybe*, *Lactarius*, *Phylloporus*, *Ramaria*, *Suillus*, and *Tylopilus*.

These forests are also the habitat for tiny, less abundant mushrooms, difficult to observe, such as *Craterellus calyculus* (= *Pseudocraterellus undulatus* (Pers.)), *Microglossum viride*, *Geoglossum difforme*, *Trichoglossum hirsutum*, *Sarcoscypha coccinea*, and *Cookeina* aff. *tricholoma*.

It is important to mention that *Cantharellus* and *Craterellus* possess a notorious diversity in Guatemala, particularly in the pine-oak forest. Flores et al. (2008a, b, c) reported the existence of at least eight species for that year, a number that has increased with the finding of *Cantharellus* cf *roseocanus* and *C.* cf *confluens*, and

Fig. 4.6. Some ectomycorrhizal mushrooms in Guatemala: (**a**) *Leccinum rugosiceps*; (**b**) *Ramaria* sp.; (**c**) *Chalciporus trinitensis*; (**d**) *Phylloporus* sp. nov.; (**e**) *Phylloporus centroamericanus*; (**f**) *Inocybe geophylla* var. *lilacina*; (**g**) *Tricholoma* aff. *caligatum*; (**h**) *Gomphus clavatus*; (**i**) *Gomphus* aff. *kauffmanii*

Fig. 4.7 Some *Cantharellus* and *Lactarius* species present in Guatemala: (**a**) *Cantharellus atrolilacinus*; (**b**) *Cantharellus cibarius complex*; (**c**) *Lactarius neo-deliciosus* sp. nov. nom. prov.; (**d**) *Lactarius indigo*; (**e**) *Lactarius* aff. *haugiae*; (**f**) *Lactarius argillaceifolius*

that could be higher once DNA analyses of the different samples collected under the name *C. cibarius* are performed.

A genus that has recently been collected in the valley of Guatemala City is *Phlebopus*, with at least two unknown species. This genus has been reported as saprophytic in other latitudes and continents, producing large fruit bodies. The specimens collected in Guatemala have been found near mature *Fraxinus* trees and also in a natural *Q. peduncularis* forest. Both species lack cystidia, a fact that could relate them to *Phlebopus mexicanus*, a species recently described by Baroni et al. (2015). However, unlike that species, the Guatemalan specimens are larger, and one of them becomes reddish after cutting and produces caps of up to 40 cm in diameter.

Table 4.1 lists the ectomycorrhizal species and morphospecies that have been found in Guatemala to date and their host trees. It was used by Rinaldi et al. (2008) as a reference about the mycorrhizal status of the cited species.

4.7 Ethnomycology

Guatemala is the second country in America with the richest ancestral knowledge and use of wild edible fungi after Mexico. This knowledge was born in the original populations that settled in the Guatemalan territory, shaping different cultures and, among these, the outstanding Mayan culture for almost 4000 years. These original populations were established in both the North and South of the country and subsequently gave rise to other ethnic groups and cultures that inhabit the central and western highlands of Guatemala, known as the K'iche', Kaqchikel, Mam, Chuj, Kanjobal, Q'eqchi', and Uspantec, among others.

The oldest elements that refer to the use of fungi in Guatemala are the mushroom-shaped stone sculptures, from the southern zone (Preclassic) and central highlands (since Preclassic to Postclassic); the clay vessels with a mushroom shape from the North (Preclassic); and the references made to mushrooms in two local texts of great historical and cultural importance: the *Popol Vuh*—the sacred book of the K'iche'-Kaqchikel people—and the *Título de Totonicapán*. Although the oldest known texts were copied at the beginning of the colonial era, their origins are far more ancient as these tell the origins, history, experiences, customs, and knowledge of the populations.

The Popol Vuh mentions the offering of mushroom caps along with other elements to deities. Unfortunately, it cannot be concluded if these were edible and/or hallucinogen mushrooms. The text mentions, in K'iche' language, the offering of *holom okox*, which means "head or cap of mushroom."

The Título de Totonicapán, initially written in K'iche' and later translated into Spanish, makes reference to the use of nine mushroom stones by the principal lords or governors, which would represent power ("and they sent me to bring the nine mushroom-stones …."). Borhegyi (1961) wrote about the possibility of an association between the number nine—in relation to the nine small mushroom figures that were found in Kaminal Juyú (actually part of Guatemala City)—and the nine Lords of the Night of the Maya pantheon, which would be also an indication of worship to the mushroom stones. However, Mayer (1977) pointed out that there was not enough evidence—in those years—to affirm that the mushroom stones were related with the use of hallucinogenic mushrooms.

These sculptures were initially studied by the Guatemalan archaeologist Guillermo Mata as well as by archaeologists Bernard Lowy, Stephan F. de Borhegyi, and Karl Mayer (Mayer 1977). They were then studied by Ohi, Torres, and

Table 4.1 List of ectomycorrhizal mushrooms collected in Guatemala until 2016

Species	*Abies*	*Pinus*	*Quercus*	*Quercus* + *Pinus*
Ascomycota				
Leotiales				
Helvella acetabulum (L.) Quél.			X	X
H. atra J. Köng	X	X		
H. crassitunicata N.S. Weber	X			
H. crispa (Scop.) Fr.		X	X	X
H. elastica Bull.		X	X	X
H. lacunosa Afzel	X	X	X	X
H. lactea Boud.			X	
H. leucomelaena (Pers.) Nannf			X	
H. macropus (Pers.) P. Karst	X	X	X	X
H. sulcata Willd.			X	X
H. villosa Schaeff.		X		
Helvella. spp.			X	X
Humaria spp.			X	X
Gyromitra aff. *ínfula*		X		
Morchellaceae				
Morchella aff *esculenta* (L.) Pers.	X	X		
Morchella spp.	X	X	X	X
Pyronemataceae				
Otidea onotica (Pers.) Fuckel			X	
Tarzetta aff. *catinus*				X
Pezizaceae				
Peziza spp.			X	X
Sarcoscyphaceae				
Wynnea americana Thaxt.		X		X
Winnea sp.		X		X
Basidiomycota				
Agaricales				
Amanita aff *atkinsoniana* (Coker) Locq.			X	
A. aff. *bisporigera* G.F. Atk.			X	X
A. brunnescens G.F. Atk.			X	X
A. caesarea (Scop.) Pers.		X	X	X
A. aff. *calyptratoides* Peck			X	X
A. calyptroderma G.F. Atk. and V.G. Ballen			X	X
A. chlorinosma (Peck) Lloyd				X
A. citrina (Schaeff.) Pers.		X	X	
A. cokeri (E.J. Gilbert and Kühner) E.J. Gilbert			X	
A. conara Tullos and Halling			X	X
A. costaricensis Singer ex Tulloss, Halling and G. M. Muell.			X	X
A. crocea (Quél.) Singer			X	X

(continued)

Table 4.1 (continued)

Species	*Abies*	*Pinus*	*Quercus*	*Quercus* + *Pinus*
A. aff. *farinosa*		X		
A. flavoconia G. F. Atk.			X	X
A. franchetti (Boud.) Fayod			X	
A. frostiana (Peck) Sacc.			X	X
A. fulva Fr.			X	X
A. gemmata (Fr.) Bertill.			X	X
A. hemibapha (Berk. and Broome) Sacc.			X	X
A. magnivelaris Peck			X	X
A. aff. *magniverrugata* Thiers & Ammirati				X
A. muscaria var. *flavivolvata* (Singer) D. T. Jenkins		X		
A. muscaria var. *persicina* Dav. T. Jenkins		X		
A. pantherina (DC.) Krombh		X		X
A. rubescens Pers.	X	X	X	X
A. vaginata (Bull.) Lam.			X	X
A. verna (Bull.) Lam.			X	X
A. xylinivolva Tulloss, Ovrebo and Halling		X		
Camarophyllus spp.		X		
Catathelasma ventricosum (Peck) Singer		X		
Cortinarius alboviolaceus (Pers.) Fr.			X	X
C. aff. *iodes* Berk. and M. A. Curtis			X	X
C. odorifer Britzelm.	X			
C. aff. *praestans* Cordier (Gillet)			X	X
C. sanguineus (Wulfen) Fr.			X	X
C. violaceus (L.) Gray	X	X		
Cortinarius spp.	X	X	X	X
Hebeloma spp.		X	X	X
Hygrophorus chrysodon (Batsch) Fr.	X	X		
H. aff. *erubescens* (Fries) Fries			X	X
H. pudorinus (Fr.) Fr.	X			
H. russula (Schaeff.) Kauffman			X	X
H. sordidus Peck			X	X
Inocybe asterospora Quél.			X	X
I. calamistrata (Fr.) Gillet			X	X
I. geophylla (Bull.) P. Kumm			X	X
I. geophylla var *lilacina* (Bull.) P. Kumm			X	X
I. rimosa (Bull.) P. Kumm.			X	X
Inocybe spp.	X	X	X	X
Laccaria amethystina (Huds.) Cooke		X	X	X
L. bicolor (Maire) P.D. Orton	X	X		
L. gomezii Singer and G.M. Muell			X	X
L. laccata (Scop.) Cooke			X	X
L. major G.M. Muell.			X	X

(continued)

Table 4.1 (continued)

Species	*Abies*	*Pinus*	*Quercus*	*Quercus + Pinus*
L. nobilis A.H. Sm.				X
L. ohiensis (Mont.) Singer			X	X
L. proxima (Boud.) Pat.				X
L. trichodermophora G.M. Muell.			X	X
Lyophyllum sp.			X	X
Tricholoma caligatum complex (Viv.) Ricken			X	X
T. flavovirens (Pers.) S. Lundell			X	X
Tricholoma spp.	X	X	X	X
Boletales				
Alpova sp.		X		
Astraeus hygrometricus (Pers.) Morgan			X	X
Aureoboletus cf *auriporus* (Peck) Pouzar			X	
Austroboletus betula (Schwein.) E. Horak			X	X
A. gracilis var. *gracilis* (Peck) Wolfe			X	X
A. subflavidus (Murrill) Wolfe			X	
A. subvirens (Hongo) Wolfe			X	X
Boletellus ananas (M.A. Curtis) Murrill			X	X
B. coccineus (Sacc.) Singer				X
B. russellii (Frost) E.-J. Gilbert			X	X
Boletinellus monticola (Singer) Watling			X	
B. rompelli (Pat. and Rick) Watling			X	
Boletus edulis Bull.	X	X		
B. floridanus (Singer) Murrill			X	
B. frostii J. L. Russell			X	X
B. guatemalensis Flores and Simonini		X		
B. luridus complex			X	X
B. luteoloincrustatus Flores and Simonini			X	
B. aff. *nobilissimus* Both and Riedel			X	
B. neo-pinophilus sp. nov. Flores and Honrubia		X		X
B. pulverulentus Opat.			X	
B. rubellus Krombholz		X	X	X
B. variipes Peck			X	
B. variipes var. *fagicola* Peck			X	
Calostoma cinnabarinum Desv.			X	
Chalciporus piperatus (Bull.) Bataille		X		
C. trinitensis (Heinem.) Singer, I.J. Araujo and M.H Ivory			X	
Chroogomphus jamaicensis (Murrill) O.K. Mill.			X	
C. rutilus (Schaeff.) O.K. Mill.			X	
C. vinicolor (Peck) O.K. Mill.			X	
Gyroporus castaneus (Bull.) Quél.			X	
G. cyanescens (Bull.) Quél.		X		

(continued)

Table 4.1 (continued)

Species	*Abies*	*Pinus*	*Quercus*	*Quercus + Pinus*
Gyroporus sp. nov.			X	
G. subalbellus Murrill		X		
G. umbrinosquamosus Murrill		X		
Gyrodon lividus (Bull.) Sacc.				Alnus
Heimioporus betula (Schwein) E. Horak			X	
Hortiboletus spp.				
Hysterangium sp.		X		
Leccinum aff. *montícola* Halling and G. M. Muell			X	X
L. rugosiceps (Peck) Singer			X	X
Leccinum spp.		X	X	X
Melanogaster sp.		X		
Octavianina sp.		X		
Phlebopus sp. nov. 1			X	
Phlebopus sp. nov. 2 (*Fraxinus*)			X	
Phlebopus sp. nov. 3			X	
Phylloporus aurantiacus Halling and G.M. Muell.				X
P. centroamericanus Singer and L.D. Gómez			X	
P. leucomycelinus Singer			X	
Phylloporus. sp. nov. 1			X	
Phylloporus. sp. nov. 2			X	
Pisolithus arhizus (Scop.) Rauschert		X	X	X
Pulveroboletus ravenelii (Berk. and M.A. Curtis) Murrill		X		X
Retiboletus griseus (Frost) Binder and Bresinsky			X	X
R. ornatipes (Perck) Binder and Bresinsky			X	X
Rhizopogon evadens A.H. Sm		X		
R. ochraceorubens A.H. Sm.		X		
Rhizopogon spp.		X	X	X
Scleroderma areolatum Ehrenb.			X	X
S. bovista Fr.			X	X
S. polyrhizum (J.F. Gmel.) Pers.		X	X	X
S. texense Berk.		X		
S. verrucosum (Bull.) Pers.			X	
Strobilomyces confusus Singer			X	X
S. strobilaceus (Scop.) Berk.		X	X	X
Suillus americanus (Peck) Snell		X		
S. bovinus (Pers.) Roussel		X		
S. brevipes (Peck) Kuntze		X		
S. decipiens (Peck) Kuntze		X		
S. granulatus (L.) Roussel		X		
S. luteus (L.) Roussel		X		
S. salmonicolor (Frost) Halling		X		

(continued)

Table 4.1 (continued)

Species	*Abies*	*Pinus*	*Quercus*	*Quercus* + *Pinus*
S. subluteus (Peck) Snell		X		
S. tomentosus (Kauffman) Singer		X		
Suillus spp.		X		
Tapinella panuoides (Batsch) E.-J. Gilbert			X	
Tylopilus aff. *alboater* (Schawein.) Murrill			X	
T. balloui (Peck) Heinem. and Rammeloo			X	X
T. chromapes (Frost) A.H. Sm. and Thiers		X		X
T. felleus (Bull.) P. Karst.			X	
T. aff. *felleus*			X	X
T. leucomycelinus (Singer and M.H. Ivory) Flores and Simonini		X	X	
T. plumbeoviolaceus (Snell and E.A. Dick) Snell and E.A. Dick		X	X	
T. porphyrosporus (Fr. and Hök) A.H. Sm. and Thiers		X		
T. violatinctus T.J. Baroni and Both		X		X
Tylopilus spp.		X	X	X
Veloporphyrellus conicus (Ravenel) B. Ortiz, Y.C. Li and Zhu L. Yang		X		
Xerocomellus chrysenteron (Bull.) Sutara		X		
Xerocomus pseudoboletinus (Murrill) Singer			X	
Xerocomus spp.			X	X
Cantharellales				
Cantharellus atrolilacinus Eyssart., Buyck and Halling			X	
C. cibarius complex Fr.	X	X	X	X
C. cinnabarinus (Schwein.) Schwein.			X	
C. aff. *confluens* Berk and M.A. Curtis			X	
C. aff. *formosus*		X		
C. ignicolor R.H. Petersen Dahlman, Danell and Spatafora		X	X	X
C. lateritius (Berk.) Singer			X	X
C. minor Peck			X	X
Clavulina cinerea (Bull.) J. Schröt		X		X
C. coralloides (L.) J. Schröt.	X	X		
Clavulina sp.	X	X		
Craterellus boyacensis Singer			X	
C. fallax A.H. Sm.			X	
C. lutescens (Fr.) Fr.			X	X
C. tubaeformis (Fr.) Quél		X	X	X
Craterellus sp nov. 1			X	
Craterellus sp nov. 2		X		
Hydnum repandum L.	X	X	X	X
H. repandum var. *album* (Quél.) Rea		X	X	X

(continued)

Table 4.1 (continued)

Species	*Abies*	*Pinus*	*Quercus*	*Quercus* + *Pinus*
H. umbilicatum Peck	X	X	X	X
Pseudocraterellus calyculus (Berk. and M.A. Curtis) D.A. Reid			X	
Gomphales				
Clavariadelphus pistillaris (L.) Donk		X	X	
C. truncatus Donk			X	
Gomphus clavatus (Pers.) Gray			X	
G. floccosus (Schwein.) Singer	X			
G. aff. *floccosus*	X			
Gomphus sp. nov 1			X	
Gomphus sp. nov 2 (*Liquidambar*)		X		
Phaeoclavulina zippelii (Lev.) Overeem			X	
Ramaria araiospora Marr and D.E. Stuntz			X	
R. botrytis (Pers.) Ricken			X	
R. flava (Schaeff.) Quél			X	
R. flavobrunnescens (G.F. Atk.) Corner			X	X
R. stricta (Pers.) Quél			X	X
R. xanthosperma (Peck) Corner		X		X
Ramaria spp.	X	X	X	X
Polyporales				
Scutiger pes-caprae (Pers.) Bondartsev and Singer			X	X
A. confluens (Alb. and Schwein.) Kotl. and Pouzar	X	X		
Laeticutis cristata (Schaeff.) Audet			X	
Coltricia cinnamomea (Jacq.) Murrill			X	X
C. montagnei (Fr.) Murrill		X		
C. perennis (L.) Murrill		X		
Russulales				
Lactarius areolatus Hesler and A.H. Sm			X	
L. argillaceifolius Hesler and A. H. Sm.			X	
L. atroviridis Peck.			X	
L. costaricensis Singer			X	
L. chrysorrheus Fr.			X	X
L. corrugis Peck			X	X
L. neo-deliciosus Flores and Honrubia sp. nov.		X		
L. fragilis (Burl.) Hesler and A.H. Sm.			X	
L. gerardii var. *subrubescens* Peck			X	
L. hygrophoroides (Berk. & M.A. Curtis) Kuntze			X	
L. indigo (Schwein.) Fr.		X	X	X
L. aff. *necator* (Bull.) Pers.			X	
L. maculatipes Burl			X	
L. mexicanus A. Kong and Estrada	X			

(continued)

Table 4.1 (continued)

Species	*Abies*	*Pinus*	*Quercus*	*Quercus* + *Pinus*
L. montanus (Hesler and A.H. Sm) Montoya and Bandala	X	X		
L. aff. *paradoxus* Beardslee and Burl.		X		X
L. peckii Burl.			X	X
L. piperatus (L.) Pers.			X	
L. psammicola f. *glaber* A.H. Sm.			X	
L. psammicola f. *psammicola* A.H. Sm.			X	
L. rimosellus Peck.			X	
L. aff. *subpurpureus* Peck		X		
L. aff. *rufus*		X		
L. salmoneus Peck		X		
L. neo-salmonicolor Flores and Honrubia sp. nov.	X			
L. subplinthogalus Cocker.			X	X
L. subvellereus Peck.			X	X
L. thyinos A. H. Sm.		X		
L. uvidus (Fr.) Fr.	X	X		
L. volemus (Fr.) Fr.		X	X	
L. aff. *yazooensis*			X	
Russula alutacea (Fr.) Fr.			X	
R. aff. *densifolia* Gill			X	X
R. brevipes Peck.		X	X	X
R. crustosa Peck.			X	
R. delica Fr.			X	X
R. emetica (Schaeff.) Pers.			X	X
R. flavida Frost			X	
R. foetens Pers.			X	
R. aff. *fragiloides* Murrill			X	
R. grata Britzelm			X	
R. lutea (Huds.) Gray			X	
R. nigricans Fr.		X	X	X
R. olivacea (Schaeff.) Fr.	X	X		X
R. queletii Fr.		X		X
R. rosea Pers.			X	X
R. sanguinaria (Schumach.) Rauschert			X	X
R. sanguinea Fr.			X	X
R. vesicatoria Murrill			X	
R. virescens (Schaeff.) Fr.			X	
Sistotrema confluens Pers.			X	
Thelephorales				
Boletopsis leucomelaena (Pers.) Fayod.			X	
Thelephora palmata (Scop.) Fr.			X	X
T. terrestris Ehrh.			X	X

(continued)

Table 4.1 (continued)

Species	*Abies*	*Pinus*	*Quercus*	*Quercus* + *Pinus*
Thelephora sp.			X	X
Sarcodon imbricatus (L.) P. Karst.			X	
Sarcodon spp.		X	X	X
Hydnellum caeruleum (Hornem.) P. Karst.			X	X
Hydnellum concrescens (Pers.) Banker			X	X

Sommekamp at the beginning of the 1980s (Ohi & Torres 1994), who came to the conclusion that there are three main styles which changed over time by political and cultural influence of the dominant populations in Mexico. The oldest have a square base and zoomorphic figures in the stipe, belonging to the Preclassic period (approximately 1000 BC); those with a tripod base and carved stipe belong to the Classic period; and those with a tripod base with simple and smooth stipe belong to the Postclassical period (1500 AD). Although some mushroom stones have been found in Mexico and El Salvador, these can be considered as an element of the Guatemalan pre-Hispanic cultures, not only by their mention in the texts already cited but also for their quantity and variety. Currently, many of these sculptures are at the National Museum of Archaeology and Ethnology (MUNAE), the Popol Vuh Museum of Francisco Marroquín University, Museo Miraflores (built in the former Kaminal Juyú) and some in museums from their original regions such as the Santa Lucía Cotzumalguapa's Museum and the Regional Museum of La Democracia, Escuintla. Many mushroom stones are in private collections and in the vaults of the National Institute of Anthropology and History (IDAEH).

Mayer (1977) mentioned the existence of two arguments against the cultural use of mushrooms in the highlands of Guatemala. First is the inexistence of hallucinogenic mushrooms in the country, a wrong argument, as Mayer not only found written evidence from the colonial period about "mushrooms that can drive a person crazy when consumed"—called *xibalbaj ocox* (Xibalbá's mushroom, or "from the hell"), *kaizalah ocox* (poisonous mushroom), and *qu'ec c'im* (mushroom that gets you drunk)—but also the recent findings of *A. muscaria* by UBIOTAH's researchers that confirmed its presence in at least seven departments of the country (Huehuetenango, San Marcos, Totonicapán, Sololá, Chimaltenango, El Progreso, and Petén), as well as *Psilocybe* spp. almost over the whole national territory. Second is the absence of evidence about the use of hallucinogenic mushrooms in Mayan documents from the highlands elaborated in the Colonial period. The mushroom stones are to this date the only clue about the use of mushrooms in any type of ritual, as shown in the already cited documents. Besides, the fact that some of them were found in burials in Kaminal Juyú shows a ritual value.

With regard to the mushroom-like clay vessels of the Preclassic period, these have a larger area of distribution—the subtropical and tropical zone of Mexico (Tabasco and Campeche), Guatemala (Petén, north of Alta Verapaz, and Escuintla), and even in El Salvador—but there are fewer records and pieces in museums. No particular function of these is known in either country. The first to describe these figures was Borhegyi (1961), in an article with photographs. Woodfill et al. (2002) found a couple of them that belonged to the Protoclassic, being intact in a burial inside a grotto in Petén, which suggests that at least those two figures had an intention of sacredness as these were turned into offering. The vessels preserved in Guatemala show a smooth surface or with fingernail incisions as external decoration. Only one vessel has been found with a painted cap, which could represent a sun, being red, with long arms and bifurcated ends, from the Carmelita region. It is worth mentioning that in some archaeological Mayan sites in Petén, large cylindrical clay vessels have been found, with lids similar to mushroom caps, but there is nothing conclusive about it.

Taking into account the unique and historical value of these figures of stone and clay, which Mayer (1977) estimated to be about 400 sculptures, the process to consider them as Cultural Heritage of Humanity by UNESCO was initiated to increase our knowledge and improve their protection.

The first reference about the species *A. muscaria* in Guatemala was documented by Lowy (1974), when he describes the history of *Maximón*—Mayan-K'iche' ancestral deity—and the presence of the thunder mushroom, *Kakuljá*. However, Lowy (1972) previously proposed the representation of this mushroom in the Madrid Codex to explain why the character who holds it in his hand subsequently self-sacrifices himself. On the other side, it is striking that *A. muscaria*, an ectomycorrhizal mushroom exclusively associated with pines in Guatemala, is represented in a Codex believed to have been elaborated in the western zone of Yucatán (García 2000), where there is no pine (Gernandt and Pérez-de la Rosa 2014). Delgado et al. (2011) note that *P. caribaea* is only found in the south of Campeche and Quintana Roo (Mexico), as well as in Petén, Belize, and Honduras. García (2000) mentions recent studies about the original Codex that point to a possible origin in Tayasal, Petén, where *P. caribaea* does exist. Besides, *A. muscaria* var. *persicina* has only been found in Poptún (Petén) and Belize, with this pine species, which gives support to García's proposal.

In sampling carried out in the highlands, it has been found that peasants who know *A. muscaria* refer to it as a mushroom that "drives people crazy" or as deadly. However, when asked about *Psilocybe*, they only recognize it because it grows in pastures or in cow's manure and because some young people seek it to "go crazy." In certain areas of Quetzaltenango and Totonicapán, the orange carpophores of *A. caesarea* complex mushrooms are not consumed because people consider them poisonous, maybe by confusing them with rain-washed *A. muscaria*. At a recent field day in San Juan Sacatepéquez, a group of people consumed a ceviche supposedly containing raw species of *A. caesarea* complex and *A. jacksonii*, reporting that after 20 min they were feeling dizzy and drunk and that the feeling lasted approximately half an hour. It is likely that some local *Amanita* species could contain substances that cause dizziness or confusion, but further study is required.

With regard to edible mushrooms, the use of more than 80 species has been documented in Guatemala (Bran et al. 2003a), of which over 40 are mycorrhizal (pictures of some of the most common species sold in markets are shown in Fig. 4.8). As already noted, the majority are consumed in the highlands, where the majority of

Fig. 4.8 Some frequent edible mycorrhizal mushrooms in Guatemala: (**a**) *Cantharellus lateritius*; (**b**) *Laccaria laccata*; (**c**) *Hydnum umbilicatum*; (**d**) *Sarcodon imbricatus*; (**e**) *Hypomyces lactifluorum* over *Russula* sp.; (**f**) *Boletus luteoloincrustatus*; (**g**) *Helvella crispa*; (**h**) *Amanita jacksonii*; (**i**) *Amanita basii*

the indigenous and mestizo population lives. Although the country is small, it can be said that the West is *mycophilic* and the East is *mycophobic*. One reason is that a large part of the population of southeastern and eastern Guatemala does not have a Mayan origin but a Spanish one (Andalusian) or is mestizo. The first Spanish settlers who established in Guatemala came from Castile, where there was no tradition of mushroom consumption as in Cataluña or Navarra. On the other hand, the local vegetation has more Neotropical species of plants and trees, although there are pine trees and oaks. In the Eastern zone, in the departments of Chiquimula, Jalapa, and Jutiapa, the most sought mycorrhizal mushroom is *C. cibarius*, with white to yellowish stipe, while those from the West tend to maintain the same coloration as the stipe. In samplings made in pine-oak forests of Jutiapa and Jalapa, it was found that very few people know the edibility of *L. indigo*, *L. deliciosus*, and *A. caesarea* and prefer instead *Pseudofistulina radicata*, a parasitic mushroom associated with *Diphysa americana* (Fabaceae), which they call "guachipilín mushroom or hoe," due to its shape and color. *Pleurotus* spp. are also consumed and are known by their generic name "ears." The limited knowledge on edible mushrooms in the Eastern zone of Guatemala has led to more cases of intoxication and deaths of children due to the consumption of poisonous mushrooms, particularly of *Amanita*.

The Western part of Guatemala offers a visual spectacle in every sense, where sales of mushrooms in municipal markets or at the Pan-American Highway cannot go unnoticed. Two departments are renowned for the sale and distribution of mushrooms in the highlands: Chimaltenango in the center of the country and Totonicapán in the West.

Chimaltenango has two main centers where the Kaqchikel population dominates: San Martín Jilotepeque and San Juan Comalapa, although the market of the departmental capital is also an important place for mushroom trading from local and nearby provenances, including the neighboring department of El Quiché. Mushrooms are sold per unit, per measure (usually a basket or plastic bowl of 10 cm in diameter), and per pound. The sale by basket is used for large volumes during the high season, particularly for the *A. caesarea* complex, *L. indigo*, and *L. deliciosus*. Among them, it is common to see the sale of *A. jacksonii*, *A. caesarea*, *Boletus variipes* group, *C. cibarius*, *C. lateritius*, *H. repandum*, *H. umbilicatum*, *L. deliciosus*, *L. indigo*, *L.* cf *paradoxus*, *R. virescens*, *R. delica*, *Suillus* spp., and occasionally *Sarcodon* spp. (pictures of wild edible mushrooms sold in markets of central and western Guatemala are shown in Figs. 4.9 and 4.10).

Totonicapán, with strong K'iche' culture, is a small mountainous department, with temperate climate. The mushrooms sold at the municipal market—which is a renowned commercial axis—come from different localities of the department, particularly from Panquix, Santa Maria Chiquimula, and Momostenango. In this market, the sale of *A. rubescens*, *C. cibarius*, *H. lacunosa*, *L. macropus*, *L. deliciosus*, *L. indigo*, *L.* aff. *salmonicolor*, *L. thyinos*, *L.* aff. *paradoxus*, *L. amethystina*, *L. laccata*, *Morchella* spp., and *Ramaria* spp., as well as the saprophytic *Ampulloclitocybe* spp. and *Collybia* sp., is obvious.

A village famous for its mushroom trading is San Juan Sacatepéquez in the department of Guatemala, just 25 km from the capital city. Although the sale of

Fig. 4.9 Edible mycorrhizal mushrooms in Guatemalan markets. (**a**) *Cantharellus lateritius* in San Juan Sacatepéquez. (**b**) *Amanita jacksonii* and *L. neo-deliciosus* in Comalapa (**c**) *Lactarius neo-deliciosus*, *L. indigo*, and *Amanita* in San Juan Sacatepéquez. (**d**) Seller of Patzún with a basket of *Ramaria* spp. and *Lactarius*. (**e**) Women buying *L. neo-deliciosus* and *A jacksonii* in Comalapa. (**f**) Sellers of fresh *Amanita basii* in Comalapa

mushrooms occurs during the rainy season, the most important date is June 24, during the patronal feast in honor of San Juan Bautista. The *A. caesarea* complex is known there as "mushroom of San Juan." The sister species, *A. jacksonii*, with a red stipe, is locally known as "mushroom of San Pedro," which is the neighboring village (San Pedro Sacatepéquez). The majority of *C. cibarius* and *C. lateritius* sold in the markets of Guatemala City are collected in the forests of this place. It is also one of the sites with abundant sales of species of *Boletus* section *Boletus*. Hernández and Mérida (2017) reported that during recent years, new mushroom species started to be sold, such as *Cortinarius* or *Calvatia*. They also mention that mushroom sales occur at different dates because of the climatic variation and rainfall patterns.

Fig. 4.10 Common species sold in markets of Guatemala. (**a**) Basket with *Cortinarius*, *Boletus*, *Sarcodon*, *Ramaria*, and other species in San Martin Jilotepeque. (**b**) *Amanita basii* in Comalapa. (**c**) Lunch of grilled *L. indigo* with tortilla. (**d**) Group of mixed milk caps: *L. indigo*, *L. neo-deliciosus*, and *L. miniatosporus*. (**e**) Basket of *Ramaria* spp. (**f**) *Amanita rubescens* with *Russula* aff. *delica*. (**g**) *A. jacksonii*, *A. garabitoana* and *L. neo-deliciosus* in Chichicastenango. (**h**) Traditional sale of *Cantharellus* in maxán leaves in Jilotepeque

An interesting find in the highland's markets is the sale of mushrooms considered toxic in other countries, such as *Gyromitra* sp. and *Cortinarius* spp. To date, no study has been conducted about the chemical content of these mushrooms, which may be toxic or may be innocuous after cooking, and/or that the population that consumes them is genetically favored with greater resistance against certain substances that can affect other population groups. Here is a glimpse at a line of research of great scientific importance about mushrooms in Guatemala.

4.8 Conclusions

Guatemala is a small country but with a great biological richness in its different life zones. One of the most interesting groups by its origin, diversity, and endemism is the mushrooms, which also bring important benefits to the ecosystems and to humans. In Guatemala, ectomycorrhizal mushrooms are associated with pine, oak, fir, and alder trees from temperate highland areas, but further research is required in warm areas where broad-leaved trees predominate, particularly leguminous.

Although the study of these mushrooms is relatively recent, a good number of species have already been identified, along with their distribution and their host plants. Other advances include the discovery of species complexes and the need to separate the different species that coexist under the same name; the synthesis of mycorrhizae using local plants and fungal strains for forestry applications; valuable information on traditional knowledge by various ethnic groups of Mayan origin in Guatemala about toxic and edible mushrooms, as well as the relationship of these organisms in the original cultures of the country since pre-Hispanic times, particularly evident in the mushroom stones. Although the reduced number of mycologists in Guatemala and financial support, the results show the enormous local fungal richness and their ecological, and economic importance. There is much left to discover and posteriorly apply in order to preserve the endemic species and their habitats, to improve their cultivation and their use with new biotechnology. The support of foreign research institutions is fundamental to reach these goals and ensure that this territory considered as a hotspot for biodiversity (Myers et al. 2000) will not be missed by social pressure and indifference.

Acknowledgments To Ana Silvia Del Valle for the translation of this document. To Faisal Abdalla Herrera for helping with the translation and finding local gray literature.

Special thanks to my friends and colleagues: Yvonne Sommerkamp, María del Carmen Bran, and Osberth Morales (Guatemala); Jesús Pérez-Moreno, Ronald Ferrera-Cerrato, Arturo Estrada-Torres, Roberto Garibay-Orijel, Xavier Parladé, Asunción Morte, and Gisela Díaz Espejo (Spain); Giampaolo Simonini, Andrea Rinaldi, Ornella Comandini, Alessandra Zambonelli, and Gloria Innocenti (Italy); Alexis Guerin-Laguette (France/New Zealand); Fuquiang Yu and Wang Yun (China); and David Pilz (USA) for helping to promote the knowledge of the Guatemalan wild edible mushrooms.

And in memory of Dr. Mario Honrubia (†), a great guide and tutor.

References

Azurdia C (2004) Priorización de la Diversidad Biológica de Guatemala en Riesgo Potencial por la Introducción y Manipulación de Organismos Vivos Modificados. Consejo Nacional de Áreas Protegidas-CONAP, Guatemala. Available from http://www.bchguatemala.gob.gt/documentos/capacitacion/modulo-informativo-sobre-biodiversidad-local-y/Priorizacion%20DB-OVMs.pdf/view. Accessed 5 May 2019

Águeda B, Parladé X, Fernández-Toirán LM et al (2006) Characterization and identification of field ectomycorrhizae of *Boletus edulis* and *Cistus ladanifer*. Mycologia 98:23–30

Águeda B, Parladé X, Fernández-Toirán LM et al (2008) Mycorrhizal synthesis between *Boletus edulis* species complex and rockroses (*Cistus* sp.). Mycorrhiza 18:443–449

Baroni TJ, Cifuentes J et al (2015) A new species of *Phlebopus* (Boletales, Basidiomycota) from Mexico. N Am Fungi 10:1–13

Boa E (2004) Wild edible fungi: A global overview of their use and importance to people. Food and Agriculture Organization of the United Nations

Borhegyi S (1961) Miniature mushroom stones from Guatemala. Am Antiq 26:498–504

Bran MC, Flores R, Rodríguez E et al (1997) Hongos ectomicorrícicos asociados a *Abies guatemalensis*, *Pinus rudis* y *P. ayacahuite* de la Sierra de los Cuchumatanes y su aprovechamiento en la producción de planta forestal micorrizada. Fase I. DIGI/Universidad de San Carlos de Guatemala, Guatemala

Bran MC, Morales O et al (2003a) Contribución al conocimiento de los hongos comestibles de Guatemala. Rev Cient 1:1–24

Bran MC, Morales O, Flores R et al (2003b) Hongos comestibles de Guatemala: diversidad, cultivo y nomenclatura vernácula. Fase III. DIGI/Universidad de San Carlos de Guatemala, Guatemala

Brown A, Kappelle M (2001) Introducción a los bosques nublados del neotrópico: una síntesis regional. In: Kappelle M, Brown A (eds) Bosques nublados del neotrópico. Instituto Nacional de Biodiversidad–INBio, Costa Rica, pp 25–40

Burrola-Aguilar C, Garibay-Orijel R, Argüelles-Moyao A (2013) *Abies religiosa* forests harbor the highest species density and sporocarp productivity of wild edible mushrooms among five different vegetation types in a neotropical temperate forest region. Agroforest Syst 87:1101–1115

Castañeda C (2008) Diversidad de Ecosistemas en Guatemala. In: Azurdia C, García F, Ríos M (eds) Guatemala y su Biodiversidad. Un enfoque histórico, cultural, biológico y económico. Documento Técnico 6-2008. Centro de Estudios Conservacionistas–CONAP, Guatemala, pp 181–229

Cáceres R (2011) Contribución al conocimiento de los hongos comestibles de la comunidad Xetonox, San Juan Comalapa, Chimaltenango. Tesis de Licenciatura. Universidad de San Carlos de Guatemala. Available from https://docplayer.es/45859402-Universidad-de-san-carlos-de-guatemala-facultad-de-ciencias-quimicas-y-farmacia.html. Accessed 3 May 2019

Comandini O, Eros-Honti Z, Jakucs E et al (2012) Molecular and morpho-anatomical description of mycorrhizas of *Lactarius rimosellus* on *Quercus* sp., with ethnomycological notes on *Lactarius* in Guatemala. Mycorrhiza 22:279–287

Delgado P, Piñero D, Rebolledo V et al (2011) Genetic variation and demographic contraction of the remnant populations of Mexican Caribbean pine (*Pinus caribaea* var. *hondurensis*: Pinaceae). Ann For Sci 68:121–128

Díaz G, Flores R, Honrubia M (2007) *Lactarius indigo* and *L. deliciosus* form mycorrhizae with Eurasian or Neotropical *Pinus* species. Sydowia 59:32–45

Díaz G, Flores R, Honrubia M (2009) Descripción de cultivos miceliares de Boletales neotropicales y europeos (*Boletus* grupo *edulis*, *Boletellus* y *Suillus*) y formación de primordios de *B. edulis* en cultivo puro. Rev Mex Mic 30:1–7

Endo N, Kawamura F, Kitahara R et al (2014) Synthesis of Japanese *Boletus edulis* ectomycorrhizae with Japanese red pine. Mycoscience 55:405–416

Farjon A (2013) *Pinus devoniana*. The IUCN Red List of Threatened Species 2013: e.T42356A2974898. Availabel from http://www.iucnredlist.org/details/42356/0 Accessed 16 March 2017

Fundación Defensores de la Naturaleza (2010) IV Actualización Plan Maestro 2010-2014: Reserva de Biósfera Sierra de las Minas. Guatemala. Available from http://www.conap.gob.gt/Documentos/SIGAP/PMR/PM%20RB%20Sierra%20de%20las%20Minas.pdf. Accessed 12 May 2019

Flores R, Bran MC (2018) Diversidad de hongos micorrícicos en bosques de *Pinus caribaea* en Poptún, Guatemala. Rev Cient 27(2):33–47. In Spanish

Flores R, Simonini G (2000). Contributo alla conoscenza delle Boletales del Guatemala. R.d.M. 2: 141-156. In Spanish

Flores R, Bran MC et al (2002) Hongos micorrícicos de bosques de pino y pinabete. DIGI/ Universidad de San Carlos de Guatemala, Guatemala

Flores R (2003) *Lactarius* sección *Dapetes y Boletus* grupo *Edulis* en Guatemala. Micorrización y estudio filogenético. Ph.D Tesis. Universidad de Murcia, España.

Flores R, Díaz G, Honrubia M (2005) Mycorrhizal synthesis of *Lactarius indigo* (Schw.) Fr. with five Neotropical pine species. Mycorrhiza 15:563–570

Flores R, Díaz G, Honrubia M (2008a) Mycorrhizal synthesis and basidioma primordium formation between *Abies guatemalensis* and *Laccaria bicolor*. Nova Hedwigia 87:153–160

Flores R, Honrubia M, Díaz G (2008b) Caracterización de cepas de *Lactarius* sección *Deliciosi* de Guatemala y su comparación con cepas europeas de *L. deliciosus*. Rev Mex Mic 26:51–55

Flores R, Ché G, García M et al (2008c) *Cantharellus* y *Craterellus* en Guatemala, diversidad y análisis microscópico de siete especies representativas en el país. Rev Cient 4:36–41

Flores R, Comandini O, Rinaldi AC (2012) A preliminary checklist of macrofungi of Guatemala, with notes on edibility and traditional knowledge. Mycosphere 3:1–21

Fulgenzi TD, Mayor JR, Henkel TW et al (2008) New species of *Boletellus* from Guyana. Mycologia 100(3):490–495

Gernandt DS, Pérez-de la Rosa JA (2014) Biodiversidad de *Pinophyta* (coníferas) en México. Rev Mex Biodivers (Supl.) 85:126–133

García A (2000) El Códice Tro-Cortesiano del Museo de América de Madrid. Rev Esp Antropol Am 30:9–25

Gentry A (2001) Patrones de diversidad y composición florística en los bosques de las montañas neotropicales. In: Kappelle M, Brown A (eds) Bosques nublados del Neotrópico. Instituto Nacional de Biodiversidad–INBio, Costa Rica, pp 85–123

Gentry A (1982) Neotropical floristic diversity: phytogeographical connections between Central and South America. Pleistoscene climatic fluctuations, or an accident of the Andean orogeny? Ann Mo Bot Gard 69:557–593

Grupe AC, Baker AD, Uehling JK et al (2015) *Sarcodon* in the Neotropics I: new species from Guyana, Puerto Rico and Belize. Mycologia 107(3):591–606

Gutiérrez-Espeleta E, Frenkel C (2005) Perspectivas de la Biodiversidad en Centroamérica 2003: Una primera aproximación al análisis de un tema prioritario. Observatorio del Desarrollo, Universidad de Costa Rica, Costa Rica

Guevara G, Bonito G et al (2008) *Tuber regimontanum*, new species of truffle from Mexico. Rev Mex Micol. 26:17–20

Halling R (1996) Boletaceae (Agaricales): latitudinal biodiversity and biological interactions in Costa Rica and Colombia. Rev Biol Trop 44:111–114

Halling RE (2001) Ectomycorrhizae: co-evolution, significance, and biogeography. Ann Mo Bot Gard 88:5–13

Halling RE, Mueller GM (2002) Agaric and Boletes of Neotropical Oakwoods. In: Watling R, Frankland J et al (eds) Tropical mycology, volume 1: macromycetes. CABI Publishing, Wallingford, pp 1–10

Halling RE, Mueller GM (2005) Common Mushrooms of the Talamanca Mountains, Costa Rica. New York Botanical Garden Press, New York, NY

Halling RE, Osmundson TW, Neves M-A (2007) Pacific boletes: implications for biogeograpic relationships. Mycol Res 112:437–447

Hasselquist NJ, Douhan GW, Allen MF (2011) First report of the ectomycorrhizal status of boletes on the Northern Yucatan Peninsula, Mexico, determined using isotopic methods. Mycorrhiza 21:465–471

Haug I, Weiß M, Homeier J et al (2005) Russulaceae and Thelephoraceae form ectomycorrhizas with members of the Nyctaginaceae (Caryophyllales) in the tropical mountain rain forest of southern Ecuador. New Phytol 165:923–936

Henkel TW, Aime MC, Uehling JK et al (2011) New species and distribution records of *Clavulina* (Cantharellales, Basidiomycota) from the Guiana Shield. Mycologia 103:883–894

Henkel TW, Aime MC, Chin MM et al (2012) Ectomycorrhizal fungal sporocarp diversity and discovery of new taxa in *Dicymbe* monodominant forests of the Guiana Shield. Biodivers Conserv 26:2195–2220

Henkel TW, Wilson AW, Aime MC et al (2014) Cantharellaceae of Guyana II: new species of *Craterellus*, new South American distribution records for *Cantharellus guyanensis* and *Craterellus excelsus*, and a key to the Neotropical taxa. Mycologia 106:307–322

Hernández M, Mérida J (2017) Etnomicología del Municipio de San Juan Sacatepéquez, Guatemala. Estudio de la diversidad, tradición, valoración y uso de los hongos locales. Tesis de Licenciatura. Universidad de San Carlos de Guatemala. Available from https://biblioteca-farmacia.usac.edu.gt/library/index.php?title=6149&query=@title=Special:GSMSearchPage@process=@autor=FLORES,%20ROBERTO%20@mode=&recnum=12

Holdridge LR, Genke W et al (1971) Forest environments in tropical life zones: a pilot study. Pergamon Press, Oxford

INAB–Instituto Nacional de Bosques (1999) Coníferas de Guatemala. Base de Datos Forestales DATAFORG. Proyecto Promoción del conocimiento y uso sostenible de las especies forestales de Guatemala. Consejo Nacional de Ciencia y Tecnología-CONCYT, Guatemala. Available from http://glifos.concyt.gob.gt/digital/multicyt/%201999.01.pdf Accessed 15 June 2017

Islebe G, Kappelle M (1994) A phytogeographical comparison between subalpine forest of Guatemala and Costa Rica. Feddes Repert 105:73–87

Islebe G, Véliz M (2001) Guatemala. In: Kappelle M, Brown A (eds) Bosques nublados del neotrópico. Instituto Nacional de Biodiversidad–INBio, Costa Rica, pp 231–242

Islebe G (1993) Will Guatemala's *Juniperus-Pinus* forests survive? Environ Conserv 20:167–168

Kropp B (2001) Familiar faces in unfamiliar places: mycorrhizal fungi associated with Caribbean pine. Mycologist 15:137–140

Lowy B (1972) Mushroom symbolism in Maya Codices. Mycologia 64:816–821

Lowy B (1974) *Amanita muscaria* and the thunderbolt legend in Guatemala and Mexico. Mycologia 66:188–191

Martínez-Peña F, Ágreda T, Águeda B et al (2012) Edible sporocarp production by age class in a Scots pine stand in Northern Spain. Mycorrhiza 22:167–174

Mayer KH (1977) The Mushroom stones of Mesoamerica. Acoma Books, California. http://www.samorini.it/doc1/alt_aut/lr/mayer-the-mushroom-stones-of-mesoamerica.pdf

Mediavilla O, Olaizola J, Santos-del Blanco L et al (2016) Mycorrhization between *Cistus ladanifer* L. and *Boletus edulis* Bull., is enhanced by the mycorrhiza helper bacteria *Pseudomonas fluorescens* Migula. Mycorrhiza 26:161–168

Morales O, Flores R, Samayoa B et al (2002) Estudio etnomicológico de la cabecera municipal de Tecpán Guatemala, Chimaltenango. Rev Cient 15(1):1–11

Morales O, Bran MC, Cáceres R (2010) Los hongos comestibles de uso tradicional en Guatemala. In: Martínez-Carrera D, Curvetto N, Sobal M et al (eds) Hacia un Desarrollo Sostenible del Sistema de Producción Consumo de los Hongos Comestibles y Medicinales en Latinoamérica: Avances y Perspectivas en el Siglo XXI. Red Latinoamericana de Hongos Comestibles y Medicinales-COLPOS-UNSCONACYT-AMC-UAEM-UPAEP-IMINAP, Puebla, México

Moyersoen B (2012) Dispersion, an Important Radiation Mechanism for Ectomycorrhizal Fungi in Neotropical Lowland Forests? In: Sudarshana P, Nageswara-Rao M, Soneji JR (eds) Tropical

Forests. IntechOpen, Rejecta. https://doi.org/10.5772/33217. Available from https://www.intechopen.com/books/tropical-forests/dispersion-an-important-radiation-mechanism-for-ectomycorrhizal-fungi-in-neotropical-lowland-forests Accessed 23 April 2019

Myers N (1998) Global biodiversity priorities and expanded conservation policies. In: Mace GM, Balmford A, Ginsberg BR (eds) Conservation in a changing world. Cambridge University Press, Cambridge, pp 273–285

Myers N, Mittermeier R, Mittermeier C et al (2000) Biodiversity hotspots for conservation priorities. Nature 403:853–858

Nuytinck J, Ammirati J (2014) A new species of *Lactarius* sect. *Deliciosi* (Russulales, Basidiomycota) from western North America. Botany 92:767–774

Ohi K, Torres M (1994) Piedras Hongo. Museo de Tabaco y Sal. Tokyo

Ortega-Martínez P, Águeda B, Fernández-Toirán LM et al (2011) Tree age influences on the development of edible ectomycorrhizal fungi sporocarps in *Pinus sylvestris* stands. Mycorrhiza 21:65–70

Ortiz–Santana B, Lodge DJ, Baroni TJ et al (2007) Boletes from Belize and the Dominican Republic. Fungal Divers 27:247–416

Pérez S, Secaira E, Macías C et al (2008) Plan de Conservación de los Bosques de Pino-Encino de Centroamérica y el Ave Migratoria *Dendroica chrysoparia*. Alianza para la Conservación de los Bosques de Pino-Encino de Mesoaméric, Serie Técnica 5. Fundación Defensores de la Naturaleza y The Nature Conservancy, Guatemala. Available from http://www.pronatura-sur.org/web/docs/PLAN_CONSERVACION_bosques_pino-encino.pdf. Accessed 12 May 2019

Palacios E (2013) Estudio de la diversidad florística de las especies arbóreas y algunas arbustivas y herbáceas de importancia socioeconómica en el área protegida Volcán y Laguna Chicabal, en el municipio de San Martín Sacatepéquez, Quetzaltenango. Tesis de licenciatura. Universidad de San Carlos de Guatemala

Porras C, Flores R (2016) Descripción microscópica de ejemplares del género *Lactarius* de las Secciones *Uvidus* y *Zonarii* en Guatemala. Rev Cient 28:2–61

Quezada M, Rodas L, Hernández B et al (2018) Encinos de Guatemala, estado de conservación y evaluación de servicios ecosistémicos. Fase III: Izabal, Zacapa y Chiquimula. DIGI/Universidad de San Carlos de Guatemala. Available from https://digi.usac.edu.gt/bvirtual/informes/puicb/INF-2017-16.pdf

Rinaldi AC, Comandini O, Flores R (2005) Guatemala: i funghi dei maya. Darwin 16:36–47

Rinaldi AC, Comandini O, Kuyper TW (2008) Ectomycorrhizal fungal diversity: separating the wheat from the chaff. Fungal Divers 33:1–45

Smith ME, Henkel TW, Uehling JK et al (2013) The Ectomycorrhizal Fungal Community in a Neotropical Forest Dominated by the Endemic Dipterocarp *Pakaraimaea dipterocarpacea*. PLOS ONE 8(1):e55160. Available from http://journals.plos.org/plosone/article?id=10.1371/journal.pone.0055160

Sommerkamp Y (1984) Estudio de los macromicetos del Biotopo Universitario "Lic. Mario Dary Rivera" para la conservación del Quetzal. Tesis de Licenciatura. Universidad de San Carlos de Guatemala

Sommerkamp Y (1990) Hongos Comestibles en los Mercados de Guatemala. Cuaderno de Investigación 3-90. DIGI/Universidad de San Carlos de Guatemala

Standley P, Steyermark J (1952) Fagaceae. In: Flora of Guatemala – Part III. Fieldiana. Bot. 24:369–396

Styles BT (1993) Genus *Pinus*: a Mexican purview. In: Ramamoorthy TP, Bye R, Lot A (eds) Biological diversity of Mexico: origin and distribution. Oxford University Press, New York, NY, pp 397–420

Styles BT, Stead JW, Rolph KJ (1982) Studies of variation in Central American pines putative hybridization between *Pinus caribaea* var. *hondurensis* and *P. oocarpa*. II. Turrialba 32:229–242

Universidad Rafael Landívar-URL (2004) Perfil ambiental de Guatemala: Informe sobre el estado del ambiente y bases para su evaluación sistemática. Instituto de Incidencia Ambiental de la Universidad Rafael Landívar, Guatemala

Universidad del Valle de Guatemala-UVG (2009) Diagnóstico Ecológico y Socioeconómico de la Ecorregión Bosques de Pino-Encino de Centroamérica. The Natural Conservancy–TNC/Centro de Estudios Ambientales y Biodiversidad, Universidad del Valle de Guatemala, Guatemala. Available from https://es.scribd.com/document/315069804/Diagnostico-Ecologico-y-Socioeconomico-de-la-Ecorregion-Bosques-de-Pino-Encino-de-Centroamerica. Accessed 23 April 2019

Valencia S (2004) Diversidad del género *Quercus* (Fagaceae) en México. Bol Soc Bot Mex 75:33–53

Van der Hammen T, Hooghiemstra H (2001) Historia y Paleoecología de los bosques montanos andinos neotropicales. In: Kappelle M, Brown A (eds) Bosques nublados del Neotrópico. Instituto Nacional de Biodiversidad–INBio, Costa Rica, p 63-84

Veblen T (1977) Las coníferas de Guatemala. Unasylva 29:25–30. Available from http://www.fao.org/3/l2015s/l2015s05.htm#las%20con%C3%ADferas%20de%20Guatemala. Accessed 23 April 2019

Véliz M, López J, Velásquez L et al (2013) Determinación, caracterización, evaluación del estado actual y uso de las especies endémicas de Guatemala. Proyecto FODECYT 02-2010. Consejo Nacional de Ciencia y Tecnología–CONCYT/Herbario BIGU, Universidad de San Carlos de Guatemala, Guatemala. Available from http://www.bchguatemala.gob.gt/documentos/capacitacion/modulo-informativo-sobre-biodiversidad-local-y/estudios-sobre-especies-endemicas-y-raras-de/Informe%20Proyecto%20FODECYT%2002-2010.pdf

Véliz M, López J, Velásquez L et al (2014) Guía para el reconocimiento de las plantas endémicas de Guatemala. Herbario BIGU. Universidad de San Carlos de Guatemala, Guatemala

Woodfill B, O'Mansky M, Spenard J (2002) Asentamientos y sitios sagrados en la región de Cancuén. In: Laporte JP, Escobedo H, Arroyo B (eds) XV Simposio de Investigaciones Arqueológicas en Guatemala. Museo Nacional de Arqueología y Etnología, Guatemala. Available from https://www.asociaciontikal.com/pdf/69.01%20-%20Woodfill%20-%20en%20PDF.pdf

Yan-Chun L, Ortiz-Santana B, Nian-Kai Z et al (2014) Molecular phylogeny and taxonomy of the genus *Veloporphyrellus*. Mycologia 106:291–306

Chapter 5
Advances in the Cultivation of *Lactarius deliciosus* (Saffron Milk Cap) in New Zealand

Alexis Guerin-Laguette, Ruth Butler, and Yun Wang

5.1 Introduction

The native flora in New Zealand is composed predominantly of endomycorrhizal plants (Orlovich and Cairney 2004). Thus, compared with the Northern Hemisphere, there are many soils in New Zealand devoid of or with less ectomycorrhizal fungi capable of outcompeting truffles and other edible mycorrhizal fungi (Hall et al. 2007). Land is available to establish edible mycorrhizal fungi orchards in regions with different soil types in temperate to subtropical climates. New Zealand has a worldwide reputation for leading and pioneering the cultivation of edible mycorrhizal fungi. In 1993, the country was second only to the United States for successfully growing the Périgord black truffle (*Tuber melanosporum*) out of its region of origin in Europe (Wang and Hall 2004). Since then, saffron milk cap (*Lactarius deliciosus*) was cultivated in New Zealand under field conditions in 2002 (Wang et al. 2002, 2013), bianchetto truffle (*T. borchii*) was cultivated in 2006 and Burgundy truffle (*T. aestivum* var. *uncinatum*) in 2007 (Guerin-Laguette et al. 2009). Brumale, the edible winter truffle (*T. brumale*) is thought to have been accidentally introduced to *T. melanosporum* truffières since at least 1995. The first brumale truffles were reported in 2010 (Guerin-Laguette et al. 2013). Two varieties of shoro, *Rhizopogon rubescens* in 2002 (Wang et al. 2002) and *R. roseolus* in 2010, were also cultivated (Visnovsky et al. 2010).

A. Guerin-Laguette (✉)
Microbial Systems for Plant Protection, The New Zealand Institute for Plant & Food Research Limited, Lincoln, New Zealand
e-mail: alexis.guerin@plantandfood.co.nz

R. Butler
Biometrics, The New Zealand Institute for Plant & Food Research Limited, Lincoln, New Zealand

Y. Wang
Plant & Food Research, Christchurch, New Zealand

J. Pérez-Moreno et al. (eds.), *Mushrooms, Humans and Nature in a Changing World*, https://doi.org/10.1007/978-3-030-37378-8_5

The cultivation of *Lactarius deliciosus* in New Zealand began in the late 1990s (Wang et al. 2013). The fruiting season usually spans from late summer (early February) to late autumn (early June). There are now several plantations producing saffron milk caps from the South to the North, including Neudorf Mushrooms in Nelson, which is commercially managed (https://www.neudorfmushrooms.co.nz/). This chapter focuses on two experimental plantations established at the Plant & Food Research (PFR) Lincoln Farm in Canterbury in December 2007. In a recent study, Guerin-Laguette et al. (2014) detailed the first 3 years (2010 to 2012) of saffron milk cap production in one of these plantations, which contains 41 *Pinus radiata* trees. This was the first intensively monitored saffron milk cap orchard in the world. The purpose of this chapter is to provide an update on the advances made in the ongoing monitoring of this trial since 2013. In addition, we present original results obtained in a similar but smaller trial (ten trees), set up on the same research campus but established with *Pinus sylvestris* (Scots pine). We compare the onset of fruiting and the mushroom yields in both sites.

Over the years, we have accumulated a large amount of saffron milk cap fruiting data over four distinct trial plantations at PFR-Lincoln, trials which were all established with pine seedlings mycorrhized by saffron milk cap under controlled conditions (see also Guerin-Laguette et al. 2000, 2014). The detailed analysis of this data is beyond the scope of this chapter, but we present several general and original observations regarding the fruiting of saffron milk cap at PFR-Lincoln. Finally, we highlight orchard design and management issues potentially affecting the cultivation of *L. deliciosus* in New Zealand and suggest the next research steps required to further improve existing cultivation practices.

5.2 Trial Plantations at PFR-Lincoln, Canterbury

5.2.1 Establishment and Description

Four saffron milk cap plantations, or plots, were established on the PFR-Lincoln Farm between 2005 and 2007. All of them are producing significant quantities of mushrooms (since 2014, the production potential of the four sites combined is ≈80 kg) and are monitored for yield data. However, for the purpose of this chapter, we will focus on two plantations only. They were established 400 m apart on lawns at the PFR-Lincoln campus in December 2007.

The first, called the "Tank" site (43°38′19.3″S 172°28′39.2″E), was initially established with 42 *P. radiata* seedlings (Figs. 5.1 and 5.2) on a silt loam soil with a clay content of ≈ 18% and a pH of 6.7 (Guerin-Laguette et al. 2014).

Of the 41 trees that survived the transplantation phase, two trees were lost to wind storms in January 2013 and March 2014, leaving the current balance of 39 trees. The site is surrounded by lawns or annual crops and is therefore exposed to the dry summer winds from the northeast and northwest. *Pinus radiata* seedlings

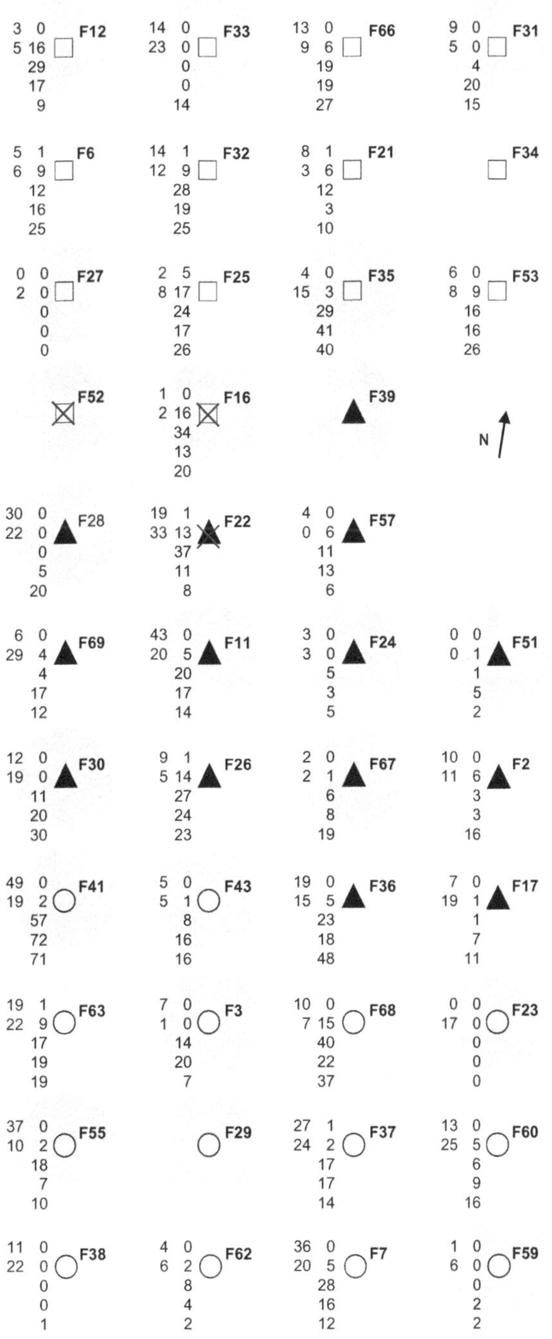

Fig. 5.1 Layout of the Tank trial site and history of mushroom production for each tree. Each square, triangle, and circle symbol indicate a *P. radiata* tree. The numbers to the left of each tree indicate, from top to bottom starting with the column to the right, the numbers of fruiting bodies harvested from that tree from 2010 to 2016. Tree F52 died during the first year after planting, while trees F22 and F16 were lost due to wind storms in 2013 and 2014, respectively. See also Guerin-Laguette et al. (2014) for the meaning of the tree symbols' shape and the light shadowed disks, which are not directly relevant to the data presented in this chapter. We kept this information for the sake of consistency between both publications

Fig. 5.2 (**a**) View of the Tank site in March 2017; (**b**) needle litter became apparent in 2015; (**c**) one-day harvest under tree F41 in March 2014; (**d**, **e**) characteristic saffron milk caps; (**f**, **g**) unprecedented winter fruiting on 4 July 2016 on tree F31

were inoculated with two *L. deliciosus* isolates selected from isolates derived from three to five mushrooms collected in a *P. sylvestris* forest at Anglesey, North Wales, in 1997 (I. Hall, pers. comm., Table 5.1). Two other PFR-Lincoln sites, "Demo" and "Barley," were established with trees inoculated with the same batch of isolates.

Table 5.1 Isolates of *Lactarius deliciosus* used in the PFR-Lincoln trial plantations with ecological information about the source basidiomata

Code	Date of isolation	Host plant	Soil ecology	Geographic origin
NZ	1997	*Pinus sylvestris*	No data	Anglesey, Wales, UK
D60	1995	*P. sylvestris*	Calcic brown soil on limestone glacial alluvium	La Joue-du-Loup, Dévoluy, Hautes-Alpes, France
D64	1996	*P. nigra*	Brown soil on shale	Olette, Pyrénées-Orientales, France
D68	1996	*P. pinaster*	Brown soil on sandstone	Banières, Gard, France
D74	1996	*P. pinaster*	Brown soil on sandstone	Banières, Gard, France

Cultures were isolated from the flesh of fresh fruiting bodies by Ian R. Hall (NZ), Alexis Guerin-Laguette (D60), and Guy Ruiz (D64, D68, and D74)

This initial pool of isolates is termed "NZ" as they were the first *L. deliciosus* isolates to be successfully cultivated in New Zealand (Wang et al. 2002, 2013). Until further genetic analysis is conducted, the number of distinct isolates present in these three sites is unknown. Given the small number of original mushrooms, the genetic variation among the "NZ" isolates is expected to be small.

The second plantation, the "House" site (43°38′19.3"S 172°28′19.6″E) (Figs. 5.3 and 5.4), was established with ten *P. sylvestris* seedlings mycorrhized by *L. deliciosus* at the same time as the Tank site. The same inoculation method (Guerin-Laguette et al. 2014) and 5 × 5 m spacing of trees (400 trees/ha) was used. However, in contrast with the Tank site, the *P. sylvestris* seedlings of the House site were each colonized by one of four distinct *L. deliciosus* isolates (Fig. 5.3, Table 5.1) originating from the south of France (Guerin-Laguette 1998). The *Lactarius*-mycorrhized pines in the trial were sheltered from the south and north by a block of another 19 trees of either *P. radiata* or *Q. robur* species which were also planted at 5 × 5 m spacings. These trees are part of a *Boletus edulis* inoculation trial (Guerin-Laguette and Wang, unpublished). The whole House site (*Lactarius* and *Boletus* trees) is sheltered on all sides but the south by mature conifer and broadleaf trees separated from the trial boundaries by 15–25 m.

5.2.2 *Management*

5.2.2.1 Planting and Irrigation

Planting holes (approximately 20 × 20 × 20 cm) were moistened with water for 24 h to soften the soil. All seedlings were irrigated after planting, especially during the first summer season. Trees in both sites received consistent irrigation provided by fixed or mobile sprinklers. The trees were irrigated during dry periods, as assessed from the visual examination of soil moisture. Irrigation ran from early summer (December) to early autumn (February), when the mushroom season started. The

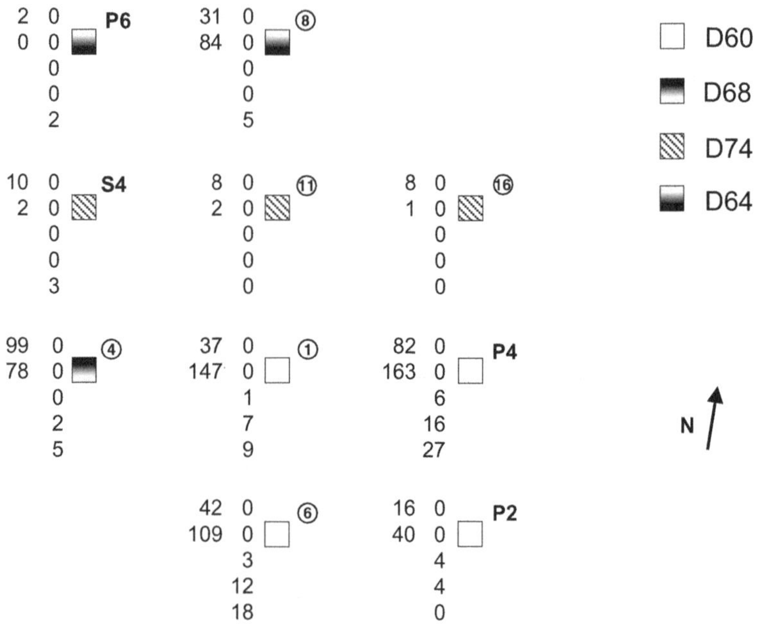

Fig. 5.3 Layout of the House trial site and history of mushroom production for each tree. Each square indicates a *P. sylvestris* tree inoculated by a distinct *L. deliciosus* isolate (see also Table 5.1). The numbers to the left of each tree indicate, from top to bottom starting with the column to the right, the numbers of fruiting bodies harvested from that tree from 2010 to 2016. Adjacent *Boletus edulis*-inoculated trees are not shown on this diagram

amount of water applied was not measured. For the first time in 2016, the irrigation was maintained throughout the autumn (March–May).

5.2.2.2 Grass Control

Grass was mown from spring until the first apparition of mushrooms. As trees developed, the ride-on mower could not access an area of 1.5 m radius around the trunk. Grass growth was controlled manually around trees to facilitate subsequent mushroom detection.

5.2.2.3 Pruning

Lower branches (first crown) of the radiata pines at the Tank site were pruned in winter 2014, i.e., 7 years after planting, to facilitate grass control and access to mushrooms below the canopy. The much slower-growing Scots pine at the House site has not been pruned. The lowest branches of *P. sylvestris* have always been in

Fig. 5.4 (**a**) Growth difference between *P. sylvestris* (small pines at the front) and *P. radiata* (arrows) 4 years after planting; (**b**) fruiting on the smallest pine (P6) in 2014; (**c**) the open canopy in March 2017; (**d**) a harvest in 2015; (**e**) mushroom cluster in 2016; and (**f, g**) a one-day harvest around tree ① (see Fig. 5.3) in 2016

contact with the soil surface, providing further shelter for mushrooms growing under the canopy.

5.2.2.4 Fruiting Body Monitoring and Yields

Every year since 2009, sites have been monitored for fruiting-body production. From the start of the mushroom season, sites were visited at least once a week, or several times a week depending on the mushroom production. On the Tank site, between 2010 (the first year of production) and 2012, the position and number of mushrooms surrounding each tree was recorded (Guerin-Laguette et al. 2014). The total mushroom weight (whole site) was also recorded. Subsequently, individual mushroom weights for each tree were recorded. On the House site, the number of mushrooms per tree and the total crop weight for the site were recorded. The mushroom survey allowed us to observe the variation of production per tree and between trees, in functions of time after planting and of changing environmental constraints. The summary of mushroom production in both sites over the first seven fruiting seasons is given in Table 5.2.

5.3 Case Study 1: Evolution of a *Pinus radiata* Plantation

This section covers the mushroom production history of the Tank site.

Table 5.2 Saffron milk cap production figures for the first seven fruiting seasons (2010 to 2016) at the Tank and House sites

	Tank					House				
Year	No. of PT	No. of FB	TFW (kg)	FW/FB (g)	FW/PT (g)	No. of PT	No. of FB	TFW (kg)	FW/FB (g)	FW/PT (g)
2009	0	0	0	0	0	0	0	0	0	0
2010	8	12	0.47	38.9	59	0	0	0	0	0
2011	28	190	12.05	63.4	430	0	0	0	0	0
2012	32	569	29.4	51.6	918	4	14	0.31	22.1	77
2013	34	536	20.18	37.6	594	5	41	0.96	23.3	191
2014	36	658	39.94	60.7	1109	7	69	4.52	65.5	646
2015	35	462	11.21	24.3	320	10	335	9.16	27.3	916
2016	36	460	14.26	31.0	419	9	626	11.04	17.6	1226

Trees were planted in December 2007

PT productive trees in the corresponding year, *FB* fruiting body, *T* total harvest, *FW* fresh weight, *FW/PT* average fruiting body yield per tree, i.e., total harvest divided per the number of productive trees in the corresponding year

5.3.1 Tree Growth and Evolution of Mushroom Production for Individual Trees

Trees grew rapidly reaching an average height of about 12 m after 9 years, as measured using an inclinometer (Clino Master, Silva) in November 2016. The history of mushroom production per tree is given in Fig. 5.1. The onset and evolution of mushroom fruiting in the first three productive years (2010–2012) has been described previously (Guerin-Laguette et al. 2014). Since 2013, 10% of trees (trees F28, F33, F38, and F59, Fig. 5.1) began to produce only as late as 5–6 years after planting (Table 5.2). Two further trees began to produce mushrooms 8 years after planting (F23 and F27, Fig. 5.1), while three others (7.3% of the number of trees at the site, F29, F34, and F39) have never produced mushrooms to date (Fig. 5.1). However, because of the indications that the root systems have begun to overlap since the seventh year after planting (please see Sect. 5.5), it is not known which tree supported the production of mushrooms that developed equidistant from trees. Therefore, the actual number of trees that failed to produce to date could be anywhere between 7.3% and 12.2% of the 41 trees successfully planted. All trees except two (F57 and F51, Fig. 5.1) have kept fruiting every year since their first year of mushroom production.

5.3.2 Saffron Milk Cap Production and the Distribution of Rainfall in 2012, 2013, and 2014

The three seasons were marked by very distinct timings of rainfall between February and May (Fig. 5.5). Figure 5.5 illustrates the effect of rainfall on the distribution and intensity of mushroom production in these years. Mushroom production clearly peaked a few weeks after significant rainfalls (Fig. 5.5). The period mid-February to mid-April was exceptionally dry in 2013 compared with 2012 and 2014 (Fig. 5.5). In 2013, the only major rainfall in early autumn (around mid-March) was $\approx$ 40 mm, or about 37% of the total rainfall developing between early February and mid-March 2012 ($\approx$107 mm, Fig. 5.5). This fall was followed by a moderate peak of mushroom production ($\approx$1.0 kg), mid-April 2013 (Fig. 5.5). It is only toward the end of the season (May 2013) that the combined significant rainfall ($\approx$90 mm) triggered the year's most abundant production peak (4.0 kg on May 20, 2013, Fig. 5.5). In 2012 and 2014, abundant rainfalls in February–March were followed by major production peaks in the first half of the fruiting season (Fig. 5.5). In 2013, late rain coincided with dropping temperatures, which in turn could have limited the formation and the development of mushrooms even when soil moisture became favorable to mushroom initiation.

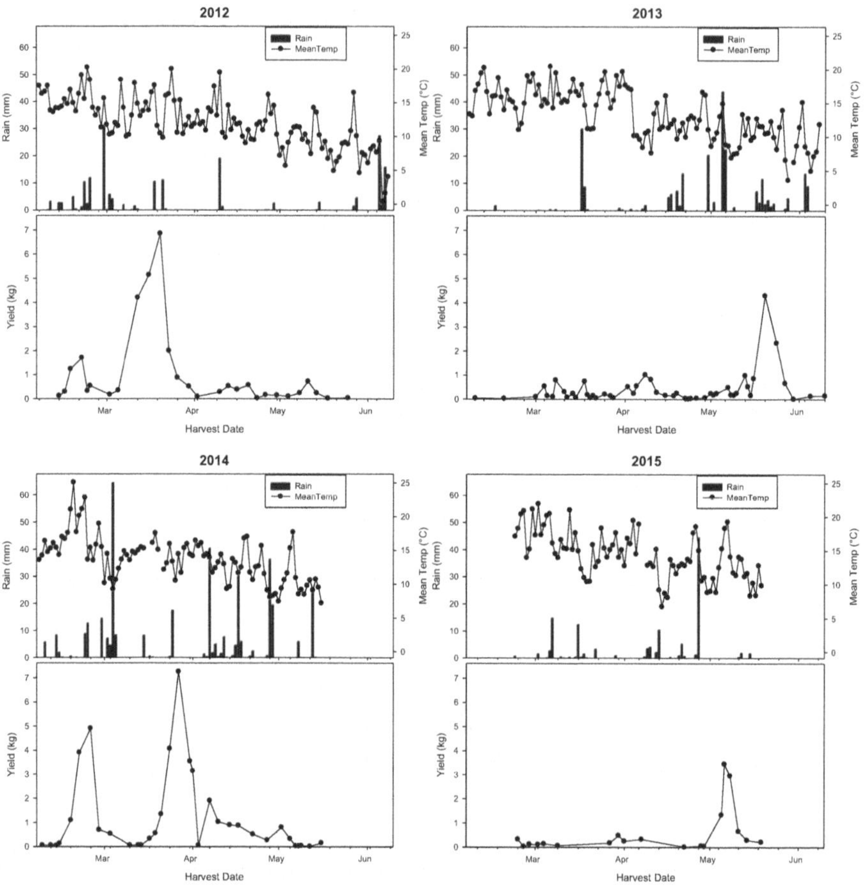

Fig. 5.5 Total yield in the Tank site at each harvest date and weather variables over 4 years from February to June in each year. Weather data was downloaded from MetWatch (http://www.hortplus.metwatch.co.nz/) for each of the four seasons 2012–2015. Data from the Lincoln Broadfield (NIWA) site was selected (43°37′34.4"S 172°28′13.4″E), as this is closest to the Tank site

5.3.3 *Production During Two Consecutive Autumn Droughts (2015 and 2016)*

In both years, rainfall was in severe deficit between February and late April compared with the previous years (see Fig. 5.5 for 2015, data not shown for 2016). In 2015, the combined rainfall during this period reached ≈70 mm (Fig. 5.5), while in 2012 and 2014, the rainfall over this time amounted to ≈130 and up to ≈310 mm, respectively (Fig. 5.5). 2013 was almost as dry as 2015 during this period (≈75 mm), but the significant rainfalls from early May had successfully triggered a peak (≈4.2 kg) production higher than the late autumn peak of 2015 (3.5 kg, Fig. 5.5). In 2016, irrigation was maintained throughout the autumn for the first time. The average weight of mushrooms produced in 2015 (24.3 g/m) and 2016 (31.0 g/m) was remarkably low in comparison with the values obtained in 2012 (51.6 g/m) and 2014 (60.7 g/m) (Table 5.2).

5.3.4 *Overview of the Production During the First Seven Mushroom Seasons*

Table 5.2 summarizes for each year the number of productive trees, mushroom numbers, total yields, and average mushroom weights on both the Tank and House sites. On the Tank site, from the third year of production (2012–2014), the average number of mushrooms was around 588 per year (≈17 mushrooms/tree). In subsequent drier autumns (2015 and 2016), the total number of mushrooms dropped by about 22%, with 461 mushrooms produced in average for 2015 and 2016 (≈13 mushrooms/tree, Table 5.2, Fig. 5.1). Furthermore, the average mushroom weight dropped by as much as 45% when comparing values between the same periods, i.e., 50 g/mushroom (g/m) in average for 2012–2014 and 27.6 g/m in 2015–2016 (Table 5.2). Thus, between 2012–2014 and 2015–2016, the average yield per tree also dropped from 0.87 to 0.37 kg/tree (Table 5.2). Compared with the 2014 highest yield (1.1 kg/tree, Table 5.2), the decrease observed in average for 2015–2016 represents a 67% fall in yield per tree. The decrease in individual mushroom weight seems to be the major reason for the observed fall in total yield (Table 5.2) during dry autumns.

5.3.5 *Outlook for the Future*

As trees grew, the canopy closed and, from 2015, a needle litter started accumulating (Fig. 5.2b). We also noticed that the production of some trees declined, with up to a 90% drop in mushroom numbers between 2014 and 2015, followed by a weak recovery in 2016 (e.g., Tree F35, Fig. 5.1). It is unknown whether this decline was due to the canopy closure, the succession of dry autumns, or the combination of both factors. The closure of the plantation canopy could be a turning point of the mushroom production because of the likely declines of light and soil temperature. We will keep monitoring this site to better understand the impact of tree ageing and canopy closure on mushroom production.

5.4 Case Study 2: Evolution of a *Pinus sylvestris* Plantation

Although the House site is smaller than the Tank site, a number of observations on the mushroom production history of the House site are very valuable and appear worthwhile as a comparison to the history on the Tank site.

5.4.1 Tree Growth and Onset of Fruiting

Pinus sylvestris seedlings grew much more slowly than the *P. radiata* seedlings planted the same year on the Tank site. Scots pines reached an average height of 3.7 m after 9 years in comparison with 12.3 m for radiata pines, i.e., only 30% of the growth of *P. radiata* (Fig. 5.4a, c). As a consequence, 10 years after planting, there is still no canopy closure and no needle litter formation (except in the vicinity of the trunk), and the space between trees is still very sunny and covered with dense grass in 2016–2017 (Fig. 5.4c, f). The first mushrooms under *P. sylvestris* appeared in 2012 or just over 4 years after planting, 2 years later than *P. radiata* planted at the same time (Table 5.2, Fig. 5.3). In the Barley site established with both pine species on the Lincoln farm (≈1 km away from the Tank site) with the same NZ inoculum as the one used in the Tank site (i.e., of similar genetic origin), we observed again that *P. sylvestris* started producing mushrooms of NZ isolates 2 years after *P. radiata* (data not shown). Therefore, it seems likely that the pine species, and its associated growth rate, is the main factor driving the delay required to start mushroom production: the slower the growth of the tree species, the later the mushroom production. Regarding the percentage of trees that became productive, the small size of the House trial limits the extent of comparison with the Tank site. However, we also noticed the progressive increase in the number of trees switching to mushroom production. Furthermore, all *P. sylvestris* trees became productive within the first four fruiting seasons (Table 5.2, Fig. 5.3). The small size of trees and the fact that mushrooms have not yet reached positions equidistant from neighboring trunks suggest that the root systems of adjacent Scots pines may not have overlapped yet. Two trees stopped fruiting: P2 in 2014 (only) and P6 in 2016 (Fig. 5.3). However, the latter is the smallest tree in this site (data not shown) and has only produced four mushrooms over two fruiting seasons so far (Fig. 5.3).

5.4.2 Overview of the Production During the First Five Mushroom Seasons, Comparison with the Tank Site, and Outlook for the Future

The number of mushrooms produced kept increasing over time (Table 5.2, Fig. 5.3). Given that the same spacing has been adopted between both sites, located only 400 m apart on a very similar lawn of the PFR campus, we assume that it is reasonable to compare the productivity of pines in both sites, i.e., the mushroom yield per productive tree over time (Fig. 5.6). We can observe that the average yield per tree on *P. sylvestris* kept increasing and has overtaken that of *P. radiata* in 2016: the production per *P. sylvestris* tree was 1.23 kg/tree compared with 0.42 kg/tree for *P. radiata* (Table 5.2, Fig. 5.6). We suggest that the following factors may help to explain the differences observed in yields between the two pine species: (1) the slow growth and compact shape of *P. sylvestris* trees (Fig. 5.4c) maintained a sunny lawn environment around *P. sylvestris* trees; (2) the smaller size of the House site achieved a more consistent, and probably more intense, irrigation; and (3) this site is more sheltered than the Tank site from Northwest and Northeast dry winds. We will keep monitoring the yield in the House site to determine the long-term evolution of yield, again in relation to the environment changes caused by tree ageing and canopy closure.

With a tree spacing of 5 × 5 m, both plantations represent a density of 400 trees per ha. Production values described here for the two most productive years, assuming that all trees are productive, represent 444 kg/ha for *P. radiata* in 2014 (7 years after planting) and 490 kg/ha for *P. sylvestris* in 2016 (9 years after planting). These values are well above maximal values (≈30 kg/ha) reported for naturally established populations of *L. deliciosus* (Martínez-Peña et al. 2012), thereby supporting the hypothesis that mushrooms could be farmed in pine plantations established with mycorrhizal seedlings.

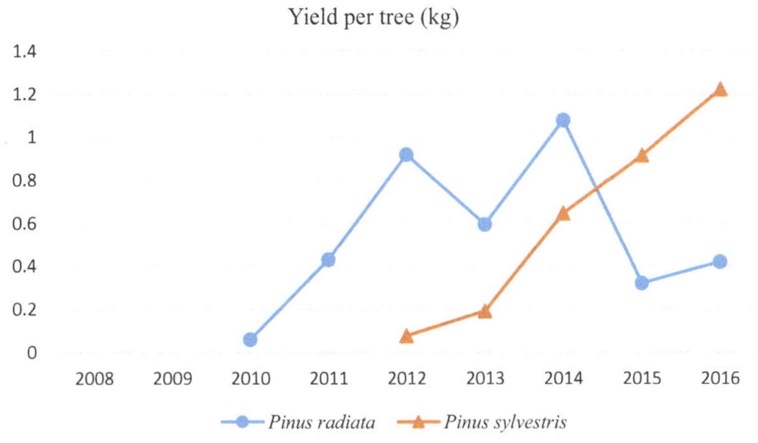

Fig. 5.6 Comparison of the yields (kg per tree) on the Tank (*Pinus radiata*) and on the House (*Pinus sylvestris*) sites since the establishment of the plantations in December 2007

5.5 Mushroom Fruiting Observations

These observations were made over the four saffron milk cap trials at PFR-Lincoln.

5.5.1 Fruiting Distribution Around Trees

Mushrooms can appear anywhere around the trunk. However, for *P. radiata* trees that are partly in full sun (i.e., edge trees not sheltered by other trees) and exposed to the wind, we noticed that the mushrooms tended to develop preferably on the shady sides of the trees (south side). Each year, for both pine species, the distance of mushrooms from the tree trunk increased, and after 7 years of fruiting, it is possible to find mushrooms up to 2.5 m from the tree trunk (over 1 m away from the drip line of trees) although mushrooms still emerged very close to the trunk. Over the years, the root systems of neighboring trees are expected to overlap each other. It is difficult to know precisely when this phenomenon started, but our field notes on the Tank site (when mushrooms started to be found at equidistant locations from trunks) suggested that such an overlap began in 2015 or ≈7 years after planting. This means that it gradually became almost impossible to assign with certainty mushrooms to particular trees, either because mushrooms grew at equidistant locations from neighbor trunks or because the underground mycelial network started to act as one individual developing multiple mycorrhizal connections with the root systems of several trees.

Mushrooms tend to grow bigger when developing within dense grass, as opposed to mushrooms forming over bare ground or below the needle litter. A grass cover is therefore desirable, but the height of the grass above the ground should preferably not exceed 15 cm. A taller grass canopy greatly increases the probability of missing mushrooms or of finding them late, when they are overmature.

5.5.2 Sustained Production from Fallen Trees

Fast-growing *Pinus radiata* trees are very sensitive to wind storms. In the Tank site to date, we have lost 6% of productive *P. radiata* trees to this phenomenon. The wind-exposed Barley site incurred similar tree losses (although none of the 11 *P. sylvestris* there were uprooted), and it is likely that more *P. radiata* trees could be lost in the future especially if a wind storm follows a rain episode that would have previously softened the ground. Interestingly, we found mushrooms fruiting all over the area in which these trees were centered up to 3 years following their death. Given the impossibility of determining precisely the date when root systems began to overlap, it is unclear how long the mycelium could fruit after the death of the associated tree. We suggest that the mushrooms could still fruit for at least one season following the tree death but cannot demonstrate this with the present data. Indeed, we cannot rule out that roots from neighboring trees have reached the area left by fallen trees and have become mycorrhizal with the mycelium that once colonized the dead tree, thus enabling the mycelium to keep fruiting.

5.5.3 Mushroom Yields per Tree

The number of mushrooms per tree can vary greatly among the trees but also over time for some trees (Figs. 5.1 and 5.3). Some trees produced just one or two mushrooms every year, while others produced tens of mushrooms every year. The production of some trees regularly increases until it reaches a plateau, while that of others oscillates or even declines sharply. In terms of the number of mushrooms produced, *P. sylvestris* became the most productive tree species after 5 years. In the House site in 2016, three trees produced over 100 mushrooms each (Fig. 5.3) with tree P4 producing up to 163 mushrooms (or 2.9 kg). In the Tank site, the most productive *P. radiata* tree (F41) achieved 5.9 kg in 2014 with only 71 mushrooms (83 g/m in average), which is less than half the number of mushrooms produced by the most productive *P. sylvestris* trees in 2016 (Fig. 5.4). The variation of yield between trees is shown for the two most productive years for both sites, 2014 for Tank and 2016 for House (Fig. 5.7). In case of *P. radiata*, two trees (F41 and F36) really stood out in terms of yield, while it appears that the majority of trees produced less than the mean value (Fig. 5.7). In the case of *P. sylvestris*, there seemed to be less variation in yield between trees. It is largely unknown why trees F41 (Tank) and P4 (House) were so productive (e.g., possible environmental reasons) or if there will be a means (e.g., genetic basis) to select for such trees, or for the right tree/isolate combination, in order to maximize yields. It may be possible to select fungus/tree partners for high mushroom yields based on genotyping; therefore, having all trees at the current spacing producing mushroom yields equivalent to the best performing trees could achieve significant yield increases 5 years after planting, equal to the 2.3 tons per ha from *P. radiata* tree F41 and 1.1 t/ha from *P. sylvestris* tree P4. These production values are not realistic targets at this stage but indicate the maximum potentials that could ultimately be achieved with each pine species. Sustained high production from *P. radiata* would require careful and heavy pruning in order to maintain a favorable mushroom environment and irrigation may also be crucial during drought-filled years as discussed below. Despite yields theoretically lower than *P. radiata*, at least over the first five fruiting seasons, slow-growing pine species such as *P. sylvestris* still appear very attractive for saffron milk cap cultivation in New Zealand conditions: they are more resistant to wind storm and will need less pruning than *P. radiata* to maintain a mushroom-friendly lawn environment. The slow pine growth delays canopy closure and needle accumulation. It also allows for more light to reach the ground in the development of the canopy. The variation of mushroom production per tree may also be related to the soil condition and water availability in addition to the tree's original mycorrhization.

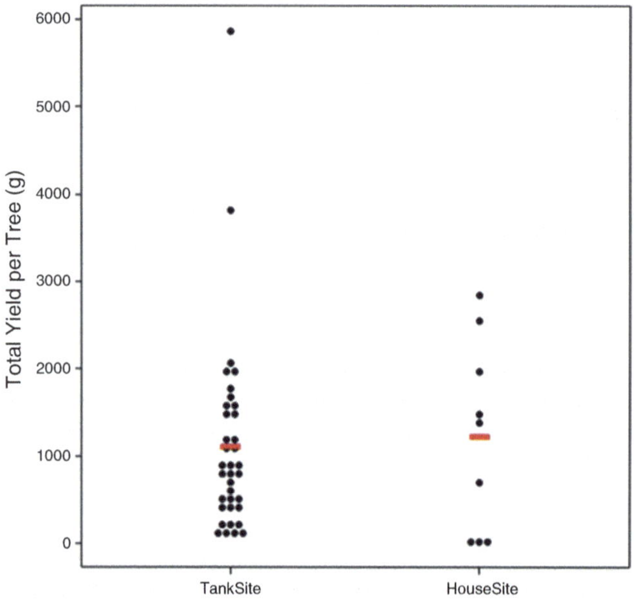

Fig. 5.7 Dot histogram of the yield per tree for the most productive fruiting seasons for the Tank (2014) and House (2016) sites. Data for the Tank site are obtained from actual individual mushroom weights. Data for the House site are inferred from the number of mushrooms per tree and the average weight of mushrooms in 2016. The red bar indicates the mean yield per tree

5.5.4 Mushroom Size

Mushrooms tended to be smaller under *P. sylvestris* than under *P. radiata* (Table 5.2). They also occurred more frequently in clusters under *P. sylvestris* (Fig. 5.4e). The largest mushroom harvested from the Lincoln trials was 250 g from *P. radiata* (Barley site). On the same host tree in Nelson, with a warmer climate than Christchurch, mushrooms have grown as big as 375 g (Theres and Hannes Krummenacher, pers. comm.). However, from a commercial point of view, smaller mushrooms may be more desirable, at least for the retail market.

5.5.5 Fruiting Season Span and Unprecedented Winter Fruiting in 2016

The mushroom season typically runs from early January until early June, with mushroom peaks distributed between February and May depending on the season. In 2016, for the first time since the mushroom surveys at Lincoln commenced in 2008, we harvested large and fresh fruiting bodies as late as July 4 from seven trees in the Tank site (Fig. 5.2) and three trees in the House site (data not shown). We suggest that the timing of the mushroom season is mainly related to the soil temperature. These observations would suggest that late June to early July 2016 temperatures were particularly mild. Indeed, between June 20 and 24, 2016, the soil temperatures 10 cm below ground were slightly higher than the maximal temperatures recorded for the same days between 2008 and 2015 and three to six degrees higher than the average temperature recorded over this period (unpublished). Soil temperatures were obtained from New Zealand's National Climate Database (https://cliflo.niwa.co.nz/) using data from the Lincoln Broadfield Electronic Weather Station (43°37′34.4″S 172°28′13.4″E), which is only ≈1.4 km away from both sites.

5.6 Cultivation Research: The Next Steps

As this chapter clearly shows, there are crucial factors in saffron milk cap production that are unknown. This is to be expected with every new crop where a large number of cultivation parameters require testing. We list below a few areas of research that could further improve *L. deliciosus* cultivation in pine orchards in New Zealand.

5.6.1 Plantation Design: Tree Composition, Plantation Layout, and Grass Cover

We recommend diversifying the pine species in order to assess the potential of each species for mushroom cultivation. New Zealand has 16 *Pinus* species, of which 13 are naturalized (http://www.nzflora.info/factsheet/Taxon/Pinus.html). It would be useful to set up plantation trials with species offering various advantages such as ornamental (*P. pinea* or *P. pinaster*), nut production (*P. pinea*), suitability to local climates (*P. canariensis*), and wind resistance (*P. sylvestris*). Depending on the location in New Zealand, pine species not causing a threat as wilding species would also be preferred. It would also be essential to compare the productivity between different pine species. Mixing pine species or pines and other tree species could also prove beneficial. Besides the tree composition, open canopy conditions should be maintained as long as possible. There are several means to achieve this, separately or in combination: (1) low density planting; (2) different planting layouts, e.g., rows or hedges instead of bidimensional planting; (3) slow-growing species; and (4) pruning. It is possible to prune pines as long as this is restricted to the active candle zones in early summer. However, the optimal pruning regime has yet to be discovered for each pine species. Finally, it is recommended that the plantation site is sheltered from dominant winds. Wind breaks could be used as long as they remain low cost. A grass cover provides an ideal micro-habitat for mushroom formation.

However, as it can generate additional cost and disturbance (soil compaction), considering naturally short grass varieties, such as turf-type tall fescues, could prove beneficial.

5.6.2 Other Management and Mushroom Monitoring Aspects

In areas with regular drought periods, such as the east of New Zealand, it is strongly advised that there is access to irrigation in order to maintain, or limit, the decline in production because of insufficient rainfall in summer and autumn. In dry areas, mulching (e.g., bark) at planting is recommended even though this may delay the onset of fruiting (Guerin-Laguette et al. 2014). The timing and quantity of irrigation and other parameters such as the water temperature or oxygenation are far from understood and would need to be thoroughly investigated. However, based on our experience during very dry spans, we recommend a few abundant water regimes, 30–50 mm every 2–3 weeks, over a regular shallow irrigation. The use of soil temperature and humidity sensors would be essential to better understand the conditions conducive to mushroom flushes. Such sensors, coupled with light intensity detectors, could monitor the changes in micro-climate resulting from plantation ageing, canopy closure, and climatic factors (drought, wind). In an intensive farming model, one would also need to consider optimizing the frequency of the mushroom checks and harvests, as well as limiting the soil compaction through repeated stepping, which seems to affect the subsequent harvest yields more than the direct act of harvesting mushrooms (Egli et al. 2006).

5.6.3 Future Perspectives

Farming edible mycorrhizal mushrooms is a very recent development in human cropping history. Many field trials will be required to improve and optimize sustainable cultural practices for new crops such as *L. deliciosus*. Research will be best achieved by a partnership between industry (farm, forestry, and mushroom industries) and science providers. The high scale supply of high-quality, controlled mycorrhizal seedlings will require technology transfer and quality control in order to avoid the pitfalls seen in the truffle industry (Murat 2015). In this chapter, we focus on the development of small relatively intensive mushroom orchards (i.e., a few well-maintained trees) as a means to diversify on-farm production while growing timber or firewood. Another possible approach would use controlled mycorrhizal seedlings for extensive large-scale production of mushrooms under a forestry model with limited tree management and natural rain. However, one difficulty of this approach is the harvest of mushrooms occurs over a short time and over a very large area.

5.7 Conclusions

The mushroom monitoring data from the Tank and House sites highlight the advances made so far in the cultivation of saffron milk cap, with annual yields of up to ≈0.5 t/ha, and strongly suggest that its efficient cultivation is achievable in the near future in the South Island of New Zealand. There are many limitations: under *P. radiata*, we know that a combination of canopy closure and dry autumns has contributed to declining yields, but we are unable to determine what the most important factor was. Irrigation seems to have tempered the effect of drought only to a very small extent. Under *P. sylvestris*, irrigation did seem to compensate for low rainfall, but the canopy was still open (favorable grass conditions) and the small number of trees enabled us to achieve a more consistent and intense irrigation. Furthermore, the House site was more sheltered than the Tank site, and the impact of this environmental context could not be measured precisely. In the case of canopy closure, it is unknown whether the main factor is over-shading (i.e., lack of light, soil temperature), increased competition between trees for limited resources (including water), or modification of the soil characteristics (needle litter, acidification, etc.). Despite these limitations, we hope that this work present enough convincing evidence that an abundant mushroom crop is already achievable and is likely to be much higher once we will have acquired the basic cultivation knowledge.

Since 2013, mushrooms from all PFR-Lincoln blocks were sold to the public and to restaurants in order to introduce them to New Zealanders (often for the first time), assess their popularity, and generate income for the trial maintenance. The feedback received from the public and the professionals has been very positive, including a constant demand from some of New Zealand's most prestigious chefs. These observations suggest that saffron milk caps could become a popular autumn crop within the next few years in New Zealand. We hope that this work can contribute to and stimulate the innovative and sustainable farming of other edible mycorrhizal mushrooms. There are over 500 of these species worldwide (Honrubia 2007) not yet studied for cultivation and which could bring multiple benefits: farm diversification and yearly income for local populations, non-meat source of protein, environment protection (preventing soil erosion and protecting the climate through carbon fixation), and health benefits such as antioxidant properties (Palacios et al. 2011). A comprehensive review of health benefits in edible ectomycorrhizal mushrooms can be found in Pérez-Moreno and Martínez-Reyes (2014).

Acknowledgments The authors are deeply grateful to Jesús Pérez-Moreno for his contribution to the present book, Ian R. Hall for having encouraged and initiated the research on cultivation of *Lactarius deliciosus* in New Zealand, Daniel Mousain for having maintained alive and generously given to us in 2005 the pure culture isolates originating from Southern France, and Shu-hong Li (Biotechnology & Germplasm Resources Institute, Yunnan, China) for his support during the 2012 harvesting season. We also would like to acknowledge all our colleagues without whose support and trust this work would have never been possible: David Hughes, Stephen Griffith, Ralph Scott, Nina Hesom-Williams, Nicholas Cummings, Ron van Toor, George Strong, and Grant Morris. Scott Thompson from Landcare Research, Lincoln, contributed to the sites' maintenance. We thank Daniel Mousain, Ron van Toor, and Andrew Pitman for helpful suggestions to improve the manuscript. Last but not least, we are grateful to all our enthusiastic saffron milk cap customers who helped us continue to pioneer the cultivation of these colorful delicacies in harsh funding times.

References

Egli S, Peter M, Buser C, Stahel W, Ayer F (2006) Mushroom picking does not impair future harvests – results of a long-term study in Switzerland. Biol Conserv 129:271–276

Guerin-Laguette A (1998) Les lactaires à lait rouge: mycorhization contrôlée des pins et caractérisation moléculaire. Application à l'étude de la compétence écologique et de la compétitivité d'isolats de *Lactarius deliciosus*. [The red milk cap mushrooms: controlled mycorrhization of pines and molecular characterization. Application to study the ecological competence and the competitiveness of isolates of *Lactarius deliciosus*.] in French with English summary. Ph.D. thesis, Ecole Nationale Supérieure Agronomique de Montpellier

Guerin-Laguette A, Plassard C, Mousain D (2000) Effects of experimental conditions on mycorrhizal relationships between *Pinus sylvestris* and *Lactarius deliciosus* and unprecedented fruit-body formation of the saffron milk cap under controlled soilless conditions. Can J Microbiol 46:790–799

Guerin-Laguette A, Hesom-Williams N, Parmenter G, Strong G, Wang Y (2009) Field research and cultivation of truffles in New Zealand: an update. In: Li DZ et al. (ed) The mycorrhizal edible mushroom resources and their sustainable utilization. The proceeding of the 5th international workshop on edible mycorrhizal mushrooms, IWEMM5, Chuxiong, 26–29 August 2007, Acta Botanica Yunnanica Suppl. XVI, pp 90–93

Guerin-Laguette A, Cummings N, Butler RC, Willows A, Hesom-Williams N, Li S, Wang Y (2014) *Lactarius deliciosus* and *Pinus radiata* in New Zealand: towards the development of innovative gourmet mushroom orchards. Mycorrhiza 24:511–523

Guerin-Laguette A, Cummings N, Hesom-Williams N, Butler R, Wang Y (2013) Mycorrhiza analyses in New Zealand truffières reveal frequent but variable persistence of Tuber melanosporum in co-existence with other truffle species. Mycorrhiza 23(2):87–98

Hall IR, Brown GT, Zambonelli A (2007) Taming the truffle – the history, lore and science of the ultimate mushroom. Timber Press, Portland, OR, 304 p

Honrubia M (2007) Wild edible fungi and rural development: a global and critical point of view. In: Abstract book of the 5th international workshop on edible mycorrhizal mushrooms, IWEMM5, Chuxiong, 26–29 August 2007, p 29

Martínez-Peña F, Ágreda T, Águeda B, Ortega-Martínez P, Fernández-Toirán LM (2012) Edible sporocarp production by age class in a Scots pine stand in Northern Spain. Mycorrhiza 22:167–174

Murat C (2015) Forty years of inoculating seedlings with truffle fungi: past and future perspectives. Mycorrhiza 25:77–81

Orlovich D, Cairney JWG (2004) Ectomycorrhizal fungi in New Zealand: current perspectives and future directions. N Z J Bot 42:721–738

Palacios I, Lozano M, Moro C, D'Arrigo M, Rostagno MA, Martínez JA, García-Lafuente A, Guillamón E, Villares A (2011) Antioxidant properties of phenolic compounds occurring in edible mushrooms. Food Chem 128:674–678

Pérez-Moreno J, Martínez-Reyes M (2014) Edible ectomycorrhizal mushrooms: biofactories for sustainable development. In: Guevara-Gonzalez R, Torres-Pacheco I (eds) Biosystems engineering: biofactories for food production in the century XXI. Springer, Heidelberg, pp 151–233 https://doi.org/10.1007/978-3-319-03880-3_6

Visnovsky SB, Guerin-Laguette A, Wang Y, Pitman AR (2010) Traceability of marketable Japanese shoro in New Zealand: using multiplex PCR to exploit phylogeographic variation among taxa in the *Rhizopogon* subgenus *roseoli*. Appl Environ Microbiol 76:294–302

Wang Y, Hall IR (2004) Edible ectomycorrhizal mushrooms: challenges and achievements. Can J Bot 82:1063–1073

Wang Y, Hall IR, Dixon C, Hance-Halloy M, Strong G, Brass P (2002) The cultivation of *Lactarius deliciosus* (saffron milk cap) and *Rhizopogon rubescens* (shoro) in New Zealand. In: Hall IR et al. (eds) Edible mycorrhizal mushrooms and their cultivation. Proceedings of the second international conference on edible mycorrhizal mushrooms, New Zealand Institute for Crop & Food Research Limited, Christchurch, New Zealand, CD-ROM, 3–6 July 2001

Wang Y, Cummings N, Guerin-Laguette A (2013) Cultivation of basidiomycete edible ectomycorrhizal mushrooms: *Tricholoma*, *Lactarius*, and *Rhizopogon* – Chapter 16. In: Zambonelli A, Bonito GM (eds) Edible ectomycorrhizal mushrooms. Soil biology, vol 34. Springer-Verlag, Berlin/Heidelberg, pp 281–304

Chapter 6
Edible Mushrooms and Their Cultural Importance in Yunnan, China

Fuqiang Yu, Alexis Guerin-Laguette, and Yun Wang

6.1 Introduction

Yunnan is a province in southwest China, which is famous for its rich biological and cultural diversity (Fig. 6.1). Northwest Yunnan is adjacent to the Eastern Himalayas and has many high snowy mountains over 5000 m in altitude and deep valleys. In the south of the province are tropical rainforests and dry hot valleys. Between these extremes there is a plateau. Elevation varies from 6740 m in the northwest to only 80 m in the south. Three rivers run north to south, and lakes are widespread. Such diverse topography created a variety of climates and nourished a rich biodiversity. Yunnan is recognized as a global biodiversity hotspot (Myers et al. 2000) The province is considered a center of distribution and divergence of many plants in the Fagaceae and the Pinaceae and of edible mushrooms such as *Tricholoma matsutake* (Zhou 1992; Li 1995; Murata et al. 2008). Visitors travelling to Yunnan during the mushroom season from June to October are impressed by the variety and the delicacy of mushrooms available in countless markets and restaurants. However, scientific knowledge of the diversity of the wild edible and medicinal mushroom in the region is still limited (Ying and Zang 1994; Wang and Liu 2002; Wang et al. 2004). More detailed and systematic surveys are needed.

F. Yu (✉)
CAS Key Laboratory for Plant Diversity and Biogeography of East Asia, Chinese Academy of Sciences, Kunming, Yunnan, China

SWFU-KIBCAS Joint Institute for Applied Mycology, Kunming, Yunnan, China
e-mail: fqyu@mail.kib.ac.cn

A. Guerin-Laguette
The New Zealand Institute for Plant and Food Research Limited, Christchurch, New Zealand
e-mail: alexis.guerin@plantandfood.co.nz

Y. Wang
Plant and Food Research, Christchurch, New Zealand

J. Pérez-Moreno et al. (eds.), *Mushrooms, Humans and Nature in a Changing World*, https://doi.org/10.1007/978-3-030-37378-8_6

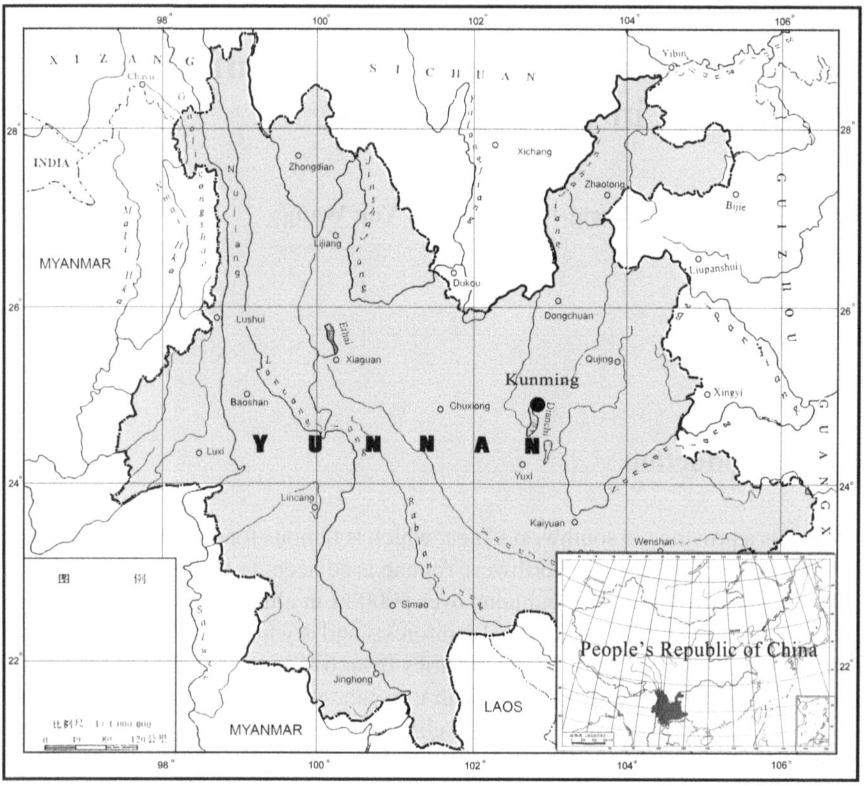

Fig. 6.1 Map of Yunnan

6.2 Diversity of Edible Mushroom in Yunnan

6.2.1 Wild Edible Mushrooms

The great variation in climate and topography has created very diverse habitats and species in Yunnan. This diversity has been sustained over geological time because of the light impact of the Quaternary glaciation on Yunnan, in comparison with its effects in Europe and North America, where the land was largely covered by glacial ice. Yunnan has the world's richest diversity of wild edible mushrooms, with over 600 edible species recorded (mycorrhizal and saprobes), which represents around 75% of the nation's total (Wang and Yang 2006). In 2011, production of edible mushrooms was over 135,000 tons, 51.85% of which were wild edible mushrooms (Tan 2012). Over 90% of these were ectomycorrhizal mushrooms. At local markets, it is possible to find 321 species of wild edible mushrooms, of which 164 are commonly traded. From 2006 to 2013, over 50,000 tons of edible mushrooms were exported annually (Sheng 2013). China is the world's largest exporter of matsutake (*T. matsutake*), and over 90% is produced in Yunnan (Wang et al. 1997). China is

Fig. 6.2 Yi people in Chuxiong, Yunnan

also emerging as a truffle-producing nation, and more than 60% of its truffles are harvested in Yunnan (Wang et al. 2008). Yunnan has 24 ethnic minorities such as Yi, Bai, Tibetan, Thai, Hani, Naxi, Miao, and Lisu (Fig. 6.2). Most of these people live in mountainous regions (Pei 2004), and wild edible mushrooms are an important livelihood in these areas (Wang and Hall 2004; Wang et al. 2008; Yang et al. 2009). Production of wild edible mushrooms has declined since large-scale commercial harvesting began in the 1990s. Protection and restoration of wild edible mushroom resources are urgent.

6.2.2 *Commercial Wild Mushrooms*

Every county in Yunnan has at least one wild mushroom market trading wild edible mushrooms harvested from surrounding forests (Fig. 6.3). At the main markets such as in Kunming and Nanhua, hundreds of tons of wild edible mushrooms change hands daily during mushroom season from June to October. A total of 321 species, belonging to 101 genera, and 47 families were identified as wild mushrooms traded in the local markets (Appendix) (Petersen and Zang 1986, 1989, 1990; Ying and Zang 1994; Wang and Liu 2002; Wang et al. 2004, 2009; Wang and Yao 2005; Yu and Liu 2005; Tang et al. 2006; Wang and Yang 2006; Wei et al. 2006, 2009; Yu et al. 2006; Zang 2006, 2013; Kirk et al. 2008; Zheng and Liu 2008; Li 2009; Li et al. 2009, 2011a, b, 2014a, b; Tian et al. 2009, 2012; Dai et al. 2010; Zhang 2010;

Fig. 6.3 Mushuihua wild edible mushroom market at Kunming

Shao 2011; Shao et al. 2011, 2012, 2014, 2016; Wang and Liu 2011; Zeng 2011, Zeng et al. 2013, 2014a, b; Cao et al. 2012; Fan and Cao 2012; Fan et al. 2012a, b, 2013, 2014; Tang 2013; Yu and Liu 2013; Tang and Yang 2014; Deng et al. 2013; Shi 2013; Zhao et al. 2013, 2015; Zhao et al. 2014a, b; Cui et al. 2015; Tang et al. 2015; Wang et al. 2015a, b; Wu et al. 2015; Yang 2015; Wang 2016). Of these commercial mushrooms, Boletaceae was the best represented family, with 27 genera and 23.05% (74 spp.) of the total species. Russulaceae, with 9.97%, includes *Russula* (15 spp.), *Lactarius* (13 spp.), and *Lactifluus* (4 spp.). Those traded in Gomphaceae, include 4 genera with 26 species, and *Ramaria* is the most species-rich genus in this family,with 22 species and 6.83% of the total. Nineteen species were in Cantharellaceae (5.92%), 16 in Tricholomataceae (4.98%), 13 in Tuberaceae (4.05%), and 11 in Lyophyllaceae (3.43%), Amanitaceae (3.12%), and Helvellaceae (3.12%). More than 164 species in the genera *Albatrellus*, *Amanita*, *Boletus*, *Cantharellus*, *Helvella*, *Lactarius*, *Morchella*, *Phylloporus*, *Ramaria*, *Russula*, *Suillus*, *Termitomyces*, *Tricholoma*, and *Tuber* were found to represent 51.09% of the total wild mushroom species in the Yunnan's local markets (Table 6.1).

We have listed 60 dominant commercial species in Yunnan belonging to the genera *Boletus*, *Cantharellus*, *Lactarius*, *Russula*, *Termitomyces*, *Tricholoma*, *Ramaria*, etc. (Table 6.2). The main commercial mushroom species sold in Yunnan and China (Boa 2004; Yu and Liu 2005), ordered by quantity and frequency of their appearance in markets, are listed in the Appendix.

Table 6.1 Species, genera, and families of commercial wild mushrooms in Yunnan

Family	No. of genera (%)	No. of species (%)
Basidiomycota		
Boletaceae	27 (27)	74 (23.05)
Russulaceae	3 (3)	32 (9.97)
Gomphaceae	4 (4)	26 (8.10)
Cantharellaceae	2 (2)	19 (5.92)
Tricholomataceae	7 (7)	16 (4.98)
Lyophyllaceae	2 (2)	11 (3.43)
Amanitaceae	1 (1)	10 (3.12)
Albatrellaceae	2 (2)	8 (2.49)
Bankeraceae	3 (3)	7 (2.18)
Ganodermataceae	2 (2)	7 (2.18)
Suillaceae	1 (1)	7 (2.18)
Polyporaceae	4 (4)	6 (1.87)
Thelephoraceae	2 (2)	6 (1.87)
Agaricaceae	3 (3)	5 (1.56)
Gomphidiaceae	2 (2)	4 (1.25)
Physalacriaceae	2 (2)	3 (0.93)
Hydnangiaceae	1 (1)	4 (1.25)
Hygrophoraceae	1 (1)	4 (1.25)
Sclerodermataceae	2 (2)	3 (0.93)
Auriculariaceae	1 (1)	3 (0.93)
Cortinariaceae	1 (1)	3 (0.93)
Hericiaceae	1 (1)	3 (0.93)
Pleurotaceae	1 (1)	3 (0.93)
Scutigeraceae	2 (2)	2 (0.62)
Gyroporaceae	1 (1)	2 (0.62)
Sparassidaceae	1 (1)	2 (0.62)
Boletinellaceae	1 (1)	1 (0.31)
Clavariaceae	1 (1)	1 (0.31)
Clavulinaceae	1 (1)	1 (0.31)
Entolomataceae	1 (1)	1 (0.31)
Hydnaceae	1 (1)	1 (0.31)
Inocybaceae	1 (1)	1 (0.31)
Meripilaceae	1 (1)	1 (0.31)
Meruliaceae	1 (1)	1 (0.31)
Omphalotaceae	1 (1)	1 (0.31)
Rhizopogonaceae	1 (1)	1 (0.31)
Schizophyllaceae	1 (1)	1 (0.31)
Tremellaceae	1 (1)	1 (0.31)
Ascomycota		
Tuberaceae	1 (1)	13 (4.05)
Helvellaceae	1 (1)	10 (3.12)

(continued)

Table 6.1 (continued)

Morchellaceae	1 (1)	7 (2.18)
Ophiocordycipitaceae	1 (1)	3 (0.93)
Cordycipitaceae	2 (2)	2 (0.62)
Leotiaceae	1 (1)	1 (0.31)
Nectriaceae	1 (1)	1 (0.31)
Shiraiaceae	1 (1)	1 (0.31)
Xylariaceae	1 (1)	1 (0.31)

6.2.3 Important Wild Edible Mushrooms

6.2.3.1 Matsutake

Tricholoma matsutake is a traditional, special delicacy in Japan, also popular in Korea and China. China is the biggest matsutake producer and exporter worldwide (Fig. 6.4). Ninety percent of Chinese matsutake are produced in Yunnan, especially in the northwest region, and exported yearly to Japan as fresh mushrooms (1000 tons, worth USD 50 million), frozen mushrooms (700 tons, USD ten million), and brined mushrooms (300 tons, USD 3.4 million), according to Wang et al. (1997).

T. matsutake is the most important species in the matsutake group. Three other related species have been found in Yunnan, including *T. bakamatsutake* Hongo, *T. fulvocastaneum* Hongo, and *T. lavendulophyllum* F. Q. Yu (Yu et al. 2006). *Tricholoma matsutake* associates with pine species in Korea, Japan, and northeast China. However, in Yunnan, in addition to growing with pines such as *Pinus yunnanensis*, it associates with many species in the Fagaceae including *Castanopsis delavayi*, *Cyclobalanopsis delavayi*, *Lithocarpus sphaerocarpus*, and *Quercus pannosa* (Fig. 6.5). Matsutake is thus called "qing-gang-jun" (evergreen oak mushroom) in southwestern China where it is harvested from oak-dominated forests (Yu 2007).

Tricholoma bakamatsutake is called "hua" (flower-like cracking pattern on cap) matsutake or fake matsutake and is only harvested by local people for domestic consumption in Yunnan (Fig. 6.6). It is quite similar to *T. matsutake* but can be distinguished by its shorter and thinner stem, dark brown zoned scales on cap, and stronger matsutake smell with a slight aroma of honey. It associates with species in the Fagaceae, such as *Quercus serrata* in Japan and northeast China, but in Yunnan, it can also be found with *Castanopsis* and *Lithocarpus* (Yu 2007).

6.2.3.2 Truffles

Yunnan has one of the richest truffle diversities in China and worldwide (Jeandroz et al. 2008; García-Montero et al. 2010; Wang and Liu 2011; Bonito et al. 2013). However, this truffle richness was unknown until the discovery, in 1989, of *Tuber sinense* (a taxon of the *Tuber indicum* complex) in Huidong County, Sichuan

Table 6.2 Dominant species of commercial wild mushrooms in Yunnan

Species name
Albatrellus ellisii
Amanita hemibapha var. *ochracea*
Auricularia delicata
Butyriboletus roseoflavus
Boletopsis grisea
Boletus bainiugan
Boletus reticuloceps
Boletus shiyong
Boletus sinoedulis
Boletus viscidiceps
Cantharellus cibarius
Cantharellus cinnabarinus
Cantharellus formosus
Catathelasma ventricosum
Craterellus tubaeformis
Cortinarius emodensis
Ganoderma cochlear
Ganoderma lingzhi
Hygrophorus russula
Laccaria laccata
Lactarius deliciosus
Lactarius hatsudake
Lactarius hygrophoroides
Lactarius vividus
Lactifluus volemus
Lanmaoa asiatica
Leccinum aurantiacum
Lentinula edodes
Lyophyllum shimeji
Morchella eximia
Morchella importuna
Morchella sextelata
Neoboletus brunneissimus
Ophiocordyceps sinensis
Phlebopus portentosus
Polypus dispansus
Ramaria eryuanensis
Ramaria hemirubella
Ramaria indoyunnaniana
Ramaria mairei
Rugiboletus extremiorientalis
Retiboletus fuscus
Russula compacta
Russula cyanoxantha

(continued)

Table 6.2 (continued)

Species name
Russula griseocarnosa
Russula virescens
Sarcodon imbricatus
Scleroderma yunnanense
Suillus pinetorum
Suterius magnificus
Termitomyces clypeatus
Termitomyces eurhizus
Termitomyces heimii
Termitomyces microcarpus
Thelephora ganbajun
Tricholoma matsutake
Tricholoma myomyces
Tuber indicum
Tuber liyuanum
Tuber pseudohimalayense

Fig. 6.4 *Tricholoma matsutake* on Shangri-La wild edible mushroom market

Province. Since then, more than 20 truffle species have been reported in Yunnan, including the commercial species *T. indicum*, *T. sinoaestivum*, and *T. pseudohimalayense* (Chen and Liu 2012; Fan and Cao 2012; Fan et al. 2012a, 2012b, 2013, 2014; Deng et al. 2013; Li et al. 2014a, b; Wan et al. 2017; Xu et al. 2017). Concurrently with increasing quantities of truffles exported to Europe, species from the *T. indicum* complex have been found in more than 20 counties in Yunnan. In

Fig. 6.5 Evergreen oak matsutake forests in northwest Yunnan

Fig. 6.6 *Tricholoma bakamatsutake* on Mushuihua wild edible mushroom market, Kunming

2006, 835 tons of fresh truffles (worth USD 26.19 million) were exported; most were from Yunnan. Recently, more than ten new white truffle species have been reported from Yunnan. Among these, *T. panzhihuanense* has been commercialized (Deng et al. 2013), and *T. liyuanum*, another newly described white truffle species,

Fig. 6.7 Truffle at Mushuihua wild edible mushroom market, Kunming

has commercial potential (Fan and Cao 2012). Harvesting and trading of truffles are quickly becoming a multimillion-dollar industry in Yunnan (Fig. 6.7).

Recent research on the *T. indicum* complex revealed that it is composed of two subspecies or species: Lineage A and Lineage B (Bonito et al. 2013; Qiao 2013; see also Chap. 2 by Wang et al. for more details). The ascocarps of *T. indicum* complex produced in dry-hot valleys have better taste than those from the plateau. The Gongshan truffles produced in the dry-hot valley of the Nu River are considered the best quality truffles of the *T. indicum* complex in Yunnan (Fig. 6.8), according to truffle dealers (Qiao et al. 2018).

6.2.3.3 Boletes

Porcini mushrooms (*Boletus* sect. *Boletus*) and closely related species are the most important wild edible mushrooms in Yunnan (Wang and Liu 2002). Around 1000 tons of dried porcini are annually exported to Europe and the USA, with a value of USD 19 million (Fig. 6.9). The majority of exported boletes are from Yunnan (Wu and Lu 2006).

Cui et al. (2015) indicated that Chinese porcini can be divided into 15 species, including nine new species, namely, *Boletus bainiugan*, *B. fagacicola*, *B. griseiceps*, *B. monilifer*, *B. sinoedulis*, *B. subviolaceofuscus*, *B. tylopilopsis*, *B. umbrinipileus*, and *B. viscidiceps*. In addition to porcini mushrooms, there are other popular edible boletoid mushrooms such as *Neoboletus brunneissimus*, *Retiboletus fuscus*, and *Rugiboletus extremiorientalis* (Wang and Yao 2005; Wu et al. 2014; Zhao et al. 2014a, b).

Fig. 6.8 The truffle habitat at Nu River valley, Gongshan, Yunnan (Photo by Shu-Hong Li)

Fig. 6.9 Dried sliced porcini at Mushuihua wild edible mushroom market, Kunming

Some boletes known as "jian-shou-qing" (turning blue when bruised or cut) are considered hallucinogenic, causing visions that Yunnan's people call "xiao-ren-ren" (little men or little people, similar to the "Lilliputian hallucinations" found in the Kuma people from New Guinea) (Arora 2008). Among these boletes are *Butyriboletus roseoflavus*, *Lanmaoa asiatica*, and *Sutorius magnificus*, all

Fig. 6.10 Jianshouqing boletes at Mushuihua wild edible mushroom market, Kunming

commonly collected in Yunnan and even more popular than porcini (Fig. 6.10) (Wang et al. 2004). Though some of these species can cause gastrointestinal distress, the local people continue to consume them (Arora 2008).

Eleven boletes, some causing confusion/uncertainty in persons and some producing toxic effects to animals in lab test, were found commonly mixed with commercial edible species in Yunnan's mushroom markets (Fig. 6.11). These are *Boletellus ananas*, *Buchwaldoboletus hemichrysus*, *Heimioporus retisporus*, *Pulveroboletus ravenelii*, *Suillellus queletii*, *Sutorius eximius*, *Tylopilus neofelleus*, *T. otsuensis*, *T. plumbeoviolaceus*, *T. virens*, and *Gyroporus castaneus* (Wang et al. 2004; Li 2009; Li et al. 2011a, b; Wu et al. 2014).

Suillus pinetorum and other six closely related species (*S. cavipes*, *S. granulatus*, *S. grevillei*, *S. luteus*, *S. pictus*, and *S. sibiricus*) are common ectomycorrhizal edible mushrooms in Yunnan (Fig. 6.12). Thirty-two species of this genus have been found including 17 new species and two new records from China (Shi 2013).

6.2.3.4 Russulaceae

In central and southern China (e.g., Hunan, Guizhou, and Yunnan Provinces), six species of *Lactarius* sect. *Deliciosi* (*L. akahatsu*, *L. deliciosus*, *L. hatsudake*, *L. hengduanensis*, *L. pseudohatsudake*, and *L. vividus*) are commonly collected, consumed, and commercialized, with various local names including "gu-shou-jun"

Fig. 6.11 Poisonous boletes mixed with other edible ones sold at the markets

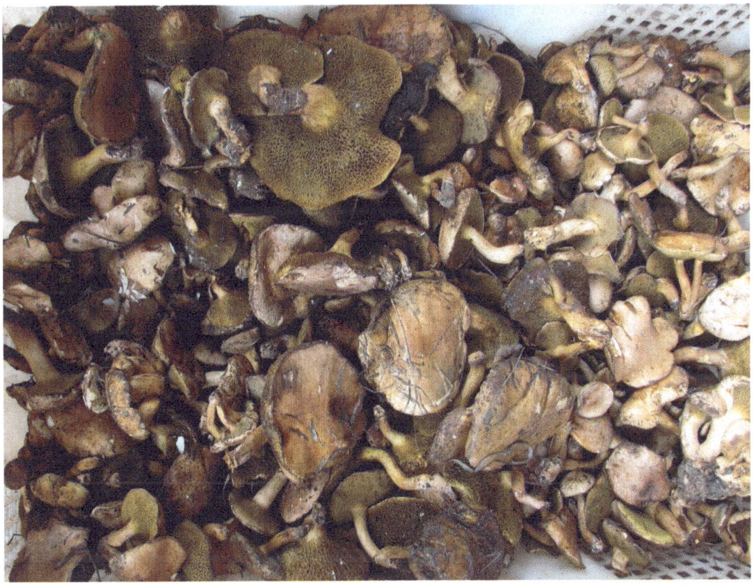

Fig. 6.12 *Suillus pinetorum* at Nanhua wild edible mushroom market

Fig. 6.13 *Lactarius* mushrooms at Ciba wild edible mushroom market, Kunming

(mushroom that fruits when the corn is ripe), "tong-lv-jun" (coppery green mushroom), and "song-jun" (pine mushroom). Because of their similar appearance and overlapping geographic distribution, *L. vividus* was misidentified as *L. akahatsu* or *L. deliciosus* in China (Wang et al. 2015a, b). *Lactarius vividus*, *L. hatsudake*, *L. deliciosus*, *Lactifluus hygrophoroides*, *L. volemus*, and other 14 milk cap mushrooms are commonly found in Yunnan markets (Fig. 6.13) (Wang et al. 2004, Wang 2016). Research on the production of mycorrhizal seedlings with *L. vividus*, *L. deliciosus*, *L. hatsudake*, and related species is now being undertaken (Wang et al. 2019). Yields of *L. volemus* were increased by field inoculation using spore inoculum in the natural habitats of this mushroom in Lancang County, Yunnan (Liu et al. 2009).

Mushrooms in the genus *Russula* are very popular in southern China. Fifteen species of *Russula* were encountered in the local markets in Yunnan, and several of these were extensively collected and sold. *Russula griseocarnosa* is the most renowned Chinese edible and medicinal mushroom. It is mainly distributed in tropical and subtropical areas and is collected, sold, and consumed under a well-known local name "da-hong-jun" (bright red mushroom) in Yunnan, Fujian, and Guangdong Provinces (Fig. 6.14) (Wang et al. 2009). *Russula virescens*, known as "qing-tou-jun" (green head mushroom), is common in central Yunnan.

6.2.3.5 *Termitomyces*

Termitomyces is known as "ji-zong" (chicken mushroom) and is very popular in Yunnan (Fig. 6.15). These grow in tropical and subtropical regions of China and are associated with termites (Fig. 6.16). Twenty-six species of *Termitomyces* were reported from China, but only 11 have been confirmed recently (Tang et al. 2006;

Fig. 6.14 *Russula griseocarnosa* at Chuxiong wild edible mushroom market, Kunming

Fig. 6.15 *Termitomyces* at Mushuihua wild edible mushroom market, Kunming

Fig. 6.16 A mushroom produced in a nest formed by *Termitomyces* and termites (Photo by Rong-Chun Li)

Wei et al. 2006, 2009). Ten *Termitomyces* species are found in Yunnan, and nine can be found in local markets, with *T. clypeatus*, *T. eurhizus*, and *T. heimii* as the dominant species (Wang et al. 2004).

6.2.3.6 *Cantharellus* and *Craterellus*

Cantharellus and *Craterellus* are commercially important genera of wild edible mushrooms, collected in Europe, Africa, Asia, and North and Central America. They are very common in the wild mushroom markets in Yunnan (Fig. 6.17). Twenty-three species of *Cantharellus* have been found in China, including four new species and four new records (Shao 2011; Shao et al. 2011, 2012, 2014, 2016; Tian et al. 2009, 2012). Twenty species have been confirmed in Yunnan, and 15 species can be encountered on the local markets, with *Cantharellus cibarius*, *C. cinnabarinus*, and *C. formosus* as the dominant species (Wang et al. 2004). *Craterellus aureus*, *C. cornucopioides*, *C. lutescens*, and *C. tubaeformis* are also common species on the markets. Due to the high species diversity and economical importance, more work on Cantharellaceae should be carried out in China.

Fig. 6.17 *Cantharellus* mushrooms at Nanhua wild edible mushroom market

6.2.3.7 Morels

Morels are usually known as "yang-du-jun" (sheep stomach mushroom) and are high-priced, popular edible mushrooms in Yunnan (Fig. 6.18). They are widely distributed throughout China. Du (2012) reported 11 new phylogenetically distinct species in China, thus establishing China as the most taxon-rich country with a total of 30 morel species. Seven species can be found at local markets, with *Morchella eximia*, *M. importuna*, and *M. sextelata* as the dominant species in Yunnan. Although cultivation of *M. rufobrunnea* and *M. importuna* has been achieved independently in the USA and China, morels collected from the wild still dominate the markets. Besides *Morchella*, ten species of *Helvella* were found on the markets in Yunnan, including four species new to science (Zhao et al. 2015).

6.2.3.8 *Amanita*

Amanita sect. *Caesareae* is a group of edible mushrooms distributed worldwide. Forty-seven species were reported from this section, including many popular species, such as *Amanita caesarea* (Caesar's mushroom), *A. caesareoides*, *A. hemibapha* var. *ochracea*, and *A. zambiana* (Tang 2013; Yang 2015). The *A. hemibapha* complex is known as "ji-dan-jun" (egg mushroom) by local people and is very popular in Yunnan (Fig. 6.19). Another good edible species is *A. yuaniana* (Yang

Fig. 6.18 Morels at Mushuihua wild edible mushroom market, Kunming

Fig. 6.19 *Amanita hemibapha* at Mushuihua wild edible mushroom market, Kunming

Fig. 6.20 *Lyophyllum shimeji* at Shangri-La wild edible mushroom market

2015). Every year, poisoning incidents with *Amanita* species are common worldwide (see for example Chap. 2 by Wang et al.) and Yunnan is no exception.

6.2.3.9 *Lyophyllum*

Lyophyllum shimeji ("honshimeji" = true shimeji in Japanese) is known as "yi-wo-ji" (a den of hens) or "yi-wo-yang" (a den of sheep) by local people and is a popular edible mushroom in Yunnan (Fig. 6.20). Based on research by Japanese mycologists, different genotypes of *L. shimeji* can adopt different trophic lifestyles. See also Chap. 2 by Wang et al. for more information about this group.

6.2.3.10 *Albatrellus*

Albatrellus species are common wild edible mushrooms sold at local markets in Yunnan. The common names are "huang-hu-zhang-jun" (yellow tiger-paw mushroom) and "di-hua-jun" (flower on earth). Zheng and Liu (2008) reported 19 species of *Albatrellus* from China, including three species new to science. *Albatrellus ellisii* is one of the 13 species of *Albatrellus* recorded in Yunnan and is the most important one found in the majority of markets (Fig. 6.21). Grilling or boiling them with water prior to cooking is known to enhance their flavor.

Fig. 6.21 *Albatrellus ellisii* at Mushuihua wild edible mushroom market, Kunming

6.2.3.11 Shoro and Other Hypogeous Fungi

The "shoro" (*Rhizopogon roseolus*) is a delicacy in Japan. In China, it is known as "ji-yao-zi" (chicken kidney) and is commonly collected and traded in Yunnan (Wang et al. 2012). The taxonomy and phylogeny of Chinese shoros has recently been revised and contains three new species (see Chap. 2 by Wang et al.) including *Rhizopogon jiyaozi* (Fig. 6.22). Two other species described from Yunnan are *R. flavidus* and *R. sinoalbidus* (Li 2014). In addition, a few species in *Gautieria, Gymnomyces, Hydnotrya, Hymenogaster, Hysterangium, Melanogaster, Rossbeevera*, and *Zelleromyces* have been found in Yunnan (Orihara et al. 2012; Li et al. 2013; unpublished data). Further research on hypogeous fungi is needed.

6.2.4 *Other Edible Wild Mushrooms*

6.2.4.1 *Ophiocordyceps sinensis*

This mushroom is called "chong-cao" (insect-fungus) or "dong-chong-xia-cao" (insect in winter and fungus in summer). This is the most expensive (on a par with gold) edible or medicinal fungus in the world (Fig. 6.23). It is considered as a tonic for both men and women, although this needs scientific confirmation, and is traditionally cooked with meat, especially with duck (it is inserted into the chest of the duck and then stewed). Southwestern China is the center of origin and differentiation of *O. sinensis*. The alpine mountains of northwest Yunnan (over 4000 m in altitude) are an important collecting region. Harvesting of *O. sinensis* provides an

Fig. 6.22 *Rhizopogon jiyaozi* at Shangri-La wild edible mushroom market

Fig. 6.23 *Ophiocordyceps sinensis* sold at Shangri-La

Fig. 6.24 *Thelephora ganbajun* at Ciba wild edible mushroom market, Kunming

important income for Tibetan populations. When snow begins to melt in early spring, it is time to search for *O. sinensis*.

6.2.4.2 *Thelephora ganbajun*

This is the world's only known edible *Thelephora* species and the most expensive edible mycorrhizal fungus in Yunnan (Fig. 6.24). In many parts of the world, *Thelephora* species are troublesome contaminants in the production of mycorrhizal seedlings. However, until recently, *T. ganbajun* has defied cultivation, and pure culture isolates have been difficult to obtain (see also Chap. 2 by Wang et al.). Protection and improvement of natural habitats are the only known means to maintain and increase yield of these economically important mushrooms. They are mainly harvested from *Pinus armandii*, *P. yunnanensis*, and *Keteleeria evelyniana* forests (unpublished data).

6.2.4.3 *Scleroderma yunnanense*

There is no mushroom book or scientific paper stating that any *Scleroderma* species is edible at a large scale. Instead, there have been quite a few reports of unpleasant results from eating *Scleroderma* species (Stevenson and Benjamin 1961; Arora 1986). Surprisingly however, a *Scleroderma* species is commonly sold at tropical and subtropical local markets in Yunnan (Fig. 6.25). This was misidentified as *S. citrinum*. It is now considered a new species, *S. yunnanense* (Zhang et al. 2013),

Fig. 6.25 *Scleroderma yunnanense* at Lancang wild edible mushroom market, Pu'er

and is associated with pine (*Pinus kesiya* var. *langbianensis*) and broadleaf (*Betula alnoides*) trees. Although local people also eat mature specimens, *S. yunnanense* is best eaten immature.

6.2.4.4 *Phlebopus portentosus*

It is a favorite edible mushroom in Xishuangbanna, the tropical region of Yunnan (Fig. 6.26). It is not an ectomycorrhizal fungus; it can be saprobic and cultivated by using saprophytic methods (Ji et al. 2011). However, in most circumstances, *Phlebopus portentosus* is associated with mealy bugs, forming fungus-insect galls on plants; more galls indicate greater productivity. Based on the fungus-insect association, *P. portentosus* can be produced by field inoculation of plants with fungal mycelia (Zhang et al. 2015).

6.2.4.5 *Schizophyllum commune*

The split gill mushroom, locally known as "bai-sheng" (white ginseng) is a popular edible mushroom in Yunnan (Fig. 6.27). Although considered a widely distributed basidiomycetous, *Schizophyllum commune* has been consumed for its nutritional value and medical efficacy in mostly Southeast Asia and is now both harvested from the wild and cultivated in Yunnan (Arbaayah and Umi 2013). It is usually cooked with eggs.

Fig. 6.26 *Phlebopus portentosus* at Jinghong wild edible mushroom market, Xishuangbanna (Photo by Chunxia Zhang)

Fig. 6.27 *Schizophyllum commune* at Lancang wild edible mushroom market, Pu'er

Fig. 6.28 Dried *Naematelia aurantialba* at a market

6.2.4.6 *Naematelia aurantialba*

An orange-red *Naematelia* species parasitizing *Stereum hirsutum* and allied species in southwestern China has recently been commercialized and cultivated (Fig. 6.28). It has traditionally been used in medicinal preparations for treatment of "lung fever," flu, coughing, asthma, and hypertension in China (Bandoni and Zang 1990).

6.2.4.7 *Oudemansiella raphanipes*

The local name for *Oudemansiella raphanipes* is "lu-shui-ji-zong" (dew termitomyces) or "cao-ji-zong" (grass termitomyces) in Yunnan. It was successfully cultivated in the 1990s (Yu et al. 2002). Until recently, this new edible mushroom, commercially called "heipijizong" or "black termite mushroom," has been widely cultivated in many parts of China (Fig. 6.29) (Hao et al. 2016).

6.3 Cultural Importance

Most species of wild edible mushrooms are eaten locally, and many are commercially harvested in Yunnan. Total quantities sold in local markets can be considerable. During the rainy season, huge quantities are collected and taken to markets in small towns and from there to larger cities. The financial contributions to rural

Fig. 6.29 Cultivation of *Oudemansiella raphanipes* in plastic greenhouse

livelihoods are not known, though the widespread sale of wild edible mushroom within Yunnan and then substantial export business demonstrate that significant amounts of money are earned. The foreign income produced from wild mushroom exportation is over USD 100 million every year. Marketing of *T. matsutake* and a few additional species, such as *O. sinensis*, *T. indicum*, and *B. bainiugan*, has significantly improved the local economy in the last few years (Wang and Yang 2006).

The majority of land in Yunnan is mountainous and home of 36 million farmers who mostly identify with ethnic minorities such as Yi, Tibetan, Hani, Naxi, Lisu, and Miao (Pei 2004). Harvesting of wild mushrooms is an important livelihood and generates 15–90% of these people's annual income. The most important commercial mushroom is probably matsutake. In the last 10 years, over 1000 tons of fresh fruit bodies of matsutake have been exported from Yunnan and Sichuan annually. More than 40 counties in Yunnan are reported to harvest matsutake. In the Shangri-La region, northwest Yunnan, harvesting matsutake can result in an annual return of over 10,000 Chinese Yuan (about USD 1500) for an average family. New houses have been constructed in Tibet using the money generated from this harvest (Fig. 6.30). Cultivation of edible mushrooms is another important means by which a farming family may transition from poverty to wealth. *Stropharia rugosoannulata* and *Phallus impudicus* are saprobic species commonly cultivated during the less busy farming seasons (Fig. 6.31) and grown in rotation with crops or under crops, fruit trees, or natural forests.

Processing edible mushrooms is a way of adding value to the harvest and increasing a family's income. Truffle wines and *Termitomyces* oils (Fig. 6.32) are special products in Yunnan and are sold nationwide.

Fig. 6.30 New Tibetan houses at Geza, Shangri-La

Fig. 6.31 Cultivation of *Stropharia rugosoannulata* in the field

Fig. 6.32 *Termitomyces* sp. in oil

6.4 Conservation

The natural production of wild edible mushrooms has declined since large-scale commercial harvesting initiated in the 1990s. Decreasing yields of wild mushrooms has been exacerbated by global warming, particularly during consecutive years of drought. A variety of efforts have been deployed to protect wild edible mushrooms. The most important has been the forest ownership reformation which occurred in 2008, giving farmers the right to manage forest products including wild mushrooms. A few regulations have been launched such as prohibiting the harvesting of immature matsutake and truffles. A few reserves for protection and study of matsutake and truffles have been established and more are planned. Experimental plantations have been set up for truffle cultivation, and the production of *T. indicum* has begun (Fig. 6.33). Other attempts to cultivate truffles (*T. borchii*, *T. melanosporum*, and *T. sinoaestivum*) and milk cap mushrooms (*L. akahatsu*, *L. deliciosus*, *L. hatsutake*, and *L. vividus*) and to understand the biology, ecology, and cultivation potential of edible mushrooms are being undertaken (Geng et al. 2009; Deng et al. 2014; Wang and Liu 2014; Wang et al. 2015a, b; Wan et al. 2016; Wang et al. 2019). However, conservation of the precious wild edible mushroom remains a vital and urgent issue.

Fig. 6.33 An experimental plantation for the cultivation of *T. indicum* at Shilin County

Acknowledgments This study has been funded by National Key Research and Development Program of China (No. 2017YFC0505206), Science and Technology Service Network Initiative, Chinese Academy of Sciences (2017), and Guizhou Science and Technology Program (No. 4002, 2018). We would like to thank Xiang-hua Wang, Rui-lin Zhao, Tie-zheng Wei, Yan-chun Li, Qi Zhao, Xiao-fei Shi, and Shan-ping Wan for their help in checking the data in the Appendix. Special thanks to Toni Atkinson for carefully reviewing the English.

Appendix 1. Commercial Wild Mushrooms in Yunnan, Their Use/Property, and Quantity Traded

Taxa	Use/property[a]	Quantity
Ascomycota		
Hypocreales		
Cordycipitaceae		
Cordyceps militaris	Medicine	I
Metacordyceps liangshanensis	Medicine NC	I
Nectriaceae		

(continued)

Pseudonectria bambusae	Medicine	I
Ophiocordycipitaceae		
Ophiocordyceps crassispora	Medicine	I
O. sinensis	Medicine	I
O. sobolifera	Medicine	III
Leotiales		
Leotiaceae		
Leotia aurantipes	Medicine NC	I
Pezizales		
Helvellaceae		
Helvella elastica	Food	I
H. involuta	Food	I
H. macropus	Food	I
H. maculatoides	Food	I
H. orienticrispa	Food	I
H. orienticrispa	Food	I
H. pseudoreflexa	Food	I
H. rugosa	Food	I
H. sublactea	Food	I
H. zhongtiaoensis	Food	II
Morchellaceae		
Morchella galilaea	Food	I
M. eohespera	Food	I
M. eximia	Food	III
M. exuberans	Food	I
M. importuna	Food	III
M. meiliensis	Food	I
M. sextelata	Food	III
Tuberaceae		
Tuber glabrum	Food	I
T. himalayense	Food	II
T. indicum	Food	IV
T. liyuanum	Food	III
T. microspermum	Food	I
T. microsphaerosporum	Food	I
T. microspiculatum	Food	I
T. microverrucosum	Food	I
T. panzhihuanense	Food	II
T. pseudohimalayense	Food	III
T. sinoaestivum	Food	II
T. sinopuberulum	Food	I
T. vesicoperidium	Food	I
Pleosporales		
Shiraiaceae		
Shiraia bambusicola	Medicine	I

(continued)

Xylariales		
Engleromyces sinensis	Medicine	II
Basidiomycota		
Agaricales		
Agaricaceae		
Agaricus flocculosipes	Food	I
A. guizhouensis	Food	I
A. macrocarpus	Food	I
Calvatia cyathiformis	Food & Medicine	I
Macrolepiota procera	Food	I
Amanitaceae		
Amanita caesareoides	Food	II
A. fritillaria	Poisonous NC	I
A. hemibapha	Food	II
A. hemibapha var. *ochracea*	Food	IV
A. javanica	Food	I
A. manginiana	Food	II
A. pseudoporphyria	Food NC	I
A. sinensis	Food NC	I
A. virgineoides	Poisonous AM	I
A. yuaniana	Food	I
Clavariaceae		
Scytinopogon echinosporus	Food	II
Cortinariaceae		
Cortinarius emodensis	Food	III
C. purpurascens	Food	I
C. tenuipes	Food	I
Entolomataceae		
Entoloma clypeatum	Food	I
Hydnangiaceae		
Laccaria alba	Food	I
L. amethystina	Food	II
Laccaria laccata	Food	III
L. vinaceoavellanea	Food	II
Hygrophoraceae		
Hygrophorus camarophyllus	Food	I
H. eburneus	Food	I
H. robustus	Food	I
H. russula	Food	III
Inocybaceae		
Inocybe flavobrunnea	Poisonous AM	I
Lyophyllaceae		
Lyophyllum fumosum	Food	II
L. shimeji	Food	IV
Termitomyces aurantiacus	Food	II

(continued)

T. bulborhizus	Food	III
T. clypeatus	Food	IV
T. eurhizus	Food	IV
T. globulus	Food	II
T. heimii	Food	III
T. mammiformis	Food	II
T. microcarpus	Food	III
T. striatus	Food	II
Omphalotaceae		
Lentinula edodes	Food	III
Physalacriaceae		
Armillaria mellea	Food	I
A. tabescens	Food	I
Oudemansiella raphanipes	Food	I
Pleurotaceae		
Pleurotus flabellatus	Food NC	I
P. ostreatus	Food	I
P. platypus	Food	I
Schizophyllaceae		
Schizophyllum commune	Food & Medicine	II
Tricholomataceae		
Catathelasma imperiale	Food	II
C. ventricosum	Food	IV
Clitocybe eccentrica	Food	I
Collybia obscura	Food	I
Lepista nuda	Food	I
Leucopaxillus tricolor	Food	I
Macrocybe gigantea	Food	I
Tricholoma bakamatsutake	Food	II
T. flavovirens	Food NC	II
T. fulvocastaneum	Food	I
T. imbricatum	Food	I
T. lavendulophyllum	Food	I
T. matsutake	Food	IV
T. myomyces	Food	III
T. robustum	Food	I
T. saponaceum	Poisonous NC	I
Auriculariales		
Auriculariaceae		
Auricularia delicata	Food	III
A. heimuer	Food	II
A. nigricans	Food	II
Boletales		
Boletaceae		
Baorangia pseudocalopus	Food	I

(continued)

Boletellus ananas	Poisonous NC	I
Boletus bainiugan	Food	IV
B. citrifragrans	Food	I
B. gertrudiae	Food	I
B. instabilis	Food	II
B. miniato-olivaceus	Food WC	I
B. obscureumbrinus	Food WC	II
B. orientialbus	Food	I
B. punctilifer	Food	I
B. reticuloceps	Food	II
B. sensibilis	Food WC	II
B. shiyong	Food	IV
B. sinoedulis	Food	IV
B. taienus	Food	I
B. tomentipes	Food	I
B. violaceofuscus	Food	I
B. viscidiceps	Food	III
B. yunnanensis	Food	I
Buchwaldoboletus hemichrysus	Poisonous NC	I
Butyriboletus roseoflavus	Food WC	IV
Caloboletus panniformis	Food	I
C. yunnanensis	Food	I
Crocinoboletus laetissimus	Food	I
Heimioporus retisporus	Poisonous AM	I
Hortiboletus rubellus	Food	II
H. subpaludosus	Food	I
Hourangia cheoi	Food	I
H. microcarpa	Food	I
H. nigropunctata	Food	I
Lanmaoa asiatica	Food WC	IV
Leccinellum crocipodium	Food	II
Leccinum aurantiacum	Food	III
L. holopus	Food	I
L. rubropunctum	Food	I
L. rugosiceps	Food	I
L. scabrum	Food	I
Neoboletus brunneissimus	Food	IV
N. thibetanus	Food	I
Phylloporus bellus	Food	I
P. brunneiceps	Food	I
P. imbricatus	Food	I
P. luxiensis	Food	I
P. maculatus	Food	I
P. pachycystidiatus	Food	I
P. rubrosquamosus	Food	I

(continued)

P. yunnanensis	Food	II
Pulveroboletus ravenelii	Poisonous NC	I
Retiboletus fuscus	Food	IV
R. griseus	Food	I
R. kauffmanii	Food	I
R. nigerrimus	Food	I
R. ornatipes	Food NC	I
Rubinoboletus balloui	Food	I
Rubroboletus sinicus	Food WC	II
Rugiboletus brunneiporus	Food	I
R. extremiorientalis	Food	IV
Strobilomyces confusus	Food NC	I
S. mirandus	Food NC	I
S. strobilaceus	Food NC	I
Suillellus luridus	Food	I
S. queletii	Poisonous NC	I
Sutorius eximius	Poisonous NC	I
S. magnificus	Food WC	IV
Tylopilus brunneirubens	Food NC	I
T. nanus	Food NC	I
T. neofelleus	Poisonous NC	I
T. otsuensis	Poisonous NC	I
T. plumbeoviolaceus	Poisonous AM	I
T. virens	Poisonous AM	I
Veloporphyrellus velatus	Food NC	I
Xerocomellus chrysenteron	Food	I
Zangia olivaceobrunnea	Food	I
Z. roseola	Food	I
Boletinellaceae		
Phlebopus portentosus	Food	IV
Gomphidiaceae		
Chroogomphus confusus	Food	I
C. filiformis	Food	I
C. orientirutilus	Food	I
Gomphidius roseus	Food	I
Gyroporaceae		
Gyroporus castaneus	Poisonous NC	I
G. longicystidiatus	Food NC	I
Rhizopogonaceae		
Rhizopogon jiyaozi	Food	I
Suillaceae		
Suillus cavipes	Food	I
S. granulatus	Food	II
S. grevillei	Food	II
S. luteus	Food	II

(continued)

S. pictus	Food NC	I
S. pinetorum	Food	III
S. sibiricus	Food	I
Sclerodermataceae		
Pisolithus arhizus	Medicine NC	I
Scleroderma flavidum	Poisonous NC	I
S. yunnanense	Food	III
Cantharellales		
Cantharellaceae		
Cantharellus appalachiensis	Food	I
C. carneoflavus	Food	I
C. cibarius	Food	IV
C. cinereus	Food	I
C. cinnabarinus	Food	III
C. formosus	Food	IV
C. hygrophorus	Food	I
C. infundibuliformis	Food	II
C. minor	Food	II
C. odoratus	Food	II
C. phloginus	Food	I
C. subalbidus	Food	II
C. tuberculosporus	Food	I
C. xanthopus	Food	III
C. yunnanensis	Food	II
Craterellus aureus	Food	II
C. cornucopioides	Food	II
C. lutescens	Food	II
C. tubaeformis	Food	III
Clavulinaceae		
Clavulina coralloides	Food	I
Hydnaceae		
Hydnum repandum	Food	II
Gomphales		
Gomphaceae		
Gomphus orientalis	Poisonous NC	I
Phaeoclavulina cyanocephala	Food	II
Ramaria asiatica	Food	II
R. botrytoides var. *microspora*	Food	II
R. brunneipes	Food	I
R. cyanocephala	Food	I
R. distinctissima	Food	II
R. eryuanensis	Food	III
R. fennica	Food	II
R. formosa	Poisonous NC	I
R. hemirubella	Food	IV

(continued)

R. hilaris	Food	II
R. indoyunnaniana	Food	III
R. laeviformosoides	Food	I
R. linearioides	Food	I
R. linearis	Food NC	I
R. mairei	Food	III
R. nanispora	Food NC	I
R. neoformosa var. *sinensis*	Food	II
R. rubriattenuipes	Food	II
R. rubricarnata var. *laeta*	Food	I
R. sanguinipes	Food	II
R. sinoconjunctipes	Food	I
R. zebrispora	Food	I
Turbinellus floccosus	Poisonous NC	II
T. fujisanensis	Poisonous NC	I
Polyporales		
Ganodermataceae		
Amauroderma rugosum	Medicine	I
Ganoderma applanatum	Medicine	II
G. cochlear	Medicine	III
G. lingzhi	Medicine	IV
G. lucidum	Medicine	II
G. mutabile	Medicine	I
G. sinense	Medicine	I
Meripilaceae		
Meripilus giganteus	Food	I
Meruliaceae		
Irpex lacteus	Food	I
Polyporaceae		
Cryptoporus sinensis	Medicine	I
Grifola frondosa	Food	I
Lentinus sajor-caju	Food	I
L. squarrosulus	Food	I
L. tuber-regium	Food	I
Wolfiporia cocos	Medicine	I
Sparassidaceae		
Sparassis latifolia	Food	I
S. subalpina	Food NC	I
Russulales		
Albatrellaceae		
Albatrellus confluens	Food	II
A. ellisii	Food	IV
A. flettii	Food	I
A. fumosus	Food	I
A. ovinus	Food	I

(continued)

A. skamanius	Food	I
A. yunnanensis	Food	I
Neoalbatrellus yasudae	Food	I
Polypus dispansus	Food NC	III
Scutiger pes-caprae	Food	I
Hericiaceae		
Hericium alpestre	Food	I
H. coralloides	Food	I
H. erinaceus	Food	I
Russulaceae		
Lactarius chichuensis	Food	I
L. cinnamomeus	Food	II
L. deliciosus	Food	IV
L. echinatus	Food	II
L. gerardii	Food	I
L. hatsudake	Food	IV
L. hengduanensis	Food	II
L. piperatus	Food WC	I
L. pseudohatsudake	Food	II
L. vividus	Food	III
L. zonarius	Food	I
Lactifluus hygrophoroides	Food	III
L. volemus	Food	IV
L. subpiperatus	Food	I
L. subpruinosus	Food	I
L. tenuicystidiatus	Food	I
L. tropicosinicus	Food	I
Russula atroaeruginea	Food	II
R. aurata	Food	I
R. brunneoviolacea	Food	I
R. compacta	Food	III
R. cyanoxantha	Food	III
R. densifolia	Food	I
R. griseocarnosa	Food & Medicine	IV
R. laurocerasi	Food	II
R. lepida	Food	I
R. melliolens	Food	I
R. nigricans	Food	I
R. ochroleuca	Food	I
R. sanguinea	Poisonous NC	I
R. virescens	Food	IV
R. viridirubrolimbata	Food	II
Thelephorales		
Bankeraceae		
Boletopsis grisea	Food	III

(continued)

Hydnellum cumulatum	Food	II
Sarcodon amarescens	Food	I
S. excentricus	Food	I
S. imbricatus	Food	IV
S. scabrosus	Food	II
S. squamosus	Food	II
Thelephoraceae		
Polyozellus multiplex	Food	I
Thelephora ganbajun	Food	IV
T. japonica	Food	I
T. palmata	Food	I
T. vialis	Food	II
T. aurantiotincta	Food	I
Tremellales		
Tremellaceae		
Naematelia aurantialba	Food & Medicine	I

[a]Food, with clear evidence that the species is used as food; Food NC, not certain/confirmed that the species is used as food; Food WC, with caution/conditions and recorded as poisonous in some references; Poisonous AM, toxic to animals in lab test, effect on humans not recorded or unknown; Poisonous NC, confusion/uncertainty; Medicine, with clear evidence that the species is used as medicine; Medicine NC, not certain/confirmed or where there is a lack of consensus; Quantity traded: I, small; II, moderate; III, large; IV, enormous quantity

References

Arbaayah HH, Umi YK (2013) Antioxidant properties in the oyster mushrooms (*Pleurotus* spp.) and split gill mushroom (*Schizophyllum commune*) ethanolic extracts. Mycosphere 4:661–673

Arora D (1986) Mushrooms demystified. Ten Speed Press, Berkeley

Arora D (2008) Xiao Ren: the 'little people' of Yunnan. Econ Bot 62:540–544

Bandoni RJ, Zang M (1990) On an undescribed *Tremella* from China. Mycologia 82:270–273

Boa E (2004) Wild edible fungi a global overview of their use and importance to people. Food and Agriculture Organization of the United Nations, Rome

Bonito G, Smith ME, Nowak M et al (2013) Historical biogeography and diversification of truffles in the Tuberaceae and their newly identified southern hemisphere sister lineage. PLoS One 8:1–15

Cao Y, Wu SH, Dai YC (2012) Species clarification of the prize medicinal *Ganoderma* mushroom "Lingzhi". Fungal Divers 56:49–62

Chen J, Liu PG (2012) Delimitation of *Tuber pseudohimalayense* and *T. pseudohimalayense* based on morphological and molecular data. Cryptogam Mycol 32:83–93

Cui YY, Feng B, Wu G et al (2015) Porcini mushrooms (*Boletus* sect. *Boletus*) from China. Fungal Divers 82:189–212

Dai YC, Zhou LW, Yang ZL et al (2010) A revised checklist of edible fungi in China. Mycosystema 29:1–21

Deng XJ, Liu PG, Liu CY et al (2013) A new white truffle species, *Tuber panzhihuanense* from China. Mycol Prog 12:557–561

Deng XJ, Yu FQ, Liu PG (2014) Contribution to confirmed & synthesized on mycorrhizae of *Tuber indicum s.l.* with two dominated & subalpine broadleaf trees in southwestern China. Am J Plant Sci 5:3269–3279

Du XH, Zhao Q, O'Donnell K et al (2012) Multigene molecular phylogenetics reveals true morels (*Morchella*) are especially species-rich in China. Fungal Genet Biol 49:455–469

Fan L, Cao JZ (2012) Two new species of white truffle from China. Mycotaxon 121:297–304

Fan L, Cao JZ, Zheng ZH et al (2012a) *Tuber* in China: *T. microspermum* and *T. microspiculatum* spp. nov. Mycotaxon 119:391–395

Fan L, Cao JZ, Li Y (2012b) *Tuber microsphaerosporum* and *Paradoxa sinensis* spp. nov. Mycotaxon 120:471–475

Fan L, Cao JZ, Yu J (2013) *Tuber* in China: *T. sinopuberulum* and *T. vesicoperidium* spp. nov. Mycotaxon 121:255–263

Fan L, Feng S, Cao JZ (2014) The phylogenetic position of *Tuber glabrum* sp. nov. and *T. sinomonosporum* nom. nov., two *Paradoxa*-like truffle species from China. Mycol Prog 13:241–246

García-Montero LG, Díaz P, Massimo GD et al (2010) A review of research on Chinese *Tuber* species. Mycol Prog 9:315–335

Geng LY, Wang XH, Yu FQ et al (2009) Mycorrhizal synthesis of *Tuber indicum* with two indigenous hosts, *Castanea mollissima* and *Pinus armandii*. Mycorrhiza 19:461–467

Hao YJ, Zhao Q, Wang SX et al (2016) What is the radicate *Oudemansiella* cultivated in China? Phytotaxa 286(1):12

Jeandroz S, Claude M, Wang YJ et al (2008) Molecular phylogeny and historical biogeography of the genus *Tuber*, the "true truffles". J Biogeogr 35:815–829

Ji KP, Cao Y, Zhang CX et al (2011) Cultivation of *Phlebopus portentosus* in southern China. Mycol Prog 10:293–300

Kirk PM, Cannon PF, Minter DW et al (2008) Dictionary of the Fungi, 10th edn. CABI Publishing, Wallingford

Li N (1995) Studies on geographical distribution, origin and dispersal of the family Pinaceae Lindl. Acta Phytotax Sin 33:105–130

Li YC (2009) Taxonomy and phylogeny of *Tylopilus s.l.* from China, with notes on the taxonomy and biogeography of *Chroogomphus*. Dissertation, Kunming Institute of Botany, Chinese Academy of Sciences

Li L (2014) Research on diversity of *Rhizopogon*, *Truncocolumella* and *Hydnotrya* in the longitudinal range gorge region, southwest China. Dissertation, Kunming University of Science and Technology

Li YC, Yang ZL, Bau T (2009) Phylogenetic and biogeographic relationships of *Chroogomphus* species as inferred from molecular and morphological data. Fungal Divers 38:85–104

Li YC, Feng B, Yang ZL (2011a) *Zangia*, a new genus of Boletaceae supported by molecular and morphological evidence. Fungal Divers 49:125–143

Li SH, Zhao YC, Yu FQ et al (2011b) Systematics of the easily confusing poisonous boletes from Yunnan wild mushroom markets. Edible Fungi China 30:34–36

Li L, Zhao YC, Zhang XL et al (2013) *Hydnotrya laojunshanensis* sp. nov. from China. Mycotaxon 125:277–282

Li SH, Heng LY, Liu CY et al (2014a) Two new truffles species, *Tuber alboumbilicum* and *Tuber pseudobrumale* from China. Mycol Prog 13:1157–1163

Li YC, Ortiz-Santana B, Zeng NK et al (2014b) Molecular phylogeny and taxonomy of the genus *Veloporphyrellus*. Mycologia 106:291–306

Liu PG, Yu FQ, Wang XH et al (2009) The cultivation of *Lactarius volemus* in China. Acta Bot Yunnanica Suppl XVI:115–116

Murata H, Babasaki K, Saegusa T et al (2008) Traceability of Asian matsutake, specialty mushrooms produced by the ectomycorrhizal basidiomycete *Tricholoma matsutake*, on the basis of retroelement-based DNA markers. Appl Environ Microbiol 74:2023–2031

Myers N, Mittermeier RA, Mittermeier CG et al (2000) Biodiversity hotspots for conservation priorities. Nature 403:853–858

Orihara T, Smith ME, Ge ZW et al (2012) *Rossbeevera yunnanensis* (Boletaceae, Boletales), a new sequestrate species from southern China. Mycotaxon 120:139–147
Pei SJ (2004) Ethnic cultural diversity and nature conservation in Yunnan. Acta Bot Yunnanica Suppl 15:1–11
Petersen RH, Zang M (1986) New and interesting clavarioid fungi from Yunnan, China. Acta Bot Yunnanica 8:281–294
Petersen RH, Zang M (1989) *Ramaria* subgenera *Ramaria* and *Laeticolora*. Acta Bot Yunnanica 11:363–369
Petersen RH, Zang M (1990) *Ramaria* subgenera *Ramaria* and *Laeticolora* in Yunnan (continued). Acta Bot Yunnanica 12:49–56
Qiao P (2013) Phylogeography and population genetics of *Tuber indicum* complex in southwest China. Dissertation, Kunming Institute of Botany, Chinese Academy of Sciences
Qiao P, Tian W, Liu PG et al (2018) Phylogeography and population genetic analyses reveal the speciation of the *Tuber indicum* complex. Fungal Genet Biol 113:14–23
Shao SC (2011) Taxonomy and phylogeny of the genus *Cantharellus* from Southwestern China with screening primers on population genetics of *C. tuberculosporus*. Dissertation, Kunming Institute of Botany, Chinese Academy of Sciences
Shao SC, Tian XF, Liu PG (2011) *Cantharellus* in southwestern China: a new species and a new record. Mycotaxon 116:437–446
Shao SC, Tian XF, Liu PG (2012) Two species with intercontinental disjunct distribution of the genus *Cantharellus*. J Yunnan Agric Univ 27:150–155
Shao SC, Buyck B, Hofstetter V et al (2014) *Cantharellus hygrophorus*, a new species in subgenus *Afrocantharellus* from tropical southwestern China. Cryptogam Mycol 35:283–291
Shao SC, Buyck B, Tian XF et al (2016) *Cantharellus phloginus*, a new-colored species from southwestern China. Mycoscience 57:144–149
Sheng J (2013) Yunnan wild edible mushroom industry. Research report. Yunnan Agricultural University/Yunnan Plateau Characteristic Agricultural Industry Research Institute, Yunnan
Shi XF (2013) Taxonomy and phylogeny of the genus *Suillus* (Boletales, Basidiomycota). Dissertation, Kunming Institute of Botany, Chinese Academy of Sciences
Stevenson JA, Benjamin CR (1961) *Scleroderma* poisoning. Mycologia 53:438–439
Tan X (2012) Studies on the development of Yunnan edible mushroom industry. Minzu Press, Beijing
Tang LP (2013). Species diversity, reticulate evolution and biogeography of *Amanita* section *Caesareae* with notes on a new species of *Rhodotus*. Dissertation, Kunming Institute of Botany, Chinese Academy of Sciences
Tang LP, Yang ZL (2014) Recent studies on fungal species diversity in the Lancang-Mekong river watershed. Resources Science 36:282–295
Tang BH, Wei TZ, Yao YJ (2006) Revision of *Termitomyces* species originally described from China. Mycotaxon 95:285–293
Tang LP, Yang ZL, Zeng NK et al (2015) Atlas of higher fungi from Lancang river valley. Yunnan Science and Technology Press, Kunming
Tian XF, Shao SC, Liu PG (2009) Two notable species of the genus *Cantharellus* Adams (Cantharellales, Basidiomycota) new to China. Edible Fungi China 28:10–11
Tian XF, Buyck B, Shao SC et al (2012) *Cantharellus zangii*, a new subalpine basidiomycete from southwestern China. Mycotaxon 120:99–103
Wan SP, Yu FQ, Tang L et al (2016) Ectomycorrhizae of *Tuber huidongense* and *T. liyuanum* with *Castanea mollissima* and *Pinus armandii*. Mycorrhiza 26:249–256
Wan SP, Xu WJ, Tang NW et al (2017) Three excavatum species *Tuber badium, T. depressum* and *T. verrucosivolvum* from Sichuan Province, China. Phytotaxa 296:228–238
Wang XH (2016) Three new species of *Lactarius* sect. *Deliciosi* from subalpine-alpine regions of central and southwestern China. Cryptogam Mycol 37:493–508
Wang Y, Hall IR (2004) Edible ectomycorrhizal mushrooms: challenges and achievements. Can J Bot 82:1063–1073

Wang XH, Liu PG (2002) Resources investigation and studies on the wild commercial fungi in Yunnan. Biodivers Sci 10:318–325

Wang Y, Liu PG (2011) Verification of Chinese names of truffles and their conservation in natural habitats. Plant Divers Resour 33:625–642

Wang SP, Liu PG (2014) Diversity of culturable bacteria associated with ascocarps of a Chinese white truffle, *Tuber panzhihuanense* (Ascomycota). Plant Divers Resour 36:29–36

Wang L, Yang ZL (2006) Wild edible fungi of the Hengduan Mountains, southwestern China. In: Christoph K, Yang YP, Weyerhäuser H et al (eds) The sustainable harvest of non-timber forest products in China. Sino-German symposium proceedings. Sino-German Center, Beijing, pp 58–65

Wang QB, Yao YJ (2005) *Boletus reticuloceps*, a new combination for *Aureoboletus reticuloceps*. Sydowia 57:131–136

Wang Y, Evans LA, Hall IR (1997) Ectomycorrhizal fungi with edible fruiting bodies. 1. *Tricholoma matsutake* and related species. Econ Bot 51:311–327

Wang XH, Liu PG, Yu FQ (2004) Color atlas of wild commercial mushrooms in Yunnan. Yunnan Science and Technology Press, Kunming

Wang Y, Liu PG, Chen J et al (2008) China–a newly emerging truffle-producing nation. In: Reynal B, Chevalier GG (eds) La Culture de la Truffe dans le Monde. Actes du Colloque, Brive-La-Gallarde, pp 35–44

Wang XH, Yang ZL, Li YC et al (2009) *Russula griseocarnosa* sp. nov. (Russulaceae, Russulales), a commercially important edible mushroom in tropical China: mycorrhiza, phylogenetic position, and taxonomy. Nova Hedwigia 88:269–282

Wang Y, Cummings N, Guerin-Laguette A (2012) Cultivation of basidiomycete edible ectomycorrhizal mushrooms: *Tricholoma*, *Lactarius*, and *Rhizopogon*. In: Zambonelli A, Bonito GM (eds) Edible ectomycorrhizal mushrooms: current knowledge and future prospects. Springer-Verlag, Berlin/Heidelberg, pp 281–304

Wang R, Liu PG, Wan SP et al (2015a) Study on mycorrhization helper bacteria (MHB) of *Tuber indicum*. Microbiol China 42:2366–2376

Wang XH, Nuytinck J, Verbeken A (2015b) *Lactarius vividus* sp. nov. (Russulaceae, Russulales), a widely distributed edible mushroom in central and southern China. Phytotaxa 231:63–72

Wang R, Guerin-Laguette A, Huang LL et al (2019) Mycorrhizal syntheses between *Lactarius* spp. section *Deliciosi* and *Pinus* spp. and effects of grazing insects in Yunnan, China. Can J For Res 49:616–627

Wei TZ, Tang BH, Yao YJ et al (2006) A revision of *Sinotermitomyces*, a synonym of *Termitomyces* (Agaricales). Fungal Divers 21:225–237

Wei TZ, Tang BH, Yao YJ (2009) Revision of *Termitomyces* in China. Mycotaxon 108:257–285

Wu JC, Lu H (2006) Prospect of wild edible mushrooms industry and suggestions for industry development. J West China Forest Sci 35:154–158

Wu G, Feng B, Xu J et al (2014) Molecular phylogenetic analyses redefine seven major clades and reveal 22 new generic clades in the fungal family Boletaceae. Fungal Divers 69:93–115

Wu G, Li YC, Zhao K et al (2015) Four new genera of the fungal family Boletaceae. Fungal Divers 81:1–24

Xu WJ, Wan SP, Huang LL et al (2017) *Tuber sinoniveum*, a new white Chinese truffle species from Yunnan, China. Phytotaxa 298:253–260

Yang ZL (2015) Atlas of the Chinese species of Amanitaceae. Science Press, Beijing

Yang XF, Wilkes A, Yang YP et al (2009) Common and privatized: conditions for wise management of matsutake mushrooms in Northwest Yunnan Province, China. Ecol Soc 14:30

Ying JZ, Zang M (1994) Economic macrofungi from southwestern China. Science Press, Beijing

Yu FQ (2007) Taxonomy and phylogeny of *Tricholoma matsutake* group (Tricholomataceae, Agaricales) and its allied species. Dissertation. Kunming Institute of Botany, Chinese Academy of Sciences

Yu FQ, Liu PG (2005) Species diversity of wild edible mushrooms from *Pinus yunnanensis* forests and conservation strategies. Biodivers Sci 13:58–69

Yu FQ, Liu PG (2013) Wild edible and medicinal mushrooms in Pu'er of Yunnan and their sustainable utilization. J Fungal Res 11:14–23

Yu FQ, Ji DG, Song MJ et al (2002) Cultivation of two varieties of *Oudemansiella furfuracea*. Edible Fungi China 21:13–15

Yu FQ, Wang Y, Liu PG (2006) *Tricholoma lavendulophyllum*, a new species from Yunnan, China. Mycotaxon 95:305–308

Zang M (2006) Flora fungorum sinicorum: Boletaceae (I). Science Press, Beijing

Zang M (2013) Flora fungorum sinicorum: Boletaceae (II). Science Press, Beijing

Zeng NK (2011) Species diversity of Boletaceae in Hainan, China, with notes on the taxonomy of *Phylloporus*. Dissertation, Kunming Institute of Botany, Chinese Academy of Sciences

Zeng NK, Tang LP, Li YC et al (2013) The genus *Phylloporus* (Boletaceae, Boletales) from China: morphological and multilocus DNA sequence analyses. Fungal Divers 58:73–101

Zeng NK, Liang ZQ, Yang ZL (2014a) *Boletus orientialbus*, a new species with white basidioma from subtropical China. Mycoscience 55:159–163

Zeng NK, Wu G, Li YC et al (2014b) *Crocinoboletus*, a new genus of Boletaceae (Boletales) with unusual boletocrocin polyene pigments. Phytotaxa 175:133–140

Zhang P (2010) A study on taxonomy of *Ramaria* from Southwestern China, with notes on the phylogeny of the genus. Dissertation, Kunming Institute of Botany, Chinese Academy of Sciences

Zhang CX, Xu XE, Liu J et al (2013) *Scleroderma yunnanense*, a new species from South China. Mycotaxon 125:193–200

Zhang CX, He MX, Cao Y et al (2015) Fungus-insect gall of *Phlebopus portentosus*. Mycologia 107:12–20

Zhao Q, Feng B, Yang ZL et al (2013) New species and distinctive geographical divergence of the genus *Sparassis* (Basidiomycota): evidence from morphological and molecular data. Mycol Prog 12:445–454

Zhao K, Feng B, Yang ZL (2014a) Molecular phylogeny of *Caloboletus* (Boletaceae) and a new species in East Asia. Mycol Prog 13:1127–1136

Zhao K, Wu G, Yang ZL (2014b) A new genus, *Rubroboletus*, to accommodate *Boletus sinicus* and its allies. Phytotaxa 188:61–77

Zhao Q, Tolgor B, Zhao YC et al (2015) Species diversity within the *Helvella crispa* group (Ascomycota: Helvellaceae) in China. Phytotaxa 239:130–142

Zheng HD, Liu PG (2008) Additions to our knowledge of the genus *Albatrellus* (Basidiomycota) in China. Fungal Divers 32:157–170

Zhou ZQ (1992) Origin, evolution and migration of *Quercus*. Acta Bot Yunnanica 14:227–236

Chapter 7
Advances in Desert Truffle Mycorrhization and Cultivation

Asunción Morte, Almudena Gutiérrez, and Alfonso Navarro Ródenas

7.1 Introduction

Desert truffles include a group of edible hypogeous fungi that belong to different genera within the order Pezizales in the Ascomycota division. The most important genera are *Terfezia*, *Picoa* and *Tirmania* (Fig. 7.1).

These fungi are frequent in acid and alkaline soils of the Mediterranean Basin, and their fructification period ranges from February to May, depending on the quantity and distribution of precipitations that occurred along the year (Honrubia et al. 2007). Desert truffles have a great interest from an ecological point of view because the group of Cistaceae host plants with which they establish a symbiotic mutualism (Fig. 7.2) is well adapted to semiarid and arid environments. Desert truffles also have an important economic interest due to their great nutritional and gastronomic values (Murcia et al. 2003) and marketable fruiting bodies (Volpato et al. 2013).

7.2 Mycorrhizal Symbiosis

First descriptions of mycorrhizae of *Terfezia* with some species from the *Helianthemum* genus were made by Awameh et al. (1979), who, after obtaining axenic fungal culture, were able to achieve the first mycorrhizal synthesis with *Terfezia*. Later on, Chevalier et al. (1984) carried out the synthesis of *Terfezia leptoderma* mycorrhizae with different *Cistus* species. At that time, this type of mycorrhiza was considered to be intermediate between endotrophic and ectotrophic.

A. Morte (✉) · A. Gutiérrez · A. N. Ródenas
Departamento Biología Vegetal (Botánica), Facultad de Biología, Universidad de Murcia, Murcia, Spain
e-mail: amorte@um.es

J. Pérez-Moreno et al. (eds.), *Mushrooms, Humans and Nature in a Changing World*, https://doi.org/10.1007/978-3-030-37378-8_7

Fig. 7.1 The most appreciated edible desert truffle species in the market: (**a**) *Terfezia claveryi*, (**b**) *Terfezia boudieri*, (**c**) *Picoa lefebvrei*, (**d**) *Tirmania nivea*, (**e**) *Tirmania pinoyi*, from alkaline soils, (**f**) *Terfezia arenaria*, (**g**) *Terfezia fanfani*, from acid soils

The first mycorrhizal synthesis of *T. claveryi* with *H. almeriense* was carried out by Cano et al. (1991). Then, Morte et al. (1994) obtained this symbiosis in vitro conditions and characterized the synthetized mycorrhizae as ectendomycorrhizae, both with intercellular and intracellular hyphae and with structures similar to coils. Furthermore, some hyphae growing around the roots were observed, but this was not considered a true mantle (Morte et al. 1994). Alsheikh (1984) differentiated this type of mycorrhiza from ecto-, endo- and ectendomycorrhizae and found similarities with some arbutoid mycorrhizae. As a result, the term "helianthemoid" was proposed, in order to describe the mycorrhiza formed between different species of *Helianthemum* and some desert truffles. Kovács et al. (2003) suggested that the term "terfezioid" was more reasonable than "helianthemoid" since they found this type of intermediate mycorrhiza in *Robinia pseudoacacia* and *Helianthemum ovatum*

Fig. 7.2 Host plant species for desert truffles: (**a**) *Helianthemum almeriense*, (**b**) *Helianthemum violaceum*, (**c**) *Helianthemum hirtum*, (**d**) *Helianthemum canariense*, (**e**) *Helianthemum lippii*, from alkaline soils, (**f**) *Tuberaria guttata*, from acid soils

and considering other associations with *Terfezia* previously described (Kagan-Zur et al. 1999). Fortas and Chevalier (1992) showed for the first time, in different growing conditions, that the type of mycorrhizae established between *Helianthemum guttatum* (=*Xolantha guttata*) and different desert truffles depends on the phosphorus content of the medium. These authors observed, in the same root, structures of both an endomycorrhiza and an ectomycorrhiza in media with low phosphorus, while only ectomycorrhizal characteristics were observed in media with high phosphorus contents. Kagan-Zur et al. (1994) showed that, at the same time, low phosphorus concentrations in the medium culture may inhibit the mycorrhization between *Helianthemum sessiliflorum* and *Terfezia leonis*, but that the same mycorrhization could be stimulated by low iron concentration.

Gutiérrez et al. (2003) described the association formed by *T. claveryi* with *H. almeriense* as an endomycorrhiza in field conditions, an ecto- and ectendomycorrhiza in greenhouse conditions and an ectomycorrhiza with mantle and Hartig net in vitro conditions (Fig. 7.3). The authors proposed that these differences probably depended on the phosphorus concentration of the medium, since in nursery conditions the phosphorus concentration was intermediate between field and in vitro conditions. Furthermore, these authors described four mycorrhizal morphotypes in roots of *H. almeriense* mycorrhized by *T. claveryi*: "club-shaped", "capitated", "moniliform" and "branched" (Gutiérrez et al. 2003). Kovács et al. (2003) observed increased fungal colonization in the in vitro association between *T. terfezioides* (=*Mattirolomyces terfezioides*) and two host plants with high inorganic phosphorus concentration in the medium culture. On the other hand, Zaretsky et al. (2006), using transformed roots of *Cistus incanus* inoculated with *T. boudieri* mycelium, proposed that the type of mycorrhiza formed is not only influenced by the fungal

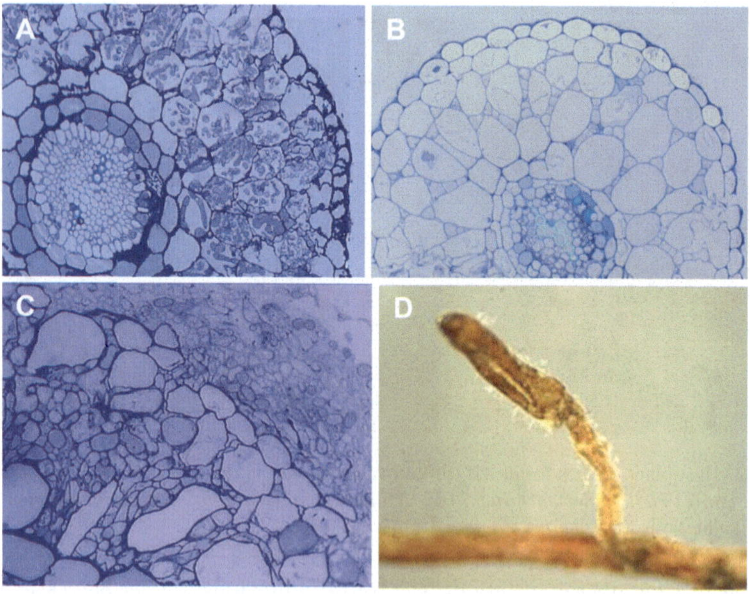

Fig. 7.3 Morphologies of the ectendomycorrhizal *continuum* between *H. almeriense* and *T. claveryi*: (**a**) inter- and intracellular hyphae, (**b**) Hartig net, (**c**) Hartig net with mantle, and (**d**) "club-shaped" root morphotype. (Photos A, C and D from Gutiérrez et al. (2003), with kind permission of Springer Science)

strain but also by the host plant's sensitivity to indoleacetic acid. This sensitivity may be related to the phosphorus and auxin exogenous levels.

Later on, Navarro-Ródenas et al. (2013) found that the mycorrhizal colonization between *H. almeriense* and *T. claveryi* increases as the availability of phosphorus and water decreases in the substrate. They found all types of colonization (intracellular, intercellular and both) in the same root, and they determined that the proportion of each one depended on the growing conditions. This led to propose the term "ectendomycorrhiza *continuum*", to define the morphology of this intermediate symbiosis (Navarro-Ródenas et al. 2013).

7.3 Mycorrhizal Plant Production

Having different in vivo and in vitro systems for plant and fungus propagation allowed to carry out several tests and combinations for the mycorrhizal synthesis between different species of *Helianthemum* with *T. claveryi* (Morte et al. 1994, 2008, 2009). In 1999, the first plantation of 60 plants of *H. almeriense* mycorrhized by *T. claveryi* was established in Zarzadilla de Totana (Murcia, Spain) (Gutiérrez 2001). The plantation produced the first truffles shortly before reaching 2 years after plantation (Honrubia et al. 2001). This plantation, the first one in the world, kept producing truffles until now, with an average of 250–400 kg/ha (Morte et al. 2008, 2009). During the last 15 years, more than 30 plantations have been established in Spain, with approximately 50,000 mycorrhizal plants of different *Helianthemum* species (*H. almeriense*, *H. violaceum* or *H. hirtum*) associated with *T. claveryi*.

The recent increasing demand for mycorrhizal plants has led to the search of new strategies in plant propagation, inoculum production (Morte et al. 2012) and the use of microorganisms present in the soil, such as plant growth-promoting rhizobacteria (PGPR) (Navarro-Ródenas et al. 2016), in order to increase the number and quality of desert truffle mycorrhizal plants.

Most of the bioassays for the production of ectomycorrhizal plants use seeds for seedling production and fungal spores as inoculum (Morte and Honrubia 2009). The same method is normally used to produce mycorrhizal seedlings of *Helianthemum x Terfezia*, but many *Helianthemum* species show erratic seed germination with high mortality of seedlings during the first 2 months after germination in nursery conditions (Morte et al. 2012). In vitro micropropagation techniques have solved this problem, getting up to 90% plant survival and a rapid micropropagation since plant multiplication, elongation and rooting occur in the same subculture, with no need of successive sub-cultivations nor addition of growth regulators (Morte et al. 2009, 2012). Moreover, a system of photoautotrophic micropropagation, based on the methodology described by Kozai (1991), has been designed for *H. almeriense* that involves a successful acclimatization from in vitro to ex vitro conditions (Andrino et al. 2012, Morte and Andrino 2014). In vitro conditions generally lead to a stomatal malfunction, poor development of epicuticular waxes, elongated and etiolated stems and poor root development (Kozai 1991, Majada et al. 2002). The methodology consists in replacing agar with sterilized perlite moistened with MS culture medium (Murashige and Skoog 1962), without carbon source, increasing photosynthetic light intensity (140–160 µmol/m^2 s) and providing an atmosphere with low relative humidity (60–70%) and 350–400 ppm of CO_2, by using plastic covers that allow gas exchange (Morte and Andrino 2014).

In relation to the inoculum of desert truffles, spore solution is commonly used due to the erratic and slow growth of the pure culture mycelium. However, a great effort has been made to improve culture conditions and nutrients and understand factors or inhibitors that may limit *Terfezia* growth. Cano et al. (1991) carried out the first pure culture isolation of *T. claveryi*, establishing that the best growing medium for this fungus was MMN (Marx, 1969) with a pH adjusted to 7.0. Later, Navarro-Ródenas et al. (2011) proved that *T. claveryi* and *P. lefebvrei* mycelia require some water potential adjustment of the in vitro culture medium, growing better with a moderate water stress of −0.45 and −0.72 kPa for *T. claveryi* and *P. lefebvrei*, respectively. At the base of that tolerance to water stress, a greater expression of the *T. claveryi* aquaporin TcAQP1 gene was found, a membrane protein that acts as a water channel and other substances facilitating the transport of water between cells (Navarro-Ródenas et al. 2013). Moreover, the use of cyclodextrins (CD), especially β-CD, could stimulate mycelial growth of *T. claveryi* until achieving a final diameter and a growth rate five times greater than that of the control without CD (López-Nicolás et al. 2013).

Recently, we have found that the true limiting factor of in vitro growth of *T. claveryi* is not nutrients, nor growing conditions, but the deficiency in certain growth factors, such as vitamins involved in glucose catabolic pathways, that the fungus may not be able to synthesize. Thus, an assay involving response surface methodology was performed using Box-Behnken design to find the optimal parameters for the high production of *T. claveryi* mycelial biomass (Arenas et al. 2018). The best results were obtained with glucose as carbon source, buffering the pH at 5 during culture, adding a pool of vitamins and adjusting the optimal concentrations of carbon and nitrogen sources of the MMN medium to 15 and 0.6 g L^{-1}, respectively. Biomass production of strain T7 in the bioreactor increased from 0.3 to 3 g L^{-1} dry weight, and productivity increased from 10.7 to 95.8 mg L^{-1} day^{-1} dry weight, thus providing a suitable amount of mycelium for large-scale mycorrhizal inoculation (Arenas et al. 2018).

Generally, the mycorrhizal symbiosis involves other soil microorganisms that are in most cases beneficial or even necessary for a right development of mycorrhizal plants (Azcón 2014). Of these microorganisms, the most studied are PGPR bacteria, a heterogeneous group of soil bacteria that can stimulate plant growth, protect them from diseases or increase their production (Bhattacharyya and Jha 2012).

In order to identify native PGPR associated with desert truffles, Navarro-Ródenas et al. (2016) isolated bacteria from the mycorrhizosphere of *H. almeriense* roots and from the peridium of *T. claveryi*, which were characterized according to several PGPR traits (auxin and siderophore productions, phosphorus solubilization and ACC deaminase activity). Furthermore, the effect of some of these bacteria at different stages of the production of mycorrhizal plants with desert truffles was evaluated. After a phylogenetic analysis of the 16S rDNA, 64 bacterial colonies

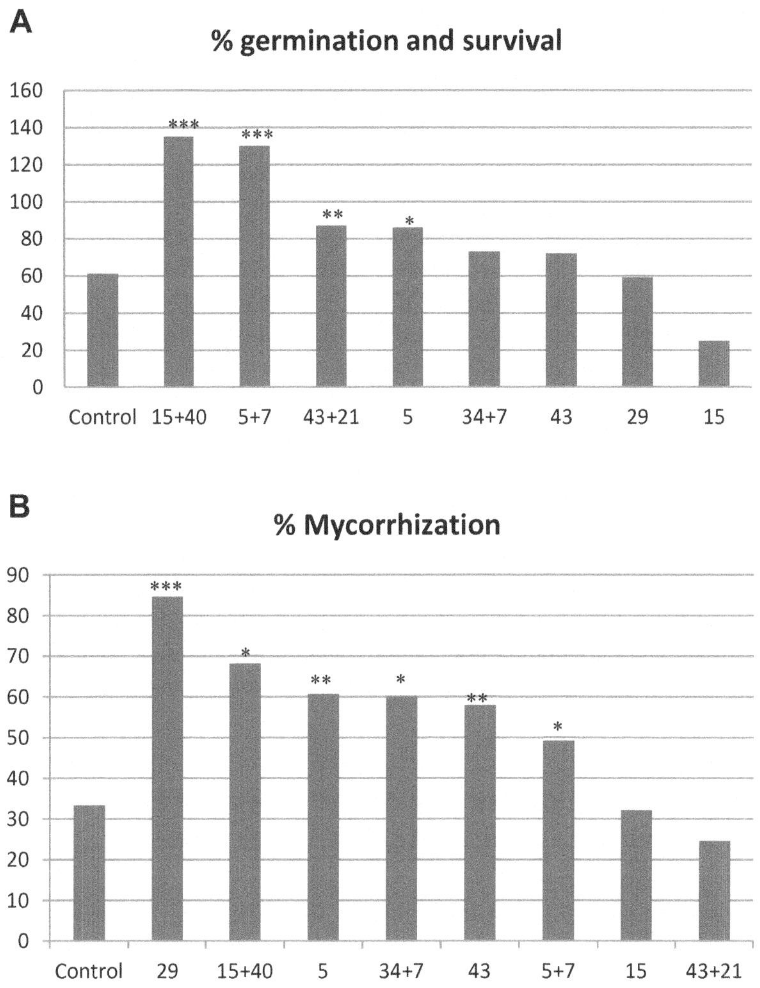

Fig. 7.4 Germination and survival (**a**) and mycorrhization (**b**) percentages of *H. almeriense* inoculated with different combinations of bacteria strains after 4 weeks in nursery conditions. 5 = *Pseudomonas fluorescens*; 15 = *Flavobacterium*; 29 = *Ps. mandelii*; 43 = *Arthrobacter* sp.; 5 + 7 = *Ps. fluorescens* + *Pseudomonas* sp.; 34 + 7 = *Ps. brenneri* + *Pseudomonas* sp.; 15 + 40 = *Flavobacterium* sp. + *Pseudomonas* sp.; 43 + 21 = *Arthrobacter* sp. + *Pseudomonas* sp. $*P \leq 0.05$ $**P \leq 0.01$ $***P \leq 0.001$; significant difference in comparison with control according to ANOVA test

were identified that were grouped in 45 strains. The 45 strains belong to 17 genera: *Achromobacter*, *Acinetobacter*, *Arthrobacter*, *Bacillus*, *Flavobacterium*, *Microbacterium*, *Microvirga*, *Novosphingobium*, *Paenibacillus*, *Phyllobacterium*, *Pseudomonas*, *Rhizobium*, *Rhodococcus*, *Sinorhizobium*, *Sphingomonas*, *Stenotrophomonas* and *Variovorax*. The most abundant genera were *Pseudomonas* (40.8%), *Bacillus* (12.2%) and *Variovorax* (8.2%). The rest of the genera did not exceed 5% of the total of strains. From the 45 strains, seven (15.6%) presented IAA (indoleacetic acid) production, ten (24.4%) produced siderophores, nine (20.0%) were able to solubilize phosphate, one strain (2.2%) showed ACC deaminase production and only one strain of *Pseudomonas fluorescens* showed three activities (P solubilizer, IAA producer and ACC deaminase activity).

Some of these bacteria had a positive effect on seed germination and plant survival, with an increase of 40–122% in comparison with the treatment without bacteria, depending on the bacterial treatment (Fig. 7.4a). The IAA-producing bacteria were particularly relevant during the mycorrhization stage increasing the root-stem ratio and colonization percentage by 47–154% in comparison with plants without bacterial inoculation, depending on the bacterial treatment (Fig. 7.4b).

Moreover, one strain of *Pseudomonas mandelii* was able to considerably increase mycorrhizal colonization but not the plant growth and could be considered as a mycorrhiza helper bacteria (MHB) (Navarro-Ródenas et al. 2016). Further studies have demonstrated that the mycorrhization percentage in plants inoculated with *Ps. mandelii* (40%) or in plants inoculated with bacteria in combination with a drought treatment (60%) was higher than control plants (20%) without bacteria inoculation or water stress (Fig. 7.5a) (Espinosa-Nicolás 2017). Moreover, this bacterium is able to mitigate the negative effect of water stress, maintaining both shoot water potentials of drought stressed and control plants at similar levels (Fig. 7.5b) (Espinosa-Nicolás 2017). *Pseudomonas mandelii* also improved the mycorrhization by *T. claveryi* in other *Helianthemum* species such as *H. violaceum* (Martínez-Ballesteros 2018).

7.4 Ecophysiological and Molecular Aspects of Desert Truffle Mycorrhizal Symbiosis Against Water Stress

In the case of desert truffles, it is essential to study the group of mechanisms that regulate the tolerance of mycorrhizal symbiosis to water stress. *Helianthemum almeriense* mycorrhizal plants with *T. claveryi* presented higher survival in drought conditions, as well as higher transpiration rates, stomatal conductance and photosynthesis than non-mycorrhizal plants, both in drought and irrigation conditions. Under water stress conditions, mycorrhizal plants accumulated higher N, P and K than non-mycorrhizal plants, which shows that the fungus is able to absorb more nutrients from the soil in drought conditions, helping the plant to keep better water and physiological levels (Morte et al. 2000). In another similar study, using desert truffle mycorrhizal plants in field conditions, it was observed that the water stress increased significantly the mycorrhizal colonization percentage up to 70%, while irrigated plants did not exceed 48% of mycorrhization (Morte et al. 2010). Furthermore, water stress induced a change in the mycorrhizal type formed, increasing intracellular colonization under drought stress (Navarro-Ródenas et al. 2013). Results obtained showed that *H. almeriense* mycorrhizal plants with *T. claveryi* maintain good physi-

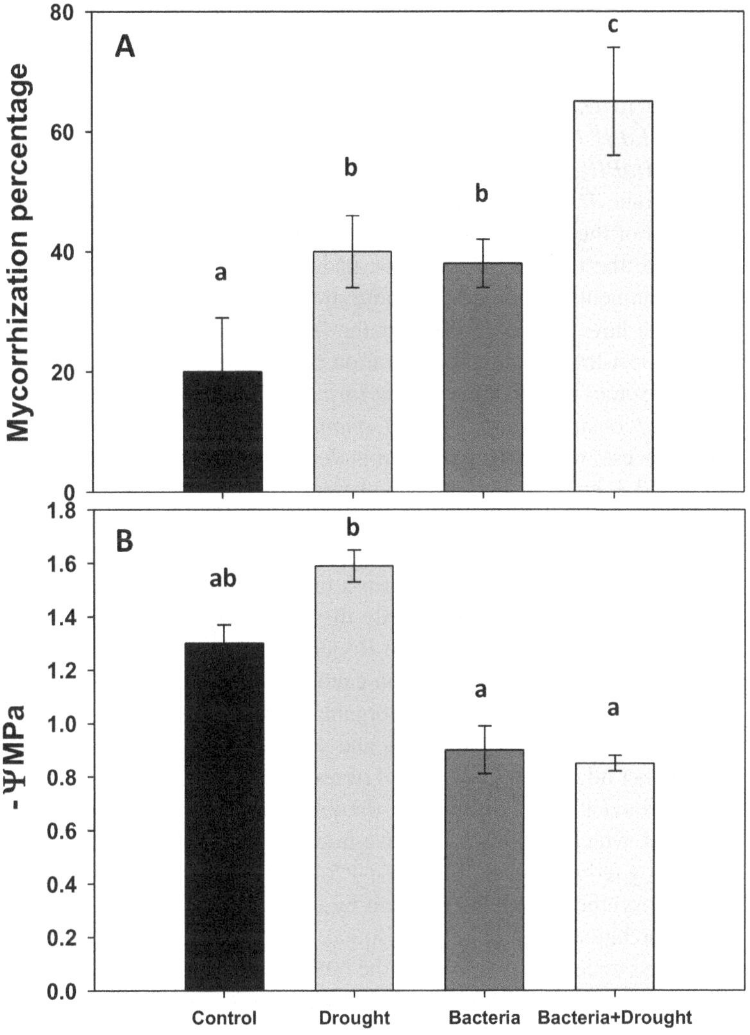

Fig. 7.5 Effect of MHB inoculation (*Pseudomonas mandelii*) and drought treatment on mycorrhization percentage (**a**) and plant water potential (**b**) of mycorrhizal *H. almeriense* plants with *T. claveryi* in nursery conditions

ological parameters with low soil matrix potential, which makes this symbiosis an alternative agricultural crop in arid and semiarid areas (Morte et al. 2010).

The molecular bases that regulate the efficient use of water in this symbiosis were analysed through the studies of aquaporin (Navarro-Ródenas et al. 2013). Aquaporins are membrane channels present in all biological kingdoms, which facilitate and regulate the passive movement of water and other small apolar molecules (Zardoya 2005). Depending on the biological group, the same organisms may present one or two aquaporin isoforms (fungi and bacteria) and up to 70 (in some plants) according to Park et al. (2010). Based on their amino acid homology, subcellular location and selectivity of different substrates, aquaporins tend to divide in different types: PIP, TIP, NIP, SIP and XIP, and the latter in other sub-groups. Five aquaporin isoforms were identified in *H. almeriense*: two belonging to the PIP1 group (*HaPIP1;1* JF49134 and *HaPIP1;1* JF491350), two belonging to the PIP2 group (*HaPIP2;1* JF491351 and *HaPIP2;2* JF491352) and one belonging to the TIP group (*HaTIP1;1* HQ234609). Some *H. almeriense* aquaporins, from leaves and/or roots, present a fine adjustment of the expression, as a function of photosynthesis or stomatal conductance, where the number of isoforms, under photosynthesis control, is greater when the environmental conditions get more stressful (Navarro-Ródenas et al. 2013).

Recently, we have also described for the first time in the Fungi Kingdom the existence of a post-transcriptional maturation of the LSU rRNA, both in ascocarps and mycelia of some species of the genera *Terfezia* (*T. claveryi*, *T. arenaria*, *T. boudieri*, *T. extremadurensis*, *T. fanfani*) and *Tirmania* (*T. nivea*) (Navarro-Ródenas et al. 2018). This process, which results in two molecules of 1.6 and 1.8 kb, corresponding to the 5′ and 3′ ends, has never been observed in fungi, although it seems to be widespread in other kingdoms, including bacteria, protozoa, worms, several arthropods, fish and rodents. During the evolution of Pezizomycetes, the introduction of sequences susceptible of cleavage to form a hidden gap must have appeared relatively late, since, as far as we know, only these two genera share the hidden gap within the class Pezizomycetes (Navarro-Ródenas et al. 2018).

The biological significance of this non-canonical post-transcriptional maturation of the rRNA LSU is still unknown in the organisms in which it has been described. It seems to be related to a better adaptation and survival in extreme conditions, correlated with a higher fidelity of translation of ribosomes, as seen in the case of the naked mole rat (Azpurua et al. 2013). In addition, the chloroplast ribosomes of an *Arabidopsis thaliana* mutant, which exhibits a defective fragmentation of chloroplast 23S rRNA, have a drastically reduced level of ribulose-1,5-bisphosphate carboxylase/oxygenase and other photosynthetic proteins encoded by chloroplasts (Nishimura et al. 2010). Another interpretation proposes a ribosomal inactivation control mechanism in response to stress conditions by dividing the 60S subunits containing 28S rRNA with a hidden gap (Nomura et al. 2016). *Terfezia* species are adapted to extreme environments and show great tolerance to drought and high temperatures (Navarro-Ródenas et al. 2011, 2013, Zambonelli et al. 2014). It is still unknown if the hidden gap of the *Terfezia* 28S rRNA influences the efficiency of translation under stress conditions and what impact it may have on its biological cycle, so it is necessary to deepen the study of the processing of the hidden gap in these desert truffles and understand if it bears any relation with the slow and erratic in vitro growth of these fungi.

Fig. 7.6 (**a**) Plantation of *Helianthemum almeriense* mycorrhizal plants with *Terfezia claveryi* in Caravaca (Murcia, Spain), (**b**) *T. claveryi* ascocarp fruiting under *H. almeriense* in April

7.5 Desert Truffle Cultivation

The first step in the establishment of a desert truffle plot is to choose suitable host plants and fungal species that are well adapted to the environmental conditions and soil characteristics (Morte et al. 2017a). Moreover, high quality mycorrhizal plants, with certified mycorrhization levels, should be selected (Honrubia et al. 2014).

Field fructifications of *T. claveryi* occur from 1 to 3 years after plantation, depending on the quality of the mycorrhizal plants, the suitability of the site, the plantation season (spring or autumn), the plantation frame and, above all, the irrigation management and the weed elimination (Morte et al. 2012). A successful plantation frame has 1.5 × 1.5 m spacing in four to five rows forming a block, with 2–3 m separation between blocks. This design produced the first ascocarps after 2 years (Fig. 7.6). The small size of these shrubs allows to arrange them closer, thus optimizing the cultivated field. This means a plantation of around 6000 plants/ha, which, while very expensive to establish, could be amortized after 5 years of cultivation if production is adequate (200–450 kg/ha) (Morte et al. 2017a).

Honrubia et al. (2014) analysed and discussed the results of 12 years of experience on desert truffle plantations and the factors that may favour truffle formation in natural production areas through "desert truffle silviculture". A successful example of this mycosilviculture in natural production areas was done in Abu Dhabi (UAE), where the production of *T. boudieri* and *T. nivea* was stimulated by spore inoculation of areas with *Helianthemum lippii* plants and by applying sprinkler irrigation and placing a fence to avoid the consumption of truffles by animals (Gouws et al. 2014).

After following the *T. claveryi* ascocarp production in a plantation with appropriate management during 15 years, a statistical correlation between the precipitation volume during autumn (September, October and November) of 1 year and the truffle production of the following year was found (Morte et al. 2012). This finding has allowed us to maintain the truffle production in dry years, adapting soil water potential to the necessary parameters to keep the mycorrhizal symbiosis productive. The average production of desert truffles in this plantation (with a plantation frame of

0.5×0.5 m) was approximately 400 kg/ha, but this production varied from 1 year to another, from a minimum of 2 kg/ha to a maximum of 1050 kg/ha obtained over the course of these 15 years (Morte et al. 2012).

Among all the agro-climatic parameters studied, the cumulative rainfall between September the first and November the 15th of the year before the collection period showed the highest correlation with the production of desert truffle throughout 15 years of study in the plantation, with a Pearson coefficient of 0.926. The next climatic parameter that showed correlation was the cumulative rainfall during March of the year of the collection period, with a Pearson coefficient of 0.643 (Morte et al. 2017b). The model that involved both precipitation period and best adjusted R^2 was a multiple linear regression model, where kg/ha = autumn precipitation∗5.11 + spring precipitation∗2.83–33.7.

Even so, more eco-physiological studies on this symbiosis are needed to evaluate a continuous production over the years, regardless of the weather. It is also necessary to identify possible competitors that, fructifying or not, can displace or cohabit with desert truffles, as well as to quantify the soil fungal biomass necessary for ascocarp formation and analyse its relationship with the vegetation cover.

As it was previously remarked, there are several productive plantations in Spain of *H. almeriense*, *H. violaceum* and *H. hirtum* mycorrhizal plants with *T. claveryi* (Fig. 7.6). Positive cultivation results have also been obtained with *T. boudieri* in Tunisia (Slama et al. 2010) and Israel (Kagan-Zur, pers. comm.). So far, all these cultivated *Terfezia* species are typical of alkaline soils. The cultivation of desert truffle species from acid soils is hindered by the fact that the host plants are annual. Numerous tests are being done to adapt perennial host plants to these soils, with edible species like *T. arenaria*, trying to benefit from previous experiences obtained on alkaline soils (Morte et al. 2017b).

Definitely, desert truffle cultivation is relatively new, with almost 20 years of experience, and of extreme complexity due to inherent aspects of fungal and plant species. The weather of the different cultivation areas is also critical. It is therefore a challenge for research to make the cultivation of desert truffles sustainable and profitable.

7.6 Conclusions

The symbiosis formed by desert truffles with Cistaceae of the genus *Helianthemum* is defined as an *ectendomycorrhizal continuum*, where same roots may present intracellular or intercellular colonization, or both, at the same time. In addition, the presence of organic phosphorus in the medium and drought conditions trigger an increase of the fungal intracellular colonization of the root, making this symbiosis tighter.

The mycelia of *T. claveryi* and *P. lefebvrei* are able to deal with a moderate water stress. At the base of this stress tolerance, there is a greater expression of aquaporin genes, both in the fungus and in the host plant of the mycorrhizal symbiosis. A high expression of fungal aquaporin is observed as the plant's photosynthesis and stomatal conductance decrease in situations of water stress. The combination of intracellular colonization together with the expression of fungal aquaporin TcAQP1 produces a morpho-physiological adaptation that favours this mycorrhizal symbiosis in arid and semiarid conditions. The beneficial bacteria of the soil, isolated from *T. claveryi* plantations, are able to increase survival and mycorrhization of nursery plants and could possibly play an important role in the biological cycle of these fungi. It is crucial to keep studying these bacteria in order to improve the field cultivation of desert truffles.

Acknowledgements This work has been partially funded by projects 20866/PI/18 (FEDER and Programa Regional de Fomento de la Investigación—Plan de Actuación 2019—de la Fundación Séneca, Agencia de Ciencia y Tecnología of the Region of Murcia, Spain) and CGL2016-78946-R (AEI/FEDER, UE). The authors thank Antonio Rodríguez for photos 1A–C and 1F–G and Maite Santiesteban Romero for the photo 2F.

References

Alsheikh A (1984) Mycorrhizae of annual *Helianthemum* species formed with desert truffles. In: Molina R (ed) Proceedings of the 6th North American conference on mycorrhizae, Oregon State University, Corvallis, Oregon, 25–29 June 1984

Andrino A, Morte A, Honrubia M (2012) Method for producing Cistaceae mycorrhized with desert truffles. Patent ES2386990, Thader Biotechnology SL, Spain

Arenas F, Navarro-Ródenas A et al (2018) Mycelium of *Terfezia claveryi* as inoculum source to produce desert truffle mycorrhizal plants. Mycorrhiza 28:691–701

Awameh MS, Alsheikh A, Al-Ghawas S (1979) Mycorrhizal synthesis between *Helianthemum ledifolium*, *H. salicifolium* and four species of the genera *Terfezia* and *Tirmania* using ascospores and mycelial cultures obtained from ascospores germination. In: Varma A (ed) Proceedings of the 4th North American conference on mycorrhizae, Colorado State University, Fort Collins, Colorado, 24–28 June 1979

Azcón R (2014) Mycorrhizosphere: the role of PGPR. In: Morte A, Varma A (eds) Root engineering: basic and applied concepts, soil biology, vol 40. Springer-Verlag, Berlin/Heidelberg, pp 107–143

Azpurua J, Ke Z, Chen IX et al (2013) Naked mole-rat has increased translational fidelity compared with the mouse, as well as a unique 28S ribosomal RNA cleavage. Proc Natl Acad Sci U S A 110(43):17350–17355

Bhattacharyya PN, Jha DK (2012) Plant growth-promoting rhizobacteria (PGPR): emergence in agriculture. World J Microbiol Biotechnol 28:1327–1350

Cano A, Honrubia M, Molina-Niñirola C (1991) Mycorrhizae in semiarid ecosystems: synthesis of mycorrhizae between *Terfezia claveryi* Chat., *Picoa juniperi* Vit. and *Helianthemum almeriense* (Cistaceae). In: Proceedings of the 3rd European symposium on mycorrhizas, University of Sheffield, Sheffield, UK, 19–23 August 1991

Chevalier G, Dupré C, Riousset L, Dexheimer J (1984) Synthese mycorhizienne entre *Terfezia leptoderma* Tul et diverses Cistaceae. Agronomie 4:210–211

Espinosa-Nicolás J (2017) Efecto de bacterias MHB en plantas micorrizadas con trufa del desierto. Bachelor thesis. University of Murcia, Spain

Fortas Z, Chevalier G (1992) Effet des conditions de culture sur la mycorrhization de l'*Helianthemum guttatum* par trois especes de terfez des genres *Terfezia* et *Tirmania* d'Algerie. Can J Bot 70:2453–2460

Gouws A, De Wet T, Abdullah F, et al (2014) Desert truffle research in U.A.E. Abstract book of second symposium on hypogeous fungi in Mediterranean basin (HYPOGES2) & fifth congress Tuber aestivum/uncinatum European Scientific Group (TAUESG5). Universitè Mohammed V. Rabat, Morocco, 9–13 April 2014

Gutiérrez A (2001) Caracterización, micorrización y cultivo en campo de las trufas de desierto. Doctoral Thesis, University of Murcia, Spain

Gutiérrez A, Morte A, Honrubia M (2003) Morphological characterization of the mycorrhiza formed by *Helianthemum almeriense* Pau with *Terfezia claveryi* Chatin and *Picoa lefebvrei* (Pat.) Maire. Mycorrhiza 13:299–307

Honrubia M, Gutiérrez A, Morte A (2001) Desert truffle plantation from south-east Spain. In: Edible mycorrhizal mushrooms and their cultivation. Proceedings of the second international conference on edible mycorrhizal mushrooms-IWEMM2, Christchurch, New Zealand, 3–6 July 2001

Honrubia M, Morte A, Gutiérrez A (2007) Las Terfezias. Un cultivo para el desarrollo rural en regiones áridas y semi-áridas. In: Reina S (coord) Truficultura. Fundamentos y Técnicas. Ediciones Mundi-Prensa, Madrid, pp 365–397

Honrubia M, Andrino A, Morte A (2014) Domestication: preparation and maintenance of plots. In: Kagan-Zur V, Roth-Bejerano N, et al (eds) Desert Truffles. Soil biology, vol 38, Chapter 22. Springer-Verlag, Berlin/Heidelberg, pp 367–387

Kagan-Zur V, Raveh E et al (1994) Initial association between *Helianthemum* and *Terfezia* is enhanced by low iron in the growth medium. New Phytol 127:567–570

Kagan-Zur V, Kuang J et al (1999) Potential verification of a host plant for the desert truffle *Terfezia pfeilii* by molecular methods. Mycol Res 103:1270–1274

Kovács G, Vágvölgyi C, Oberwinkler F (2003) *In vitro* interaction of the truffle *Terfezia terfezioides* with *Robinia pseudoacacia* and *Helianthemum ovatum*. Folia Microbiol 48:369–378

Kozai T (1991) Photoautotrophic micropropagation. In Vitro Cell Dev Biol Plant 27:47–51

López-Nicolás JM, Pérez-Gilabert M et al (2013) Mycelium growth stimulation of the desert truffle *Terfezia claveryi* Chatin by β-cyclodextrin. Biotechnol Prog 29(6):1558–1564

Majada JP, Fall MA et al (2002) Effects of natural ventilation on leaf ultrastructure of *Dianthus caryophyllus* L. cultured *in vitro*. In Vitro Cell Dev Biol Plant 38:272–278

Martínez-Ballesteros A (2018) Evaluación de los mecanismos de acción de la MHB *Pseudomonas mandelii* #29 sobre plantas micorrizadas con trufa de desierto. Bachelor thesis, University of Murcia, Spain

Marx D (1969) The influence of ectotrophic fungi on the resistance of pine roots to pathogenic infections. I. Antagonism of mycorrhizal fungi to root pathogenic fungi and soil bacteria. Phytopathology 59:153–163

Morte A, Andrino A (2014) Domestication: preparation of mycorrhizal seedlings. In: Kagan-Zur V, Roth-Bejerano N et al (eds) Desert truffles. Soil biology, vol 38, Chapter 21. Springer-Verlag, Berlin/Heidelberg, pp 343–365

Morte A, Honrubia M (2009) Biotechnology for the industrial production of ectomycorrhizal inoculum and mycorrhizal plants. In: Varma A, Verma N (eds) Text book on molecular biotechnology. I. K. International Publishing House Pvt. Ltd, New Delhi, pp 691–704

Morte A, Cano A et al (1994) *In vitro* mycorrhization of micropropagated *Helianthemum almeriense* plantlets with *Terfezia claveryi* (desert truffle). Agric Sci Finl 3:309–314

Morte A, Lovisolo C, Schubert A (2000) Effect of drought stress on growth and water relations of the mycorrhizal associations *Helianthemum almeriense-Terfezia claveryi*. Mycorrhiza 10:115–119

Morte A, Honrubia M, Gutiérrez A (2008) Biotechnology and cultivation of desert truffles. In: Varma A (ed) Mycorrhiza: genetics and molecular biology, eco-function, biotechnology, eco-physiology, structure and systematics, 3rd edn. Springer-Verlag, Berlin/Heidelberg, pp 467–483

Morte A, Navarro-Ródenas A, Nicolás E (2010) Physiological parameters of desert truffle mycorrhizal Helianthemun almeriense plants cultivated in orchards under water deficit conditions. Symbiosis 52(2):133–139

Morte A, Zamora M et al (2009) Desert truffle cultivation in semiarid Mediterranean areas. In: Azcón-Aguilar C, Barea JM et al (eds) Mycorrhizas: functional processes and ecological impact. Springer-Verlag, Berlin/Heidelberg, pp 221–234

Morte A, Andrino A et al (2012) *Terfezia* cultivation in arid and semiarid soils. In: Zambonelli A, Bonito GM (eds) Edible ectomycorrhizal mushrooms, Soil biology, vol 34. Springer-Verlag, Berlin/Heidelberg, pp 241–263

Morte A, Pérez-Gilabert M et al (2017a) Basic and applied research for desert truffle cultivation. In: Varma A, Prasad R, Tuteja N (eds) Mycorrhiza: ecophysiology, secondary metabolites, nanomaterials. Springer, Berlin, pp 23–42
Morte A, Andrino A et al (2017b) Turmicultura en España, oportunidades y necesidades del sector. In: Sociedad Española de Ciencias Forestales (ed) Gestión del monte: servicios ambientales y bioeconomía. 7CFE01-543, pp 1–9
Murashige T, Skoog F (1962) A revised medium for rapid growth and bio assays with tobacco tissue cultures. Physiol Plant 15:473–497
Murcia MA, Martínez-Tomé M et al (2003) Effect of industrial processing on desert truffles *Terfezia claveryi* Chatin and *Picoa juniperi* Vittadini: proximate composition and fatty acids. J Sci Food Agric 83:535–541
Navarro-Ródenas A, Lozano-Carrillo MC et al (2011) Effect of water stress on *in vitro* mycelium cultures of two mycorrhizal desert truffles. Mycorrhiza 21:247–253
Navarro-Ródenas A, Bárzana G et al (2013) Expression analysis of aquaporins from desert truffle mycorrhizal symbiosis reveals a fine-tuned regulation under drought. Mol Plant-Microbe Interact 26(9):1068–1078
Navarro-Ródenas A, Berná LM et al (2016) Beneficial native bacteria improve survival and mycorrhization of desert truffle mycorrhizal plants in nursery conditions. Mycorrhiza 26:769–779
Navarro-Ródenas A, Carra A, Morte A (2018) Identification of a hidden gap in 26S rRNA of desert truffles uncovers an alternative rRNA posttranscriptional maturation in the Kingdom Fungi. Front Microbiol 9:996
Nishimura K, Ashida H et al (2010) A DEAD box protein is required for formation of a hidden break in *Arabidopsis* chloroplast 23S rRNA. Plant J 63:766–777
Nomura T, Ito M, Kanamori M, Shigeno Y et al (2016) Characterization of silk gland ribosomes from a bivoltine caddisfly, *Stenopsyche marmorata*: translational suppression of a silk protein in cold conditions. Biochem Biophys Res Commun 469(2):210–215
Park W, Scheffler BE et al (2010) Identification of the family of aquaporin genes and their expression in upland cotton (*Gossypium hirsutum* L.). BMC Plant Biol 10:142
Slama A, Fortas Z et al (2010) Cultivation of an edible desert truffle (*Terfezia boudieri* Chatin). Afr J Microbiol Res 4:2350–2356
Volpato G, Rossi D, Dentoni D (2013) A reward for patience and suffering: Ethnomycology and commodification of desert truffles among Sahrawi refugees and nomads of Western Sahara. Econ Bot 67(2):147–160
Zambonelli A, Donnini D et al (2014) Hypogeous fungi in Mediterranean maquis, arid and semi-arid forests. Plant Biosyst 148(2):392–401
Zardoya R (2005) Phylogeny and evolution of the major intrinsic protein family. Biol Cell 97:397–414
Zaretsky M, Kagan-Zur V et al (2006) Analysis of mycorrhizal associations formed by *Cistus incanus* transformed root clones with *Terfezia boudieri* isolates. Plant Cell Rep 25:62–70

Part III
Case Studies

Chapter 8
Diversity, Biogeographic Distribution, Ecology, and Ectomycorrhizal Relationships of the Edible Porcini Mushrooms (*Boletus* s. str., Boletaceae) Worldwide: State of the Art and an Annotated Checklist

Matteo Gelardi

8.1 Introduction

Porcini species, also known as king boletes, penny bun, ceps, or cepes (*Boletus* s. str., Boletaceae), are among the most widely priced and sought-after wild edible mushrooms in the world, and, given the high commercial value, their trade plays an important economic role as a source of relevant income for local communities (Hall et al. 1998a; Sabra and Walter 2001; Boa 2004; Sitta et al. 2007a; Arora 2008; Sitta and Floriani 2008; Mello 2012; Mortimer et al. 2012; Feng et al. 2012; Sitta and Davoli 2012; Dentinger and Suz 2014; Cui et al. 2016).

Boletus s. str. is a group of pileate/stipitate to rarely sequestrate mushroom-forming fungi. This group is circumscribed and separated from the other boletoid genera on account of the following unifying key features: (1) medium-large to large sized, fleshy basidiomes; (2) initially white and then yellowish and, in most species, finally olive-brown hymenophore; (3) the presence of a whitish thin veil-like hyphal structure occluding tube mouth ("stuffed pores") in the first developmental stages; (4) stipe nearly always enlarged toward the base and more or less distinctly reticulate; (5) white context and basal mycelium; (6) tissues unchanging or only occasionally turning light blue or brownish upon bruising and on exposure; (7) persistent fragrant, nutty odor and mild taste; (8) olive-brown spore print; (9) ellipsoid-fusiform to occasionally ovoid basidiospores; (10) pileipellis structure variously arranged (trichoderm, ixotrichoderm, physalo-palisadoderm, hymeniform, tricho-epithelium); (11) hymenophoral trama bilateral divergent of the "*Boletus* type"; (12) fertile caulohymenium; (13) lateral stipe stratum of the "boletoid type"; (14) inamyloid hyphae with Melzer's reagent; (15) the absence of clamp connections;

M. Gelardi (✉)
Via dei Barattoli 3a, I-00061, Anguillara Sabazia (RM), Italy

J. Pérez-Moreno et al. (eds.), *Mushrooms, Humans and Nature in a Changing World*, https://doi.org/10.1007/978-3-030-37378-8_8

and (16) seemingly gymnocarpic ontogenetic development (Coker and Beers 1943; Smith and Thiers 1971; Corner 1972; Singer 1986; Muñoz 2005; Šutara 2005, 2014; Watling 2008; Cui et al. 2016). *Boletus* s. str. is monophyletic only when restricted to the species orbiting around *B. edulis* Bull.: Fr. (= *Boletus* sect. *Boletus* or *Boletus* sect. *Edules* Fr.) (Binder and Hibbett 2006; Dentinger 2007; Watling 2008; Desjardin et al. 2009; Dentinger et al. 2010; Li et al. 2011, 2014a, b; Feng et al. 2012; Nuhn et al. 2013; Arora and Frank 2014a; Wu et al. 2014, 2016; Cui et al. 2016; Gelardi et al. 2015; Halling et al. 2015; Smith et al. 2015; Henkel et al. 2016; Ortiz-Santana et al. 2016; Chakraborty et al. 2017; Orihara and Smith 2017; Farid et al. 2018; Parihar et al. 2018; Bozok et al. 2019, 2020; Crous et al. 2019; Loizides et al. 2019; Kuo and Ortiz-Santana 2020).

Several multi-locus molecular phylogenetic studies carried out across the northern hemisphere in the last 20 years, initially at regional level and later at a global scale, have revealed a number of new species-level lineages within *Boletus* s. str., and many other previously described taxa have been confirmed with robust statistical supports (Simonini et al. 2001; Moor et al. 2002; Iotti et al. 2005; Leonardi et al. 2005; Mello et al. 2006; Zhao et al. 2006; Vizzini et al. 2007; Beugelsdijk et al. 2008; Lian et al. 2008; Korhonen et al. 2009; Dentinger et al. 2010; Yan et al. 2010; Feng et al. 2012; Arora and Frank 2014b; Dentinger and Suz 2014; Fedosova and Kovalenko 2014; Zeng et al. 2014; Cui et al. 2016; Chakraborty et al. 2017; Sarwar et al. 2018; Crous et al. 2019), suggesting that this group is particularly diverse and ecologically important for forest ecosystems. Furthermore, the high number of recently detected phylogenetic species still lacking a scientific name prove that there are likely additional undescribed taxa yet to be discovered and documented, especially from temperate eastern and tropical southeastern Asia and Australasia (Feng et al. 2012; Wu et al. 2014; Cui et al. 2016).

Molecular phylogenies indicate *Boletus* s. str. to be divided into five distinct lineages currently defined with provisional names: "porcini s. str.," "*Obtextiporus*," "*Inferiboletus*," "*Alloboletus*," and "*Orientiboletus*" (Dentinger et al. 2010; Feng et al. 2012; Cui et al. 2016). "*Alloboletus*" occupies the most basal position within *Boletus* s. str. and may represent the evolutionary relict of the genus (Feng et al. 2012). "Porcini s. str." is in turn further subdivided into three phylogenetic clades, namely, "Edulis clade," "Aereus clade," and "Variipes clade," plus a fourth, yet unnamed clade which is close to the Aereus clade and consisting of three recently described Chinese species (Feng et al. 2012; Cui et al. 2016), that might be called "Botryoides clade". It is not yet clear whether these informal groups should be retained at the generic, subgeneric, or sectional rank, and further analysis on a more inclusive number of taxa will be required in order to elucidate their definitive taxonomic placement.

8.2 Materials and Methods

The following checklist is divided into three different sections according to the taxonomic status of the treated species: (1) species validly published and with molecular phylogenetic support; (2) species validly published but without molecular phylogenetic support; and (3) species not yet validly published. In this latter section, species preceded by an asterisk ("*") have not been sequenced to date. Names in bold refer to taxa considered valid at the specific rank (either confirmed or not by molecular analyses). Names in normal refer to species, subspecies, varieties, and forms related to the pertinent species listed above. The symbol "≡" refers to homotypic synonyms, while the symbol "=" refers to heterotypic synonyms. Invalid taxa are preceded by an en dash ("–"). Taxa in normal preceded by a question mark ("?") are dubitatively considered synonyms of the aforementioned species in bold face. Wherever the basionym is not indicated, it is understood its coincidence with the main taxon in bold face. Nomenclatural issues follow the rules of the ICN (International Code of Nomenclature for algae, fungi, and plants, Shenzhen Code 2018). All taxa are enumerated in alphabetic order.

8.3 Results

8.3.1 A Checklist of Porcini Mushrooms

8.3.1.1 Species Validly Published and Molecularly Supported

- ***Boletus aereus*** Bull., Herb. France 9: tab. 385. 1789: Fr. Syst. Mycol. 1: 393. 1821 (Fig. 8.1a).

  ≡ *Tubiporus aereus* (Bull.: Fr.) P. Karst., Bidr. Känn. Finl. Nat. Folk 37: 5. 1882.
  ≡ *Dictyopus aereus* (Bull.: Fr.) Quél., Enchir. Fung.: 159. 1886.
  ≡ *Suillus aereus* (Bull.: Fr.) Kuntze, Revis. Gen. Pl.: 535. 1898.
  ≡ *Tubiporus aereus* (Bull.: Fr.) Ricken, Vadem. Pilzf.: 206. 1918.
  ≡ *Boletus edulis* subsp. *aereus* (Bull.: Fr.) Konrad & Maubl., Icon. Sel. Fung. 6: 454. 1935.
  ≡ *Tubiporus edulis* subsp. *aereus* (Bull.: Fr.) Maire, Publ. Instit. Botàn. 3: 45. 1935.
  ≡ *Boletus edulis* f. *aereus* (Bull.: Fr.) Vasilkov, Belyj Grib: 19. 1966.
  = *Tubiporus ustulatus* Paulet, Traité Champ. 2: 368, pl. 168. 1793.
  = *Boletus cepa* Thore, Essai Chloris: 482. 1803.
  = *Boletus aereus* var. *cepa* (Thore) DC., in de Candolle & Lamarck, Fl. Franç., Edn 3. 6: 42. 1815.
  = *Boletus aereus* var. *leucoporus* Pers., Mycol. Eur. 2: 138. 1825.
  = *Boletus sykorae* Smotlacha, Čas. Českoslov. Houb. 15: 19. 1935.
  = *Boletus edulis* f. *subaereus* Vasilkov, Belyj Grib: 19. 1966.
  = *Boletus subaereus* Pilát, Bull. Soc. Mycol. France 86 (4): 881. 1971.
  = *Boletus mamorensis* Redeuilh, Bull. Soc. Mycol. France 94 (3): 303. 1978.
  = *Boletus aereus* var. *mamorensis* (Redeuilh) Guinberteau, Bull. FAMM 55 (1): 26. 2019.
  – *Boletus edulis* f. *aereus* (Bull.: Fr.) Vasilkov, Out. Geogr. Inv. Cap fungi U.S.S.R.: 32. 1955. (nom. inval., art. 41.5, basionym not cited)

? – *Boletus aereus* var. *squarrosus* Rezende-Pinto, Brotéria N.S. 9 (3): 93. 1940. (nom. inval., art. 39.1, no Latin diagnosis)

Nomenclatural comments: "*Boletus aeneus* Bull." in Rostkovius (Deutschl. Fl., *in* Sturm, Die Pilze Deutschl. 3: 109. 1844) is an evident typographical error for *Boletus aereus* Bull.: Fr.

- Known distribution: Europe (mainly southern, including Corsica, Sardinia, Naxos, Cyprus Islands, etc.), northern Africa (Morocco, Algeria), Asia Minor and Middle East (Turkey, Israel), and southwestern and central Asia (Ukraine, Russia, Azerbaijan and Kazakhstan), exotic in South Africa.
- Habitat and mycorrhizal association: *B. aereus* is a xero-thermophilic southern species mainly restricted to the Mediterranean Basin or to warm temperate regions on dry soil; forming ECM with broad-leaved trees and Mediterranean shrubs such as *Quercus* spp., including evergreen *Q. ilex* L., *Q. suber* L., *Q. alnifolia* Poech, and probably also *Q. coccifera* L., *Castanea sativa* L., *Fagus sylvatica* L. (Fagaceae), *Carpinus betulus* L. (Betulaceae), and *Tilia cordata* Mill. (Tiliaceae); in Mediterranean low maquis type with *Erica arborea* L., *Arbutus unedo* L., and *A. andrachne* L. (Ericaceae), *Juniperus* spp. (Cupressaceae), and *Cistus* spp. and *Halimium lasianthum* Spach (Cistaceae), occasionally with *Betula aetnensis* Raf. (Betulaceae) and *Pinus nigra* subsp. *laricio* (Poir.) Maire (Pinaceae) and possibly also with exotic *Eucalyptus* spp. (Myrtaceae); seldom above 1000 m alt. (up to 1500 m alt.).
- Relevant bibliography (selected): Singer (1967), Pilát and Dermek (1974), Alessio (1978, 1985), Redeuilh (1978 as "*B. mamorensis*"), Engel et al. (1983), Redeuilh and Simonini (1993), Lannoy and Estadès (2001), Estadès and Lannoy (2004), Muñoz (2005), van der Linde and Noordeloos (2005), Watling and Hills (2005), Calzada (2007), Klofac (2007), Knudsen and Taylor (2008), Šutara et al. (2009), Manavella (2010), Kibby (2011), Fedosova and Kovalenko (2014), Rodà (2012), Galli (2013), Vasquez (2014), Mikšík (2017), Noordeloos et al. (2018), Guinberteau (2019 as "*B. aereus* var. *mamorensis*"), Loizides et al. (2019).
- Notes: Information about the occurrence of *B. aereus* (commonly known as the "black king bolete") in northern Africa can be found in Bertault (1979), El-Assfouri et al. (2006), Bakkali et al. (2009), Haimed et al. (2013, 2015), Nounsi et al. (2014), and Outcoumit et al. (2014). *B. aereus* also occurs in Azerbaijan (Mustafabayli and Aghayeva 2019) and appears to be uncommon to rare in Ukraine and Russia (Akulov and Leontyev 2008; Heluta 2012; Fedosova and Kovalenko 2014), it has also been introduced in South Africa (Goldman and Gryzenhout 2019). All records of this species outside the geographic range reported above (i.e., in Australia; see Watling and Gregory 1988, 1989; Watling and Li 1999) should be critically assessed. Moreover, any reports of *B. aereus* from the Pacific Coast of western North America (Smith 1975; Thiers 1975; Arora 1986, 1991; Bessette et al. 2000) must be referred to *B. regineus*.
- *Boletus sykorae* Smotlacha from the Czech Republic (Smotlacha 1935) is just a later synonym of *B. aereus*.

- Pilát (1970) described *B. subaereus* Pilát from the Czech Republic; however, it is nothing more than a phenotype of *B. aereus* devoid or nearly so of the reticulum on the stipe.

- *Boletus mamorensis* Redeuilh (Redeuilh 1978; Outcoumit et al. 2006) has been confirmed as contaxic with *B. aereus* by means of molecular analysis (Dentinger et al. 2010; Guinberteau 2019; Loizides et al. 2019).

- De Rezende-Pinto (1940) invalidly described *B. aereus* var. *squarrosus* from Brazil; however, based on the available data, it is unclear whether this infraspecific taxon is a member of *Boletus* s. str. and is therefore only tentatively included as a synonym of *B. aereus*.
- ***Boletus albobrunnescens*** Desjardin, Dentinger & D. Arora Mycologia 106 (4): 830. 2014.
- Known distribution: Southeastern Asia (Thailand).
- Habitat and mycorrhizal association: Occurring in tropical environment in mixed evergreen and deciduous forests with *Pinus* spp. (Pinaceae) and *Castanopsis* spp. (Fagaceae).
- Relevant bibliography: Pukahuta et al. (2009, as "*Boletus pallidus*"), Halling et al. (2014).
- Notes: This species was previously misidentified as *Boletus pallidus* Frost by Pukahuta et al. (2009).
- ***Boletus austroedulis*** Halling & Fechner Mycologia 106 (4): 832. 2014 (Fig. 8.1b).
- Known distribution: Eastern Australia (Queensland).
- Habitat and mycorrhizal association: Occurring with Myrtaceae, especially *Corymbia intermedia* (R.T. Baker) K.D. Hill & L.A.S. Johnson and *Eucalyptus* spp., *Allocasuarina torulosa* (Aiton) L.A.S. Johnson (Casuarinaceae), and *Acacia* spp. (Fabaceae).
- Relevant bibliography: Halling et al. (2014).
- Notes: *B. austroedulis* is the only member of *Boletus* s. str. native to Australia known to date (Halling et al. 2014) and one of the very few endemic species known to occur in the southern hemisphere.
- ***Boletus bainiugan*** Dentinger *in* Dentinger & Suz Index Fungorum 29: 1. 2013. (Corresponding to ***Boletus* sp. nov. 2** in Dentinger et al. 2010 and ***Boletus* sp. 6** in Feng et al. 2012) (Fig. 8.1c).
- Known distribution: Central and southwestern China (Henan, Sichuan, and Yunnan Provinces).
- Habitat and mycorrhizal association: Occurring in subtropical environment in mixed forests with *Pinus yunnanensis* Franch., *P. kesiya* var. *langbianensis* (A. Chev.) Silba (Pinaceae), and *Castanea mollissima* Blume (Fagaceae).
- Relevant bibliography: Wang et al. (2004, as "*B. edulis*"), Dentinger (2013), Dentinger and Suz (2014), Cui et al. (2016).
- Notes: Microscopic features reported in the original description by Dentinger (2013) are wrong since they were taken with a 40× objective (Dentinger in litt.). The color image of *Boletus* sp. in Sitta et al. (2007a) might well be *B. bainiugan*.
- ***Boletus barrowsii*** Thiers & A.H. Smith Mycotaxon 3 (2): 262. 1976.

- Known distribution: North America, west of Rocky Mountains (western Canada as far north as British Columbia and western USA, reported from Washington, Oregon, Idaho, Colorado, California, Arizona, New Mexico south to Mexico).
- Habitat and mycorrhizal association: Occurring in temperate montane coniferous forests with *Pinus ponderosa* Douglas ex C. Lawson, *Abies* spp., and *Picea* spp. (Pinaceae) but also in mixed woods in warm Pacific coastal regions under evergreen *Quercus agrifolia* Née (Fagaceae) and *Arbutus menziesii* Pursh (Ericaceae).
- Relevant bibliography: Thiers (1976), Arora (1986, 1991, 2008), States (1990), Phillips (1991), Both (1993), Evenson (1997), Bessette et al. (2000), Miller and Miller (2006), Kuo (2007), Trudell and Ammirati (2009), Scates (2011), Camacho and Bruns (2008), Siegel and Schwarz (2016).
- Notes: Popularly known in the western USA as the "white king bolete," *B. barrowsii* has recently turned out to be a collective species and should therefore be split in two different, independent taxa, one occurring in western North America (*B. barrowsii* s. str.), the other in eastern North America (see next species) (Feng et al. 2012).
- Information about the occurrence of *B. barrowsii* from Mexico is found in Villareal and Perez-Moreno (1989) and García-Jiménez (2013). The species has also been reported from Pakistan (Sarwar and Khalid 2014), but, as for many other American species said to occur in the Old World, even in this case, the identification is likely mistaken.
- ***Boletus botryoides*** B. Feng, Yang Y. Cui, J.P. Xu & Zhu L. Yang Fungal Div. 81 (1): 195. 2016. (Corresponding to ***Boletus* sp. nov. 2** in Feng et al. 2012)
- Known distribution: Central China (Hunan Province).
- Habitat and mycorrhizal association: Occurring in subtropical environment with members of the Fagaceae.
- Relevant bibliography: Cui et al. (2016).
- ***Boletus edulis*** Bull., Herb. Fr. 2: tab. 60. 1782: Fr. Syst. Mycol. 1: 392. 1821 (Fig. 8.1d,e and f).

≡ *Leccinum edule* (Bull.: Fr.) Gray, Nat. Arr. Brit. Pl. 1: 647. 1821.
≡ *Tubiporus edulis* (Bull.: Fr.) P. Karst., Rev. Mycol. 3: 16. 1881.
≡ *Dictyopus edulis* (Bull.: Fr.) Quél., Enchir. Fung.: 159. 1886.
= *Boletus bulbosus* Schaeff., Fung. Bavar. Palat. Nasc. 4: 87. 1774.
= *Suillus bulbosus* (Schaeff.) Kuntze, Revis. Gen. Pl. 3 (2): 535. 1898.
= *Boletus elephantinus* With., Arr. Brit. Pl. (Edn. 3) 4: 277. 1776.
= *Boletus edulis* var. *elephantinus* (With.) Pers., Syn. Meth. Fung. 2: 510. 1801.
= *Leccinum elephantinum* (With.) Gray, Nat. Arr. Brit. Pl. 1: 648. 1821.
= *Boletus edulis* var. *tuberosus* Pers., Syn. Meth. Fung. 2: 511. 1801.
= *Boletus edulis* var. *communis* Alb. & Schwein., Consp. Fung.: 240. 1805.
= *Tubiporus esculentus* Paulet, Icon. Champ.: pl. 168 bis, fig. 1. 1808-1835.
= *Tubiporus albus* Paulet, Icon. Champ.: pl. 177, fig. 1. 1808-1835.
= *Boletus albus* (Paulet) Pers., Traité Champ. Com.: 233. 1818. (nom. illeg., art. 53.1, non Schaeff., 1774).

= *Boletus esculentus* var. *albus* (Paulet) Pers., Mycol. Eur. 2: 132. 1825.
= *Boletus edulis* var. *albus* (Paulet) E.-J. Gilbert, Les Bolets 3: 153. 1931.
= *Boletus edulis* f. *albus* (Paulet) J.A. Muñoz, Fungi Europ. 2: 314. 2005.
= *Boletus persoonii* Bon, Doc. Mycol. 19 (74): 61. 1988. [nom. nov. based on *Boletus esculentus* var. *albus* (Paulet) Pers.]
= *Boletus solidus* Sowerby, Col. Fig. Engl. Fung. Mushr. 4: tab. 419. 1809.
= *Boletus citrinus* A. Venturi, I Miceti dell'Agro Bresciano: 45, tab. 59, fig. 1. 1860. (nom. illeg., art. 53.1, non Withering, 1776).
= *Suillus citrinus* Kuntze, Rev. Gen. Pl. 3 (2): 535. 1898. (nom. nov., art. 58.1).
= *Boletus edulis* f. *citrinus* (Kuntze) J.A. Muñoz, Fungi Europ. 2: 316. 2005. (nom. illeg., art. 53.1, non Vasilkov, 1966).
= *Boletus venturii* Bon, Doc. Mycol. 17 (65): 51. 1986. (nom. nov. based on *Boletus citrinus* A. Venturi).
= *Boletus filiae* Gill., Tab. Anal. Hym.: 143. 1884.
= *Gyrodon filiae* (Gill.) Sacc. & Cub., Syll. Fung. 6: 53. 1888.
? = *Boletus crassus* Battarra ex Jacz., Opred. Grib. 1: 594. 1913. (nom. illeg., art. 53.1, non Massee, 1892).
= *Boletus edulis* var. *clavipes* Peck, Bull. N.Y. St. Mus. 2 (8): 133. 1889.
= *Boletus edulis* subsp. *clavipes* (Peck) Singer, Am. Midl. Nat. 37: 25. 1947.
= *Boletus clavipes* (Peck) Pilát & Dermek, Hríbovite Huby: 97. 1974.
= *Boletus edulis* var. *laevipes* Massee, Brit. Fung. Fl. 1: 284. 1892.
= *Boletus edulis* var. *pseudopurpureus* J. Murr, Öst. Bot. Z. 67 (10–12): 348. 1918.
= *Boletus edulis* f. *pseudopurpureus* (J. Murr) Vasilkov, Belyj Grib: 15. 1966.
= *Boletus edulis* var. *abietis* Schiemek, Pilz- und Kräuterfreund 4: 170. 1921.
? = *Boletus stellenbossiensis* van der Byl, South Afr. J. Sci. 22: 167. 1925.
= *Boletus edulis* var. *citrinus* E.-J. Gilbert, Les Bolets 3: 153. 1931.
= *Boletus edulis* f. *citrinus* (E.-J. Gilbert) Vasilkov, Belyj Grib: 13. 1966.
= *Boletus edulis* subsp. *euedulis* Maire, Treb. Mus. Ciènc. Nat. Barcelona, sér. bot. 15 (2): 41. 1933.
= *Boletus slovenicus* Smotl., Čas. Československ. Houb. 28 (7–8): 104. 1951.
? = *Boletus edulis* subsp. *aurantioruber* E.A. Dick & Snell, Mycologia 57 (3): 454. 1965.
? = *Boletus edulis* f. *aurantioruber* (E.A. Dick & Snell) Vasilkov, Belyj Grib: 15. 1966.
? = *Boletus aurantioruber* (E.A. Dick & Snell) Both, Bessette & W.J. Neill, Mushrooms of Cape Cod and the National Seashore: 32. 2001.
= *Boletus edulis* f. *quercicola* Vasilkov, Belyj Grib: 12. 1966.
= *Boletus quercicola* (Vasilkov) Singer, Sydowia 30 (1–6): 257. 1978.
= *Boletus edulis* f. *betulicola* Vasilkov, Belyj Grib: 13. 1966.
= *Boletus betulicola* (Vasilkov) Pilát & Dermek, Hríbovite Huby: 96. 1974.
= *Boletus edulis* f. *praecox* Vasilkov, Belyj Grib: 14. 1966.
= *Boletus edulis* f. *arcticus* Vasilkov, Belyj Grib: 16. 1966.
= *Boletus edulis* f. *roseipes* Vasilkov, Belyj Grib: 16. 1966.
= *Boletus edulis* f. *tardus* Vasilkov, Belyj Grib: 17. 1966.
= *Boletus carpinicola* Pilát, Česká Mykol. 21 (2): 61. 1967.

= *Boletus chippewaensis* A.H. Sm. & Thiers, Boletes of Michigan: 366. 1971.
= *Boletus edulis* var. *ochraceus* A.H. Sm. & Thiers, Boletes of Michigan: 373. 1971.
= *Boletus edulis* subsp. *trisporus* Watling, Notes R. Bot. Gdn. Edinb. 33 (2): 326. 1974.
= *Boletus edulis* var. *arenarius* H. Engel, Krieglst. & Dermek in Engel et al., Dickröhrlinge. Die Gattung Boletus in Europa: 45. 1983.
= *Boletus edulis* var. *pusteriensis* Ferrarese & Simonini, Micol. Veget. Medit. 11 (2): 132. 1996.
= *Boletus fulvomaculatus* Estadès & Lannoy, Doc. Mycol. 31 (121): 60. 2001.
= *Boletus edulis* f. *fulvomaculatus* (Estadès & Lannoy) Lécuru, Index Fungorum 384: 1. 2019.
= *Boletus edulis* var. *grandedulis* D. Arora & Simonini, Econ. Bot. 62 (3): 374. 2008.
= *Boletus pinetorum* M. Korhonen, Karstenia 49 (2): 44. 2009.
? – *Ceriomyces crassus* Battarra, Fung. Arim. Hist.: 62, pl. 29, fig. A-B. 1775. [no binomial system].
– *Boletus edulis* var. *piceicola* Vasilkov, Ed. Pois. fung. Europ. Part USSR: 40. 1948. (nom. inval., art. 39.1, no Latin diagnosis)
– *Boletus edulis* var. *quercicola* Vasilkov, Ed. Pois. Fung. Europ. Part USSR: 40. 1948. (nom. inval., art. 39.1, no Latin diagnosis)
– *Boletus edulis* var. *betulicola* Vasilkov, Ed. Pois. Fung. Europ. Part USSR: 46. 1948. (nom. inval., art. 39.1, no Latin diagnosis)
– *Boletus edulis* f. *laevipes* (Massee) Vasilkov, Belyj Grib: 13. 1966. (nom. inval., art. 41.5, basionym not cited)
? – *Boletus edulis* [var. *aurantioruber*] f. *roseus* H.V. Smith & A.H. Smith, Non-Gilled Fleshy Fungi: 251. 1973. (nom. inval., art. 39.1, no Latin diagnosis)
– *Boletus edulis* subsp. *betulicola* ("betulicolus") (Vasilkov) Hlaváček, Mykol. Sborn. 71 (1): 8. 1994. (nom. inval., art. 41.5, basionym not cited)
– *Boletus edulis* subsp. *slovenicus* (Smotl.) Hlaváček, Mykol. Sborn. 71 (1): 6. 1994. (nom. inval., art. 41.5, basionym not cited)
– *Boletus edulis* var. *arcticus* (Vasilkov) Hlaváček, Mykol. Sborn. 71 (1): 9. 1994. (nom. inval., art. 41.5, basionym not cited)
– *Boletus reticulatus* var. *citrinus* (A. Venturi) Hlaváček, Mykol. Sborn. 71 (4): 113. 1994. (nom. inval., art. 41.5, basionym not cited)
– *Boletus reticulatus* var. *albus* (Pers.) Hlaváček, Mykol. Sborn. 71 (4): 114. 1994. (nom. inval., art. 41.5, basionym not cited)
– *Boletus edulis* f. *nigricans* Lannoy & Estadès, Les bolets, Mém Hors Sér. 6: 96. 2001. (nom. inval., art. 39.1, no Latin diagnosis)

- Known distribution: Apparently throughout North America (Canada, the USA including Alaska, Mexico), Europe, northern Africa (Morocco, Algeria), Asia Minor and Middle East (Turkey, Israel), and Asia (Russia, including Siberia, northern and northeastern China including Inner Mongolia, Heilongjiang, and Jilin Provinces, Japan), reported, although not molecularly confirmed, from Guatemala, Pakistan, Iran, Nepal, India, and Thailand, exotic in Chile, sub-Saharan Africa (South Africa, Swaziland, Zimbabwe, Kenya, Malawi, Mozambique), Australia, New Zealand, and perhaps central South America (Peru and Brazil).

- Habitat and mycorrhizal association: *B. edulis* is a widespread and quite common species with a preference for hygrophilous and cool temperate, mesophilic woodlands on acid soil but also extending into boreal and subarctic environments in the northern hemisphere; forming ECM with broad-leaved and conifer trees such as *Quercus* spp. [including deciduous *Q. lobata* Née, *Q. robur* L., *Q. petraea* (Mattuschka) Liebl., *Q. cerris* L., *Q. pyrenaica* Willd., *Q. rubra* L., etc., and live *Q. suber*, *Q. ilex*, *Q. agrifolia*, and *Q. parvula* var. *shrevei* (C.H. Mull.) Nixon], *Castanea sativa*, *Fagus sylvatica* (Fagaceae), *Corylus avellana* L., *Carpinus betulus*, *Betula* spp. (*B. aetnensis*, *B. alba* L., *B. ermanii* Cham., *B. papyrifera* Marshall, *B. pendula* Roth) (Betulaceae), *Tilia cordata* (Tiliaceae) and probably *Fraxinus* spp. (Oleaceae); *Picea* spp. [*P. abies* (L.) H. Karst., *P. sitchensis* (Bong.) Carr.], *Abies* spp. (*A. alba* Mill., *A. cephalonica* Loudon), *Pinus* spp. [*P. attenuata* Lemmon, *P. contorta* Douglas ex Loudon, *P. densiflora* Siebold & Zucc., *P. hartwegii* Lindl., *P. muricata* D. Don, *P. nigra* J.F. Arnold, *P. patula* Schiede ex Schltdl. & Cham., *P. pinea* L., *P. radiata* D. Don, *P. strobus* L., *P. sylvestris* L. s.l., *P. thunbergii* Parl. and probably also *P. mugo* Turra, *P. cembra* L., *P. uncinata* Mill., *P. resinosa* Sol. ex Aiton, *P. banksiana* Lamb., etc.], *Tsuga canadensis* (L.) Carrière and *Pseudotsuga menziesii* (Mirb.) Franco (Pinaceae), *Erica arborea* L. (Ericaceae), and *Populus trichocarpa* Torr. & A.Gray ex. Hook. (Salicaceae) (Camacho and Bruns 2008), exceptionally with *Cistus* spp., *Helianthemum nummularium* (L.) Mill., and *Halimium lasianthum* (Cistaceae) in coastal sand dunal habitats in southern Europe (Lavorato 1991; Águeda et al. 2006, 2008; Oria de Rueda et al. 2008; Ponce et al. 2011; Mediavilla et al. 2016), with *Arbutus menziesii* (Ericaceae) along the Pacific West Coast in North America (Siegel and Schwarz 2016), and with exotic *Eucalyptus* spp. plantations (Myrtaceae) in Mediterranean environment (Alvarez 2008; pers obs); up to 2500 m altitude with *Arctostaphylos uva-ursi* (L.) Spreng. (Ericaceae), *Salix herbacea* L. and *S. repens* L. (Salicaceae) in European alpine meadows and in the arctic belt (Watling 2002; Manavella 2007) or even at higher altitudes in North America (Hall et al. 2003).
- Relevant bibliography (selected): Coker and Beers (1943), Singer (1947), Dick and Snell (1965, as "*B. edulis* subsp. *aurantioruber*"), Vasilkov (1966), Singer (1967), Snell and Dick (1970), Smith and Thiers (1971), Pilát and Dermek (1974), Grund and Harrison (1976), Pegler and Young (1981), Alessio (1978, 1985, 1991), Zerova and Rozhenko (1988), Breitenbach and Kränzlin (1991), Phillips (1991), Both (1993), Redeuilh and Simonini (1993), Bessette et al. (2000, 2016), Lannoy and Estadès (2001), Roody (2003), Estadès and Lannoy (2004), Muñoz (2005), van der Linde and Noordeloos (2005), Watling and Hills (2005), McNeil (2006), Calzada (2007), Klofac (2007), Manavella (2007, 2010), Arora (2008), Camacho and Bruns (2008), Knudsen and Taylor (2008), Korhonen et al. (2009), Šutara et al. (2009), Kuo (2010c), Kibby (2011), Mello (2012),

Rodà (2012), Galli (2013), Endo et al. (2014), Fedosova and Kovalenko (2014), Ivanov (2014), Vasquez (2014), Cui et al. (2016), Siegel and Schwarz (2016), Mikšík (2017), Noordeloos et al. (2018).

- Notes: This iconic bolete is the type species of the genus *Boletus* and appears to be holarctic, being widespread all over the northern hemisphere. According to the present state of knowledge and supported by molecular phylogenetic analysis (Dentinger et al. 2010; Feng et al. 2012; Fedosova and Kovalenko 2014; Cui et al. 2016), it is undoubtedly the porcini mushroom with the widest geographic distribution range.
- The occurrence of *B. edulis* in northern Africa, Turkey and Israel has been signaled by Binyamini (1975), Bertault (1979), El-Assfouri et al. (2006), Larouz (2007), Haimed et al. (2013), Nounsi et al. (2014), Outcoumit et al. (2014), Ben Hassine Ben Ali and Stephenson (2016b), and Bozok et al. (2020). The species has also been reported from Guatemala (Bran et al. 2002; Flores Arzù et al. 2002, 2012), Pakistan (Razaq and Shahzad 2013; Razaq et al. 2014; Sarwar and Khalid 2014), Iran (Asef 2008), Nepal (Adhikari and Durrieu 1996; Giri and Rana 2007), India (Boruah et al. 1996; Lakhanpal 1996; Adhikary et al. 1999; Anand et al. 2014; Wani et al. 2013 as "*Boletus edulu*"; Das and Das 2014; Karun and Sridhar 2014; Verma and Pandro 2018), Thailand (Soytong 1994; Chandrasrikul 1996; Thongklam 2008), and New Guinea (Hongo 1973), but in all cases, molecular confirmation is required. Furthermore, reports on the occurrence of *B. edulis* from tropical China (Chiu 1948; Bi et al. 1990, 1994, 1997; Zang 1995, 2006; Li and Song 2000; Wu et al. 2011) appear to be unlikely. *Boletus edulis* is allochthonous and associated with artificial exotic pine or broadleaved plantations in Peru (Boa 2004), Brazil (Sobestiansky 2005; Sulzbacher et al. 2013; Putzke and Putzke 2019), Chile (Chung et al. 2010), Kenya (Hall et al. 1998b; Mello et al. 2006), Malawi (Boa 2004), Mozambique (Uaciquete et al. 1996; Boa 2004), South Africa, Swaziland, and Zimbabwe (Pearson 1950; Stephens and Kidd 1953; Marais and Kotzè 1977; van der Westhuizen and Eicker 1994; Masuka 1996; Mello et al. 2006; Gryzenhout 2010; Coetzee 2011; Goldman and Gryzenhout 2019; Dentinger in litt.), Australia (Catcheside and Catcheside 2012), and New Zealand (Wang et al. 1995; Stringer et al. 2002; Orlovich et al. 2004; Wang and Hall 2004; Mello et al. 2006; Horak pers. comm.).
- *Boletus edulis* has been separated into dozens distinct infraspecific taxa over time, such as subspecies, varieties, and forms, mainly correlated to particular morphological (size, shape), chromatic (depigmented, albinotic, xanthoid, etc.), ecological (related to ECM host tree or soil composition), or aberrant (i.e., *B. edulis* var. *pusteriensis*) morphotypes. Molecular studies (Simonini et al. 2001; Moor et al. 2002; Iotti et al. 2005; Leonardi et al. 2005; Mello et al. 2006; Zhao et al. 2006; Vizzini et al. 2007; Beugelsdijk et al. 2008; Dentinger et al. 2010) indicate these taxa to represent mere variability within *B. edulis* and, accordingly, not to deserve recognition at species level unlike previously supposed.
- The North American species *B. clavipes* and *B. chippewaensis* have been molecularly demonstrated as later heterotypic synonyms of *B. edulis* (Dentinger et al.

2010). Likewise, *B. aurantioruber* appeared to be conspecific with *B. edulis* in Dentinger et al. (2010); however, judging from the phylogenetic tree shown in more recent publications (Nuhn et al. 2013; Halling et al. 2015; Crous et al. 2019), this taxon would rather be an independent species. More investigation and sequence data are required to solve the matter (Halling pers. comm.).

- *Boletus slovenicus* Smotlacha from the Czech Republic (Smotlacha 1951) appears to be another albinotic, slender form of *B. edulis*.
- The real identity of *B. stellenbossiensis* from South Africa (van der Bijl 1925; Doidge 1950; Pearson 1950; Reid 1973; Coetzee 2011) is still uncertain, and it is not to be excluded that this species will in the future be merged with *B. edulis*.
- ***Boletus fagacicola*** B. Feng, Yang Y. Cui, J.P. Xu & Zhu L. Yang Fungal Div.: 81 (1): 197. 2016. (Corresponding to **Boletus sp. 9** in Feng et al. 2012)
- Known distribution: Central, southern, and southwestern China (Yunnan, Hunan, and Hainan Provinces).
- Habitat and mycorrhizal association: Occurring in subtropical broad-leaved forests with members of the Fagaceae.
- Relevant bibliography: Cui et al. (2016).
- ***Boletus fibrillosus*** Thiers California Mushrooms: 45. 1975 (Fig. 8.1g).
- Known distribution: North America, west of Rocky Mountains (southwestern Canada and western USA, British Columbia south to Washington, Oregon and northern California and east to Montana, Idaho, and Wyoming).
- Habitat and mycorrhizal association: Occurring in temperate environment in coastal and montane Pacific coniferous or dense mixed forests with *Pseudotsuga menziesii* (Mirb.) Franco, *Abies* spp. [*A. grandis* (Douglas ex D. Don) Lindley], *Pinus* spp., *Picea* spp., *Tsuga* spp. (Pinaceae), and *Quercus* spp. (Fagaceae).
- Relevant bibliography: Thiers (1975), Arora (1986, 2008), Both (1993), Bessette et al. (2000 as "*Boletus mottiae*"), Scates (2011), Siegel and Schwarz (2016).
- Notes: As already pointed out by Arora (2008), the original description of *B. fibrillosus* by Thiers (1975) provided misleading information concerning the color of the pores, which were mistakenly defined as yellow from the very first developmental stage of the basidiomes, determining subsequent erroneous interpretations and classification of this uncommon species.
- ***Boletus gertrudiae*** Peck Bull. N. Y. State Mus. 150: 50. 1911.
- Known distribution: North America, east of Rocky Mountains (southeastern Canada and eastern USA, New York south to North Carolina, west to Indiana and Tennessee, western limits not yet established, perhaps also in Mexico).
- Habitat and mycorrhizal association: Occurring in temperate environment in broad-leaved forests or in mixed woods under *Quercus* spp. and *Fagus* spp. (*F. grandifolia* Ehrh.) (Fagaceae).
- Relevant bibliography: Peck (1911), Snell (1936), Snell and Dick (1970), Smith and Thiers (1971), Smith and Smith (1973), Singer (1977), Both (1993), Bessette et al. (2000, 2016).
- Notes: The alliance of *B. gertrudiae* with *B. separans* and *B. violaceofuscus* within the "*Alloboletus*" clade was already inferred by Simonini et al. (2001) and

later strengthened by Desjardin et al. (2009, 2011), Nuhn et al. (2013), and Trappe et al. (2013).

- The species has been reported from Mexico (García-Jiménz 2013). Likewise, as many other American species, *B. gertrudiae* has been also cited from India (Lakhanpal 1996) and various times from China (Chiu 1948, 1957; Li and Song 2000; Wang 2004; Zang 2006; Dai et al. 2010), but the name seems to have been misapplied.
- ***Boletus griseiceps*** B. Feng, Yang Y. Cui, J.P. Xu & Zhu L. Yang Fungal Div. 81 (1): 198. 2016. (Corresponding to collection **HKAS 71346** in Feng et al. 2012)
- Known distribution: Southeastern China (Fujian Province).
- Habitat and mycorrhizal association: Occurring with members of the Fagaceae in subtropical environment.
- Relevant bibliography: Cui et al. (2016).
- ***Boletus himalayensis*** S. Jabeen, S. Sarwar & A.N. Khalid Turkish J. Bot. 42: 2. 2018.
- Known distribution: Pakistan.
- Habitat and mycorrhizal association: Forming ECM with subalpine coniferous trees such as *Cedrus deodara* (Roxb. ex D. Don) G. Don and *Pinus wallichiana* A.B. Jacks. (Pinaceae).
- Relevant bibliography: Sarwar et al. (2018).
- ***Boletus hiratsukae*** Nagasawa Proc. Japan Acad. 70, Ser. B: 10. 1994 (Fig. 8.1h).
- Known distribution: Eastern Asia (eastern Russia, Japan).
- Habitat and mycorrhizal association: Occurring in temperate environment in mixed coniferous and deciduous forests with *Abies* spp. (*A. firma* Siebold & Zucc., *A. sachalinensis* F. Schmidt), *Pinus* spp. (*P. densiflora*, *P. thunbergii*), *Picea jezoensis* var. *hondoensis* (Mayr) Rehder (Pinaceae), *Castanopsis cuspidata* (Thunb.) Schottky, and *Quercus serrata* Thunb. (Fagaceae), also with the presence of *Celtis sinensis* var. *japonica* (Planch.) Nakai and *Larix kaempferi* (Lamb.) Carr.
- Relevant bibliography: Nagasawa (1994), Fedosova and Kovalenko (2014).
- ***Boletus indoedulis*** D. Chakraborty, K. Das, A. Baghela, S. Adhikari & Halling Nova Hedwigia 105 (1–2): 4. 2017.
- Known distribution: India (Sikkim Himalaya).
- Habitat and mycorrhizal association: Forming ECM with *Lithocarpus* spp. (Fagaceae).
- Relevant bibliography: Chakraborty et al. (2017).
- ***Boletus meiweiniuganjun*** Dentinger *in* Dentinger & Suz Index Fungorum 29: 1. 2013. (Corresponding to ***Boletus* sp. nov. 3** in Dentinger et al. 2010 and ***Boletus* sp. 7** in Feng et al. 2012)
- Known distribution: Northeastern and southwestern China (Heilongjiang, Yunnan, and perhaps Inner Mongolia Provinces) and Korea.
- Habitat and mycorrhizal association: Forming ECM with *Pinus* spp. (*P. yunnanensis*, *P. kesiya* var. *langbianensis*), *Abies* spp. (Pinaceae), and possibly *Quercus* spp. (Fagaceae).
- Relevant bibliography: Dentinger (2013), Dentinger and Suz (2014).

- Notes: Molecularly recognized by Feng et al. (2012) and formally described by Dentinger (2013), this species has later been treated as a synonym of *B. bainiugan* by Cui et al. (2016). No morphological characters can actually be used to distingiush *B. meiweiniuganjun* from *B. bainiugan* (Feng, in litt.). Further insights on the taxonomic status of *B. meiweiniuganjun* are required.
- Microscopic features reported in the original description by Dentinger (2013) are wrong since they were taken with a 40× objective (Dentinger, in litt.).
- Basidiomes of a porcini species putatively referable to *B. meiweiniuganjun* have also been recorded from Inner Mongolia (Sitta in litt.).
- ***Boletus monilifer*** B. Feng, Yang Y. Cui, J.P. Xu & Zhu L. Yang Fungal Div. 81 (1): 199. 2016. (Corresponding to ***Boletus* sp. 1** in Feng et al. 2012)
- Known distribution: Southwestern China (Yunnan Province).
- Habitat and mycorrhizal association: Forming ECM with broad-leaved trees such as *Lithocarpus* spp., *Castanopsis* spp., and *Quercus* spp. (Fagaceae) in subtropical montane environment.
- Relevant bibliography: Cui et al. (2016).
- ***Boletus nobilis*** Peck Bull. N. Y. State Mus. 94 (8): 48. 1905 (Fig. 8.1i).
- Known distribution: North America, east of Rocky Mountains (southeastern Canada, eastern USA west to Michigan and Illinois, south to Mississippi and Florida). Distribution limits presently unclear.
- Habitat and mycorrhizal association: Occurring in temperate environment in open deciduous forests with *Fagus* spp. and *Quercus* spp. (Fagaceae).
- Relevant bibliography: Peck (1905), Snell and Dick (1970), Smith and Thiers (1971 as "*B. separans*"), Smith and Smith (1973 as "*B. separans*"), Grund and Harrison (1976 as "*B. separans*"), Smith et al. (1981 as "*B. separans*"), Both (1993), Bessette et al. (2000, 2016, 2019), Roody (2003).
- Notes: Smith and Thiers (1971) relegated this species in synonymy with *B. separans*, but molecular analysis (Dentinger et al. 2010; Feng et al. 2012; Nuhn et al. 2013) has revealed the autonomy of the two species.
- *B. nobilis* has recently been reported from Thailand (Thongklam 2008) but was likely misidentified.
- ***Boletus nobilissimus*** Both & R. Riedel North American Boletes: 132. 2000.
- Known distribution: North America, east of Rocky Mountains (northeastern USA, New England, New York, Pennsylvania, west to Wisconsin and Minnesota, south to North Carolina).
- Habitat and mycorrhizal association: Occurring in temperate environment in mixed broad-leaved and coniferous forests under *Quercus rubra* L. (Fagaceae) with the presence of *Pinus strobus* L. (Pinaceae).
- Relevant bibliography: Bessette et al. (2000, 2016).
- Notes: Presently reported only from the type locality in the Western New York state and from New England, this species is reasonably widespread throughout northeastern North America (see mycoportal.org).
- *B. nobilissimus*, however, is morphologically very similar to *B. atkinsonii*, and it is not to be excluded that the two species could be the same (since the latter has

not been sequenced to date) or at most separate but very closely related to each other.

- ***Boletus orientialbus*** N.K. Zeng & Zhu L. Yang Mycoscience 55 (3): 160. 2014 (Fig. 8.1j) (Corresponding to ***Alloboletus* sp. 2** in Feng et al. 2012).
- Known distribution: Southeastern China (Fujian Province).
- Habitat and mycorrhizal association: Occurring in subtropical environment with *Lithocarpus* spp. and *Castanopsis* spp. (Fagaceae).
- Relevant bibliography: Huang (1999 as "*Boletus albus*"), Zeng et al. (2014), Cui et al. (2016).
- Notes: Cui et al. (2016) refer to this species as "*Boletus* sp. 14" in Feng et al. (2012); however, there appears to be no "*Boletus* sp. 14" in the latter publication.
- ***Boletus pinophilus*** Pilát & Dermek Ceská Mykol. 27 (1): 6. 1973 (Fig. 8.1k).

= *Boletus edulis* var. *pinicola* Vittad., Fung. Mang.: 168. 1835.
= *Boletus pinicola* (Vittad.) A. Venturi, I Miceti dell'Agro Bresciano: 39. 1860. (nom. illegit., art. 53.1) [non Swartz 1810 = *Fomitopsis marginata* (Pers.: Fr.) P. Karst. 1881].
= *Boletus aestivalis* var. *pinicola* (Vittad.) Sacc., Syll. Fung. 19: 154. 1910.
= *Boletus edulis* subsp. *pinicola* (Vittad.) Konrad & Maubl., Icon. Sel. Fung. 6: pl. 453. 1935.
= *Tubiporus edulis* subsp. *pinicola* (Vittad.) Maire, Publ. Inst. Bot. 3 (4): 45. 1937.
= *Boletus edulis* f. *pinicola* (Vittad.) Vasilkov, Belyj Grib: 14. 1966.
? = *Boletus rubiginosus* Fr., Obs. Mycol. 2: 245. 1818. (nom. illeg., art. 53.1) (non Retzius 1769).
= *Dyctiopus edulis* var. *fuscoruber* Quél., Compt. Rend. Assoc. Franç. Avancem. Sci. 18 (2): 5. 1890.
= *Boletus edulis* var. *fuscoruber* (Quél.) Bataille, Bolets: 14. 1908.
= *Boletus pinophilus* f. *fuscoruber* (Forq. in Quél.) Estadès & Lannoy, Doc. Mycol. 31 (121): 60. 2001.
= *Boletus edulis* subsp. *subhepaticus* Fayod, Ann. R. Soc. Agr. Torino 35: 104. 1893.
? = *Boletus reticulatus* var. *rubiginosus* Pelt. in E.-J. Gilbert, Les Bolets 3: 154. 1931.
= *Boletus vinosulus* Kluzák, Papoušek & Šutara, Sbor. Jihočes. Muz. v Čes. Budějovicích Přir. Vědy 32: 43. 1992.
= *Boletus pinophilus* var. *viridicaerulescens* Estadès & Lannoy, Doc. Mycol. 31 (121): 60. 2001.
– *Boletus pinicola* var. *fuscoruber* (Quél.) Angarano, Boll. Gr. Mic. Bres. 30 (1–6): 25. 1987. (nom. inval., art. 41.5, basionym not cited)
– *Boletus pinophilus* var. *fuscoruber* (Quél.) Cetto, Fung. Vero 1: 264. 1987. (nom. inval., art. 41.5, basionym not cited)

- Known distribution: Europe, Asia Minor (Turkey), southwestern Asia (Ukraine, Russia), exotic in eastern North America, Guatemala, Chile, and New Zealand, apparently also in Mexico and South Africa.

- Habitat and mycorrhizal association: widespread but less common than *B. edulis* and *B. reticulatus*, growing preferably in hygrophilous and cool, mesophilic habitat on acid soil; forming ECM with broad-leaved trees such as *Quercus* spp., *Castanea sativa*, *Fagus sylvatica* (Fagaceae), and maybe *Carpinus* spp. and *Betula* spp. (Betulaceae) but especially with conifers like *Pinus* spp. (*Pinus cembra*, *P. nigra*, *P. nigra* subsp. *laricio*, *P. pinaster* Aiton, *P. radiata*, *P. sylvestris*, *P. uncinata*, etc.), *Picea abies*, and *Abies alba* (Pinaceae).
- Relevant bibliography (selected): Singer (1967), Pilát and Dermek (1974), Alessio (1978, 1985), Engel et al. (1983), Breitenbach and Kränzlin (1991), Kluzák (1992 as "*B. vinosulus*"), Redeuilh and Simonini (1993, 1999), Lannoy and Estadès (2001), Estadès and Lannoy (2004), Muñoz (2005), van der Linde and Noordeloos (2005), Watling and Hills (2005), Calzada (2007), Klofac (2007), Knudsen and Taylor (2008), Šutara et al. (2009), Manavella (2010), Kibby (2011), Rodà (2012), Galli (2013), Fedosova and Kovalenko (2014), Ivanov (2014), Vasquez (2014), Mikšík (2017), Noordeloos et al. (2018).
- Notes: Known as the "red king bolete," *B. pinophilus* is the least common and culinary appreciated species among the European porcini mushrooms. Its occurrence seems to be restricted to temperate Europe; however, it has also been reported (either under the name *B. pinicola* or *B. pinophilus*) from Asia Minor (Bozok et al. 2020), western Asia (Heluta 2012; Fedosova and Kovalenko 2014; Ivanov 2014) and in association with exotic pine trees (such as *P. sylvestris*) from eastern North America (Both 1993), Mexico (Ibarra et al. 1999; Montoya-Esquivel et al. 2001, 2012; Hernández and López 2006; García-Jiménez 2013), Guatemala (Bran et al. 2003a; Che and Flores Arzù 2010; Flores Arzù et al. 2012), Chile (Chung et al. 2010), South Africa (Pearson 1950; Stephens and Kidd 1953; Coetzee 2011), and New Zealand (Horak pers. comm.). Reports of this taxon from North America (Bessette et al. 2000) and China (Mao 2000; Zang 2006; Thongklam 2008) seem to be mistaken according to recent morphological and molecular investigations (Arora 2008; Dentinger et al. 2010; Feng et al. 2012).
- The binomial *Boletus pinicola* (Vittad.) A. Venturi is illegitimate since it is a later homonym of the validly published *B. pinicola* Sw. (art. 53.1), now *Fomitopsis marginata* (Pers.: Fr.) P. Karsten.
- *Boletus vinosulus* Kluzák, Papoušek & Šutara from the Czech Republic (Kluzák 1992) is undoubtedly the same species as *B. pinophilus*. The synonymy has also been confirmed by Josef Šutara, one of *B. vinosulus*' authors (Mikšík in litt.).
- ***Boletus pseudopinophilus*** A.R. Bessette, Bessette, J. Craine & J.L. Frank Persoonia 42: 387. 2019.
- Known distribution: North America, east of Rocky Mountains (southeastern USA, from Virginia south and west into Georgia, Mississippi, and Texas). Distribution limits yet to be assessed.
- Habitat and mycorrhizal association: Occurring in temperate environment under *Pinus elliottii* Engelm. and *P. palustris* Mill. (Pinaceae).
- Relevant bibliography: Weber & Smith 1985; as "*Boletus pinophilus*"; Bessette et al. 2000, 2007 both as "*B. pinophilus*," 2016, 2019), Crous et al. (2019).

- Notes: *B. pseudopinophilus* appears to be molecularly different from the morphologically similar European *B. pinophilus* (Bessette et al. 2016, 2019; Crous et al. 2019) and with the latter species has been previously confused by American authors (Weber & Smith 1985; Bessette et al. 2000, 2007).
- ***Boletus quercophilus*** Halling & G.M. Mueller Mycologia 91 (5): 898. 1999 (Fig. 8.2l).
- Known distribution: Central America (Costa Rica).
- Habitat and mycorrhizal association: Forming ECM with *Quercus* spp. (*Q. copeyensis* C.H. Mull., *Q. seemannii* Liebm., etc.) (Fagaceae) in tropical montane environment up to 3000 m alt.
- Relevant bibliography: Halling and Mueller (1999, 2005).
- Notes: As for *B. luteoloincrustatus*, this species might be widespread in neotropical montane cloud forests of Central America dominated by Fagaceae, but its distribution limits are presently unknown.
- ***Boletus regineus*** D. Arora & Simonini Econ. Bot. 62 (3): 374. 2008 (Fig. 8.2m).
- Known distribution: North America, west of Rocky Mountains (from central California north to Oregon and Washington).
- Habitat and mycorrhizal association: Occurring mainly in the coastal Pacific regions with hardwoods such as *Notholithocarpus densiflorus* (Hook. & Arn.) Manos, Cannon & S.H. Oh, *Chrysolepis chrysophylla* (Dougl. ex Hook.) Hjelmq., and *Quercus* spp. (*Fagaceae*) but also with *Arbutus menziesii*, *Arctostaphylos* spp. (Ericaceae), and more rarely conifers (Pinaceae).
- Relevant bibliography: Smith (1975), Thiers (1975), Arora (1986, 1991, 2008), Bessette et al. (2000) (all as "*Boletus aereus*"), Scates (2011), Siegel and Schwarz (2016).
- Notes: Known along the Pacific coast as the "queen bolete," *B. regineus* has long been misidentified by American authors as the European *B. aereus*, but the two species are definitely distinct based on morphological and molecular analysis (Arora 2008; Dentinger et al. 2010; Feng et al. 2012).
- Camacho and Bruns (2008) provided a description of *B. mottiae* which is probably to be referred to *B. regineus*.
- ***Boletus reticulatus*** Schaeffer Fung. Bavar. Palat. Nasc. 4: 78. 1774 (Fig. 8.2n). [non *B. reticulatus* (Hoffm.: Fr.) Pers. 1801 (nom. illeg., art. 53.1)]

  ≡ *Suillus reticulatus* (Schaeff.) Kuntze, Revis. Gen. Pl., 3 (2): 535. 1898.
  ≡ *Boletus edulis* var. *reticulatus* (Schaeff.) Bataille, Bolets: 14. 1908.
  ≡ *Boletus edulis* subsp. *reticulatus* (Schaeff.) Konrad & Maubl., Icon. Sel. Fung. 4 (fasc. 2): pl. 398. 1926.
  ≡ *Tubiporus edulis* ssp. *reticulata* (Schaeff.) Maire, Publ. Instit. Botan. 3 (4): 45. 1937.
  ≡ *Boletus edulis* f. *reticulatus* (Schaeff.) Vasilkov, Belyj Grib: 18. 1966.
  ? = *Boletus mutabilis* Batsch, Elenchus Fung.: 99. 1783.
  ? = *Boletus subfuscus* With., Bot. Arr. Brit. Pl., Edn 3 (London) 2: 405. 1792.
  = *Boletus reticulatus* var. *minor* Alb. & Schwein., Consp. Fung.: 240. 1805.
  = *Tubiporus aestivalis* Paulet, Icon. Champ.: pl. 170. 1808-1835.
  = *Boletus aestivalis* (Paulet) Fr., Epicr. Syst. Mycol.: 422. 1838.

Fig. 8.1 (**a**) *Boletus aereus* (**b**) *Boletus austroedulis* (photo R.E. Halling) (**c**) *Boletus* cf. *bainiugan* (photo N. Sitta) (**d**) *Boletus edulis* (**e**) *Boletus edulis* f. *citrinus* (**f**) *Boletus edulis* f. *albus* (**g**) *Boletus fibrillosus* (photo J.L. Frank) (**h**) *Boletus hiratsuke* holotypus (photo E. Nagasawa) (**i**) *Boletus nobilis* (photo A.E. Besette) (**j**) *Boletus orientialbus* (photo N.K. Zeng) (**k**) *Boletus pinophilus*. Photos by M. Gelardi unless otherwise stated

= *Versipellis aestivalis* (Paulet) Quél., Enchir. Fung.: 158. 1886.
= *Suillus aestivalis* (Paulet) Kuntze, Revis. Gen. Pl. 3 (2): 535. 1898.
= *Boletus luteus* Ventenat, Hist. champ. France: index, t. 4. 1812. (nom. illeg., art. 53.1).
= *Boletus carpinaceus* Velen., Novitates Mycologicae: 158. 1939.
? = *Boletus reticulatus* var. *xanthoporus* Loup & Poluzzi, Scheiz. Z. Pilzk. 8: 128. 1939.
– *Boletus septimus* Schaeff., Fung. Bavar. Palat. Nasc. 2: tab. 108. 1763. (nom. inval., art. 23.6, see ex 13).
– *Boletus reticulatus* subsp. *carpinaceus* (Velen.) Hlaváček, Mykologický Sborník 71 (2): 54. 1994. (nom. inval., art. 41.5, basionym not cited)
– *Boletus subreticulatipes* Mendaza & Ruiz, Orden Boletales en España, Tomo 1: 58. 1997. (nom. inval. art. 39.1, no Latin diagnosis)

- Known distribution: Europe (including Cyprus and Azores Islands), northern Africa (Morocco, Algeria, Tunisia), Asia Minor and Middle East (Turkey, Jordan, Israel), southwestern and central Asia (Ukraine, Russia, and most likely Azerbaijan and Kazakhstan), exotic in South Africa.
- Habitat and mycorrhizal association: *B. reticulatus* is a widespread xerothermophilic species commonly found in warm temperate regions of the Mediterranean Basin with a preference for acid soil; forming ECM with broadleaved trees, deciduous and evergreen *Quercus* spp., *Castanea sativa*, *Fagus sylvatica* (Fagaceae), *Corylus avellana*, *Carpinus betulus*, *Betula pendula* (Betulaceae), *Tilia cordata* (Tiliaceae) and perhaps *Alnus glutinosa* (L.) Gaertn. (Betulaceae), *Juniperus communis* L. (Cupressaceae), and perhaps *Cistus* spp. (Cistaceae), occasionally with conifers such as *Picea abies* (L.) H. Karst., *Abies alba* Mill., *Pinus sylvestris* (Pinaceae), and probably other pine species, like the introduced Mexican species *P. patula* Schiede ex Schltdl. & Cham. in South Africa; at least in Italy and Spain possibly with exotic *Eucalyptus* spp. plantations (especially *E. camaldulensis* Dehnh. and *E. globulus* Labill.) (Myrtaceae) (Alvarez 2008; pers obs); up to 2350 m altitude with *Arctostaphylos uva-ursi* (Ericaceae) in Italian northwestern alpine meadows (Manavella 2007).
- Relevant bibliography (selected): Singer (1967), Pilát and Dermek (1974), Alessio (1978, 1985), Breitenbach and Kränzlin (1991), Redeuilh and Simonini (1993, 1999), Lannoy and Estadès (2001), Estadès and Lannoy (2004), Muñoz (2005), van der Linde and Noordeloos (2005), Watling and Hills (2005), Calzada (2007), Klofac (2007), Manavella (2007, 2010), Knudsen and Taylor (2008), Šutara et al. (2009), Kibby (2011), Rodà (2012), Galli (2013), Fedosova and Kovalenko (2014), Ivanov (2014), Vasquez (2014), Cancarè (2015), Mikšík (2017), Noordeloos et al. (2018).
- Notes: There still exist conflicting nomenclatural views between European mycologists about the use of binomials *B. aestivalis* (Paulet) Fr. 1838 vs *B. reticulatus* Schaeff. 1774 as the correct name to apply to the present species. *Boletus reticulatus* Schaeff. could be treated as doubtful and has been considered by several authors (i.e., Redeuilh 1988; Redeuilh and Simonini 1993; Lannoy and

Estadès 2001; Muñoz 2005; Galli 2013) not applicable to a porcini mushroom since the original diagnosis by Schaeffer described a cyanescent context. Indeed, Alessio (1991) pointed out that such a character is reported in the German diagnosis but not in the Latin one (Schaeffer provided a double diagnosis, in German and Latin, for all species described in his work). Moreover, the bluing context is not visible in the accompanying original plate 108 (Schaeffer 1763), or, whenever detectable, it is so faint to be irrelevant (Alessio 1991). Fries (1821: 392) treated *B. reticulatus* as a variant of *B. edulis*, suggesting that Schaeffer's plate represented for him a non-bluing taxon and that he considered it as a member of the porcini mushroom consortium. There also exists the homonym *B. reticulatus* (Hoffm.: Fr.) Pers. 1801, based on *Mucilago reticulata* Hoffm. 1796: Fr. ["*Mucilaga*" in the text, but *Mucilago* in the illustration], now *Ceriporia reticulata* (Hoffm.: Fr.) Domański 1963, combined by Fries (1815) as *Polyporus reticulatus* (Hoffm.: Fr.) Fr. The latter name was sanctioned by himself in 1821. According to the rules of the ICN, Fries sanctioning protects the name resulting from Persoon's combination only against heterotypic synonyms but not against heterotypic homonyms as is the case of *B. reticulatus* Schaeff. As a result, *B. reticulatus* (Hoffm.: Fr.) Pers. is an illegitimate combination (art. 53.1), since *B. reticulatus* (Hoffm.: Fr.) Pers. ["*reticulata*"] is not a sanctioned name and is a later homonym of Schaeffer's name. *B. reticulatus* Schaeff. is therefore a prior and legitimate name.

- On the other hand, the genus *Tubiporus* Paulet (1808) was considered an invalid nomen nudum (Donk 1955). However, Donk's statement is wrong because *Tubiporus* is a valid although illegitimate name since Paulet erected it as a substitute for *Boletus* L. As a matter of fact, he wrote on page 45 of his Prospectus (1808): "...on a prefère...celui de *Tubiporus* (pour dire à tubes et à pores), à la place du suillus de Micheli ou du boletus de Linné" [...we prefer... that of *Tubiporus* (meaning with tubes and pores) instead of *Suillus* Micheli or *Boletus* Linné], making, as a result, an original diagnosis unnecessary. Moreover, in agreement with Redeuilh (1990: 31), we believe that Paulet indeed provided a brief original diagnosis by writing "*Tubiporus*, champignons poreux ou tubuleux, cèpe" [*Tubiporus*, mushrooms with tubes or pores, ceps] to distinguish it from other genera. Soon after, he validly and legitimately published the name *Tubiporus aestivalis* Paulet (1808–1835) in his Iconographie des champignons. Paulet linked the plate to his earlier work Traité des champignons (1793) with a reference to page 371 of the second volume where the description of this taxon is found under the vernacular name "Le grand Mouffeux." Finally, Fries (1838) combined the epithet under *Boletus* as *B. aestivalis* (Paulet) Fr. However, it might be considered a doubtful name when referred to the current interpretation of this species, since in Paulet's plate the stipe is smooth and no mention is done on the presence of a reticulum in the original description.
- Consequently, since both *B. reticulatus* and *B. aestivalis* are valid and legitimate names from the nomenclatural viewpoint but with discordant or ambiguous taxonomic interpretations, we currently refrain from calling this species *B. aestivalis*,

and, despite not accepted by some authors, we prefer to use the older binomial *B. reticulatus* to name it.

- Information about the occurrence of the "summer king bolete" in northern Africa can be retrieved in Bertault (1979), Larouz (2007), Nounsi et al. (2014), Outcoumit et al. (2014), Haimed et al. (2015), and Ben Hassine Ben Ali and Stephenson (2016a, b).
- Records from Azerbaijan under the names *B. edulis* var. *arenarius* and *B. varipes* (Mustafabayli and Aghayeva 2019) are almost certainly to be referred to *B. reticulatus*, whereas Fedorenko (2019) reported this species from Kazakhstan, which would currently make up the easternmost geographic limit for *B. reticulatus*.
- Extralimital records of this species, either under the name *B. reticulatus* or *B. aestivalis*, include northeastern USA (Snell and Dick 1970; Bessette et al. 2000), Mexico (Hernández and López 2006), South Africa (van der Westhuizen and Eicker 1994; Coetzee 2011; Goldman and Gryzenhout 2019), Pakistan (Razaq and Shahzad 2013; Karun and Sridhar 2014; Sarwar and Khalid 2014), China (including Taiwan) (Yeh and Chen 1981; Li and Song 2000; Zang 2006; Mao 2009; Dai et al. 2010), Japan (Yamanaka et al. 2000), India (Lakhanpal and Sagar 1989; Lakhanpal 1996; Anand et al. 2014), Thailand (Soytong 1994; Thongklam 2008), and Australia (Watling and Gregory 1988, 1989; Watling & Li 1999), but in all cases, with the only possible exception of South African collections, these names have probably been misapplied.
- ***Boletus reticuloceps*** (M. Zang, M.S. Yuan & M.Q. Gong) Q.B. Wang & Y.J. Yao Sydowia 57 (1): 132. 2005 (Fig. 8.2o).

  ≡ *Aureoboletus reticuloceps* M. Zang, M.S. Yuan & M.Q. Gong, Acta Mycol. Sin. 12: 277. 1993. (basionym)

- Known distribution: East Asia, Nepal, central, southwestern, and southeastern China (Tibet, Yunnan, Sichuan, Hubei, and Taiwan Provinces).
- Habitat and mycorrhizal association: Forming ECM with subalpine conifers such as *Tsuga* spp., *Abies* spp. [*A. georgei* Orr, *A. faxoniana* Rehder & E.H. Wilson, *A. kawakamii* (Hayata) T. Ito, *A. squamata* Masters], and *Picea* spp. [*P. likiangensis* (Franch.) E. Pritz., *P. spinulosa* (Griff.) A. Henry] (Pinaceae), also in mixed woods with *P. densata* Mast. and with broad-leaved trees such as *Quercus aquifolioides* Rehder & E.H. Wilson, *Betula delavayi* Franch. (Betulaceae), and *Salix* spp. (Salicaceae); also found in association with an herbaceous plant *Kobresia capillifolia* (Decne.) C.B. Clarke in alpine meadows of southwestern China (Gao and Yang 2010); up to 4000 m alt.
- Relevant bibliography: Zang et al. (1993), Ying and Zang (1994), Wang (2004), Wang et al. (2004), Wang and Yao (2005), Zang (2006), Yuan and Sun (2007), Cui et al. (2016), Feng et al. (2017).
- Notes: *B. reticuloceps* likely evolved during the rising of Himalayan range (Yang 2005) and may be a relict species in nature (Zang et al. 1993, Zang 1995). Its dispersal from the Hengduan Mountains to Taiwan through central China may have occurred during the Quaternary glaciations (Feng et al. 2017).

- ***Boletus rex-veris*** D. Arora & Simonini Econ. Bot. 62 (3): 373. 2008. (Corresponding to ***Boletus* sp. nov.** in Camacho and Bruns 2008) (Fig. 8.2p).
- Known distribution: North America, west of Rocky Mountains (southwestern Canada and western USA, Washington east to Idaho and south to Oregon and California, perhaps extending into Mexico).
- Habitat and mycorrhizal associations: Occurring in temperate montane coniferous forests with *Abies* spp. [particularly *A. concolor* (Gordon & Glend.) Hildebr. and *A. grandis*] and *Pinus* spp. [especially *P. contorta* subsp. *murrayana* (Balfour) Engelmann and *P. ponderosa*] (Pinaceae) during the spring or early summer (April–July); up to 2130 m alt.
- Relevant bibliography: Thiers (1975 as "*B. edulis*"), Arora (2008), Camacho and Bruns (2008 as "*Boletus* sp. nov."), Scates (2011).
- Notes: Known as the "spring king bolete," this taxon has long been misinterpreted in the Pacific Northwest either as *B. edulis* or *B. pinophilus*. However, it is an independent phylogenetic species that has recently been revealed by means of molecular research (Camacho and Bruns 2008) and formally established as a novel species by Arora and Simonini (Arora 2008).
- *Xerocomus tumidus* (Fr.) E.-J. Gilbert as reported by López and García (2009) from Mexico might instead represent *B. rex-veris*.
- ***Boletus rubriceps*** D. Arora & J.L. Frank North American Fungi 9 (6): 3. 2014.
- Known distribution: North America, west of Rocky Mountains (western to southwestern USA, Montana, Wyoming, Colorado south to Arizona and New Mexico and possibly extending to northern Mexico).
- Habitat and mycorrhizal association: Occurring in temperate environment in mesophilic montane coniferous forests, associated with *Picea* spp. (*P. engelmannii* Parry ex Engelm.), *Pinus* spp., and perhaps *Abies* spp. (Pinaceae).
- Relevant bibliography: Snell and Dick (1958, 1970 as "*Boletus edulis* subsp. *pinicola*?"), Arora (1986 as "*Boletus edulis*"), States (1990 as "*Boletus edulis*"), Evenson (1997 as "*Boletus edulis*"), Kuo (2010d as "*Boletus* cf. *pinophilus*"), Arora and Frank (2014b).
- Notes: *B. rubriceps* is commonly called the "Rocky Mountain red-capped or ruby-capped king bolete" and appears to be very strictly allied to *B. edulis* by means of morphological and molecular analysis (Arora and Frank 2014b). As a matter of fact, *B. rubriceps* has long been confused by western American mycologists (Snell and Dick 1958, 1970; Arora 1986; States 1990; Evenson 1997; Kuo 2010d) with the type species of the genus *Boletus* up to very recent times.
- ***Boletus semigastroideus*** M. Nuhn, Manfr. Binder, A.F.S. Taylor, Halling & Hibbett Fung. Biol. 117 (7-8): 493. 2013.

Nom. nov. for *Secotium areolatum* G. Cunn., New Zealand J. Sci. Technol. 23: 172B. 1942. Non *Boletus areolatus* Berkeley, Hooker's J. Bot. Kew Gard. Misc. 4: 138. 1852 [= *Tylopilus areolatus* (Berk.) Henn. 1898].
= *Notholepiota areolata* (G. Cunn.) E. Horak, New Zealand J. Bot. 9: 479. 1971.

- Known distribution: New Zealand.

- Habitat and mycorrhizal association: Occurring in temperate environment under *Kunzea* spp. and *Leptospermum* spp. (*Myrtaceae*).
- Relevant bibliography: Cunningham (1942, 1944 as "*S. areolatum*"), Horak (1971 as "*N. areolata*"), Nuhn et al. (2013).
- Notes: *B. semigastroideus* is an overall whitish secotioid porcini mushroom and one of the very few *Boletus* s. str. species native to the southern hemisphere. It was originally described as *Secotium areolatum* by Cunningham (1942) and then recombined by Horak (1971) as *Notholepiota areolata*, the type species of the genus. Recent molecular investigation (Vellinga 2004; Nuhn et al. 2013) indicated it to belong in *Boletus* s. str.; however, the name *Boletus areolatus* resulted pre-occupied and it was consequently renamed as *B. semigastroideus*.
- ***Boletus separans*** Peck Bull. Buffalo Soc. Nat. Sci. 1: 59. 1873 (Fig. 8.2q).

  ≡ *Suillus separans* (Peck) Kuntze, Revis. Gen. Pl., 3 (2): 536. 1898.
  ≡ *Boletus edulis* subsp. *separans* (Peck) Singer, Amer. Midl. Nat. 37: 26. 1947.
  ≡ *Boletus edulis* f. *separans* (Peck) Vasilkov, Belyj Grib: 15. 1966.
  ≡ *Xanthoconium separans* (Peck) Halling & Both, Bull. Buffalo Soc. Nat. Sci. 36: 240. 1998.
  = *Boletus pseudoseparans* Grand & A.H. Smith, Mycologia 63 (1): 115. 1971.
  – *Boletus reticulatus* subsp. *separans* (Peck) Hlaváček, Mykol. Sborn. 71 (2): 52. 1994. (nom. inval., art. 41.5, basionym not cited)

- Known distribution: North America, east of Rocky Mountains (eastern Canada and eastern USA, west to Tennessee, Michigan, and Texas, south to Florida and probably Mexico), also reported from northern South America (Colombia).
- Habitat and mycorrhizal association: Occurring in temperate environment in broad-leaved or mixed forests with *Quercus* spp. (*Q. humboldtii* Bonpl. in Colombia) (Fagaceae) and occasionally with *Picea* spp. (artificial plantations of *P. abies*) and *Pinus* spp. (i.e., *P. elliottii*) (Pinaceae).
- Relevant bibliography: Peck (1872 as "*Boletus edulis*," 1873, 1889), Saccardo (1888), Coker and Beers (1943), Singer (1947), Snell and Dick (1961, 1970), Grand and Smith (1971 as "*Boletus pseudoseparans*"), Pomerleau (1980), Smith et al. (1981), Phillips (1991), Both (1993), Halling and Both (1998), Bessette et al. (2000 as "*Xanthoconium separans*," 2016, 2019), Simonini et al. (2001), Roody (2003 as "*Xanthoconium separans*"), McNeil (2006 as "*Xanthoconium separans*"), Kuo and Methven (2014), Kuo (2015).
- Notes: Commonly known in the USA as the "lilac bolete," *B. separans* was placed into the genus *Xanthoconium* due to some morphological and macrochemical peculiarities (Both 1993; Halling and Both 1998). Preliminary molecular analysis suggested a possible placement of *B. separans* and allied "*Alloboletus*" species outside *Boletus* s. str. (Simonini et al. 2001), but recent and inclusive investigation based on multi-locus molecular data has, by contrast, indicated a natural affiliation of these species with the porcini mushrooms (Dentinger et al. 2010; Feng et al. 2012; Nuhn et al. 2013) and have returned *X. separans* to *Boletus* s. str., being with certainty a porcini mushroom.

- According to Singer (1977), the descriptions of *B. separans* provided by Smith and Thiers (1971), Smith and Smith (1973), Grund and Harrison (1976), and Smith et al. (1981) are likely to be referred to *B. nobilis*.
- A collection of *B. separans* from Colombia (which represents the southernmost limit of the distribution range for this species) is microscopical dissimilar from the North American samples (Simonini et al. 2001) and may indeed be a different species. Unfortunately, molecular data are missing and the taxonomic status of this sample currently remains unresolved.
- Villareal and Perez-Moreno (1989) and García-Jiménez (2013) reported on the presence of this species from Mexico. *B. separans* has also been repeatedly cited from Europe (Pilát and Dermek 1974; Angarano 1989; Cetto 1989; Knudsen 1992; Olsson 1997; Lannoy and Estadès 2001; Estadès and Lannoy 2004; Perič and Perič 2006; Kibby 2011) and China (Zang 2006), but molecular analysis has never, as yet, demonstrated evidence on the occurrence of this North American species in the Old World. As a result, the aforementioned records should be considered nothing more than atypical phenotypes of either *B. pinophilus* or *B. reticulatus* (Simonini et al. 2001) in Europe and a misidentification of any of the several Chinese porcini mushrooms in the Far East.
- Razaq et al. (2014) noticed the presence of *B. separans* in Pakistan, but their species is obviously different from the American one.
- A putative additional species that resembles *B. separans* has recently been found in eastern Australia and provisionally named *B. austroseparans*, differing from the former in a spore print and a hymenophore colored like members of *Tylopilus* but with identical macrochemical reactions (Halling in litt.). Supplementary investigation is obviously required to assess this taxon.
- *Boletus pseudoseparans* Grand & A.H. Smith is practically indistinguishable from *B. separans* and is currently considered a later synonym of the former species (Halling and Both 1998; Bessette et al. 2000, 2016; Simonini et al. 2001).
- ***Boletus shiyong*** Dentinger *in* Dentinger & Suz Index Fungorum 29: 1. 2013. (Corresponding to ***Boletus*** **sp. 5** in Feng et al. 2012)
- Known distribution: Southwestern China (Yunnan Province).
- Habitat and mycorrhizal association: Occurring in subalpine environment in mixed forests with *Pinus* spp. (*P. densata* Mast.), *Picea* spp. (Pinaceae), and *Quercus* spp. (*Q. aquifolioides* Rehder & E.H. Wilson) (Fagaceae).
- Relevant bibliography: Dentinger (2013), Dentinger and Suz (2014), Cui et al. (2016).
- Notes: Microscopic features reported in the original description by Dentinger (2013) are wrong since they were taken with a 40× objective (Dentinger in litt.).
- ***Boletus sinoedulis*** B. Feng, Yang Y. Cui, J.P. Xu & Zhu L. Yang Fungal Div. 81 (1): 203. 2016. (Corresponding to ***Boletus*** **sp. 10** in Feng et al. 2012)
- Known distribution: Southwestern China (Sichuan and Yunnan Provinces).
- Habitat and mycorrhizal association: Forming ECM with subalpine coniferous trees such as *Picea* spp. (especially *P. likiangensis*) and *Abies* spp. (Pinaceae).
- Relevant bibliography: Cui et al. (2016).

- ***Boletus subalpinus*** (Trappe & Thiers) M. Nuhn, Manfr. Binder, A.F.S. Taylor, Halling & Hibbett Fung. Biol. 117 (7–8): 493. 2013.

  ≡ *Gastroboletus subalpinus* Trappe & Thiers, Brittonia 21: 245. 1969. (basionym)

- Known distribution: North America, west of Rocky Mountains (Washington and Oregon south to California).
- Habitat and mycorrhizal association: Occurring in temperate environment in montane and subalpine coniferous forests with *Pinus* spp. (*P. albicaulis* Engelm.), *Abies* spp. (*A. magnifica* A. Murray), and *Tsuga mertensiana* (Bong.) Carr. (Pinaceae).
- Relevant bibliography: Thiers and Trappe (1969), Thiers (1975, 1989), Arora (1986, 2008), Both (1993), Bessette et al. (2000), Nouhra et al. (2002), Trappe et al. (2007), Scates (2011), Camacho and Bruns (2008) (all as "*Gastroboletus subalpinus*"), Kuo (2013).
- Notes: Initially assigned to the polyphyletic genus *Gastroboletus*, *B. subalpinus* is one of the few secotioid porcini species and grows subhypogeus to partially emergent. Only in very recent times it has formally been transferred to *Boletus* s. str., being supported by molecular evidence (Nuhn et al. 2013).
- ***Boletus subcaerulescens*** (E.A. Dick & Snell) Both, Bessette & A.R. Bessette *in* Bessette, Roody & A.R. Bessette North American boletes: 161. 2000 (Fig. 8.2r).

  ≡ *Boletus edulis* subsp. *subcaerulescens* E.A. Dick & Snell, Mycologia 57 (3): 455. 1965. (basionym)
  ≡ *Boletus edulis* f. *subcaerulescens* (E.A. Dick & Snell) Vasilkov, Belyj Grib: 16. 1966.
  ≡ *Boletus separans* var. *subcaerulescens* (E.A. Dick & Snell) A.H. Smith & Thiers, Boletes of Michigan: 365. 1971.

- Known distribution: North America, east of Rocky Mountains (southeastern Canada and northeastern USA, New England, New York, Michigan, south to South Carolina and west to Wisconsin and Minnesota).
- Habitat and mycorrhizal association: Occurring in temperate environment in pure stand or in mixed coniferous forests with *Pinus* spp. (especially the invasive *P. sylvestris* but also with *P. rigida* Mill. and *P. banksiana*), *Picea* spp. (artificial plantations of *P. abies*) (Pinaceae), and *Betula* spp. (Betulaceae).
- Relevant bibliography: Dick and Snell (1965 as "*B. edulis* subsp. *subcaerulescens*"), Snell and Dick (1970 as "*B. edulis* subsp. *subcaerulescens*"), Smith and Thiers (1971 as "*B. separans* subsp. *subcaerulescens*"), Both (1993), Bessette et al. (2000, 2016), Simonini et al. (2001 as "*B. separans* subsp. *subcaerulescens*").
- Notes: Formerly described as a subspecies of *B. edulis* (Dick and Snell 1965) and later recombined as a variety of *B. separans* (Smith and Thiers 1971), this porcini mushroom, commonly called "almost bluing king bolete," was then raised to the specific rank (Bessette et al. 2000), and such an autonomous position has been subsequently confirmed by molecular studies (Dentinger et al. 2010).

- Considering the low phylogenetic variation among the ITS sequences of *B. subcaerulescens*, *B. subalpinus*, *B. regineus*, and *B. pinophilus*, Feng et al. (2012) suggested to merge all these taxa into a single species; however, given the clear-cut morphological differences, we prefer to retain them separate and independent from one another for the time being.
- ***Boletus subviolaceofuscus*** B. Feng, Yang Y. Cui, J.P. Xu & Zhu L. Yang Fungal Div. 81 (1): 204. 2016. (Corresponding to ***Boletus violaceofuscus* 2** W.F. Chiu in Feng et al. 2012)
- Known distribution: Southwestern China (Yunnan Province).
- Habitat and mycorrhizal association: Occurring in subtropical environment with *Lithocarpus* spp. (Fagaceae).
- Relevant bibliography: Cui et al. (2016).
- ***Boletus tylopilopsis*** B. Feng, Yang Y. Cui, J.P. Xu & Zhu L. Yang Fungal Div. 81 (1): 206. 2016.
- Known distribution: Southwestern China (Yunnan Province) and most likely Laos (based on photographic material in Læssøe et al. 2018).
- Habitat and mycorrhizal association: Occurring in subtropical environment with members of the Fagaceae.
- Relevant bibliography: Cui et al. (2016).
- ***Boletus umbrinipileus*** B. Feng, Yang Y. Cui, J.P. Xu & Zhu L. Yang Fungal Div. 81 (1): 206. 2016. (Corresponding to ***Boletus* sp. 3** in Feng et al. 2012)
- Known distribution: Southwestern China (Yunnan Province).
- Habitat and mycorrhizal association: Occurring in subtropical environment with members of the Fagaceae.
- Relevant bibliography: Cui et al. (2016).
- ***Boletus variipes*** Peck Rep. (Annual) Trustees State Mus. Nat. Hist. 41: 76. 1888. (Corresponding to ***B. variipes*-1** in Feng et al. 2012) (Fig. 8.2s).

  = *Boletus variipes* var. *fagicola* A.H. Smith & Thiers, Boletes of Michigan: 370. 1971.
  = *Boletus variipes* var. *albipes* Peck, Rep. (Annual) Trustees State Mus. Nat. Hist. 41: 76. 1888.
  = *Boletus variipes* var. *pallidipes* Peck, Rep. (Annual) Trustees State Mus. Nat. Hist. 41: 76. 1888.
  = *Boletus variipes* var. *tenuipes* Peck, Rep. (Annual) Trustees State Mus. Nat. Hist. 41: 76. 1888.

- Known distribution: North America, east of Rocky Mountains (eastern Canada and northeastern USA west to Minnesota and Iowa and perhaps Missouri). Southern distribution limits yet to be established.
- Habitat and mycorrhizal association: Occurring in temperate environment in broad-leaved forests with *Quercus* spp. (*Q. rubra*), *Fagus* spp. (*F. grandifolia*) (Fagaceae), *Populus* spp. (Salicaceae), and *Acer* spp. (Sapindaceae), also reported in mixed coniferous forests with *Tsuga* spp., *Pinus* spp., and *Picea* spp. (artificial plantations of *P. abies*) (Pinaceae).

- Relevant bibliography: Peck (1888, 1889), Coker and Beers (1943), Snell and Dick (1970), Smith and Thiers (1971), Grund and Harrison (1976), Pomerleau (1980), Smith et al. (1981), Phillips (1991), Both (1993), Horn et al. (1993), Bessette et al. (2000, 2016), Roody (2003), McNeil (2006), Binion et al. (2008), Kuo (2010e), Scates (2011).
- Notes: Molecular phylogenetic analysis (Dentinger et al. 2010; Feng et al. 2012) has clearly demonstrated that *B. variipes* is a collective species that should be split into two distinct taxa which are mainly characterized by different distribution patterns. *B. variipes* s. str. appears to be restricted to northeastern North America, while the other cryptic species (see below) occurs in southeastern North America and Central America. The biogeographic range of both species, however, is at present only partly known, and a more comprehensive sampling will be necessary to understand their entire distribution.
- The three varieties described by Peck (1888) should be considered no more than aspects of the infraspecific morphological variability of *B. variipes*.
- *Boletus variipes* has also been reported from India (Lakhanpal and Sagar 1989; Lakhanpal 1996) and China (Li and Song 2000; Mao 2000, 2009; Zang 2006; Dai et al. 2010) but almost certainly misidentified like most of the extralimital bolete species said to occur in the Far East. Furthermore, *B. variipes* s. Watling from the Philippines as defined in Dentinger et al. (2010) is a new species (see below).
- ***Boletus violaceofuscus*** W.F. Chiu Mycologia 40 (2): 210. 1948 (Fig. 8.2t).
- Known distribution: Eastern and southeastern Asia, China (Yunnan and Taiwan Provinces), Laos, eastern Russia, Japan.
- Habitat and mycorrhizal association: Occurring in subtropical to temperate regions in mixed broad-leaved and coniferous forests under *Quercus* spp. (*Q. serrata*, *Q. variabilis* Blume) (Fagaceae), *Pinus* spp. (*P. densiflora*, *P. thunbergii*) (Pinaceae), and *Betula* spp. (Betulaceae).
- Relevant bibliography: Chiu (1948, 1957), Hongo (1960), Ying et al. (1987), Imazeki et al. (1988), Imazeki and Hongo (1987), Bi et al. (1990, 1994), Chen et al. (1997), Borghi et al. (2000), Floriani et al. (2000), Mao (2000, 2009), Simonini et al. (2001), Wang (2004), Wang et al. (2004), Zang (2006), Sitta et al. (2007b), Wu et al. (2011), Cui et al. (2016), Læssøe et al. (2018).
- Notes: The presence of *B. violaceofuscus* in eastern Russia has been molecularly confirmed by Fedosova and Kovalenko (2014).
- ***Boletus viscidiceps*** B. Feng, Yang Y. Cui, J.P. Xu & Zhu L. Yang Fungal Div. 81 (1): 208. 2016. (Corresponding to ***Boletus* sp. 4-1** and ***Boletus* sp. 4-2** in Feng et al. 2012)
- Known distribution: Southwestern China (Yunnan Province).
- Habitat and mycorrhizal association: Occurring in subtropical environment with *Lithocarpus* spp. and *Quercus* spp. (Fagaceae).
- Relevant bibliography: Feng et al. (2012), Cui et al. (2016).
- Notes: This species was considered as a probable synonym of *B. viscidiceps* (Cui et al. 2016), and this identity has now been confirmed (Feng in litt.).

8.3.1.2 Species Validly Published but Without Molecular Support

- ***Boletus atkinsonii*** Peck Bull. N. Y. State Mus. 94 (8): 20. 1905.
- Known distribution: North America, east of Rocky Mountains (southeastern Canada and northeastern USA south to North Carolina, western limits yet to be established), also reported from Mexico.
- Habitat and mycorrhizal association: Occurring in temperate environment in mixed broad-leaved forests with *Quercus* spp. and *Fagus* spp. (Fagaceae).
- Relevant bibliography: Peck (1905), Snell and Dick (1970), Smith and Thiers (1971), Singer (1977), Smith et al. (1981), Both (1993), Bessette et al. (2000, 2016), Kuo (2010a).
- Notes: *B. atkinsonii* has been reported from Mexico (Montoya-Esquivel 1998) and various times from China (Bi et al. 1984, 1994; Li and Song 2000; Zhuang 2001; Zang 2006), but in all cases, the identity of these collections with North American material should be confirmed by means of molecular methods. *B. atkinsonii* s. Sims et al. (1997) and Watling (2001) from the Philippines represents a new, yet unnamed species (see below).
- ***Boletus castanopsidis*** Hongo in Kobayasi et al. Bull. Natn. Sci. Mus., Tokyo 16 (3): 548. 1973.
- Known distribution: New Guinea (West Sepik).
- Habitat and mycorrhizal association: Growing in tropical environment under *Castanopsis* spp. and *Quercus* spp. (Fagaceae) with the presence of *Araucaria* spp. (Araucariaceae)
- Relevant bibliography: Hongo (1973).
- Notes: *B. castanopsidis* is one of the very few endemic porcini species occurring in the southern hemisphere. Molecular research is required to confirm its status.
- ***Boletus frustulosus*** Peck Bull. Torrey Bot. Club 24 (3): 146. 1897.
- ≡ *Ceriomyces frustulosus* (Peck) Murrill, Mycologia 1 (4): 145. 1909.
- Known distribution: North America, east of Rocky Mountains (southeastern USA, Mississippi and Alabama). Distribution limits unknown.
- Habitat and mycorrhizal association: Occurring in temperate environment, in open ground on clay banks. ECM symbiosis undetermined.
- Relevant bibliography: Peck (1897), Murrill (1909, 1914 as "*Ceriomyces frustulosus*"), Singer (1947), Both (1993).
- Notes: *B. frustulosus* is a critical porcini mushroom species due to shortage of information available. Singer (1947) believed it to be either *B. edulis* or an old *B. aereus* and Both (1993) to be a probable representative of *Gastroboletus*. A modern re-description based on fresh material will be necessary to clarify its taxonomy, ecology, phylogeny, and distribution boundaries.
- ***Boletus gigas*** Berk. Hooker's J. Bot. Kew Gard. Misc. 4: 138. 1852.
- ≡ *Suillus gigas* (Berk.) Kuntze, Revis. Gen. Pl. 3 (2): 535. 1898.
- Known distribution: South and southeastern Asia, India (Sikkim Himalaya) and perhaps China.

- Habitat and mycorrhizal association: Forming ECM with *Betula* spp. (Betulaceae) and *Andromeda* spp. (Ericaceae).
- Relevant bibliography: Berkeley (1852), Horak (1980), Lakhanpal (1996).
- Notes: Little information is known upon this species, and although also reported from China (Li and Song 2000; Zang 2006), its taxonomic circumscription and distribution limits are presently unknown.
- ***Boletus insuetus*** A.H. Smith & Thiers Boletes of Michigan: 367. 1971.
- Known distribution: North America, east of Rocky Mountains (northeastern USA, Michigan). Distribution limits unknown.
- Habitat and mycorrhizal association: Occurring in temperate environment in mixed forests. ECM relationships unknown.
- Relevant bibliography: Smith and Thiers (1971), Smith and Smith (1973), Smith et al. (1981), Both (1993).
- Notes: Other than the type description, there seems to be no other significant information on this taxon. It is suggested a morphological affinity of *B. insuetus* with *B. gertrudiae* (Smith and Thiers 1971), but the taxonomic circumscription of the former species is presently indefinite.
- ***Boletus luteoloincrustatus*** R. Flores & Simonini Riv. Micol. 43 (2): 125. 2000 (Fig. 8.2u).
- Known distribution: Central America (Guatemala).
- Habitat and mycorrhizal association: Forming ECM with *Quercus* spp. (Fagaceae), also with the presence of *Pinus* spp. (Pinaceae) in tropical montane environment up to 2300 m alt.
- Relevant bibliography: Flores Arzù and Simonini (2000), Bran et al. (2003b), Morales et al. (2010), Flores Arzù et al. (2012).
- Notes: This species might be widespread in montane cloud forests of Central America dominated by Fagaceae, but its distribution limits are presently unknown.
- ***Boletus mottiae*** Thiers California Mushrooms (as "*B. mottii*"): 36. 1975.
- Known distribution: North America, west of Rocky Mountains (southwestern Canada and western USA, British Columbia south to Washington, Oregon, and California, west to Montana, Idaho, and Colorado), also reported from Europe (Austria).
- Habitat and mycorrhizal association: Occurring in temperate environment in coastal and montane Pacific forests with *Quercus* spp. (Fagaceae).
- Relevant bibliography: Thiers (1975), Both (1993), Bessette et al. (2000), Arora (2008), Klofac (2007, 2010), Scates (2011).
- Notes: *B. mottiae* appears to be morphologically nearly identical to *B. fibrillosus*. Furthermore, their symbiotic hosts and distribution range seem to overlap. Accordingly, *B. mottiae* and *B. fibrillosus* may well be conspecific as suggested by Arora (2008). Unfortunately, the type specimen of *B. mottiae* is too old and in poor condition for a successful DNA extraction (Arora 2008; Dentinger in litt.). Fresh material will be necessary to elucidate its taxonomic status and relationships with the similar *B. fibrillosus*.
- In a recent molecular work retrieved on the web (Camacho and Bruns 2008), *B. mottiae* and *B. fibrillosus* are considered separate, but it is unclear where the

samples were collected and, most of all, whether their *B. mottiae* is the same as Thiers' species. *B. mottiae* s. Camacho and Bruns might indeed be *B. regineus*.

- *Boletus mottiae* has also been reported from central Europe (Austria), and the identification of the European material was made by Thiers himself, who examined and confirmed the samples and also rectified upon the putative ectomycorrhizal association of the species, claiming that it is only linked with *Quercus* and not with conifers also as previously assumed in the protologue (Klofac 2007, 2010). If such an observation turned out to be correct, it might at least in part explain misidentifications and general confusion between *B. mottiae* and *B. fibrillosus*.
- ***Boletus multipunctus*** Peck Bull. N. Y. State Mus. 54 (5): 952. 1902.
- Known distribution: North America, east of Rocky Mountains (USA, New York, Connecticut, and Pennsylvania). Distribution limits unknown.
- Habitat and mycorrhizal association: Occurring in temperate environment in deciduous forests with *Quercus* spp. (Fagaceae).
- Relevant bibliography: Peck (1902), Snell and Dick (1970), Smith and Thiers (1971), Both (1993).
- Notes: *B. multipunctus* has also been recognized from China (Li and Song 2000; Wang 2004; Zang 2006), but the generic placement of this species is still debated; Peck (1902), Snell and Dick (1970), and Singer (1986) placed it in *Boletus* s. str., whereas Murrill (1909, 1910) and Smith and Thiers (1971) indicated it as a synonym of or a closely related species to *B. roxanae* Frost (now a member of *Aureoboletus* Pouzar). Molecular analysis on fresh material is desperately required to ascertain its status and phylogenetic relationships.
- ***Boletus occidentalis*** B. Ortiz & T.J. Baroni Fungal Div. 27: 314. 2007 (Fig. 8.2v).
- Known distribution: Central America (Dominican Republic).
- Habitat and mycorrhizal association: Occurring in tropical environment with *Pinus occidentalis* Sw. (Pinaceae).
- Relevant bibliography: Ortiz-Santana et al. (2007).
- ***Boletus subreticulatus*** Corner Boletus in Malaysia: 129. 1972.
- Known distribution: Insular Malaysia (Borneo).
- Habitat and mycorrhizal association: Occurring in tropical montane environment under *Quercus* spp. (Fagaceae).
- Relevant bibliography: Corner (1972), Horak (2011).
- Notes: *B. subreticulatus* is a critical taxon, and further morphological and molecular examinations on fresh material are needed to assess its taxonomic position, ECM partnerships, and biogeographic distribution patterns.

8.3.1.3 Species Yet to Be Validly Published

- ***Boletus barrowsii*** Thiers & A.H. Smith s. auct. (Corresponding to ***Boletus barrowsii* 2** in Feng et al. 2012)

- Known distribution: North America, east of Rocky Mountains. Distribution limits unknown.
- Habitat and mycorrhizal association: Unknown.
- Relevant bibliography: Feng et al. (2012).
- Notes: This species has recently been separated from *B. barrowsii* s. str. by means of molecular tools but still remains undescribed due to shortage of data.
- ***Boletus* cf. *pinophilus***
- Known distribution: Mexico (Oaxaca).
- Habitat and mycorrhizal association: Unknown.
- Relevant bibliography: Crous et al. (2019).
- ***Boletus* cf. *reticulatus***
- Known distribution: North America, east of Rocky Mountains (Illinois). Distribution limits unknown.
- Habitat and mycorrhizal association: Growing in temperate environment in mixed deciduous woodlands under *Quercus* spp. (especially *Q. alba* L.) (Fagaceae) and *Carya* spp. (Juglandaceae).
- Relevant bibliography: Snell and Dick (1970 as "*B. edulis* subsp. *reticulatus*"), Bessette et al. (2000 as "*B. reticulatus*"), Kuo (2010b).
- Notes: According to Kuo (2010b), preliminary unpublished molecular analysis carried out by a team of Swedish and British mycologists in 2004 indicated this species to be strictly allied to the European *B. reticulatus* although different from the latter. Additional collections and further phylogenetic investigation will, of course, be necessary to formally describe this new species.
- ****Boletus leptocephalus*** Peck Bull. Torrey Bot. Club 25: 371. 1898. (nom. illeg., art. 53.1) [non *Boletus leptocephalus* Jacq., Miscell. Austriaca 1: 142, t. 12. 1778. = *Cerioporus leptocephalus* (Jacq.) Zmitr.].
- Known distribution: North America, east of Rocky Mountains (southeastern USA, Alabama). Distribution limits unknown.
- Habitat and mycorrhizal association: Occurring in temperate environment in dry, open woods under *Pinus* spp. (Pinaceae).
- Relevant bibliography: Peck (1898), Murrill (1909, 1910), Both (1993).
- Notes: *B. leptocephalus* is another critical North American species for which very little information is at present available. Murrill (1909, 1910) and Both (1993) considered it the same as *B. edulis*. In any case, *B. leptocephalus* Peck is invalid as the name was pre-occupied.
- ****Boletus phaeocephalus*** Pat. & C.F. Baker s. Corner pro parte J. Straits Brch. R. Asiat. Soc. 78: 70. 1918. Corner, Boletus in Malaysia: 125. 1972.
- ? = *Boletus rufobadius* Bres., Bull. Soc. Mycol. Fr. 6: 37. 1890.
- – "*Tylopilus phaeocephalus*" (Pat. & C.F. Baker) Singer *in* Singer, J. García & L.D. Gómez ad interim, Nova Hedwigia 102: 51. 1991. (nom. inval., art. 41.5, basionym not cited)
- Known distribution: Singapore, Peninsular and Insular Malaysia (Borneo).

- Habitat and mycorrhizal association: Occurring in mixed tropical forests. ECM relationships unknown.
- Relevant bibliography: Corner (1972), Singer et al. (1991 as "*Tylopilus phaeocephalus*"), Lee et al. (2002), Horak (2011).
- Notes: Originally described by Patouillard and Baker (1918), this species has later been recombined by Singer firstly in *Pulveroboletus* (Singer 1947) and then in *Xerocomus* (Singer et al. 1983). However, *B. phaeocephalus* as interpreted by Corner (1972) is clearly different from Patouillard and Baker's species and presumably represents a mixture of species (Singer 1981; Singer et al. 1983), one of which from Borneo could be a porcini mushroom (see also Watling 2000, 2001), not a *Tylopilus* as conjectured by Singer (1981) and Singer et al. (1983, 1991). Hence, Corner's bolete is apparently undescribed and would require a proper name being a new species.
- *Boletus phaeocephalus* has also been reported from China (Li and Song 2000; Zang 2006), but its occurrence north of Malaysia should be confirmed by further investigation.
- *Boletus olivaceobrunneus* Zeller & F.D. Bailey from North America is neither a later synonym of *B. edulis* nor of *B. phaeocephalus* (in either the original interpretation or Corner's interpretation). It should not be a member of the porcini mushrooms at all as previously supposed by Singer (1947), Stuntz and Isaacs (1962), Corner (1972), and Horak (2011) but is rather a member of the currently circumscribed genus *Porphyrellus* (Thiers 1975) and according to C.B. Wolfe a later synonym for *Porphyrellus atrofuscus* E.A. Dick & Snell (Halling in litt.). The use of the epithet "*olivaceobrunneus*" to describe olive-brown pileal forms of *B. edulis* (such as in Mikšík 2017: 51) should therefore be considered a misapplication.
- Finally, mention should be made of *B. umbrinellus* Pat. & C.F. Baker from Singapore and *B. rufobadius* Bres. from Cameroon. According to Singer's revision of the type of *B. umbrinellus*, it appears to be conspecific with part of the material of *B. phaeocephalus* s. Corner and is currently considered a member of *Boletellus* (Singer et al. 1983). Instead, the identity of *B. rufobadius* with *B. phaeocephalus* s. Corner as formerly suggested by Corner (1972) and subsequently by Horak (2011) is yet to be determined.
- ***Boletus* sp.** (Corresponding to **"*Alloboletus*" sp. 1** in Feng et al. 2012)
- Known distribution: Southeast Asia (Bangladesh).
- Habitat and mycorrhizal associations: Forming ECM with *Shorea robusta* Roth (Dipterocarpaceae) in tropical environment.
- Relevant bibliography: Feng et al. (2012).
- Notes: Yet undescribed due to the paucity of fresh material available, this species is the only known porcini mushroom apparently associated to Dipterocarpaceae (Feng et al. 2012).
- ***Boletus* sp. nov. 1**
- Known distribution: Southeast Asia (Philippines).
- Habitat and mycorrhizal associations: Forming ECM with *Pinus kesiya* Royle ex Gordon in tropical environment.

- Relevant bibliography: Sims et al. (1997 as "*Boletus atkinsonii*"), Watling (2001 as "*Boletus atkinsonii*"), Dentinger et al. (2010 as "*Boletus* sp. nov. 1" or "*Boletus variipes*" s. Watling).
- Notes: This species has been variously interpreted in the literature, but molecular analysis (Dentinger et al. 2010) clearly indicates it is a new, yet unnamed species.
- ***Boletus* sp. 8**
- Known distribution: Central China (Hunan Province).
- Habitat and mycorrhizal associations: Unknown.
- Relevant bibliography: Feng et al. (2012).
- ***Boletus* sp. 11**
- Known distribution: Eastern China (Shandong Province).
- Habitat and mycorrhizal associations: Unknown.
- Relevant bibliography: Feng et al. (2012).
- ***Boletus* sp. 12**
- Known distribution: Eastern China (Jiangsu Province).
- Habitat and mycorrhizal associations: Unknown.
- Relevant bibliography: Feng et al. (2012).
- ****Boletus subvariipes*** T.H. Li *in* Bi et al. A preliminary agaric flora of Hainan Province: 303. 1997. (nom. inval., art. 39.1, original diagnosis absent)
- – *Boletus subvariipes* T.H. Li *in* Bi et al. ex Q.B. Wang & Y.J. Yao in Q.B. Wang, Taxonomy and molecular systematics of Boletus in China: 77. 2004. (nom. inval., art. 39.1, no Latin diagnosis)
- Known distribution: Southeastern China (Hainan Province).
- Habitat and mycorrhizal association: Occurring in tropical environment under unknown host trees.
- Relevant bibliography: Bi et al. (1997), Li and Song (2000), Wang (2004).
- Notes: *B. subvariipes* has never been validly published and still remains a taxon in desperate need of better taxonomic circumscription. According to the description provided by Li and Song (2000), this species might be an intermediate form between a pileate-stipitate and a secotioid porcini mushroom.
- ***Boletus variipes*** Peck s. auct. (Corresponding to ***Boletus* sp. 4** in Dentinger et al. 2010 and to ***B. variipes*-2** in Feng et al. 2012) (Fig. 8.2w).
- = *Boletus variipes* var. *fagicola* A.H. Smith & Thiers s. auct.
- = *Xerocomus phaeocephalus* (Pat. & C.F. Baker) Singer s. Singer in Singer, Araujo & Ivory, Beih. Nova Hedwigia 77: 83. 1983.
- Known distribution: North and Central America, east of Rocky Mountains (southeastern USA, North Carolina west to Texas, extending southward into the neotropics through Mexico, Belize, and Guatemala and south to Costa Rica), probably reaching Northern South America (Colombia). Northern distribution limits yet to be established.
- Habitat and mycorrhizal association: Occurring in temperate to tropical environment in broad-leaved forests with *Quercus* spp. (*Q. oleoides* Schltdl. & Cham.) and *Fagus* spp. (*F. grandifolia*) (Fagaceae), sometimes with the presence of *Pinus* spp. (*P. caribaea* Morelet) (Pinaceae).

Fig. 8.2 (**l**) *Boletus quercophilus* (photo R.E. Halling) (**m**) *Boletus regineus* (photo J.L. Frank) (**n**) *Boletus reticulatus* (photo M. Gelardi) (**o**) *Boletus reticuloceps* (photo B. Feng) (**p**) *Boletus rex-veris* (photo J.L.Frank) (**q**) *Boletus separans* (**r**) *Boletus subcaerulescens* (**s**) *Boletus variipes* s.str. (**t**) *Boletus violaceofuscus* (photo Y. Taneyama) (**u**) *Boletus luteoloincrustatus* (photo R. Flores Arzu) (**v**) *Boletus occidentalis* (photo T.J. Baroni) (**w**) *Boletus variipes* s.auct. (photo R.E. Halling). Photos by A.E. Bessette unless otherwise stated

- Relevant bibliography: Weber and Smith (1985), Ortiz-Santana et al. (2007, as "*B. variipes* var. *fagicola*"), Dentinger et al. (2010), Feng et al. (2012); probably also Bessette et al. (2019).
- Notes: This unnamed taxon has been mentioned from Central America under the misapplied names *B. variipes* by Ibarra et al. (1999), García-Jiménez (2013), Robles-García et al. (2018) (Mexico), Flores Arzù et al. (2012) (Guatemala), and Winkler (2016) (Colombia) and *B. variipes* var. *fagicola* by Ortiz-Santana et al. (2007) (Belize) and Halling (in litt.) (Costa Rica). Moreover, according to Dentinger et al. (2010), *Xerocomus phaeocephalus* s. Singer (Singer et al., 1983) should be referred to the present species. Being a novel species, this taxon deserves a new name to science.

8.4 Conclusions

The real number of porcini species occurring in the world seems to be far beyond the 19 taxa supposed by Dentinger et al. (2010), the 20/25 hypothesized by Watling (2008), or the 30 conjectured by Nuhn et al. (2013). As a matter of fact, according to the present list, *Boletus* s. str. currently comprises a total of more than 60 species detected by means of morphological and/or molecular studies, 12 out of which still lack a scientific binomial (including *B. barrowsii* s. auct., *B.* cf. *pinophilus*, *B.* cf. *reticulatus*, *B. leptocephalus*, *B. phaeocephalus* s. Corner, *B. subvariipes*, and *B. variipes* s. auct.), and additional members are expected to be uncovered in the near future, especially from the pantropical belt (Feng et al. 2012; Flores Arzù and Vizzini pers. comm.). At least two species are secotioid (*B. semigastroideus* and *B. subalpinus*), while the remaining are typically boletoid. "Porcini s. str." includes the largest number of taxa within *Boletus* s. str. (presumably more than 40) and is followed by "*Alloboletus*" with nine species (*Boletus* sp. nov., *B. gertrudiae*, *B. nobilis*, *B. orientialbus*, *B. semigastroideus*, *B. separans*, *B. subviolaceofuscus*, *B. tylopilopsis*, *B. violaceofuscus*), "*Orientiboletus*" (one species, *B. griseiceps*), "*Obtextiporus*" (one species, *B. albobrunnescens*), and "*Inferiboletus*" (one species, *B. austroedulis*). Within "porcini s. str.," the "Edulis clade" includes 14 species (*B. edulis*, *B. fagacicola*, *B. fibrillosus*, *B. himalayensis*, *B. pinophilus*, *B. pseudopinophilus*, *B. regineus*, *B. reticuloceps*, *B. rex-veris*, *B. rubriceps*, *B. sinoedulis*, *Boletus* sp. 8, *B. subalpinus*, *B. subcaerulescens*), the "Aereus clade" includes 15 species (*B. aereus*, *B. barrowsii*, *B. barrowsii* s. auct., *B. bainiugan*, *Boletus* cf. *reticulatus*, *B. indoedulis*, *B. meiweiniuganjun*, *B. nobilissimus*, *B. quercophilus*, *B. reticulatus*, *B. shiyong*, *Boletus* sp. nov. 1, *Boletus* sp. 11, *Boletus* sp. 12, *Boletus viscidiceps*), the "Variipes clade" includes three species (*B. hiratsukae*, *B. variipes*, *B. variipes* s. auct.), and the "Botryoides clade" includes three species (*B. botryoides*, *B. monilifer*, *B. umbrinipileus*). The American species *B. atkinsonii* and *B. mottiae* might reasonably be additional members of the "Aereus clade" and the "Edulis clade," respectively.

Biogeographic distribution of porcini mushrooms is presently poorly known, and for several species, it is only speculative; however, their overall endemism is restricted to the northern hemisphere where species are mostly associated to representatives of the Fagaceae and Pinaceae. Based on the present state of knowledge, only four endemic species occur in Europe and western Asia (including Asia Minor) (*B. aereus*, *B. edulis*, *B. pinophilus*, *B. reticulatus*), three in northern Africa and Middle East (*B. aereus*, *B. edulis*, *B. reticulatus*), 24 in North America (*B. atkinsonii*, *B. barrowsii*, *B. barrowsii* s. auct., *Boletus* cf. *pinophilus*, *Boletus* cf. *reticulatus*, *B. edulis*, *B. fibrillosus*, *B. frustulosus*, *B. gertrudiae*, *B. insuetus*, *B. leptocephalus*, *B. mottiae*, *B. multipunctus*, *B. nobilis*, *B. nobilissimus*, *B. pseudopinophilus*, *B. regineus*, *B. rex-veris*, *B. rubriceps*, *B. separans*, *B. subalpinus*, *B. subcaerulescens*, *B. variipes*, *B. variipes* s. auct.), and four in Central America (but this area is severely understudied!) (*B. luteoloincrustatus*, *B. occidentalis*, *B. quercophilus*, *B. variipes* s. auct.), whereas more than 29 species are recorded from eastern and southeastern Asia (*B. albobrunnescens*, *Boletus* "*Alloboletus*" sp. nov., *B. bainiugan*, *B. botryoides*, *B. edulis*, *B. fagacicola*, *B. gigas*, *B. griseiceps*, *B. himalayensis*, *B. hiratsukae*, *B. indoedulis*, *B. meiweiniuganjun*, *B. monilifer*, *B. orientialbus*, *B. phaeocephalus*, *B. reticuloceps*, *B. shiyong*, *B. sinoedulis*, *Boletus* sp. nov. 1, *Boletus* sp. 8, *Boletus* sp. 11, *Boletus* sp. 12, *B. subreticulatus*, *B. subvariipes*, *B. subviolaceofuscus*, *B. tylopilopsis*, *B. umbrinipileus*, *B. violaceofuscus*, *B. viscidiceps*). No endemic porcini mushrooms have as yet been described from central and southern South America and sub-saharan Africa (although at least *B. aereus*, *B. edulis*, *B. pinophilus*, and *B. reticulatus* are found introduced adventitiously and associated with non-native, exotic pine, oak, or birch plantations throughout the temperate regions of the southern hemisphere), but a few species occur in Australasia, such as *B. castanopsidis* reported from New Guinea (Hongo 1973), *B. austroedulis* recently described from Australia (Halling et al. 2014), and an additional secotioid species recently recombined into *Boletus* (viz., *B. semigastroideus*) from the genus *Notholepiota* E. Horak that would seem native to New Zealand (Nuhn et al. 2013). The main hotspot for porcini diversity is southeastern Asia, and such a richness is consistent with the paleotropic origin of *Boletus* s. str., which might have arisen and subsequently migrated with several events of radiation and vicariance/speciation from Australasia or the Indo-Malayan area toward Europe and the Americas (Dentinger et al. 2010; Feng et al. 2012).

Supplementary indigenous species not yet formally described have been located from the USA (Dentinger et al. 2010); Mexico (Crous et al. 2019); China (Feng et al. 2012; Cui et al. 2016), including Hainan island (Bi et al. 1997; Li and Song 2000; Wang 2004); Vietnam, Laos, and Cambodia (Mortimer et al. 2012; Læssøe et al. 2018); Bangladesh (Feng et al. 2012); Philippines (Sims et al. 1997; Watling 2001; Dentinger et al. 2010); Sarawak (Borneo) (Dentinger in litt.); Russia (Fedosova and Kovalenko 2014); Guatemala (with at least four new taxa) (Flores Arzù and Vizzini pers. comm.); and, in the southern hemisphere, from central-south Africa (Dentinger et al. 2010) and New Guinea (Hongo 1973).

It is undeniable that porcini species are among the most sought-after, tasty, popular, and largely appreciated forest-occurring fungi (Yun and Hall 2004); nowadays, they can probably be considered the most iconic wild edible mushrooms worldwide and they are commonly used as an exquisite ingredient of a large variety of processed foods. The longest culture of consumption, preservation, and trade of porcini mushrooms is from Europe (especially Italy) and is dated back to the ancient Greeks and Romans (Buller 1914), although only in recent times it has achieved a much larger, international scale (Sitta et al. 2007a; Sitta and Davoli 2012). The practice of collecting porcini mushrooms for culinary purposes in the New World is historically and intimately connected to the Italian immigrants who settled in North America at the beginning of the twentieth century (Arora 2008), starting a mycophilic tradition that was previously practically unknown. Along with other members of the family Boletaceae, porcini mushrooms are traditionally harvested by rural communities and offered for sale in local roadside food markets as well as in stores in eastern and southeastern Asia, particularly China (especially Yunnan Province), in order to earn additional income to supplement the household economy. In addition, at least four Chinese species (*B. bainiugan*, *B. meiweiniuganjun*, *B. sinoedulis*, *B. shiyong*) are commonly exported dried or otherwise preserved to Europe due the increasing demand of this gourmet delicacy (Sitta et al. 2007b; Sitta and Floriani 2008; Feng et al. 2012; Dentinger and Suz 2014; Feng and Yang 2018). Similarly, allochthonous porcini species introduced to South Africa have long been exported to Europe (Sitta et al. 2007b). However, as far as the endemic southern hemisphere species (*B. castanopsidis*, *B. austroedulis*, *B. semigastroideus*) are concerned and based on available knowledge, no tradition of human consumption appears to exist.

Even though the present work is the most inclusive list of porcini mushrooms to date, further investigation will be needed in order to assess the definitive number of species worldwide and to better elucidate their ecology, biogeographic distribution, and evolutionary patterns.

Acknowledgments The present author is indebted to L.A. Parra (Aranda de Duero, Spain) for his invaluable revision of the nomenclature. Special thanks are due to R.E. Halling (New York, USA), B.T.M. Dentinger (London, England), E. Horak (Innsbruck, Austria), and G. Simonini (Reggio Emilia, Italy) for their comments upon taxonomy and distribution range of porcini mushrooms. T. Svetasheva (Tula, Russia), V. Janda (Prague, Czech Republic), G. Simonini, L.A. Parra, M. Mikšík (Prague, Czech Republic), W. Klofac (Michelbach, Austria), S. Nelles (Brisbane, Australia), I. Hosen and M. Zhang (Guangdong, China), and A. Vizzini (Turin, Italy) also provided precious bibliographic sources. I am also grateful to B. Feng (Yunnan, China) for reviewing the text and for his useful suggestions and comments for the improvement of the manuscript. Finally, I am obliged to T.J. Baroni (Cortland, USA), A.E. Bessette (Syracuse, USA), B. Feng (Kunming, China), R. Flores-Arzù (Guatemala City, Guatemala), J.L. Frank (Ashland, USA), R.E. Halling (New York, USA), E. Nagasawa (Tottori, Japan), N. Sitta (Lizzano in Belvedere, Italy), Y. Taneyama (Nagano, Japan), and N.-K. Zeng (Haikou, China) for sending colored pictures of several porcini mushrooms from around the world.

References

Adhikari MK, Durrieu G (1996) Ethnomycologie Nepalaise. Bull Soc Mycol France 112:31–41

Adhikary RK, Baruah P, Kalita P et al (1999) Edible mushrooms growing in the forests of Arunachal Pradesh. Adv Hort Forest 6:119–123

Akulov OY, Leontyev DV (2008) Fungi of the Red Data Book of Ukraine from the National Nature Park Gomilshanski Lisy. Ukr J Bot 65(4):586–589

Alessio CL (1978) Il porcino. Saggio sulle specie di boleti incluse in questa denominazione. Edagricole, Bologna

Alessio CL (1985) Boletus Dill. ex L. Fungi Europaei 2. Giovanna Biella, Saronno

Alessio CL (1991) Boletus Dill. Ex L. (Supplemento). Fungi Europaei 2a. Giovanna Biella, Saronno

Alvarez ML (2008) Micoflora (Basidiomycota) de los eucaliptales del NO de la Península Ibérica. Guineana 14:1–502

Anand N, Mathur A, Chowdhry PN (2014) First report on survey of macro fungal biodiversity in Rajouri Dist. (J&K), India. World J Pharm Sci 3(12):1385–1412

Angarano B (1989) Boletus separans Peck. Boll Gr Micol Bres 32(5-6):34–35

Arora D (1986) Mushrooms demystified, 2nd edn. Ten Speed Press, Berkeley

Arora D (1991) All that the rain promises, and more: a hip pocket guide to western mushrooms. Ten Speed Press, Berkeley

Arora D (2008) California porcini: three new taxa, observation on their harvest, and the tragedy of no commons. Economic Bot 62(3):356–375

Arora D, Frank JL (2014a) Clarifying the butter boletes: a new genus, Butyriboletus, is established to accommodate Boletus sect. Appendiculati, and six new species are described. Mycologia 106(3):464–480. https://doi.org/10.3852/13-052

Arora D, Frank JL (2014b) Boletus rubriceps, a new species of porcini from the southwestern USA. North American Fungi 9(6):1–11. https://doi.org/10.2509/naf2014.009.006

Asef MR (2008) Macrofungi of Arasbaran (N.W. Iran) 2. Boletes (families *Boletaceae* and *Suillaceae*). Rostaniha 9(2):210–229

Águeda B, Parladé J, de Miguel AM et al (2006) Characterization and identification of field ectomycorrhizae of Boletus edulis and Cistus ladanifer. Mycologia 98(1):23–30. https://doi.org/10.3852/mycologia.98.1.23

Águeda B, Parladé J, Fernández-Toirán LM et al (2008) Mycorrhizal synthesis between Boletus edulis species complex and rockroses (Cistus sp.). Mycorrhiza 18(8):443–449. https://doi.org/10.1007/s00572-00

Bakkali YSE, Kerdouh B, Mousain D et al (2009) Phylogenetic diversity of Moroccan cork oak woodlands fungi. Biotechnol Agron Soc Environ 13(4):521–528

Ben Hassine Ben Ali M, Stephenson SL (2016a) A report on the ethnomycological use of species of mushrooms found in northwestern Tunisia. Micol Veget Medit 31(1):53–60

Ben Hassine Ben Ali M, Stephenson SL (2016b) Preliminary checklist of the macrofungi from northwestern Tunisia. Curr Res Environ Appl Mycol 6(1):35–44. https://doi.org/10.5943/cream/6/1/4

Berkeley MJ (1852) Decades of fungi. Decades XXXIX-XL. Sikkim and Khassya (Indian) fungi, collected by Dr. J.D. Hooker. Hooker's J Bot Kew Gard Misc 4:130–142

Bertault R (1979) Bolets du Maroc. Bull Soc Mycol France 95(3):297–318

Bessette AE, Roody WC, Bessette AR (2000) North American boletes. A color guide to the fleshy pored mushrooms. Syracuse University Press, Syracuse

Bessette AE, Roody WC, Bessette AR, Dunaway DL (2007) Mushrooms of the southeastern United States. Syracuse University Press, Syracuse

Bessette AE, Roody WC, Bessette AR (2016) Boletes of eastern North America. Syracuse University Press, Syracuse

Bessette AE, Bessette AR, Lewis DP (2019) Mushrooms of the Gulf Coast States: A field guide to Texas, Louisiana, Missisipi, Alabama, and Florida. University of Texas Press, Austin

Beugelsdijk DCM, van der Linde S, Zuccarello GC et al (2008) A phylogenetic study of Boletus section Boletus in Europe. Persoonia 20:1–7

Bi ZS, Li TH, Zheng GY et al (1984) Basidiomycetes from Dinghu Mountain of China III. Some species of *Boletaceae* (2). Acta Mycol Sin 3(4):199–206; (in Chinese)
Bi ZS, Zheng GY, Li TH (1990) Macrofungus flora of the mountainous district of north Guangdong. Guangdong Science & Technology Press, Guangzhou; (in Chinese)
Bi ZS, Zheng GY, Li TH (1994) Macrofungus flora of Guangdong Province. The Chinese University Press, Hongkong; (in Chinese)
Bi ZS, Li TH, Zhang WM et al (1997) A preliminary agaric flora of Hainan Province. Guangdong Higher Education Press, Guangzhou; (in Chinese)
Binder M, Hibbett DS (2006) Molecular systematics and biological diversification of Boletales. Mycologia 98(6):971–981. https://doi.org/10.3852/mycologia.98.6.971
Binion DE, Burdsall HH Jr, Stephenson SL et al (2008) Macrofungi associated with oaks of eastern North America. West Virginia University Press, Morgantown
Binyamini N (1975) Fleshy fungi of Israel (Agaricales). Hakibbutz Hameuchad Publishing House, Tel Aviv; (in Hebrew)
Boa ER (2004) Wild edible fungi: a global overview of their use and importance to people. Non-wood forest products, vol 17. Food and Agriculture Organization of the United Nations, Rome
Borghi E, Galli F, Rinaldi L (2000) Boletus violaceofuscus. Atti Centro Studi Flora Medit 12:1–20
Boruah P, Adhikary RK, Kalita P et al (1996) Some edible fungi growing in the forest of East Khasi Hills (Meghalaya). Adv For Res India 14:214–219
Both EE (1993) The boletes of North America. A compendium. Buffalo Museum of Science, Buffalo
Bozok F, Assyov B, Taşkin H (2019) First records of Exsudoporus permagnificus and Pulchroboletus roseoalbidus (Boletales) in association with non-native Fagaceae, with taxonomic remarks. Phytologia Balcanica 25(1):13–27
Bozok F, Assyov B, Taşkın H, Doğan HH, Büyükalaca S (2020) Molecular phylogenetic studies of Turkish boletes with emphasis on some recently described species. Nova Hedwigia (in press) https://doi.org/10.1127/nova_hedwigia/2019/0563
Bran MC, Morales O, Cáceres R et al (2002) Hongos comestibles de Guatemala: Diversidad, cultivo y nomenclatura vernácula. (fase II). Informe de Avance. Dirección General de Investigación, Guatemala City
Bran MC, Morales O, Flores Arzù R et al (2003a) Hongos comestibles de Guatemala: diversidad, cultivo y nomenclatura vernácula (Fase III). Informe Técnico Final, Dirección General de Investigación. Universidad de San Carlos de Guatemala, Guatemala City
Bran MC, Morales O, Cáceres R et al (2003b) Contribución al conocimiento de los hongos comestibles de Guatemala. IIQB–USAC. Revista Científica. (Special issue):1–24
Breitenbach J, Kränzlin F (1991) Fungi of Switzerland. Vol. 3. Boletes and agarics. Verlag Mykologia, Luzern, p 1
Buller AHR (1914) The fungus lore of the Greeks and Romans. Trans Br Mycol Soc 5:21–66
Calzada DAC (2007) Guía de los boletos de España y Portugal. Náyade Editorial, Medina del Campo
Camacho FJ, Bruns T (2008) The molecular systematics of the Western U.S. edible ectomycorrhizal mushrooms in the Boletus edulis complex, Boletus section Edules. Department of Plant and Microbial Biology University of California, Berkeley. http://nature.berkeley.edu/brunslab/ftp/WesternEdulisGroup.pdf. Accessed 12 July 2016
Cancarè F (2015) Primi ritrovamenti di Boletus aestivalis in eucalipteti dell'areale mediterraneo. Riv Micol 58(1):53–60
Catcheside PS, Catcheside DEA (2012) Boletus edulis (Boletaceae), a new record for Australia. J Adel Bot Gard 25:5–10
Cetto B (1989) I Funghi dal vero, vol 6. Saturnia, Trento
Chakraborty D, Das K, Baghela A et al (2017) A new species of porcini mushroom from India with morphology and phylogeny. Nova Hedwigia 105(1–2):197–204. https://doi.org/10.1127/nova_hedwigia/2017/0405
Chandrasrikul A (1996) Mushrooms of Thailand. Thai Watthanapanich Publishing, Bangkok; (in Thai)

Che G, Flores Arzù R (2010) Descripción y comparación de estructuras microscópicas de Boletus pinophilus Pilát & Dermek, de distintas procedencias de Guatemala, respecto a las descritas para Europa. Rev Cient 19:86–91

Chen CM, Huang HW, Yeh KW (1997) The boletes of Taiwan (VII). Taiwania 42:174–179

Chiu WF (1948) The boletes of Yunnan. Mycologia 40(2):199–231

Chiu WF (1957) Atlas of the Yunnan boletes - bolete flora of Yunnan. Science Press, Beijing; (in Chinese)

Chung P, Pinilla JC, Casanova K et al (2010) Incorporación de Boletus edulis y Boletus pinicola en plantaciones de Pinus radiata en Chile. In: Martínez-Carrera D, Curvetto N, Sobal M et al (eds) Hacia un desarrollo sostenible del sistema de producción-consumo de los hongos comestibles y medicinales en Latinoamérica: Avances y perspectivas en el siglo XXI. La red latinoamericana de hongos comestibles y medicinales, Capítulo 9. Puebla, pp 125–140

Coetzee JC (2011) Preliminary checklist of the poroid mushrooms ('Boletes', Boletales) recorded from South Africa. MycoAfrica 4(2):1–4

Coker WC, Beers AH (1943) The Boletaceae of North Carolina. University of North Carolina Press, Chapel Hill

Corner EJH (1972) Boletus in Malaysia. Government Printing Office, Singapore

Crous PW, Carnegie AJ, Wingfield MJ et al (2019) Fungal Planet description sheets: 868–950. Persoonia 42:291–473. https://doi.org/10.3767/persoonia.2019.42.11

Cui YY, Feng B, Wu G et al (2016) Porcini mushrooms (Boletus sect. Boletus) from China. Fungal Div 81(1):189–212. https://doi.org/10.1007/s13225-015-0336-7

Cunningham GH (1942) Two additional New Zealand Gasteromycetes. New Zealand J Sci Technol 23:172B–173B

Cunningham GH (1944) The Gasteromycetes of Australia and New Zealand. John McIndoe, Dunedin

Dai YC, Zhou LW, Yang ZL et al (2010) A revised checklist of edible fungi in China. Mycosystema 29(1):1–21; (in Chinese)

Das AK, Das N (2014) Diversity of mycoflora in coniferous forests of munsiary and its adjoining areas of Uttarakhand, India. Int J Pl Anim Environ Sci 4(4):21–24

De Rezende-Pinto MC (1940) Notas micológicas. Brotéria 9(3):91–93

Dentinger BTM (2007) Systematics and evolution of porcini and clavarioid mushrooms. PhD thesis, University of Minnesota

Dentinger BTM (2013) Nomenclatural novelties. Index Fungorum 29:1

Dentinger BTM, Suz LM (2014) What's for dinner?: undescribed species of porcini in a commercial packet. PeerJ 2:e570. https://doi.org/10.7717/peerj.570

Dentinger BTM, Ammirati JF, Both EE et al (2010) Molecular phylogenetics of porcini mushrooms (Boletus section Boletus). Mol Phyl Evol 57(3):1276–1292. https://doi.org/10.1016/j.ympev.2010.10.004

Desjardin DE, Binder M, Roekring S et al (2009) Spongiforma, a new genus of gastroid boletes from Thailand. Fungal Div 37:1–8

Desjardin DE, Peay KG, Bruns TD (2011) Spongiforma squarepantsii, a new species of gasteroid bolete from Borneo. Mycologia 103(5):1119–1123

Dick EA, Snell WH (1965) Notes on boletes. XV. Mycologia 57(3):448–458

Doidge EM (1950) The South African fungi and lichens to the end of 1945. Bothalia 5:1–1094

Donk MA (1955) The generic names proposed for Hymenomycetes. IV. Boletaceae. Reinwardtia 3(2):275–313

El-Assfouri A, Ouazzani-Touhami A, Douira A (2006) Inventaire des champignons comestibles de la forêt de la Mamora (Maroc). Bull Ass Entr Mycol Bot Appl 46:5–9

Endo N, Kawamura F, Kitahara R et al (2014) Synthesis of Japanese Boletus edulis ectomycorrhizae with Japanese red pine. Mycoscience 55(1):405–416. https://doi.org/10.1016/j.myc.2013.11.008

Engel H, Krieglsteiner GJ, Dermek A et al (1983) Dickröhrlinge. Die Gattung Boletus in Europa. Verlag Heinz Engel, Weidhausen bei Coburg

Estadès A, Lannoy G (2004) Les bolets européens. Bull Féd Mycol Dauphinée–Savoie 174(3):3–79

Evenson VS (1997) Mushrooms of Colorado and the southern Rocky Mountains. Denver Botanic Press, Denver

Farid A, Gelardi M, Angelini C et al (2018) Phylloporus and Phylloboletellus are no longer alone: Phylloporopsis gen. nov. (Boletaceae), a new smooth-spored lamellate genus to accommodate the American species Phylloporus boletinoides. FUSE 2:341–359. https://doi.org/10.3114/fuse.2018.02.10

Fedorenko VA (2019) Annotated checklist of Basidiomycota new to Republic of Kazakhstan. Curr Res Environ Appl Mycol (J Fungal Biol) 9:271–287. https://doi.org/10.5943/cream/9/1/23

Fedosova AG, Kovalenko AE (2014) Section Boletus of the genus Boletus in Russia: Species diversity, morphology, ecology and phylogeny. Mycol Phytopat 48(2):86–106; (in Russian)

Feng B, Yang ZL (2018) Studies on diversity of higher fungi in Yunnan, southwestern China: a review. Plant Div 40:165–171

Feng B, Xu J, Wu G et al (2012) DNA sequence analyses reveal abundant diversity, endemism and evidence for Asian origin of the porcini mushrooms. PLoS One 7(5):e37567. https://doi.org/10.1371/journal.pone.0037567

Feng B, Liu JW, Xu JP et al (2017) Ecological and physical barriers shape genetic structure of the Alpine porcini. Mycorrhiza 27(3):261–272. https://doi.org/10.1007/s00572-016-0751-y

Flores Arzù R, Simonini G (2000) Contributo alla conoscenza delle Boletales de Guatemala. Riv Micol 43(2):121–145

Flores Arzù R, Bran MC, Rodríguez E et al (2002) Hongos micorrícicos de bosques de pino y pinabete. Dirección General de Investigación – DIGI, Universidad de San Carlos de Guatemala, Guatemala City

Flores Arzú R, Comandini O, Rinaldi AC (2012) A preliminary checklist of macrofungi of Guatemala, with notes on edibility and traditional knowledge. Mycosphere 3(1):1–21

Floriani M, Simonini G, Sitta N (2000) Boletus violaceofuscus, una specie orientale commercializzata in Italia. Boll Gr Micol Bres 43(3):4–15

Fries EM (1815) Observationes mycologicae, vol 1. Gehr. Bonnier, Copenhagen

Fries EM (1821) Systema Mycologicum, vol 1. Ernest Mauritius, Greifswald

Fries EM (1838) Epicrisis systematis mycologici seu synopsis hymenomycetum. Typographia Academica, Uppsala

Galli R (2013) I Boleti, Atlante pratico-monografico per la determinazione dei boleti, 4th edn. Micologica, Vergiate

Gao Q, Yang ZL (2010) Ectomycorrhizal fungi associated with two species of Kobresia in an alpine meadow in the eastern Himalaya. Mycorrhiza 20(4):281–287. https://doi.org/10.1007/s00572-009-0287-5

García-Jiménez J (2013) Diversidad de macromicetos en el estado de Tamaulipas, México. PhD en Ciencias Naturales, Universidad Autónoma de Nuevo León, Linares

Gelardi M, Vizzini A, Ercole E et al (2015) Circumscription and taxonomic arrangement of Nigroboletus roseonigrescens gen. et sp. nov., a new member of Boletaceae from tropical south-eastern China. PloS One 10(8):e0134295. https://doi.org/10.1371/journal.pone.0134295

Giri A, Rana P (2007) Some higher fungi from Sagarmatha National Park (SNP) and its adjoining areas, Nepal. Sci World 5(5):67–74. https://doi.org/10.3126/sw.v5i5.2659

Goldman GB, Gryzenhout M (2019) Field guide to mushrooms & other fungi of South Africa. Struik Nature, Cape Town

Grand LF, Smith AH (1971) A previously unrecognized southern species of Boletus. Mycologia 63(1):114–117

Grund DW, Harrison AK (1976) Nova Scotian Boletes. Bibliotheca Mycologica 47:1–283

Gryzenhout M (2010) Mushrooms of South Africa. In: Pocket guide. Struik Nature, Cape Town

Guinberteau J (2019) Découverte de Boletus mamorensis en Corse 1ère récolte non marocaine. Bull FAMM 55(1):25–32

Haimed M, Nmichi A, Touhami AO et al (2013) Bibliographic inventory of Moroccan Central Plateau fungi. J Anim Plant Sci 18(2):2723–2749

Haimed M, El-Kholfy S, El-Assfouri A et al (2015) Inventory of Basidiomycetes and Ascomycetes harvested in the Moroccan Central Plateau. Int J Pure Appl Bio 3(1):100–108

Hall IR, Buchanan PK, Wang Y et al (1998a) Edible and poisonous mushrooms: an introduction. New Zealand Institute for Crop and Food Research Limited, Christchurch

Hall IR, Lyon AJE, Wang Y et al (1998b) Ectomycorrhizal fungi with edible fruiting bodies – 2. Boletus edulis. Econ Bot 52(1):44–56. https://doi.org/10.1007/BF02861294

Hall IR, Stephenson S, Buchanan P et al (2003) Edible and poisonous mushrooms of the world. Timber Press, Portland

Halling RE, Both EE (1998) Generic affinity of Boletus separans. Bull Buff Soc Nat Sci 36:239–243

Halling RE, Mueller GM (1999) New boletes from Costa Rica. Mycologia 91(5):893–899

Halling RE, Mueller GM (2005) Common mushrooms of the Talamanca mountains, Costa Rica. Memoirs of the New York Botanical Garden, vol 90. New York Botanical Garden Press, Bronx

Halling RE, Desjardin DE, Fechner N et al (2014) New porcini (Boletus sect. Boletus) from Australia and Thailand. Mycologia 106(4):830–834. https://doi.org/10.3852/13-340

Halling RE, Fechner N, Nuhn M et al (2015) Evolutionary relationships of Heimioporus and Boletellus (Boletales), with an emphasis on Australian taxa including new species and new combinations in Aureoboletus, Hemileccinum and Xerocomus. Austr Syst Bot 28:1–22. https://doi.org/10.1071/SB14049

Heluta VP (2012) An overview of species of the genus Boletus L. as candidates for inclusion in the Red Data Book of Ukraine. http://www.researchgate.net/publication/285417703. Accessed 25 July 2016

Henkel TW, Obase K, Husbands D et al (2016) New Boletaceae taxa from Guyana: Binderoboletus segoi gen. and sp. nov., Guyanaporus albipodus gen. and sp. nov., Singerocomus rubriflavus gen. and sp. nov., and a new combination for Xerocomus inundabilis. Mycologia 108(1):157–173. https://doi.org/10.3852/15-075

Hernández SG, López AR (2006) Encuesta sobre los hongos comestibles de la región del Cofre de Perote, Ver. México. http://www.scribd.com/doc/53617/encuesta-sobrelos-hongos-del-cofre-de-perote. Accessed 13 July 2016

Hongo T (1960) The Agaricales of Japan 1–2. Rhodophyllaceae, Paxillaceae, Gomphidiaceae, Boletaceae and Strobilomycetaceae. Acta Phyto Geo 18(4):97–112

Hongo T (1973) Enumeration of the Hygrophoraceae, Boletaceae and Strobilomycetaceae. Mycological reports from New Guinea and the Solomon Islands (16–21). Bull Nat Sci Mus Tokyo 16(3):537–557

Horak E (1971) Contributions to the knowledge of the Agaricales s.l. (Fungi) of New Zealand. New Zealand J Bot 9:463–493

Horak E (1980) Indian Boletales and Agaricales. Revisions and new taxa. Sydowia 33:88–110

Horak E (2011) Revision of Malaysian species of Boletales s. l. (Basidiomycota) described by E. J. H. Corner (1972, 1974). Malayan Forest Records 51, Kuala Lumpur

Horn B, Kay R, Abel D (1993) A guide to Kansas mushrooms. Kansas UP

Huang NL (1999) Colored illustrations of macrofungi (mushrooms) of China. China Agricultural Press, Beijing; (in Chinese)

Ibarra LV, Dávalos LG, Luz AK (1999) Inventario preliminar de los hongos macroscópicos de la comunidad Wirrárixa (Huichol) de Tateikie (San Andrés Cohamiata), Jalisco, México. In: Ibarra LV (ed) Etnomicología de la etnia Wirrárixa (Huichol), Jalisco, México. Tesis de Maestría, Capítulo 2, Universidad de Guadaljara, pp 27–49

Imazeki R, Hongo T (1987) Coloured illustrations of mushrooms of Japan, vol 1. Hoikusha Publishing, Osaka; (in Japanese)

Imazeki R, Otani Y, Hongo T (1988) Fungi of Japan. Yama-kei Publishers, Tokyo; (in Japanese)

Iotti M, Barbieri E, Stocchi V et al (2005) Morphological and molecular characterisation of mycelia of ectomycorrhizal fungi in pure culture. Fungal Div 19:51–68

Ivanov AI (2014) Agaricomycetes of Volga upland. Order Boletales.. Privately published (in Russian)

Karun NC, Sridhar KR (2014) A preliminary study on macrofungal diversity in an arboretum and three plantations of the southwest coast of India. Curr Res Environ Appl Mycol 4(2):173–187. https://doi.org/10.5943/Cream/4/2/5

Kibby GG (2011) British boletes with keys to species. Privately published, London

Klofac W (2007) Schlüssel zur Bestimmung von Frischfunden der europäischen Arten der Boletales mit röhrigem Hymenophor. Öster Z Pilzk 16:187–279

Klofac W (2010) The genus Aureoboletus, a world-wide survey. A contribution to a monographic treatment. Öster Z Pilzk 19:133–174

Kluzák Z (1992) Nový druh hřibu, hřib vínový – Boletus vinosulus. Sbor Jihočes Muz v Čes Budějovicích, Přír Vědy 32:43–47

Knudsen H (1992) Boletus Dill. ex L.: Fr. In: Hansen L, Knudsen H (eds) Polyporales, Boletales, Agaricales, Russulales. Nordic macromycetes, vol 2. Nordswamp, Copenhagen

Knudsen H, Taylor AFS (2008) Boletales E.-J. Gilbert. In: Knudsen H, Vesterholt J (eds) Funga Nordica – Agaricoid, boletoid and cyphelloid genera. Nordsvamp, Copenhagen

Korhonen M, Liimatainen K, Niskanen T (2009) A new boletoid fungus, Boletus pinetorum, in the Boletus section Boletus from Fennoscandia (Basidiomycota, Boletales). Karstenia 49(2):41–60

Kuo M (2007) Boletus barrowsii. http://www.mushroomexpert.com/boletus_barrowsii.html. Accessed 12 July 2016

Kuo M (2010a) Boletus atkinsonii. http://www.mushroomexpert.com/boletus_atkinsonii.html. Accessed 10 July 2016

Kuo M (2010b) Boletus cf. reticulatus. http://www.mushroomexpert.com/boletus_cf_reticulatus.html. Accessed 11 July 2016

Kuo M (2010c) Boletus cf. edulis. http://www.mushroomexpert.com/boletus_cf_edulis.html

Kuo M (2010d) Boletus cf. pinophilus. http://www.mushroomexpert.com/boletus_cf_pinophilus.html. Accessed 11 July 2016

Kuo M (2010e) Boletus variipes. http://www.mushroomexpert.com/boletus_variipes.html. Accessed 13 July 2016

Kuo M (2013) Boletus subalpinus. http://www.mushroomexpert.com/boletus_subalpinus.html. Accessed 14 July 2016

Kuo M (2015) Boletus separans. http://www.mushroomexpert.com/boletus_separans.html. Accessed 15 July 2016

Kuo M, Methven AS (2014) Mushrooms of the Midwest. University of Illinois, Champaign

Kuo M, Ortiz-Santana B (2020) Revision of leccinoid fungi, with emphasis on North American taxa, based on molecular and morphological data. Mycologia 112:197–211. https://doi.org/10.1080/00275514.2019.1685351

Læssøe T, Pedersen OS, Sysouphanthong P (2018) An introduction to the edible, poisonous and medicinal fungi of northern Laos. Vientiane

Lakhanpal TN (1996) Mushrooms of India: Boletaceae. In: Mukherji KG (ed) Studies in cryptogamic botany, vol 1. APH Publishing Corporation, New Delhi

Lakhanpal TN, Sagar A (1989) Fleshy fungi of N.W. Himalayas – XIV. Three species of Boletus new to India. Kavaka 17(1–2):32–37

Lannoy G, Estadès A (2001) Flore mycologique d'Europe 6 – Les bolets. Doc Mycol Mém hors série 6, Lille

Larouz B (2007) Biodiversité fongique du Maroc: Etude des champignons supérieur du Moyen Atlas. Thèse de Doctorat, Université Ibn Tofaïl

Lavorato C (1991) Chiave analitica e note bibliografiche della micoflora del cisto. Boll AMER 24(3):16–45

Lee SS, Watling R, Noraini-Sikin Y (2002) Ectomycorrhizal basidiomycota fruiting in lowland rain forests of peninsular Malaysia. Bois et Forêts des Tropiques 274(4):33–43

Leonardi M, Paolocci F, Rubini A et al (2005) Assessment of inter-and intra-specific variability in the main species of Boletus edulis complex by ITS analysis. FEMS Microbiol Lett 243:411–416

Li TH, Song B (2000) Chinese boletes: a comparison of boreal and tropical elements. In: Walley AJS (ed) Tropical Mycology 2000. BMS The Millennium Meeting on Tropical Mycology. Liverpool John Moores University, Liverpool, p 1

Li YC, Feng B, Yang ZL (2011) Zangia, a new genus of Boletaceae supported by molecular and morphological evidence. Fungal Div 49(1):125–143. https://doi.org/10.1007/s13225-011-0096-y

Li YC, Ortiz-Santana B, Zeng NK et al (2014a) Molecular phylogeny and taxonomy of the genus Veloporphyrellus. Mycologia 106(2):291–306. https://doi.org/10.3852/106.2.291
Li YC, Li F, Zeng NK et al (2014b) A new genus Pseudoaustroboletus (Boletaceae, Boletales) from Asia as inferred from molecular and morphological data. Mycol Progr 13(4):1207–1216. https://doi.org/10.1007/s11557-014-1011-1
Lian B, Zang JP, Hou WG et al (2008) PCR-based sensitive detection of the edible fungus Boletus edulis from rDNA ITS sequences. Electr J Biotechnol 11(3):1–8
Loizides M, Bellanger J-M, Assyov B et al (2019) Present status and future of boletoid fungi (Boletaceae) on the island of Cyprus: Cryptic and threatened diversity unravelled by ten-year study. Fungal Ecol 41:65–81. https://doi.org/10.1016/j.funeco.2019.03.008
López AR, García JA (2009) Xerocomus tumidus. Funga Veracruzana 109:1–2
Manavella M (2007) Funghi d'alta quota nelle Alpi Cozie – Boletus edulis e Boletus aestivalis in zona alpina – 3° contributo. Il Micologo 26(119):3–15
Manavella M (2010) Sulla presenza dei Boleti del "Gruppo"di Boletus edulis Bull.: Fr. nel Piemonte sud-occidentale. Il Micologo 35(128):3–25
Mao XL (2000) The Macrofungi of China. Henan Science and Technology Press, Zhengzhou; (in Chinese)
Mao XL (2009) Macromycetes of China. Science Press, Beijing; (in Chinese)
Marais LJ, Kotzé JM (1977) Notes on ectotrophic mycorrhizae of Pinus patula in South Africa. South Afr For J 100:61–71. https://doi.org/10.1080/00382167.1977.9629436
Masuka AJ (1996) Dynamics of mushroom (Boletus edulis) production in pine plantations in Zimbabwe. J Appl Sci South Africa 2(2):69–76. https://doi.org/10.4314/jassa.v2i2.16877
McNeil R (2006) Le grand livre des champignons du Quebec et de l'est du Canada. Editions Michel Quintin, Waterloo
Mediavilla O, Olaizola J, Santos-del-Blanco L et al (2016) Mycorrhization between Cistus ladanifer L. and Boletus edulis Bull is enhanced by the mycorrhiza helper bacteria Pseudomonas fluorescens Migula. Mycorrhiza 26(2):161–168
Mello A (2012) State of the art of the research on Boletus edulis. In: Zambonelli A, Bonito GM (eds) Edible ectomycorrhizal mushrooms. Springer-Verlag, Berlin/Heidelberg, pp 73–82
Mello A, Ghignone S, Vizzini A et al (2006) ITS primers for identification of marketable boletes. J Biotechnol 121:318–329
Mikšik M (2017) Hřibovité Houby Evropy. Vydalo nakladatelství Svojtka & Co., Praha; (in Czech)
Miller HR, Miller OK (2006) North American mushrooms: a field guide to edible and inedible Fungi. Falcon Guide, Guilford
Montoya-Esquivel A (1998) Ethnomycology of Tlaxcala, Mexico. McIlvainea 13(2):6–12
Montoya-Esquivel A, Estrada-Torres A, Kong A et al (2001) Commercialization of wild mushrooms during market days of Tlaxcala, Mexico. Micol Apl Int 13:31–40
Montoya-Esquivel A, Torres-García EA, Kong A et al (2012) Gender differences and regionalization of the cultural significance of wild mushrooms around La Malinche Volcano, Tlaxcala, Mexico. Mycologia 104(4):826–834
Moor D, Brodmann P, Nicholas G et al (2002) Polymerase chain reaction (PCR) for the detection of king bolete (Boletus edulis) and slippery jack (Suillus luteus) in food samples. Eur Food Res Tech 214(4):340–345
Morales O, Bran MC, Cáceres R (2010) Los hongos comestibles de uso tradicional en Guatemala. In: Martínez-Carrera D, Curvetto N, Sobal M et al (eds) Hacia un desarrollo sostenible del sistema de producción-consumo de los hongos comestibles y medicinales en Latinoamérica: Avances y perspectivas en el siglo XXI. La red latinoamericana de hongos comestibles y medicinales, Capítulo 25. Puebla, pp 437–464
Mortimer PE, Karunarathna SC, Li QH et al (2012) Prized edible Asian mushrooms: ecology, conservation and sustainability. Fungal Div 56(1):31–47
Muñoz JA (2005) Boletus s.l. (excl. Xerocomus). Fungi Europaei, vol 2. Edizioni Candusso, Alassio
Murrill WA (1909) The Boletaceae of North America – II. Mycologia 1(4):140–158. https://doi.org/10.2307/3753125

Murrill WA (1910) Boletaceae. North American flora. Agaricales. Family 6. Bull N Y Bot Gard 9(3):133–161
Murrill WA (1914) American Boletes. Privately published, New York
Mustafabayli EH, Aghayeva DN (2019) New records of mushrooms for the mycobiota of Azerbaijan. Ukrainian Bot J 76(4):356–361
Nagasawa E (1994) A new species of Boletus sect. Boletus from Japan. Proc Japan Acad 70(Ser B):10–14
Noordeloos ME, den Bakker HC, van der Linde S (2018) Order Boletales. In: Noordeloos ME, Kuyper TW, Somhorst I et al (eds) Flora Agaricina Neerlandica, vol 7. Candusso Editrice, Origgio, pp 1–225
Nouhra ER, Castellano MA, Trappe JM (2002) NATS truffle and truffle-like fungi 9: Gastroboletus molinae sp. nov (Boletaceae, Basidiomycota), with a revised key to the species of Gastroboletus. Mycotaxon 83:409–414
Nounsi A, Outcoumit A, Selmaoui K et al (2014) Inventaire des champignons ectomycorhiziens du Maroc. J Appl Biol 79:6826–6854
Nuhn ME, Binder M, Taylor AFS et al (2013) Phylogenetic overview of the Boletineae. Fungal Biol 117(7–8):479–511. https://doi.org/10.1016/j.funbio.2013.04.008
Olsson U (1997) Boletus separans Peck – en ny ädelsopp I Sverige. Jordstjärnan 18(1):23–25
Oria de Rueda JA, Martìn-Pinto P, Olaizola J (2008) Bolete productivity of cistaceous scrublands in northwestern Spain. Econ Bot 62(3):323–330
Orihara T, Smith ME (2017) Unique phylogenetic position of the African truffle-like fungus, Octaviania ivoryana (Boletaceae, Boletales), and the proposal of a new genus, Afrocastellanoa. Mycologia 109(2):323–332. https://doi.org/10.1080/00275514.2017.1301750
Orlovich D, Stringer A, Yun W et al (2004) Boletus edulis Bull. ex Fries in New Zealand. Austr Mycol Soc News 1(1):6
Ortiz-Santana B, Lodge DJ, Baroni TJ et al (2007) Boletes from Belize and the Dominican Republic. Fungal Div 27:247–416
Ortiz-Santana B, Bessette AE, McConnell OL (2016) Boletus durhamensis sp. nov. from North Carolina. Mycotaxon 131(3):703–715. https://doi.org/10.5248/131.703
Outcoumit A, Yamni K, Ouazzani TA et al (2006) Boletus mamorensis, le cèpe de la Mamora (Maroc). Bull Soc Linn Bordeaux 141(34):169–174
Outcoumit A, El-kholfy S, Ouazzani TA et al (2014) Bibliographic inventory of Tangier fungi: catalogue of the Basidiomycetes fungal flora. Int J Plant Animal Environ Sci 4(1):205–256
Parihar A, Hembrom M, Vizzini A et al (2018) Indoporus shoreae gen. et sp. nov. (Boletaceae) from tropical India. Cryptogamie Mycologie 39(4):447–466. https://doi.org/10.7872/crym/v39.iss4.2018.1
Patouillard N, Baker CF (1918) Some Singapore Boletineae. J Straits Brch R Asiat Soc 78:67–72
Paulet JJ (1793) Traité des Champignons, vol 2. Imprimerie royale, Paris
Paulet JJ (1808) Prospectus du traité historique, graphique, culinaire et medical des champignons. Paris
Paulet JJ (1808–1835) Iconographie des Champignons. Paris
Pearson AA (1950) Cape agarics and boleti. Trans Br Mycol Soc 33:276–316
Peck CH (1872) Report of the botanist (1870). Ann Rep N Y State Mus. Nat Hist 24:41–108
Peck CH (1873) V. Description of new species of fungi. Bull Buffalo Soc Nat Sci 1:41–72
Peck CH (1888) Report of the Botanist (1887). Rep Trustees State Mus Nat Hist 41:49–122
Peck CH (1889) Boleti of the United States. Ann Rep N Y State Mus. Nat Hist 2(8):73–166
Peck CH (1897) New species of fungi. Bull Torrey Bot Club 24(3):137–147
Peck CH (1898) New species of Alabama fungi. Bull Torrey Bot Club 25:368–372
Peck CH (1902) Report of the State Botanist (1901). Bull N Y State Mus 54(5):931–991
Peck CH (1905) Report of the State Botanist (1904). Bull N Y State Mus 94(8):1–58
Peck CH (1911) Report of the State Botanist (1910). Bull N Y State Mus 150:1–100
Pegler DN, Young TWK (1981) A natural arrangement of the Boletales, with reference to spore morphology. Trans Br Mycol Soc 76(1):103–146. https://doi.org/10.1016/S0007-1536(81)80013-7
Perič B, Perič O (2006) Contribution to the study of the genus Boletus s.l. in Montenegro. Mycol Montenegrina 9:35–54

Phillips R (1991) Mushrooms of North America. Little Brown Company, Boston
Pilát A (1970) Boletus subaereus species nova ex affinitate Boleti aerei Bull. ex Fr. Bull Soc Mycol Fr 86(4):881–882
Pilát A, Dermek A (1974) Hríbovité huby. Československé hríbovité a sliziakovité huby (Boletaceae – Gomphidiaceae). Veda, Bratislava
Pomerleau R (1980) Flore des champignons au Quebec et regions limitrophes. Les Editions la Presse, Montreal
Ponce RA, Águeda B, Agreda T et al (2011) Rockroses and Boletus edulis ectomycorrhizal association: realized niche and climatic suitability in Spain. Fung Ecol 4(3):224–232. https://doi.org/10.1016/j.funeco.2010.10.002
Pukahuta C, Palasarn W, Pimmongkol A (2009) Taxonomic study of wild edible pore fungi found in Ubon Ratchathani. J Micr Soc Thailand 23(1):30–33
Putzke J, Putzke MTL (2019) Cogumelos (fungos Agaricales) no Brasil - Ordens Boletales (Boletaceae e Paxillaceae), Polyporales (Polyporaceae/Lentinaceae), Russulales (Russulaceae) e Agaricales (Cortinariaceae, Inocybaceae, Pluteaceae e Strophariaceae), vol 2. e-book, São Gabriel
Razaq A, Shahzad S (2013) Newly recorded species of Boletaceae form Pakistan. Pakistan J Bot 45(3):1473–1476
Razaq A, Shahzad S, Ali H et al (2014) New reported species of macro fungi from Pakistan. J Agri Food Appl Sci 2(3):67–71
Redeuilh G (1978) Boletus mamorensis nov. sp. Bull Soc Mycol France 94(3):299–303
Redeuilh G (1988) Etudes nomenclaturales sur les bolets. III. Critique nomenclatural des noms de bolets in "Boletus" (Fungi Europaei) par C.L. Alessio – Saronno 1985. Doc Mycol 18(72):28–39
Redeuilh G (1990) Etudes nomenclaturales sur les bolets. VI. Corrections, additions et commentaries aux etude I–V (in D.M. fasc. 72: 13–49, 1988). A. – Première tranche. Doc Mycol 20(79):25–46
Redeuilh G, Simonini G (1993) Comité pour l'unification des noms de bolets européens. Fiches du groupe I ("groupe edulis"). Bull Soc Mycol France 109(2):77–91
Redeuilh G, Simonini G (1999) Comité pour l'unification des noms de bolets européens. Fiches du groupe IV. Bull Soc Mycol France 115(4):435–453
Reid DA (1973) A reappraisal of type and authentic specimens of basidiomycetes in the Van der Bijl herbarium, Stellenbosch. J South Afr Bot 39:141–178
Robles-García D, Suzán-Azpiri H, Montoya-Esquivel A et al (2018) Ethnomycological knowledge in three communities in Amealco, Quéretaro, México. J Ethnobiol Ethnomed 14:art 7. https://doi.org/10.1186/s13002-017-0202-7
Rodà P (2012) Funghi aspromontani comparati. Boletales. AZ Editrice, Reggio Calabria
Roody WC (2003) Mushrooms of west Virginia and the central Appalachians. University of Kentucky Press, Lexington
Sabra A, Walter S (2001) Non-wood forest products in the Near East: a regional and national overview. Working paper FOPW/01/2. FAO, Rome
Saccardo PA (1888) Sylloge hymenomycetum, Vol. II. Polyporeae, Hydneae, Thelephoreae, Clavarieae, Tremellineae. Sylloge Fungorum 6:1–928
Sarwar S, Khalid AN (2014) Addition to the boletes: Boletus pakistanicus sp. nov. from the coniferous forests of Pakistan. Int J Agric Biol 16:663–667
Sarwar S, Jabeen S, Ahmad I et al (2018) Boletus himalayensis (Basidiomycota; Boletales), a new porcini species from Pakistan. Turkish J Bot 42(6):790–800. https://doi.org/10.3906/bot-1711-19
Scates K (2011) Trial field key to boletes in the Pacific Northwest. http://www.svims.ca/council/Bolete.htm. Accessed 30 July 2016
Schaeffer CJ (1763) Fungorum qui in Bavaria et Palatinatu circa Ratisbonam nascuntur icones nativis coloribus expressae. Tomus secundus, Regensburg
Siegel N, Schwarz C (2016) Mushrooms of the Redwood Coast. A comprehensive guide to the fungi of coastal northern California. Ten Speed Press, Berkeley

Simonini G, Floriani M, Binder M et al (2001) Two close extraeuropean boletes: Boletus violaceofuscus and Boletus separans. Micol Veget Medit 16(2):148–170
Sims KP, Watling R, de la Cruz R et al (1997) Ectomycorrhizal fungi of the Philippines: a preliminary survey and notes on the geographic biodiversity of the Sclerodermatales. Bio Cons 6:45–58
Singer R (1947) The Boletoideae of Florida. The Boletineae of Florida with notes on extralimital species III. Am Mid Nat 37(1):1–135. https://doi.org/10.2307/2421647
Singer R (1967) Die Röhrlinge. Teil II. Die Boletoideaea und Strobilomycetaceae. Die Pilze Mitteleuropas, Band 6. Verlag Julius Klinkhardt, Bad Heilbrunn
Singer R (1977) The Boletineae of Florida. Bibliotheca Mycologica 58:1–312
Singer R (1981) Notes on bolete taxonomy – III. Persoonia 11(3):269–302
Singer R (1986) The Agaricales in modern taxonomy, 4th edn. Koeltz Scientific Books, Koenigstein
Singer R, Araujo I, Ivory MH (1983) The ectotrophically mycorrhizal fungi of the neotropical lowlands, especially central Amazonia (litter decomposition and ectomycorrhiza in Amazonian forests 2). Beih Nova Hedwigia 77:1–352
Singer R, García J, Gómez LD (1991) The Boletineae of Mexico and Central America III. Beih Nova Hedwigia 102:1–99
Sitta N, Davoli P (2012) Edible ectomycorrhizal mushrooms: International markets and regulations. In: Zambonelli A, Bonito GM (eds) Edible ectomycorrhizal mushrooms. Springer, Heidelberg, pp 355–380
Sitta N, Floriani M (2008) Nationalization and globalization trends in the wild mushroom commerce of Italy with emphasis on porcini (Boletus edulis and allied species). Econ Bot 62(3):307–322
Sitta N, Davoli P, Fontana P et al (2007a) I funghi spontanei nel commercio e nell'alimentazione umana. In: Parliamo di funghi – II. Tossicologia, commercializzazione, legislazione, vol 7. Giunta della Provincia Autonoma di Trento, Trento, pp 159–236
Sitta N, Togni N, Zotti M (2007b) Guida alla conoscenza e all'analisi dei funghi secchi. In: Parliamo di funghi – II. Tossicologia, commercializzazione, legislazione, vol 7. Giunta della Provincia Autonoma di Trento, Trento, pp 237–318
Smith AH (1975) A field guide to western mushrooms. University of Michigan Press, Ann Arbor
Smith AH, Smith HV (1973) Notes on Michigan Boletaceae. Persoonia 7:321–331
Smith AH, Thiers HD (1971) The boletes of Michigan. University of Michigan Press, Ann Arbor
Smith AH, Smith HV, Weber NS (1981) How to know the non-gilled mushrooms. Wm C Brown, Dubuque
Smith ME, Amses KR, Elliott TF et al (2015) New sequestrate fungi from Guyana: Jimtrappea guyanensis gen. sp. nov., Castellanea pakaraimophila gen. sp. nov., and Costatisporus cyanescens gen. sp. nov. (Boletaceae, Boletales). IMA Fungus 6(2):297–317. https://doi.org/10.5598/imafungus.2015.06.02.03
Smotlacha F (1935) Boletus sýkorae Sm. – Hřib černohnědý. Čas Čes Houb 15:19–21
Smotlacha F (1951) Hřib slovenský či šaštínský. Čas Čes Houb 28(7–8):102–104
Snell WH (1936) Notes on boletes. IV. Mycologia 28(1):13–23
Snell WH, Dick EA (1958) Notes on boletes. X. A few miscellaneous discussions and a new subspecies. Mycologia 50(1):57–65
Snell WH, Dick EA (1961) Notes on boletes. XIV. Mycologia 53(3):228–236
Snell WH, Dick EA (1970) The boleti of northeastern North America. Cramer, Lehre
Sobestiansky G (2005) Contribution to a macromycete survey of the states of Rio Grande do Sul and Santa Catarina in Brazil. Braz Arch Biol Tech 48:437–457
Soytong K (1994) Mushrooms and macrofungi in Thailand. Sirithum offset Press, Ubon Rachthani; (in Thai)
States JS (1990) Mushrooms and truffles of the Southwest. University of Arizona Press, Tucson
Stephens EL, Kidd MM (1953) Some South African edible fungi. Longmans Southern Africa Ltd, Cape Town
Stringer A, Wang Y, Bulman S et al (2002) Boletus edulis sensu lato down under. In: Hall IR, Wang Y, Zambonelli A et al (eds) Edible mycorrhizal mushrooms and their cultivation. Proceedings

of the Second International Conference on Edible Mycorrhizal Mushrooms. New Zealand Institute for Crop and Food Research Limited, Christchurch, p 5
Stuntz DE, Isaacs BF (1962) Pacific Northwestern fungi. I. Mycologia 54:272–296
Sulzbacher MA, Grebenc T, Jacques RJS et al (2013) Ectomycorrhizal fungi from southern Brazil – a literature-based review, their origin and potential hosts. Mycosphere 4(1):61–95
Šutara J (2005) Central European genera of the Boletaceae and Suillaceae, with notes on their anatomical characters. Czech Mycol 57(1–2):1–50
Šutara J (2014) Anatomical structure of pores in European species of genera Boletus s.str. and Butyriboletus (Boletaceae). Czech Mycol 66(2):157–170
Šutara J, Mikšík M, Janda V (2009) Hřibovité houby. Academia, Praha; (in Czech)
Thiers HD (1975) California mushrooms. A field guide to the boletes. Afner Press, New York
Thiers HD (1976) Boletes of the southwestern United States. Mycotaxon 3(2):261–273
Thiers HD (1989) Gastroboletus revisited. In: Samuels GJ (ed) Mycological contributions celebrating the 70th birthday of Clarck T. Rogerson. Memoirs of the New York Botanical Garden, vol 49. New York Botanical Garden Press, Bronx, pp 355–359
Thiers HD, Trappe JM (1969) Studies in the genus Gastroboletus. Brittonia 21(3):244–254
Thongklam S (2008) Diversity of boletes in some national parks of upper northern Thailand. PhD Dissertation, Chiang Mai University
Trappe JM, Evans F, Trappe JM (2007) Field guide to North American truffles: hunting, identifying, and enjoying the world's most prized fungi. Ten Speed Press, Berkeley
Trappe JM, Castellano MA, Halling RE et al (2013) Australasian sequestrate fungi. 18: Solioccasus polychromus gen. & sp. nov., a richly colored, tropical to subtropical, hypogeous fungus. Mycologia 105(4):888–895. https://doi.org/10.3852/12-046
Trudell S, Ammirati J (2009) Mushrooms of the Pacific Northwest. Timber Press, Portland
Uaciquete A, Dai MDL, Motta H (1996) Distribução, valor economico e uso sustentavel do cogumelo comestível em Moçambique. Grupo de Trabalho Ambiental, Maputo
van der Bijl PA (1925) Descriptions of some previously unnamed South African fungi. South Afr J Sci 22:167–169
van der Linde S, Noordeloos ME (2005) Hoe raak ik thuis in de boleten – 5: Het Eekhoorntjesbrood en verwanten (Boletus sectie Boletus). Coolia 48(2):59–70
van der Westhuizen GCA, Eicker A (1994) Mushrooms of southern Africa (field guide). Struik Publishers, Cape Town
Vasilkov BP (1966) Belyj grib – opyt monografii odnogo vida. Nauka Publishing House, Moskva-Leningrad; (in Russian)
Vasquez G (2014) Indagini micologiche sulle Boletales epigee del terittorio siciliano – Mappatura e censimento delle specie. Tesi di dottorato, Università degli Studi di Catania
Vellinga EC (2004) Genera in the family Agaricaceae: evidence from nrITS and nrLSU sequences. Mycol Res 108:354–377
Verma B, Pandro V (2018) Distribution of boleteceous mushrooms in India, some new records from sal forest of central India. Int J Curr Micr App Sci 7(6):1694–1713
Villareal L, Perez-Moreno J (1989) Los hongos comestibles silvestres de Mexico, un enfoque integral. Micol Neo Apl 2:77–114
Vizzini A, Mello A, Ghignone S et al (2007) Boletus edulis complex: From phylogenetic relationships to specific primers. Pagine Micol 30:48–52
Wang QB (2004) Taxonomy and molecular systematics of Boletus in China. PhD dissertation, Institute of Microbiology Chinese Academy of Sciences (in Chinese)
Wang Y, Hall IR (2004) Edible ectomycorrhizal mushrooms: challenges and achievements. Can J Bot 82(8):1063–1073
Wang QB, Yao YJ (2005) Boletus reticuloceps, a new combination for Aureoboletus reticuloceps. Sydowia 57(1):131–136
Wang Y, Sinclair L, Hall IR et al (1995) Boletus edulis sensu lato: a new record for New Zealand. N Z J Crop Hort Sci 23(2):227–231. https://doi.org/10.1080/01140671.1995.9513892

Wang XH, Liu PG, Yu FQ (2004) Color atlas of wild commercial mushrooms in Yunnan. Yunnan Science and Technology Press, Kunming; (in Chinese)
Wani MP, Bhat MY, Wani AH (2013) Mushroom flora of District Kupwara (Jammu and Kashmir). J Himalayan Ecol Sustian Dev 7:20–27
Watling R (2000) Bresadola, Cesati and Patouillard's old world boletes. In: Papetti C, Consiglio G (eds) Micologia 2000. Associazione Micologica Bresadola, Trento, p 577
Watling R (2001) The relationships and possible distributional patterns of boletes in south-east Asia. Mycol Res 105(12):1440–1448
Watling R (2002) Mycota of some British shrub-plant communities. Feddes Repertorium 113(1–2):161–164
Watling R (2008) A manual and source book on the boletes and their allies. Synopsis fungorum, vol 24. Fungiflora, Oslo
Watling R, Gregory NM (1988) Observations on the boletes of the Cooloola Sandmass, Queensland and notes on their distributions in Australia. Part 2A. Smooth spored taxa – Introduction, keys and references. Proc Royal Soc Queensland 99:45–63
Watling R, Gregory NM (1989) Observations on the boletes of the Cooloola Sandmass, Queensland and notes on their distributions in Australia. Part 2D: Smooth spored taxa – Boletaceae, Xerocomaceae. Proc Royal Soc Queensland 100:31–47
Watling R, Hills AE (2005) Boletes and their allies – Boletaceae, Strobilomycetaceae, Gyroporaceae, Paxillaceae, Coniophoraceae, Gomphidiaceae (revised and enlarged edition). In: Henderson DM, Watling R (eds) British fungus flora, agarics and boleti, vol 1. HMSO, Edinburgh, pp 1–174
Watling R, Li TH (1999) Australian boletes. A preliminary survey. Royal Botanic Garden Edinburgh, Edinburgh
Weber NS, Smith AH (1985) A field guide to southern mushrooms. University of Michigan Press, Ann Arbor
Winkler D (2016) MushRoaming Colombia. http://mushroaming.com/content/colombia-tour. Accessed 02 Aug 2016
Wu XL, Dai YC, Li TH et al (2011) Fungi of tropical China. Science Press, Beijing; (in Chinese)
Wu G, Feng B, Xu JP et al (2014) Molecular phylogenetic analyses redefine seven major clades and reveal 22 new generic clades in the fungal family Boletaceae. Fungal Div 69(1):93–115. https://doi.org/10.1007/s13225-014-0283-8
Wu G, Li YC, Zhu XT et al (2016) One hundred noteworthy boletes from China. Fungal Div 81(1):25–188. https://doi.org/10.1007/s13225-016-0375-8
Yamanaka K, Namba K, Tajiri A (2000) Fruit body formation of Boletus reticulatus in pure culture. Mycoscience 41:189–191
Yan J, Li T, Ma SB et al (2010) Isolation and culture of two bolete strains (Boletus edulis). Acta Edulis Fungi 17:26–31
Yang ZL (2005) Diversity and biogeography of higher fungi in China. In: Xu JP (ed) Evolutionary genetics of fungi. Horizon Bioscience, Norfolk, p 35
Yeh KW, Chen ZC (1981) The boletes of Taiwan (II). Taiwania 26:100–115
Ying JZ, Zang M (1994) Economic macrofungi from southwestern China. Science Press, Beijing; (in Chinese)
Ying JZ, Mao XL, Ma QM et al (1987) Icons of medicinal fungi from China. Science Press, Beijing; (in Chinese)
Yuan MS, Sun PQ (2007) The coloured atlas of mushrooms (macrofungi) in China. Sichuan Science and Technology Press, Chengdu; (in Chinese)
Yun W, Hall IR (2004) Edible ectomycorrhizal mushrooms: challenges and achievements. Can J Bot 82(8):1063–1073
Zang M (1995) Observation on the genera Boletus and Sinoboletus from southwestern China. Bull Nat Sci Mus Tokyo Ser Bot 21(2):67–74
Zang M (2006) Boletaceae (I). Flora fungorum sinicorum, vol 22. Science Press, Beijing; (in Chinese)

Zang M, Yuan MS, Gong MQ (1993) Notes on and additions to Chinese members of the Boletales. Acta Mycol Sin 12(4):275–282; (in Chinese)
Zeng NK, Liang ZQ, Yang ZL (2014) Boletus orientialbus, a new species with white basidioma from subtropical China. Mycoscience 55(3):159–163. https://doi.org/10.1016/j.myc.2013.07.004
Zerova MJ, Rozhenko GL (1988) Atlas s'edobnykh i yadovitykh gribov (Atlas of edible and poisonous fungi – Ukraine). Radyans'ka shkola, Kiev; (in Russian)
Zhao YC, Fang L, Zhang X et al (2006) The structural characteristics of ITS region of Boletus edulis (Boletoueae) in Yunnan. Acta Bot Yunnan 28(6):575–580; (in Chinese)
Zhuang WY (2001) Higher fungi of tropical China. Mycotaxon, Ithaca

Chapter 9
Recent Insights in the Phylogeny, Species Diversity, and Culinary Uses of Milkcap Genera *Lactarius* and *Lactifluus*

Jorinde Nuytinck, Eske De Crop, Lynn Delgat, Quinten Bafort, Mauro Rivas Ferreiro, Annemieke Verbeken, and Xiang-Hua Wang

9.1 Introduction

Milkcaps belong to the Russulales, one of the larger orders within the Agaricomycotina with an estimated diversity of approximately 4000 species (Miller et al. 2006). The order represents a remarkable diversity in basidiocarp (fruiting body) morphologies including corticioid, polyporoid, coralloid, discoid, hydnoid, agaricoid, pleurotoid, and gasteroid species. Equally diverse are the life strategies, with white rot saprotrophic, ectomycorrhizal (EM), parasitic, endophytic, and entomosymbiotic species. From all families within the Russulales, the Russulaceae has the highest species richness and is most diverse both in fruiting body morphologies and life strategies. Apart from the Albatrellaceae, it is the only family in the Russulales in which an EM symbiosis has evolved (Hibbett et al. 2014). The family originally encompassed only the genera *Lactarius* Pers. and *Russula* Pers., characterized by the agaricoid habit and the presence of large isodiametric cells called sphaerocytes (Miller et al. 2006). But with the molecular revolution in taxonomy, our concept of Russulaceae has fundamentally changed, and the family now includes species previously classified in Corticiaceae (Larsson and Larsson 2003),

J. Nuytinck (✉)
Naturalis Biodiversity Center, Leiden, The Netherlands

Research Group Mycology, Department of Biology, Ghent University, Ghent, Belgium
e-mail: jorinde.nuytinck@naturalis.nl

E. De Crop · L. Delgat · Q. Bafort · A. Verbeken
Research Group Mycology, Department of Biology, Ghent University, Ghent, Belgium

M. Rivas Ferreiro
Naturalis Biodiversity Center, Leiden, The Netherlands

X.-H. Wang
CAS Key Laboratory for Plant Diversity and Biogeography of East Asia, Kunming Institute of Botany, Chinese Academy of Sciences, Kunming, People's Republic of China

J. Pérez-Moreno et al. (eds.), *Mushrooms, Humans and Nature in a Changing World*, https://doi.org/10.1007/978-3-030-37378-8_9

Elasmomycetaceae (Miller et al. 2001; Buyck et al. 2008; Nuytinck et al. 2003; Desjardin 2003), and some pleurotoid and annulate taxa (Miller et al. 2001; De Crop et al. 2017).

Perhaps the most striking insight is the abandoning of the strict *Lactarius* (with milk and pseudocystidia) - *Russula* (without milk or pseudocystidia) bifurcation (Buyck et al. 2008). This traditional partition was largely based on European species and does not hold when including species from the tropics (Verbeken and Nuytinck 2013). Nowadays, four agaricoid[1] genera are distinguished. The traditional genus *Russula* is quasi monophyletic, only a small clade: *Russula* subsect. *Ochricompactae* Bills and O.K. Mill. was left out and is now forming, together with the former species *Lactarius furcatus* Coker and *Lactarius stenophyllus* Berk., the new genus *Multifurca* Buyck and V. Hofst. (Buyck et al. 2008; Wang et al. 2018). *Multifurca* can be recognized by the orange spore print, forking gills, and a characteristic zonation of the context (Buyck et al. 2008). Latex can be present or absent. *Multifurca* is by far the less diverse of the four agaricoid genera, with only twelve described species (Wang et al. 2018; Verbeken et al. 2018).

Lactarius in a traditional sense consisted of two reciprocal monophyletic clades: a smaller clade including the type species for the genus *Lactarius* (*Lactarius piperatus* (L.: Fr.) Pers.) and a clade containing the majority of the known *Lactarius* species (Buyck et al. 2008). Because of the quantitative difference in species richness, it was proposed to conserve the name *Lactarius* for the largest clade with type species *Lactarius torminosus* (Schaeff.: Fr.) Pers., and to change the smaller clade into the genus *Lactifluus* (Pers.) Roussel (a former taxonomic synonym of *Lactarius* with type species *Lactifluus volemus* (Fr.) Kuntze) (Buyck et al. 2010). The proposal was accepted (Barrie 2011).

There are no true synapomorphies characterizing each of the two milkcap genera. In general, *Lactifluus* shows the tendency to have a hymenophoral trama composed of sphaerocytes and often possesses thick-walled elements in the stipiti- or pileipellis or lamprocystidia in the hymenium, while most *Lactarius* species lack these features (Verbeken and Nuytinck 2013). *Lactifluus* is also generally characterized by the complete absence of zonate and viscid to glutinous caps, while it contains many species with veiled and velvety caps. In *Lactarius* many, but not all species, have zonate and sticky to slimy caps (Verbeken and Nuytinck 2013). Pleurotoid milkcaps are so far only known in *Lactifluus* (Buyck et al. 2008; Verbeken and Nuytinck 2013), while gasteroid and secotioid species are only known within *Lactarius*. Besides these morphological trends, the genera also differ in distribution, with *Lactarius* mainly distributed in the Northern hemisphere, while *Lactifluus* has its main range in the tropics. These trends might be helpful when identifying milkcap species, but they are not exclusive. There are species, especially in the tropics, in which a DNA marker is needed to determine which genus they belong to (De Crop 2016).

---

[1]These genera are almost entirely agaricoid (pileate-stipitate with lamellae), but *Russula* and *Lactarius* include some angiocarpic representatives, and *Lactifluus* includes species with a pleurotoid morphology.

This contribution highlights our most recent insights in the phylogenic relationships and evolutionary history of the milkcap genera. We also summarize knowledge on the edibility of some well-known milkcap taxa.

9.2 *Lactarius*

The most comprehensive molecular phylogenies of *Lactarius* s.s. that have up to now been published comprise only a fraction of the ±420 species that have been described. Buyck et al. (2008) included 12 species and classified them within the three traditionally recognized subgenera: *L.*[2] subg. *Plinthogalus* (Burl.) Hesler and A.H. Sm., *L.* subg. *Russularia* (Fr. ex Burl.) Kauffman, and *L.* subg. *Piperites* (Fr. ex J. Kickx f.) Kauffman. Since the type species of the genus changed to *L. torminosus*, we should now use the name *L.* subg. *Lactarius* for the latter taxon. Eberhardt and Verbeken (2004) and Geml et al. (2009) included more species in their studies, but still placed all of them in the same three traditional subgenera. Verbeken et al. (2014) showed in their phylogeny that several tropical *Lactarius* species form a diverging grade to the recognized subgenera. Figure 9.1 gives insight in our current understanding at subgenus level of the phylogeny of *Lactarius*. Several new tropical lineages need to be described at subgenus level.

Other studies have focused on subgenera or sections in *Lactarius* in more detail. Much work has been done on *L.* sect. *Deliciosi* (Fr.) Redeuilh, Verbeken, and Walleyn. Nuytinck et al. (2007) and Nuytinck and Verbeken (2007) included all up to then described species in a two-locus worldwide phylogeny and concluded that intercontinental overlap is much smaller than previously assumed. Several new species have since been described and illustrate that our knowledge of the genus in North America and Asia still has gaps (Nuytinck and Ammirati 2014; Wang et al. 2015b; Wang 2016; Nuytinck et al. 2017).

Wisitrassameewong et al. (2014a, b, 2015, 2016) explored the diversity of *L.* subg. *Russularia* focusing mainly on Southeast Asia. They revealed 23 new species and explored the historical biogeography of the subgenus. The European taxa of *L.* subg. *Plinthogalus* and American varieties of *L. lignyotus* Fr. were re-evaluated in Stubbe and Verbeken (2012). Some tropical species of the same subgenus were explored in Stubbe et al. (2008) and Verbeken et al. (2008).

The diversity in Europe is rather well understood, although an ongoing study reveals cryptic diversity in some species (Nuytinck et al., unpublished results). To better understand the worldwide diversity and evolution of *Lactarius*, we need to focus our attention on the following four questions: (1) Where did the genus originate and can we revise our Eurocentric view on the classification of the genus by including more tropical taxa? (2) What are the important drivers of diversification in *Lactarius*? (3) We understand the distribution of circum-arctic *Lactarius* species but

---

[2] We use *L.* as the abbreviation of *Lactarius* and *Lf.* to abbreviate *Lactifluus*.

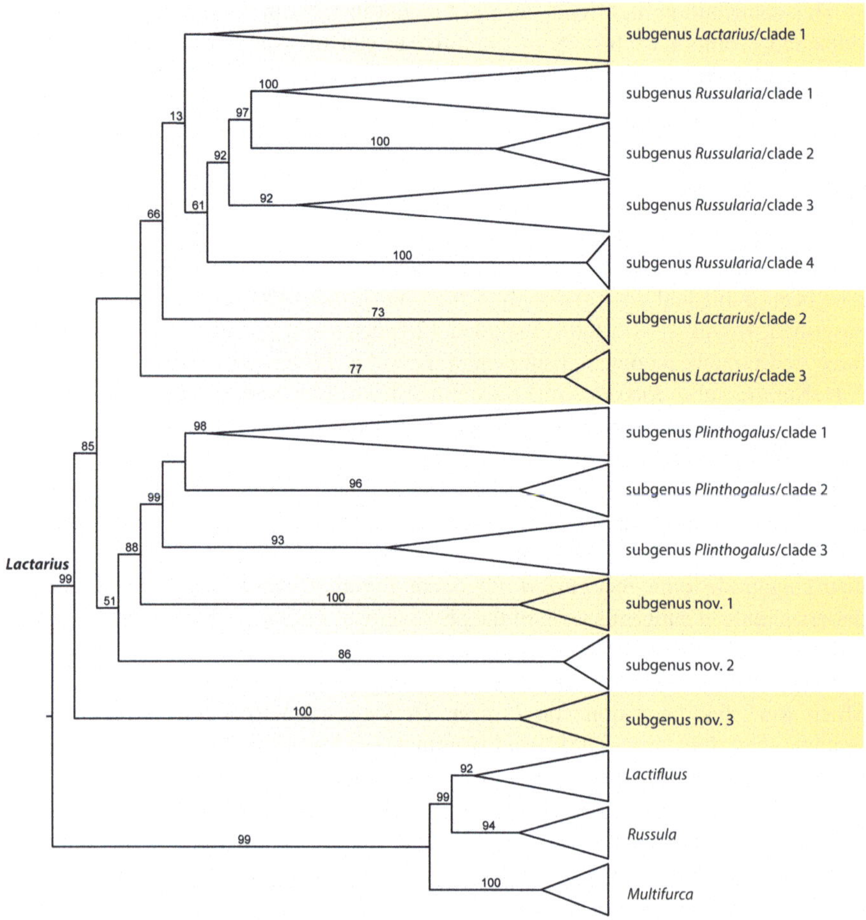

Fig. 9.1 Phylogenetic overview of the genus *Lactarius*, based on an unpublished Maximum Likelihood analysis of a dataset of four markers (ITS, LSU, *rpb1*, *rpb2*)

which boreal and temperate taxa have an intercontinental distribution? How and when did they migrate and what is the role of their EM partner therein? and (4) Can our understanding of the delimitation of and relationships between the many sections and subsections in *L.* subg. *Lactarius* be improved? The delimitation of *L.* subg. *Russularia* from *L.* subg. *Lactarius* is not straightforward (Fig. 9.1). Morphology and phylogeny are in many instances in conflict. A well-resolved phylogeny based on a balanced worldwide sampling will allow to estimate divergence times for the origin and diversification of *Lactarius* and reconstruct the geographical history of the genus. More sampling in North America and in the tropics deserves special attention.

9.3 *Lactifluus*

Lactifluus appears to be one of the dominant EM genera in the tropics (Tedersoo et al. 2010, 2011), yet until recently a robust phylogenetic framework and classification for the genus was lacking. De Crop et al. (2017) assembled a multi-gene dataset representing 80% of the described *Lactifluus* species diversity and 30% of the type specimens. This multi-gene phylogeny was combined with a morphological study, focusing on five important characteristics (basidiocarp type, presence of a secondary velum, color reaction of the latex/context, pileipellis type and presence of true cystidia). The results show that *Lactifluus* can be subdivided in four well-supported subgenera (Fig. 9.2). Two of the traditional, morphology-based subgenera were retained but amended (*Lf.* subg. *Lactariopsis* (Henn.) Verbeken and *Lf.* subg. *Lactifluus*), and two new subgenera were proposed (*Lf.* subg. *Gymnocarpi* (R. Heim ex Verbeken) De Crop and *Lf.* subg. *Pseudogymnocarpi* (Verbeken) De Crop). Within these subgenera, many sections could be delimited. Some of them were confirmed in their traditional delimitation, others were either amended or synonymized. At least ten additional clades were revealed that could represent new sections (Fig. 9.2).

The majority of *Lactifluus* species remains undescribed. There are currently 191 named *Lactifluus* species, 25 of which are not present in any DNA sequence database. In the ITS dataset De Crop (2016) assembled, she delimited 371 species, many thus were labelled as *Lactifluus* sp. Using a species accumulation curve, she estimated that there might be approximately 530 (95% Confidence Interval: 461–601) *Lactifluus* species on earth (Fig. 9.3).

Considering this large amount of unknown species, it is not surprising that new lineages are still discovered, an example of which was published in Wang et al. (2015a). This study, based on 28 samples from southern China, revealed three new lineages of *Lactifluus*. Two of these lineages were described as new sections, *Lf.* sect. *Ambicystidiati* X.H. Wang and *Lf.* sect. *Tenuicystidiati* X.H. Wang and Verbeken. From a morphological point of view, *Lf. ambicystidiatus* X.H. Wang is a striking new species, with both lamprocystidia and macrocystidia in the hymenium, a unique combination within the Russulaceae. Furthermore, only remnants of lactiferous hyphae are present in *Lf. ambicystidiatus*, suggesting that the ability to form a lactiferous system has been lost in this lineage (Wang et al. 2015a).

A striking observation is the occurrence of several large, cryptic or semi-cryptic species complexes in *Lactifluus*. Stubbe et al. (2010, 2012) examined the lineage of *Lf. gerardii* (Peck) Kuntze. At the start of this study, only a handful of species were known; in the end more than 30 phylogenetic species were discovered. Based on the current data, merely two-thirds are morphologically identifiable species. In the complex around *Lf. volemus*, Van de Putte et al. (2010, 2012, 2016) applied phylogenetic species recognition and discovered about 45 different clades. Some of them could be morphologically distinguished and were described as new species. Others remain cryptic; i.e., no morphological differences were found up to now. *Lactifluus piperatus* is also part of a species complex, estimated to contain over 30 clades, the

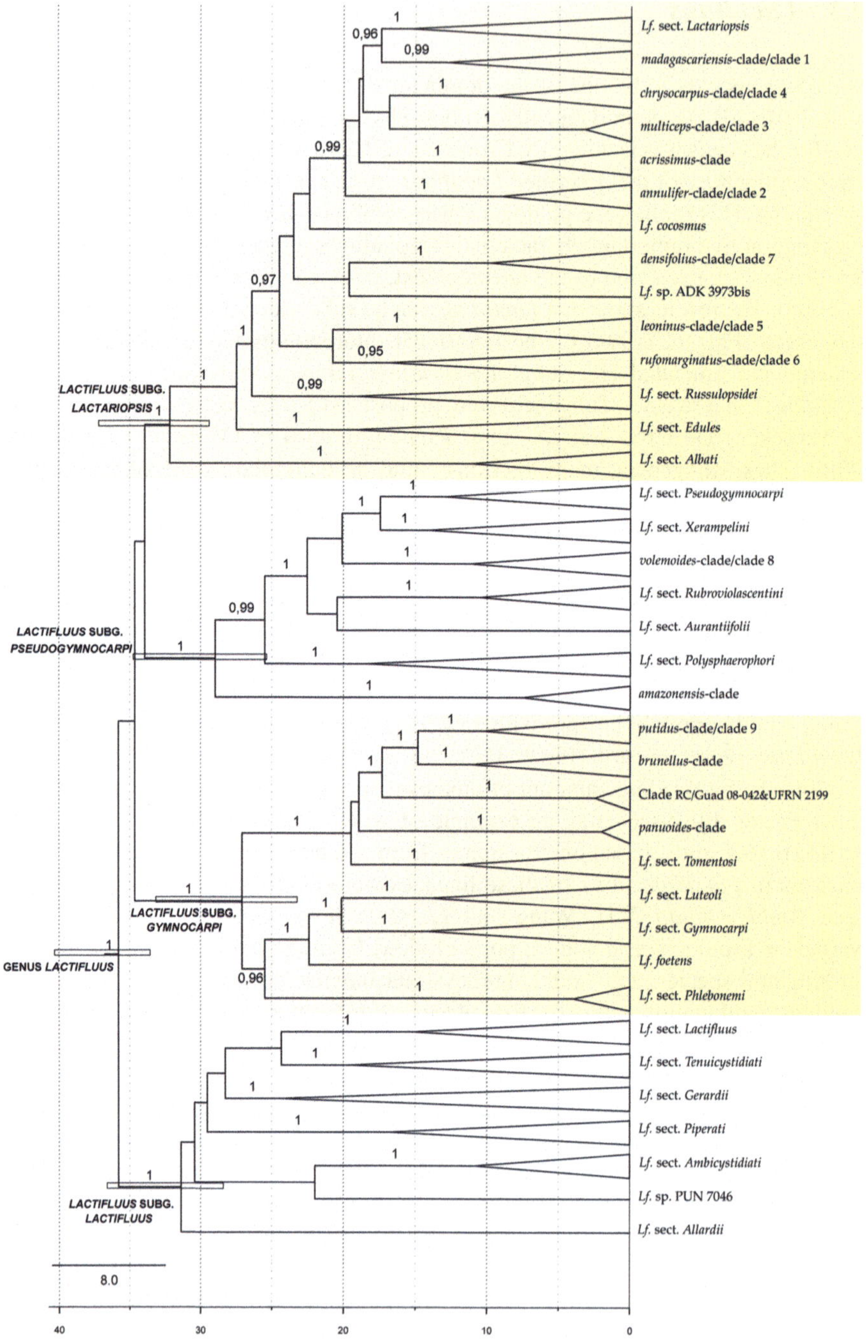

Fig. 9.2 Phylogenetic overview of the genus *Lactifluus*, inferred from a dated BEAST phylogeny presented in De Crop (2016). Time scale = million years. Undescribed clades are named after one representative inside that clade and received consecutive clade numbers

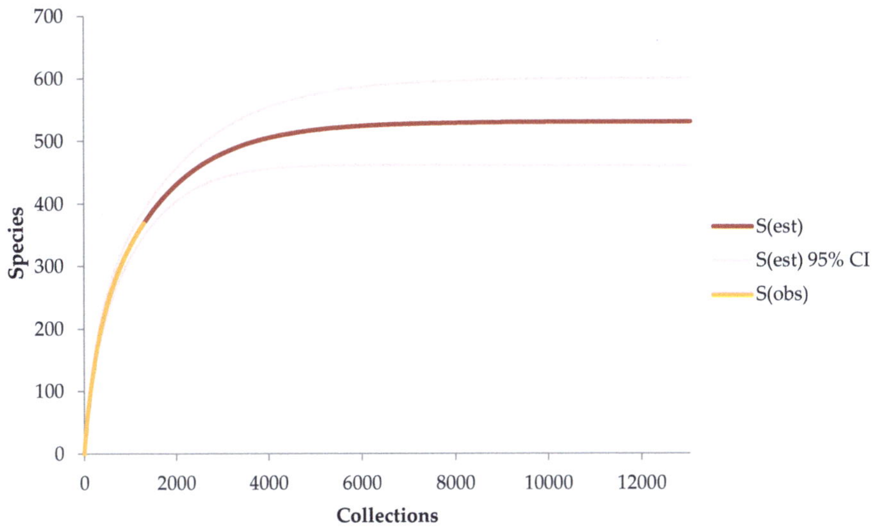

Fig. 9.3 Species accumulation curve of observed (obs) and estimated (est) species richness (S) of the genus *Lactifluus*. Species richness was estimated by extrapolation of the rarefaction curve, with 95% confidence interval (pink lines)

majority being undescribed (De Crop et al. 2014; De Crop 2016). Apart from these species complexes, several other species are assumed to be part of species complexes (De Crop 2016; Wang et al. 2015a). For example, the African *Lf. gymnocarpoides* (Verbeken) Verbeken, *Lf. pumilus* (Verbeken) Verbeken, and *Lf. longisporus* (Verbeken) Verbeken all have similar morphological characteristics and are hard to distinguish in the field. In the temperate regions, *Lf. vellereus* (Fr.) Kuntze and *Lf. bertillonii* (Neuhoff ex Z. Schaef.) Verbeken are also assumed to be part of a species complex.

9.4 Culinary Uses

In many parts of the world, mushrooms are of great importance as seasonal food source. Which species are consumed and the way they are prepared differ according to cultural habits. Russulaceae species are eaten in many parts of Africa, Asia, Europe, and Central and North America. Milkcaps can hardly be confused with deadly poisonous mushrooms because of their latex, and they often fruit in large numbers, which makes them popular at markets. The only studied species that is potentially harmful is *Lactarius necator* (Bull.: Fr.) Pers. It is reported to contain the mutagen necatorin (=7-hydroxycoumaro(5,6-c)cinnoline, von Wright and Suortti (1983); Suortti (1984)). Necatorin has been shown to be comparable to aflatoxin B1 in the strength of its mutagenic activity. Boiling reduces the concentration of this compound, but does not effectively eliminate it (Suortti 1984). Probably due to the acrid taste, most western European authorities classify this mushroom as inedible or poor. However, it has popularly been used (after pickling or parboiling) as a spice in mushroom dishes in northern and eastern Europe and Siberia.

Several European countries have a strong tradition in collecting wild edible fungi, and milkcaps are often very popular. Due to cultural differences, the same species can be considered a delicacy in northern Europe and an indigestible poisonous terror in the south (e.g., *L. torminosus* and *L. necator*; Basso 1999). As the name of the section and its most popular species give away, many species in *Lactarius* sect. *Deliciosi* are edible and delicious. In fact, all species are probably edible, but some are very inferior in taste (e.g., *L. salmonicolor*; Heim and Leclair 1950). Much depends however on the way of preparation; in several regions, a mix of any *Lactarius* sect. *Deliciosi* species is collected or sold (e.g., in Slovakia and Barcelona, own observations). The culinary properties of these orange milkcaps are especially appreciated in southern Europe (e.g., in Catalunya, but also in the rest of Spain and southern France, and on a smaller scale in Greece and Italy), eastern Europe, and Russia. People from Catalunya seem to prefer the taste of *L. sanguifluus* and even more the taste of *L. vinosus*. *Lactarius vinosus* is said to be most tasteful when greenish stains have developed on the pileus by the first frosts. In Catalunya, these species are commercially the second most important, after the *Tuber* P. Micheli ex F.H. Wigg. species. *Lactarius deliciosus* must have been popular since long in the Mediterranean region. A fresco from Herculaneum, a city that was buried during the same series of eruptions that destroyed Pompeii in the year 79, accurately depicts several basidiocarps of *L. deliciosus* together with pheasants (Ainsworth 1976). *Lactarius deliciosus* and relatives are being sold freshly or canned and are eaten roasted, prepared in a dish or in a cold salad. Eating *L. deliciosus* (and perhaps other species with orange latex) may cause the urine to be blood red, while this is not the case for red-milked species such as *L. sanguifluus* (Heim and Leclair 1950; Marchand 1971; Sterner and Anke 1995). This is by no means harmful. In a Russian study, the protein content of *L. deliciosus* was determined (Stankyavichene and Urbonas 1988). Nineteen amino acids, including all essential ones, were found, and these mushrooms were recommended as valuable food. The lipid content of *L. deliciosus* was shown to be only 1.02% (Ondrušek and Proštenik 1978). *Lactarius deliciosus* and relatives have also been used in several studies focusing on the concentrations of trace elements such as Cd, Co, Cu, Fe, Hg, Mn, Ni, Pb, and Zn in edible mushrooms (Falandysz et al. 2002; Isiloglu et al. 2001). For example, *Lactarius sanguifluus* was shown to accumulate easily several of these metals (a.o. Cd, Mn, Pb, and Zn), especially when growing close to a road (Isiloglu et al. 2001).

Some European *Lactifluus* species are edible and consumed in several countries. *Lactifluus volemus*, *Lf. oedematopus*, and *Lf. subvolemus*, for example, are popular species as they produce many large fruiting bodies that are easily identified and have an excellent taste (Van de Putte 2012). A recent Turkish report discusses two cases of acute pancreatitis caused by the consumption of *Lf. subvolemus* however (Karahan et al. 2016). The European large and white species, *Lf. vellereus*, *Lf. bertillonii* (Neuhoff ex Z. Schaef.) Verbeken, *Lf. piperatus*, and *Lf. glaucescens* (Crossl.) Verbeken, have an acrid taste and are only eaten in some regions, where they are

Fig. 9.4 Edible *Lactifluus* species on African markets: (**a**) cooked *Lactifluus* species for sale on Foumban market (Cameroon); (**b**) a local guide with a basket full of *Lactifluus* species (Foumban, Cameroon); (**c**) *Lactifluus* species for sale on Kigoma market (Tanzania); (**d**) *Lf. rubroviolascens* collected for consumption (Foumban, Cameroon); (**e**) cooked *Lactifluus* species (Foumban, Cameroon) (Photographs by E. De Crop)

parboiled or preserved with salt before consumption to remove the acrid taste (Heilmann-Clausen et al. 1998).

In African countries with woodlands and riparian forests, fungi, and especially EM fungi appear in great numbers at the beginning of the rain season. Mainly women and children go out in the forests to collect edible species, which are then sold at the local markets and along roadsides, either fresh, dried, or boiled (Fig. 9.4). Many *Lactarius*, *Lactifluus*, *Russula*, *Amanita* Pers., *Cantharellus* Juss., and *Termitomyces* R. Heim species are considered edible mushrooms (Härkönen et al. 2003). Milkcap species often found at the market are *Lf.* cf. *rubroviolascens* (R. Heim) Verbeken, *Lf. denigricans* (Verbeken and Karhula) Verbeken, *Lf. gymnocarpus* (R. Heim ex Singer) Verbeken, *Lf. albomembranaceus* De Wilde and Van de Putte, *Lf. densifolius* (Verbeken and Karhula) Verbeken, *Lf. edulis* (Verbeken and Buyck) Buyck, *Lf. gymnocarpoides*, *Lf. volemoides* (Karhula) Verbeken, *Lf. xerampelinus* (Karhula and Verbeken) Verbeken, and *L. kabansus* Pegler and Piearce (Rammeloo and Walleyn 1993; Verbeken and Walleyn 1999; De Kesel et al. 2002; Härkönen et al. 2003; Kinge et al. 2011; Sharp 2011, 2014).

Milkcap species are among the favorite edible mushrooms for local mushroom pickers in North America, and among them are several *Lactifluus* species. In Pennsylvania, for example, some locals go on "milkie mushroom" hunting trips, especially to collect milkcaps (Russell 2006). *Lactifluus* cf. *volemus*, *Lf. corrugis* (Peck) Kuntze and *Lf. hygrophoroides* (Berk and Curtis) Kuntze are the most famous ones, with *Lf.* cf. *volemus* recurrently being reported as the best and most flavorful milkcap (Peck 1885; Metzler and Metzler 1992; Roody 2003; Russell 2006; Lincoff 2010; Van de Putte 2012). Some authors also mention *Lactifluus gerardii* and *Lf. luteolus* (Peck) Verbeken as edible (Roody 2003; Bessette 2007), while only a minority likes to eat the milkcaps with peppery latex, such as *Lf.* cf. *piperatus*, *Lf.*

cf. *glaucescens*, *Lf. deceptivus* (Peck) Kuntze, *Lf. subvellereus* (Peck) Nuytinck, and *Lf. subgerardii* (Hesler and Sm.) Stubbe. Some authors even report some of these peppery tasting species as being poisonous (Bessette 2007). For other species, such as *Lf. allardii*, the edibility is unknown (Bessette 2007). Also famous is a *Hypomyces* (Fr.) Tul. and C. Tul. species (*H. lactifluorum* (Schwein.) Tul. and C. Tul.) that parasitises milkcaps and *Russula* species, turning them into "lobster mushrooms." The infected basidiocarps that show a red "crust" on the outside (the *H. lactifluorum* subiculum) and are white inside are considered a delicacy. Members of *L.* subg. *Russularia* are not often consumed, possibly due to their inferior taste, their fragile texture and small size. Additionally, the odor is rather unpleasant in some species (bug-like). A notable exception is the "candy caps," e.g., *L. rubidus* (Hesler and Sm.) Methven and *L. rufulus* Peck, which have a delightful sweetish and strong fragrance and are used to flavor desserts and cookies, but also smoked meats.

In Central America the tradition of eating wild fungi was already present centuries ago, in cultures like the Aztec and the Mayans. Today, the tradition persists in Mexico and Guatemala. Several milkcap species, like *L. indigo*, *L.* cf. *deliciosus*, *L.* cf. *salmonicolor*, *Lf. Deceptivus*, and *Lf.* cf. *volemus* are known to be sold on local and regional markets and to be collected for self-consumption (Kong Luz 1995; Montoya and Bandala 1996; Montoya et al. 1996; Comandini et al. 2012; Van de Putte 2012). Lobster fungus is also very popular in this area. However, south of these countries, in South America and the rest of Central America, the custom of eating wild fungi appears to be mostly absent (Boa 2004). To our knowledge no Russulaceae species are being consumed there, even though edible species do occur.

In Asia, members of *Lf.* sect. *Lactifluus* are known to be collected and eaten by local people of China, Japan, and Thailand (Wang et al. 2004; Le 2007; Lincoff 2010; Van de Putte et al. 2012). For example, in China, Yunnan province harbors the biggest number of ethnic minorities and most diverse habitats, and might be the place with the highest diversity of commercialized milkcaps in the country. Milkcaps are one of the most popular fungal groups collected, commercialized, and consumed by the local people there. These milkcaps are often sold freshly in summer (June to October), but dried products are also seen in supermarkets. Local people fry them with green chili. Wang (2000) reported 13 species of milkcaps from the local markets in Yunnan, accounting for 6.7% of the total number of commercialized fungi in the province (Wang and Liu 2002). The most common species among them are *L. deliciosus*, *L. hatsudake* Tanaka, and several species of *Lactifluus*, e.g., *Lf.* cf. *volemus*, *Lf.* cf. *hygrophoroides*, *Lf. echinatus* (basionym: *L. echinatus* Thiers), and *Lf.* cf. *gerardii*. These *Lactifluus* species turn out to be species complexes in molecular phylogenies (Stubbe et al. 2010; Van de Putte 2012; Van de Putte et al. 2016; Wang et al. 2015a). Therefore, the real number of commercialized milkcaps is even higher. Later, Wang et al. (2004) added *Lf.* cf. *rugatus* and the peppery *Lf.* cf. *piperatus* to the list of commercialized milkcaps in Yunnan. The Yunnan *Lactifluus rugatus* turned out to be a new species *L. tenuicystidiatus* X.H. Wang and A. Verbeken described by Wang and Verbeken (2006). The most recent molecular phylogeny shows that it represents a species complex including at least four species (Wang et al. 2015a) with an additional edible species related to it, *Lf. subpruinosus* X.H. Wang. In central and eastern China, one of the most famous edible mushrooms is *L. vividus* X.H. Wang, Nuytinck, and Verbeken. This mushroom, which is called "pine mushroom," has passed under the names *L. deliciosus*, *L. akahatsu*, and *L. salmonicolor* in Chinese literature for many years and has only recently been formally named (Wang et al. 2015b). A paper published in 2019 investigates the chemical composition, antioxidant, and antihyperglycemic activities of *L. vividus*, but wrongly identifies the species as *L. deliciosus* (Xu et al. 2019). The study concludes that it is a valuable wild mushroom with high protein, carbohydrate, and dietary fiber contents, while low in fat and calorie, extensive unsaturated fatty acids contents, with negligible health risks about harmful metal elements (Xu et al. 2019). In total, the collected and consumed milkcaps of *L.* sect. *Deliciosi* in subtropical parts of China include at least five species: *L. akahatsu* Tanaka, *L. deliciosus*, *L. hatsudake*, *L. vividus*, and *L. subindigo*. Very recently Wang (2016) found that some dealers transport subalpine *L. pseudohatsudake* X.H. Wang and *L. hengduanensis* X.H. Wang to markets in central Yunnan and sell them using the same local names for *L. hatsudake* and *L. deliciosus*. In western Sichuan, where spruces grow, *L. deterrimus* is one of the most commonly collected edible mushrooms (our observation). Although this species also grows in northwestern China (Gansu and Qinghai Provinces), local people do not collect it. The same situation occurs with *L. deliciosus* in northeastern China (Inner Mongolia): it fruits luxuriantly in *Pinus sylvestris* var. *mongolica* forests, but is not touched by local people (our observation). They do not recognize it as an edible mushroom. Apparently people in different regions in China have different traditions to recognize milkcaps as culinary material or not.

References

Ainsworth GC (1976) Introduction to the history of mycology. Cambridge University Press, Cambridge

Barrie FR (2011) Report of the general committee: 11. Taxon 60:1211

Basso MT (1999) *Lactarius* Pers, vol 7. Fungi Europaei, Mykoflora, Alassio

Bessette A (2007) Mushrooms of the Southeastern United States. Syracuse University Press, Syracuse, NY

Boa E (2004) Wild edible fungi. A global overview of their use and importance to people. Non-wood forest products 17. Food and Agriculture Organization of the United Nations (FAO), Rome

Buyck B, Hofstetter V, Eberhardt U, Verbeken A, Kauff F (2008) Walking the thin line between *Russula* and *Lactarius*: the dilemma of *Russula* subsect. *Ochricompactae*. Fungal Divers 28:15–40

Buyck B, Hofstetter V, Verbeken A, Walleyn R (2010) Proposal 1919: to conserve *Lactarius* nom. cons. (Basidiomycota) with a conserved type. Mycotaxon 111:504–508

Comandini O, Eros-Honti Z, Jakucs E, Arzu RF, Leonardi M, Rinaldi AC (2012) Molecular and morpho-anatomical description of mycorrhizas of *Lactarius rimosellus* on *Quercus* sp., with ethnomycological notes on *Lactarius* in Guatemala. Mycorrhiza 22(4):279–287

De Crop E (2016) Global phylogeny and evolutionary history of the genus *Lactifluus*. PhD, Ghent University, Belgium

De Crop E, Nuytinck J, Van de Putte K, Lecomte M, Eberhardt U, Verbeken A (2014) *Lactifluus piperatus* (Russulales, Basidiomycota) and allied species in Western Europe and a preliminary overview of the group worldwide. Mycol Prog 13(3):493–511

De Crop E, Nuytinck J, Van de Putte K, Wisitrassameewong K, Hackel J, Stubbe D, Hyde KD, Roy M, Halling RE, Moreau PA, Eberhardt U, Verbeken A (2017) A multi-gene phylogeny of *Lactifluus* (Basidiomycota, Russulales) translated into a new infrageneric classification of the genus. Persoonia 38:58–80

De Kesel A, Codjia JTC, Yorou NS (2002) Guide des champignons comestibles du Bénin. Jardin botanique national de Belgique, Meise, Belgium

Desjardin DE (2003) A unique ballistosporic hypogeous sequestrate *Lactarius* from California. Mycologia 95:148–155

Eberhardt U, Verbeken A (2004) Sequestrate *Lactarius* species from tropical Africa: *L. angiocarpus* sp. nov. and *L. dolichocaulis* comb. nov. Mycol Res 108:1042–1052

Falandysz J, Gucia M, Skwarzec B, Frankowska K, Klawikowska K (2002) Total mercury in mushrooms and underlying soil substrate from the Boreka Forest, northeastern Poland. Arch Environ Contam Toxicol 42:145–154

Geml J, Laursen GA, Timling I, McFarland JM, Booth MG, Lennon N, Nusbaum C, Taylor DL (2009) Molecular phylogenetic biodiversity assessment of arctic and boreal ectomycorrhizal *Lactarius* Pers. (Russulales; Basidiomycota) in Alaska, based on soil and sporocarp DNA. Mol Ecol 18(10):2213–2227

Härkönen M, Niemelä T, Mwasumbi L (2003) Tanzanian mushrooms. Edible, harmful and other fungi, vol 10. Norrlinia, Helsinki

Heilmann-Clausen J, Verbeken A, Vesterholt J (1998) The genus *Lactarius* Vol. 2—fungi of Northern Europe. Svampetryk: Danish Mycological Society. 287 p, vol 2. Svampetryk, Greve

Heim R, Leclair A (1950) Notes systématiques sur les Champignons du Perche. II. Les Lactaires à lait rouge (Stirpe deliciosus). Revue Mycologique 15(2):65–78

Hibbett DS, Bauer R, Binder M, Giachini AJ, Hosaka K, Justo A, Larsson E, Larsson KH, Lawrey JD, Miettinen O, Nagy LG, Nilsson RH, Weiss M, Thorn RG (2014) Agaricomycetes. In: McLaughlin DJ, Spatafora JW (eds) The Mycota part A. Systematics and evolution, vol 7. 2 edn. Springer Verlag, Berlin, pp 373–429

Isiloglu M, Yilmaz F, Merdivan M (2001) Concentrations of trace elements in wild edible mushrooms. Food Chem 73:169–175

Karahan S, Erden A, Cetinkaya A, Avci D, Ortakoyluoglu AI, Karagoz H, Bulut K, Basak M (2016) Acute pancreatitis caused by mushroom poisoning: a report of two cases. J Investig Med High Impact Case Rep 4(1):1–3

Kinge TR, Tabi EM, Mih AM, Enow EA, Njouonkou L, Nji TM (2011) Ethnomycological studies of edible and medicinal mushrooms in the mount cameroon region (Cameroon, Africa). Int J Med Mushrooms 13(3):299–305

Kong Luz A (1995) Estudio taxonomico sobre el genero *Lactarius* (Russulales, Mycetae) en el volcan la Malintzi, Tlaxcala. Universidad Nacional Autonoma de Mexico, Mexico

Larsson E, Larsson KH (2003) Phylogenetic relationships of russuloid basidiomycetes with emphasis on aphyllophoralean taxa. Mycologia 95(6):1037–1065

Le HT (2007) Biodiversity of the genus *Lactarius* (Basidiomycota) in northern Thailand. PhD dissertation, Chiang Mai University, Thailand

Lincoff G (2010) The complete mushroom hunter: an illustrated guide to finding, harvesting, and enjoying wild mushrooms. Quarry Books, London

Marchand A (1971) Champignons du nord et du midi 1. Société mycologique des Pyrénées méditerranéennes, Perpignan

Metzler S, Metzler V (1992) Texas mushrooms, 1st edn. University of Texas Press, Austin, TX

Miller SL, Larsson E, Larsson KH, Verbeken A, Nuytinck J (2006) Perspectives in the new Russulales. Mycologia 98(6):960–970

Miller SL, McClean TM, Walker JF, Buyck B (2001) A molecular phylogeny of the Russulales including agaricoid, gasteroid and pleurotoid taxa. Mycologia 93(2):344–354

Montoya L, Bandala V (1996) Additional new records on *Lactarius* from Mexico. Mycotaxon 57:425–450
Montoya L, Bandala VM, Guzmán G (1996) New and interesting species of *Lactarius* from Mexico including scanning electron microscope observations. Mycotaxon 57:411–424
Nuytinck J, Ammirati JF (2014) A new species of *Lactarius* sect. *Deliciosi* (Russulales, Basidiomycota) from western North America. Botany 92(10):767–774
Nuytinck J, Verbeken A (2007) Species delimitation and phylogenetic relationships in *Lactarius* section *Deliciosi* in Europe. Mycol Res 111:1285–1297
Nuytinck J, Verbeken A, Delarue S, Walleyn R (2003) Systematics of European sequestrate Lactarioid Russulaceae with spiny spore ornamentation. Belg J Bot 136(2):145–153
Nuytinck J, Verbeken A, Miller SL (2007) Worldwide phylogeny of *Lactarius* section *Deliciosi* inferred from ITS and glyceraldehyde-3-phosphate dehydrogenase gene sequences. Mycologia 99(6):820–832
Nuytinck J, Verbeken A, Saar I, Lambert H, Bérubé J, Voitk A (2017) *Lactarius splendens*, a second species with white latex in *Lactarius* section *Deliciosi*. Botany 95(8):859–863
Ondrušek V, Proštenik M (1978) Sphingolipids in fruit bodies of the basidiomycete *Lactarius deliciosus*. Exp Mycol 2:156–160
Peck CH (1885) Report of the Botanist (1884). Ann Rep New York State Mus Nat Hist 38:77–138
Rammeloo J, Walleyn R (1993) The edible fungi of Africa south of the Sahara. Scripta Botanica Belgica 5:1–62
Roody WC (2003) Mushrooms of West Virginia and the Central Appalachians. University Press of Kentucky, Lexington
Russell B (2006) Field guide to wild mushrooms of Pennsylvania and the Mid-Atlantic. Pennsylvania State University Press, University Park
Sharp C (2011) A pocket guide to mushrooms in Zimbabwe, vol 1. Directory Publishers, Bulawayo
Sharp C (2014) A pocket guide to the mushrooms of Zimbabwe, vol 2. TWP Sdn. Bhd., Malaysia
Stankyavichene DR, Urbonas VA (1988) Biochemical characterisation of edible mushrooms in the Lithuanian SSR (USSR): 7. Protein substance content in the fruit bodies of *Agaricus bisporus*, *Lactarius deliciosus*, *Lactarius torminosus*, *Lactarius necator*, and *Lactarius rufus*. Liet TSR Mokslu Akad Darbai Ser C Biol Mokslai (4):109–115
Sterner O, Anke H (1995) Toxic terpenoids isolated from higher fungi. Czech Mycol 48(1):39–52
Stubbe D, Le HT, Wang XH, Nuytinck J, Van de Putte K, Verbeken A (2012) The Australasian species of *Lactarius* subgenus *Gerardii* (Russulales). Fungal Divers 52(1):141–167
Stubbe D, Nuytinck J, Verbeken A (2008) *Lactarius* subgenus *Plinthogalus* of Malaysia. Fungal Divers 32:125–156
Stubbe D, Nuytinck J, Verbeken A (2010) Critical assessment of the *Lactarius gerardii* species complex (Russulales). Fungal Biol 114(2–3):271–283
Stubbe D, Verbeken A (2012) *Lactarius* subg. *Plinthogalus*: the European taxa and American varieties of *L. lignyotus* re-evaluated. Mycologia 104(6):1490–1501
Suortti T (1984) Stability of necatorin, a highly mutagenic compound from *Lactarius necator* mushroom. Food Chem Toxicol 22(7):579–581
Tedersoo L, Bahram M, Jairus T, Bechem E, Chinoya S, Mpumba R, Leal M, Randrianjohany E, Razafimandimbison S, Sadam A, Naadel T, Koljalg U (2011) Spatial structure and the effects of host and soil environments on communities of ectomycorrhizal fungi in wooded savannas and rain forests of Continental Africa and Madagascar. Mol Ecol 20(14):3071–3080
Tedersoo L, Nilsson RH, Abarenkov K, Jairus T, Sadam A, Saar I, Bahram M, Bechem E, Chuyong G, Koljalg U (2010) 454 Pyrosequencing and Sanger sequencing of tropical mycorrhizal fungi provide similar results but reveal substantial methodological biases. New Phytol 188(1):291–301
Van de Putte K (2012) Hidden diversity exposed: a case study of *Lactifluus volemus* sensu lato. Ghent University, Ghent
Van de Putte K, Nuytinck J, Das K, Verbeken A (2012) Exposing hidden diversity by concordant genealogies and morphology-a study of the *Lactifluus volemus* (Russulales) species complex in Sikkim Himalaya (India). Fungal Divers 55(1):171–194

Van de Putte K, Nuytinck J, De Crop E, Verbeken A (2016) *Lactifluus volemus* in Europe: three species in one – revealed by a multilocus genealogical approach, Bayesian species delimitation and morphology. Fungal Biol 120(1):1–25

Van de Putte K, Nuytinck J, Stubbe D, Huyen TL, Verbeken A (2010) *Lactarius volemus* sensu lato (Russulales) from northern Thailand: morphological and phylogenetic species concepts explored. Fungal Divers 45(1):99–130

Verbeken A, Nuytinck J (2013) Not every milkcap is a *Lactarius*. Scr Bot Belg 51:162–168

Verbeken A, Stubbe D, Nuytinck J (2008) Two new *Lactarius* species from Cameroon. Cryptogam Mycol 29(2):137–143

Verbeken A, Stubbe D, van de Putte K, Eberhardt U, Nuytinck J (2014) Tales of the unexpected: angiocarpous representatives of the Russulaceae in tropical South East Asia. Persoonia 32:13–24

Verbeken A, Walleyn R (1999) Studies in tropical African *Lactarius* species 7. A synopsis of the section *Edules* and a review on the edible species. Belg J Bot 132(2):175–184

Verbeken A, Wisitrassameewong K, Kim NK, Lee H, Lim YW, Taylor A, Nuytinck J (2018) Nomenclatural novelties. Index Fungorum 366

von Wright A, Suortti T (1983) Preliminary characterization of the mutagenic properties of 'necatorin', a strongly mutagenic compound of the mushroom *Lactarius necator*. Mutat Res Lett 121:103–106

Wang XH (2000) A taxonomic study on some commercial species in the genus *Lactarius* (Agaricales) from Yunnan province, China. Acta Bot Yunnanica 22(4):419–427

Wang XH (2016) Three new species of *Lactarius* sect. *Deliciosi* from subalpine-alpine regions of Central and Southwestern China. Cryptogam Mycol 37(4):493–508

Wang XH, Buyck B, Verbeken A (2015a) Revisiting the morphology and phylogeny of *Lactifluus* with three new lineages from southern China. Mycologia 107(5):941–958

Wang XH, Halling RE, Hofstetter V, Lebel T, Buyck B (2018) Phylogeny, biogeography and taxonomic re-assessment of *Multifurca* (Russulaceae, Russulales) using three-locus data. PLoS One 13(11):e0205840

Wang XH, Liu PG (2002) Resources investigation and studies on the wild commercial fungi in Yunnan. Biodivers Sci 10(3):318–325

Wang XH, Liu PG, Yu FQ (2004) Color atlas of wild commercial mushrooms in Yunnan. Yunnan Science and Technology Press, Kunming

Wang XH, Nuytinck J, Verbeken A (2015b) *Lactarius vividus* sp. nov. (Russulaceae, Russulales), a widely distributed edible mushroom in central and southern China. Phytotaxa 231(1):63–72

Wang XH, Verbeken A (2006) Three new species of *Lactarius* subgen. *Lactiflui*. Nova Hedwigia 83:167–176

Wisitrassameewong K, Looney B, Le HT, De Crop E, Das K, Van de Putte K, Eberhardt U, Jiayu G, Stubbe S, Hyde KD, Verbeken A, Nuytinck J (2016) *Lactarius* subgenus *Russularia* (Basidiomycota, Russulales): biodiversity, molecular phylogeny and evolutionary relationships. Fungal Biol 120:1554–1581

Wisitrassameewong K, Nuytinck J, Hampe F, Hyde KD, Verbeken A (2014a) *Lactarius* subgenus *Russularia* (Russulaceae) in South-East Asia: 2. Species with remarkably small basidiocarps. Phytotaxa 188(4):181–197

Wisitrassameewong K, Nuytinck J, Hyde KD, Verbeken A (2014b) *Lactarius* subgenus *Russularia* (Russulaceae) in Southeast Asia: 1. Species with very distant gills. Phytotaxa 158(1):23–42

Wisitrassameewong K, Nuytinck J, Le HT, De Crop E, Hampe F, Hyde KD, Verbeken A (2015) *Lactarius* subgenus *Russularia* (Russulaceae) in South-East Asia: 3. New diversity in Thailand and Vietnam. Phytotaxa 207(3):215–241

Xu Z, Fu L, Feng S, Yuan N, Huang Y, Liao J, Zhou L, Yang H, Ding C (2019) Chemical composition, antioxidant and antihyperglycemic activities of the wild *Lactarius deliciosus* from China. Molecules 24:1357

Chapter 10
Advances in the Cultivation of Truffles in Canada

Shannon M. Berch

10.1 Introduction

Canadian truffle cultivation began about 15 years ago in BC when news articles about truffle cultivation caught the attention of certain farmers and they contacted professional agrologists with questions. Although we have come a long way since then, having overcome the major hurdle of demonstrating that truffle production is possible here, many of the issues that face other truffle-producing regions of the world remain. Of the ten Canadian provinces and three territories, truffle cultivation activity is currently found only in four provinces: BC, Ontario, Quebec, and Nova Scotia. In this article, I provide an overview of the truffle industry in Canada and highlight our successes and challenges.

10.2 British Columbia

The Truffle Association of British Columbia (TABC; http://www.bctruffles.ca) was established in 2004 by a group of producers, harvesters, mycologists, professional agrologists, and truffle enthusiasts. Membership is small (about 25), but recent successes in harvesting native truffles and cultivated truffles have spurred an increase in interest and queries to the Association. In 2014, TABC produced a guidance document entitled "Mediterranean Truffle Orchard Best Management Practices" that covers the key issues in truffle cultivation (http://www.bctruffles.ca/Home/publications) for growers in BC. The purpose of the document is to provide information to those who are considering truffle cultivation and to the provincial government in

S. M. Berch (✉)
British Columbia Ministry of Environment, Victoria, BC, Canada

J. Pérez-Moreno et al. (eds.), *Mushrooms, Humans and Nature in a Changing World*, https://doi.org/10.1007/978-3-030-37378-8_10

their assessment of farm operations for "qualifying agricultural use" and the more favorable farm tax status.

To the best of my knowledge, TABC is the only truffle association in Canada. The North American Truffle Growers Association (NATGA; http://trufflegrowers.com/) was established in North Carolina in 2005 and, according to their web site, has members in Tennessee, Virginia, South Carolina, Iowa, Kentucky, Arkansas, Idaho, and Oregon including 51 active member farms. The North American Truffling Society (NATS) has been in existence for decades, helping to enhance the scientific knowledge of North American truffles. Quite recently, the Western Truffle Growers Association and the Idaho Truffle Farmers group have been initiated to bring together growers sharing similar geography, issues, and approaches.

It is currently estimated that there are 10 ha of truffle-inoculated trees known to be cultivated in the southwest coastal and Okanagan regions of BC (unpublished report on the Market Potential of the Culinary Truffle Industry of British Columbia); *Tuber melanosporum, Tuber aestivum, Tuber borchii*, and *Tuber macrosporum* are the cultivated species. Because this estimate is based on information from current members of the Truffle Association of BC and there are growers who are not members of the Association, we speculate that the actual number of hectares under Mediterranean truffle cultivation in BC could be two or three times this estimate.

Périgord black truffles were first produced in BC at the Stewart farm in the Lower Fraser Valley in the winter of 2013 (http://www.vancouversun.com/life/first+p%C3%A9rigord+truffle+worth+kilo+found+abbotsford/8155234/story.html). Tragically, since then that orchard has been devastated by Eastern Filbert blight (BC Ministry of Agriculture 2016; *Anisogramma anomala*), and truffle production has ceased. The next truffle orchard to produce in BC is located on Vancouver Island, and the first truffle produced there was *T. aestivum* in August of 2013. In the winter of 2014, the Périgord black truffle orchard at the same farm began to produce. In part because Eastern filbert blight is not yet as much of a problem on Vancouver Island and because the orchard also contains oaks, the truffle production trajectory of this orchard is very encouraging with the number of producing trees and total truffle production increasing each year over the past four harvest seasons (Fig. 10.1). Some of the truffle orchards in BC are yet too young to be expected to produce truffles, but a few others are old enough to produce truffles but have remained unproductive to date.

In the early days of the truffle industry in BC, a few growers worked with a truffle expert to inoculate seedlings at their truffle orchard locations. Soon, nurseries in BC and in the USA began providing truffle-inoculated seedlings to growers. Quality control issues in one of those nurseries lead some growers to plant seedlings that may have had little or no ectomycorrhiza formation by the target *Tuber* species; examination of the roots of some of these trees after out-planting detected little or no *T. melanosporum* ectomycorrhizas (Berch and Bonito 2014). Quality control issues in another nursery led to the accidental introduction of *T. borchii* (bianchetto truffle; Berch and Bonito 2014) and *Tuber lyonii* (pecan truffle; unpublished information) into one truffle orchard in the Lower Fraser Valley, BC; the good news is that bianchetto truffles were harvested from this nursery in February 2016 (Fig. 10.2),

Fig. 10.1 Some of the Périgord black truffles harvested in BC in February, 2016

Fig. 10.2 The first bianchetto truffle harvested in BC, January 2016

proved to be popular with chefs and consumers (http://www.langleyadvance.com/community/367579001.html) and have continued to produce each winter since then.

Bianchetto truffles have been harvested at a truffle orchard in Idaho, USA, since 2012 (http://www.truffletree.com/new-idaho-tuber/), but this was the first Canadian harvest of this truffle. Quality control issues in a third nursery led to the introduction of *Tuber brumale* into a different *T. melanosporum* truffle orchard, this one on Vancouver Island, BC (Berch and Bonito 2014), but the management of this orchard has been neglected, and no truffles have yet been harvested.

At present, there appears to be only one nursery in BC actively working on providing truffle-inoculated seedlings for purchase by truffle growers. Inspired by the devastation of the hazelnut industry in the Lower Fraser Valley of BC by Eastern filbert blight, caused by the fungus *Anisogramma anomala*, Thom Odell and Hayley Argen of Nature Tech Nursery now offer tissue culture-originated hazelnut trees that are resistant to the blight and were developed by the Oregon State University breeding program. This nursery is currently developing methods to inoculate these disease-resistant varieties with *T. melanosporum* and *T. aestivum*. This work is critical to the future of the truffle industry in BC as all existing hazelnut-based truffle orchards in the Lower Fraser Valley, including the Stewart farm, winner of the TABC prize for first independently confirmed *T. melanosporum* truffle production in BC, are now dead or dying due to this pathogen. Oaks serving as truffle hosts in BC orchards are unaffected by this fungus. Hazelnut-based truffle orchards on Vancouver Island and the Okanagan Valley are not yet affected by Eastern Filbert blight.

Charles Lefevre of New World Truffieres (NWT; http://www.truffletree.com/) has been a supporter of the BC truffle industry since its inception, donating inoculated trees for the TABC demonstration truffière at UBC in Vancouver and participating in almost every major truffle event in the province. NWT has a dozen or more clients in Canada, mostly with small truffle orchards in BC and one in Ontario. Due to restrictions by the Canadian Food Inspection Agency (http://www.inspection.gc.ca/plants/plant-pests-invasive-species/imports/permit-usa/eng/1343427289390/1343427392092) on importation from the USA and movement from province to province of hazelnut plants (except sterile tissue cultured plantlets for planting) and seeds, into BC, NWT provides primarily *Quercus robur* seedlings to growers in Canada. As of late winter 2019, the Canadian Food Inspection Agency has identified *Anisogramma anomala* for deregulation which may open the borders to the movement of hazelnut plant materials. Two earlier attempts at using *Quercus ilex* in BC gave mixed results and indicated that this species is not reliably hardy here. The first plantation of *T. melanosporum* on *Q. ilex* at UBC farm in Vancouver suffered frost damage due to a cold spell in November of their first year in the ground and, although all of the plants sprouted from the base again the next growing season, further dieback the next winter convinced us to replant with *Q. robur*. However, a truffle orchard on Vancouver Island is producing truffles under *Q. ilex*, *Q. robur*, and *Corlyus avellana*; the *Q. ilex* trees have been frost-nipped but continue to grow and support truffle production.

Canadian research on truffle cultivation has been carried out primarily in British Columbia (BC) where the truffle industry is furthest developed. TABC has been instrumental in carrying out this research, and the Investment Agriculture Foundation (IAF) has contributed funding for three truffle projects with matching funds from TABC and industry. In the first, investigation of 8 truffle orchards in 3 truffle growing regions of the province indicated that in most orchards, the target truffle fungus (*T. melanosporum* or *T. aestivum*) was detectable as ectomycorrhizal morphotypes on the root systems of host trees (*C. avellana* and *Q. robur*) up to 5 years after planting (Berch and Bonito 2014). In other orchards, where the inoculated seedlings all originated in one specific truffle tree nursery, no ectomycorrhizas of the target *Tuber* species were found. A surprise finding of this work was the detection of 3 species of noncommercial native truffles (*Tuber anniae*, *Tuber beyerlei*, and *Tuber menseri*) forming ectomycorrhizas on host trees in a number of the orchards.

To evaluate the context, current status, and future of the truffle industry in BC, TABC worked with IAF and a private consultant to recently produce an unpublished, draft report on the Market Potential of the Culinary Truffle Industry of British Columbia. Some of the major conclusions of that work are:

- The greatest challenges for truffle growers are maintaining high soil pH through initial and repeated addition of lime and maintaining the commitment to truffière management through the waiting years when there are no truffles produced and no financial return on their investment.
- The native truffle industry is still in its infancy and is limited by the small number of dog handlers and competent truffle dogs and the lack of understanding of ecological characteristics of productive truffle sites.
- The truffle industry can benefit by providing a truffle experience to the consumer in addition to providing truffles.
- The quality control issues that challenge the truffle industry include ensuring that only high quality seedlings colonized by only the target *Tuber* species are sold and that truffle dogs teams providing services to growers are competent and reliable.

As has been demonstrated in many truffle-growing areas, in BC as well, truffle growers may have to wait 7 or more years after planting for the production of their first truffles; understandably, some orchardists begin to wonder whether the truffle orchard they are so carefully managing still has the target *Tuber* species in it. Sampling roots and identifying the ectomycorrhizas formed by the target *Tuber* species is difficult, time-consuming, and expensive and not easily available to the orchardist. A project jointly funded by IAF, TABC, NATGA, and a Canadian biotechnology company aimed to fill this gap by adapting the methods developed and applied in Spain (Suz et al. 2006; Fischer and Colinas 2007; Adamo 2014) to conditions in North America and thereby provide a reliable and cost-effective service to truffle growers to detect *Tuber* species in their orchard soils.

In order to improve awareness of truffles and the truffle industry and to raise funds to continue supporting truffle research, TABC successfully hosted the first truffle festival in Canada in February of 2016. The festival consisted of three sold-out events: a combined truffle educational interaction and truffle dog demonstration at the University of British Columbia/TABC demonstration truffle orchard; a locally harvested truffle-themed dinner at a venue in Vancouver, BC; and a locally harvested truffle-themed lunch at a venue in Abbotsford, BC, combined with a native truffle foray and a search of a truffle orchard led by teams of truffle dogs and their handlers. Each year since then, TABC has put on some truffle-focused events to raise funds in support of the demonstration orchard at UBC Farm and to raise awareness about our local and cultivated truffles.

Truffle growers in BC have experienced many of the quality assurance issues encountered in other truffle-growing areas of the world including the accidental introduction of nontarget *Tuber* species in commercial nurseries. Although third-party assessment of inoculated seedlings prior to planting can help address this issue, it cannot solve the problem because there are very few competent professionals offering this service and it is not possible to examine every ECM of every plant so it is always possible that some contamination will escape detection. To ensure that accidental introduction of nontarget *Tuber* species does not happen, the onus is on the nursery to use both morphological and molecular methods to assess each truffle or truffle piece used as inoculum and to manage stock in the nursery to avoid seedling to seedling movement of different truffle species.

The high cost of real estate, including farmland, might be a hurdle to the development of truffle cultivation in the province. To illustrate, in the summer of 2016 one of BC's producing truffières was put up for sale. The whole parcel is 4 ha (10 acres) with about 1.2 ha (3 acres) planted to truffle trees (about 350 hazelnuts and 150 oaks). The trees are now 11 years old, a lot of the oaks are showing brulés, and bianchetto truffles have been produced. The sale price was CAN$1.7 million. As Charles Lefevre has suggested to me, perhaps this high land cost could be considered an incentive to growing a high-value crop like truffles.

Recent work by Berch and Bonito (2016) has confirmed what truffle dog teams have been demonstrating for the last few years, which is that the important culinary truffles, *Tuber gibbosum*, *Tuber oregonense* (referred to as Oregon white truffles), and *Leucangium carthusianum* (referred to as Oregon black truffles) (Fig. 10.3), occur in coastal BC associated with Douglas-fir. Other native species of *Tuber* were detected in BC, *Tuber* sp. 44, *Tuber* sp. A, *T. anniae* species complex, *T. beyerlei*, and *T. menseri*, but none of these are currently commercially harvested. The native culinary truffles are not yet cultivated in truffle orchards even though the common habitat for them is orchard-like in many ways, i.e., young Douglas-fir plantations that have become spontaneously colonized perhaps from spores in the soil or in small mammal droppings from adjacent mature stands.

10.3 Ontario

Paul Thomas and Robert Chang of the American Truffle Company are establishing truffle orchards with partners in the USA and Canada, with most of their activities centered on the USA. They have one client with an established *T. aestivum* truffière in Ontario and are in discussions with the owners of more properties. Their business model has a partnership structure in which guidance is provided and payment for

Fig. 10.3 Native truffles harvested by truffle dog teams in British Columbia. (**a**) *Tuber oregonense* harvested by Brooke Fochuk and Dexter. (**b**) Shannon Berch appreciating the aroma of *Leucangium carthusianum*

expertise is required only after truffles are successfully produced (http://www.americantruffle.com/).

A graduate student, Charles Shearer, at the University of Guelph, bitten by the truffle and hazelnut bug, spent his spare time developing a protocol for producing oak and hazelnut seedlings inoculated with *T. aestivum* (Burgundy truffle) using spores collected from northern populations of truffles produced in a truffle plantation on the island of Gotland, Sweden. He focused on getting the process right and perhaps planting his own truffle orchard. Recently, a plant nursery in Dunnville, ON, started offering Burgundy truffle-inoculated oak and hazelnut seedlings for sale, and at least one truffle orchard has been established near Guelph, ON (Drake 2015).

10.4 Quebec

Jérôme Quirion of ArborInnov (http://arborinnov.com/en/truffle-cultivation/) is developing a wide array of exotic and native *Tuber*-inoculated plants (e.g., various oaks, hazelnut, pine, hickory, and hornbeam) for sale to truffle growers. Currently, he is monitoring 2000 trees in test orchards of *T. aestivum*, *T. canaliculatum*, and *T. borchii* in southern Quebec. His work on *T. canaliculatum*, the Appalachian truffle, appears to be the first attempt to cultivate this native, endemic, truffle species, and he reports that the species grows quite aggressively on host tree roots forming abundant ectomycorrhizas and producing brulés within the first 2–3 years in the orchard. Truffles have yet to be harvested, but fruiting is expected to occur September through mid-December. The first two commercial *T. canaliculatum* truffle orchards were planted by customers, one in Québec and the other in Vermont, USA. ArborInnov has collaborated on a number of research and development projects on truffle production over the past few years, but, as the work was proprietary, results are not published in the scientific literature.

10.5 Nova Scotia

Alexander J. Wilson, Wilsonia Golden Truffles (http://users.eastlink.ca/~wilsonia/truffles/about.html), is currently growing English oak (*Q. robur*) seedlings inoculated with spores from Burgundy truffle (*T. aestivum*) and has established a half hectare English oak/Burgundy truffle orchard. Inoculated plants are not yet available for sale.

10.6 Final Thoughts

A quality assurance issue that has arisen in BC only now that truffles are being harvested is ensuring that truffle dog teams are competent and reliable. No grower wants to pay for a truffle dog service that fails to find all of the marketable truffles in the orchard but how is the truffle grower able to assess the quality of the team? Scent detection canine handlers and their dogs are already certified by independent certifying bodies to find termites, narcotics, mold, explosives, and bed bugs. Truffle detection is fundamentally no different from finding anything else, and the competence of the canine team is just as important to truffle growers as it is to security officers searching for explosives. Certification might provide truffle growers with a measure of assurance that if truffles are present, they will be found.

In BC, we currently have two excellent truffle dog teams with more teams in training. Brooke Fochuk and her puggle, Dexter, and John Kelly and his Lagotto Romagnolo, Macchi, (Fig. 10.4) have demonstrated their remarkable ability to find cultivated truffles in truffières and native truffles in forests. By finding truffles in

a.

b.

Fig. 10.4 Highly successful truffle dog teams in BC. (**a**) Brooke Fochuk and her puggle, Dexter, in a Burgundy truffle orchard; (**b**) John Kelly and his Lagotto Romagnolo, Macchi, in an orchard producing bianchetto truffles

truffières and forest at the same time and same place that others teams have failed, these two teams have proven that well-trained and highly competent truffle dog teams are key to the success of the truffle industry in BC and throughout Canada.

References

Adamo R (2014) Detection of fungal DNA in forest soils: optimization of molecular techniques for high-throughput sequencing. https://pdfs.semanticscholar.org/5196/6639a41d3e7cdd36b252299a9fe01e5587cb.pdf

Berch SM, Bonito G (2014) Cultivation of Mediterranean species of *Tuber* (Tuberaceae) in British Columbia, Canada. Mycorrhiza 24(6):473–479

Berch SM, Bonito G (2016) Truffle diversity (*Tuber*, Tuberaceae) in British Columbia. Mycorrhiza 26(6):587–594

British Columbia Ministry of Agriculture (2016) Eastern filbert blight. http://www2.gov.bc.ca/assets/gov/farming-natural-resources-and-industry/agriculture-and-seafood/animal-and-crops/plant-health/phu-easternfilbertblightss.pdf

Drake E (2015) Hazelnuts and truffles a winning combination. Ontario Agricultural College, University of Guelph. https://www.uoguelph.ca/oac/news/hazelnuts-and-truffles-winning-combination

Fischer C, Colinas C (2007) Revised methodology for the certification of *Quercus ilex* seedlings inoculated with *Tuber melanosporum* for commercial application. http://trufflegrowers.com.au/wp-content/uploads/2014/01/Fischer_Colinas_Methodology-for-certification-Spanish-of-Quercus-ilex-seedlings.pdf

Suz LM, Martın MP, Colinas C (2006) Detection of *Tuber melanosporum* DNA in soil. FEMS Microbiol Lett 254:251–257

Chapter 11
Diversity and Ecology of Edible Mushrooms from Patagonia Native Forests, Argentina

Carolina Barroetaveña and Carolina V. Toledo

11.1 The Macrofungi and the Patagonian Andean Forest

The eastern slope of the Andean-Patagonian forests covers a wide altitudinal, latitudinal, soil, and rainfall variation, conditioned by a marked decrease of the latter toward the east, with annual average values varying from 3500 mm to 500 mm in few kilometers. This variability offers a broad spectrum of "niches" for mushrooms. There is a great diversity of species of macroscopic fungi with large proportion of endemisms (Singer 1969; Moser and Horak 1975; Horak 1979; Garrido 1986; Rajchenberg 2006), including some edible species with interesting organoleptic characteristics and nutritional value (Gamundi and Horak 1993; Barroetaveña and Toledo 2016a; Barroetaveña et al. 2016; Toledo et al. 2016a, b). In a similar way, the Chilean Patagonian region, with higher precipitations, presents a high fungal diversity with several reported edible species (some shared with those of the Eastern side of the Andes), from the Agaricales, Boletales, Russulales, Aphyllophorales, Gasteromycetes, and Ascomycetes, associated with different trees and forests (Valenzuela-Flores 1993, 2003).

Temperate forests of southern Argentina begin in the North of Neuquén Province, at 36° 25′ South Latitude, extending 2088.7 km to Tierra del Fuego Province, at 54°

C. Barroetaveña (✉)
Centro de Investigación y Extensión Forestal Andino Patagónico CIEFAP, Esquel, Chubut, Argentina

Universidad Nacional de la Patagonia San Juan Bosco, Facultad de Ingeniería, Esquel, Chubut, Argentina

Consejo Nacional de Investigaciones Científicas y Tecnológicas (CONICET), Esquel, Argentina
e-mail: cbarroetavena@ciefap.org.ar

C. V. Toledo
Centro de Investigación y Extensión Forestal Andino Patagónico CIEFAP, Esquel, Chubut, Argentina

J. Pérez-Moreno et al. (eds.), *Mushrooms, Humans and Nature in a Changing World*, https://doi.org/10.1007/978-3-030-37378-8_11

53′SL. It has a maximum width of 107.56 km in the continental sector, and 225.9 km in Tierra del Fuego, covering a total area of 3,350,786 ha (SAyDS 2007). They form pure or mixed forests. The predominant tree species belong to the genus *Nothofagus*, such as Coihue (*N. dombeyi* (Mirb.) Oerst.), Ñire (*N. antarctica* (G.Forst.) Oerst.), Lenga (*N. pumilio* (Poepp. and Endl.) Krasser), Raulí (*N. alpina* (Poepp. and Endl.) Oerst.), Pellín oak (*N. obliqua* (Mirb.) Oerst.), and Guindo (*N. betuloides* (Mirb.) Oerst.), along with conifers *Austrocedrus chilensis* (D.Don) Pic.Serm. and Bizzarri and *Araucaria Araucana* (Mol.) C. Koch.

11.2 Why Is It Interesting to Meet New Wild Edible Mushrooms?

Wild edible mushrooms (WEM) currently have a huge gastronomic interest as they are foods that can be easily certified as ecological and organic, given their provenance of ecosystems with little or no tillage. In addition, they are collected and processed for sale at very low cost. On the other hand, they are categorized by those people who appreciate delicate tastes and flavors as "delicatessen" products, for which a higher price can be paid (Deschamps 2002). Additionally, consumption of edible fungi has increased in recent years mainly because they are considered natural and healthy foods, due to their high protein and low calories content (Barroetaveña and Toledo 2016b; Toledo et al. 2016b; Barros et al. 2007). Also, several mushrooms species are being considered functional foods, those food included in the diet which has one or more targeted functions in the human body, being able to reduce/prevent the likelihood of health disorders (Martins et al. 2017; Morris et al. 2017). Several mushrooms present proven functional properties related to the improvement of digestive and immune functions, reducing risk of cardiovascular disorders and certain tumors, antioxidant properties, antidiabetic effects, and neurogenerative potential. Bioactive compounds responsible for these beneficial effects can be extracted, isolated, and/or purified to prepare dietary supplements known as nutraceuticals, leading to the new era in medicine and healthcare (Morris et al. 2017).

To determine the edibility of a particular species, several works in the field of ethnomycology, which address aspects related to cultural perception, classification and traditional use of fungi, show that effective consumption is the given evidence of edibility (e. g. Ruan-Soto et al. 2007; Garibay-Orijel et al. 2006, 2007) and possibly potential for consumption, if the registry is very isolated. Boa (2004), in his extensive work of global collection on WEM species, uses the same criteria to consider its edibility: "The only reliable guide to edibility is the knowledge that someone has eaten a particular type and survived". However, information available in different field guides in relation to the species reported as edible can be contradictory. As an example, *Gyromitra esculenta* (Pers.) Fr., considered a delicacy in Finland (Härkönen 1998) and Chile (Valenzuela-Flores 2003) is considered toxic and even lethal if consumed raw in the United States (Hall et al. 2003). The knowledge on different ways of preparation or cooking might also determine that the same species is edible or not.

11.3 Wild Edible Mushroom Species from Patagonian Forests

Mushroom species with good possibilities to be used as food, fruit on different substrates or on the forest floor, and establish different ecological associations with trees, either as parasites, symbionts, lignicolous, or humicolous. Their fruit bodies are of appropriate size, good availability, with attractive organoleptic characteristics in relation to color, flavor, and texture (Toledo et al. 2014, 2016a). Fruiting and development of WEM in forest environments is conditioned by various environmental factors, such as exposure on the slopes, tree age and cover, tree, and understory species, all of which will determine the amount of light that enter the forest as well as temperature, content of organic matter, and soil moisture (Vogt et al. 1992; Bergemann and Largent 2000; Pilz et al. 2004). Knowing these environmental characteristics, along with fruiting phenology, finding frequency, availability, and biomass production (Straatsma et al. 2001; Garibay-Orijel et al. 2009; Toledo et al. 2014), allows to evaluate their economically and environmentally sustainable harvest and, eventually, suggest management options to increase their productivity within the framework of what has been called mycosilviculture (Savoie and Largeteau 2011). Studying the ecology of the highly valued *Cantharellus cibarius* Fr. and *Tricholoma magnivelare* (Peck) Redhead from western USA forests (Pilz et al. 1998, 1999, 2004) has allowed to generate tools and policies for sustainable use and monitoring actions (Pilz and Molina 2002). In the Andean Patagonian region of Argentina, the harvest and consumption of WEM has been largely limited almost exclusively to *Morchella* spp. species and, on a much smaller scale, *Cyttaria darwinii* Berk. and *C. hariotii* E. Fisch. (Deschamps 2002; Boa 2004; Albertó et al. 2010) and *Grifola gargal* Singer, by the Mapuche settlers in the San Martin de los Andes region (Neuquén, Argentina). Another species widely consumed in the region is *Suillus luteus* (L.) Roussel, associated with plantations of exotic conifers (Barroetaveña 2006; De Michelis and Rajchenberg 2007). Reports that address the traditional use of WEM as a food resource by indigenous communities include *Cyttaria darwinii*, *C. hariotii*, *C. hookeri* Berk., *C. berteroi* Berk. and *C. espinosae* Lloyd (Espinosa 1926; Martínez-Crovetto 1968; Vera 1991; Smith-Ramírez 1997; Molares et al. 2019), *Grifola gargal* Singer (Baeza and Manuel 1930; Gunckel 1959; Molares et al. 2019), *Morchella intermedia* Boud., *Fistulina antarctica* (as *F. hepatica* Martínez-Crovetto 1968; Smith-Ramírez 1997; Molares et al. 2019), *Lycoperdon* sp. (Zárraga and Vogel 2005), *Calvatia bovista* (L.) T. Macbr. and *C. lilacina* (Mont. and Berk.) Henn. (Martínez-Crovetto 1968; Vera 1991). However, there are other edible species that have been reported as of great edible interest in the region, such as *Grifola sordulenta* (Mont.) Singer, *Fistulina endoxantha* Speg., *Aleuria aurantia* (Pers.) Fuckel, *Cortinarius xiphidipus* M.M. Moser and E. Horak, *Lepista fibrosissima* Singer, and the coraloid fungus *Ramaria patagonica* (Speg.) Corner (Gamundi and Horak 1993; Toledo 2016; Toledo et al. 2016a). *Aleurodiscus vitellinus* (Lev.) Pat. also constitutes a species with potential gastronomic value, given its particular shape, bright color, good texture, and mild flavor that make it interesting for consumption (Gamundi and Horak 1993; Gorjón et al. 2013). Of lower edible quality, *Cortinarius magellanicus* Speg.complex, *Camarophyllus adonis* Singer, *Hydropus dusenii* (Bres.) Singer, and *Tricholoma fusipes* E. Horak (Gamundi and Horak 1993) have been cited. Field observations during three seasons showed that *G. sordulenta*, *A. aurantia*, *L. fibrosissima*, *C. adonis*, and *T. fusipes* appeared very rarely in their natural habitat, while species such as *C. magellanicus* complex, *C. xiphidipus*, *G. gargal*, *C. hariotii*, *F. antarctica*, *F. endoxantha*, *H. dusenii*, *R. patagonica*, and *A. vitellinus* (all endemic to the Patagonian Andean forest) together with other cosmopolitan species such as *Macrolepiota procera* (Scop.) Singer, *Lycoperdon perlatum* Pers. and *Lepista nuda* (Bull.) Cooke did it frequently, which allowed going further into the study of those variables that favor its fruiting. To achieve this, we worked on 11 *Nothofagus* sp. stands located in a geographical gradient from 43° 20′08″ S and 71° 45′06″ W to 39° 25′28″ S and 71° 19′00″ W, in territories of Chubut, Río Negro and Neuquén provinces (Toledo et al. 2014). The species studied and their characteristics are presented below:

11.3.1 Aleurodiscus vitellinus. *(Lév.) Pat*

Phylum: Basidiomycota
Class: Agaricomycetes
Order: Russulales
Family: Steraceae

Description: Easily recognized by its cup-shaped gelatinous fruitbodies (Fig. 11.1), 2–6 cm in diameter and up to 4 cm height, orange or pink color, attached

Fig. 11.1 Fruitbodies of *A. vitellinus* over a *Nothofagus dombeyi* branch

by a central, short, ribbed stem (Toledo et al. 2016a; Gorjón et al. 2013). On *N. dombeyi*, *N. betuloides* and *N. pumilio*.

Lifestyle and habit: Lignicolous. Fructifies abundantly on bark over dead or alive branches and trunks. Prefers open places, with low tree cover.

Organoleptic characteristics: Texture and color varies with its water content, being firm and of pink orange color when young, becoming gelatinous and strong orange at maturity. Aroma and fungal taste very pleasant.

11.3.2 Hydropus dusenii *(Bres.) Singer*

Phylum: Basidiomycota
Class: Agaricomycetes
Order: Agaricales
Family: Marasmiaceae

Description: Funned-shaped fructifications with dry appearance, similar to a "trumpet" (Fig. 11.2), whitish to pale ochraceous in color (Toledo et al. 2016a; Singer 1978). Moderately abundant species mainly associated with *N. dombeyi*, and also *N. betuloides* and *N. pumilio* (Toledo et al. 2016a; Gamundi and Horak 1993).

Lifestyle and habit: Lignicolous. Fructifies on fallen logs with advanced degradation state. Usually in groups, rarely solitary.

Organoleptic characteristics: Very thin context, with a cartilaginous texture. Mild fungal aroma and taste.

Fig. 11.2 Fruitbodies of *H. dusenii* over fallen *Nothofagus dombeyi* wood

Fig. 11.3 Fruitbody of *F. antarctica* on *Nothofagus dombeyi* trunk

11.3.3 Fistulina antarctica *Speg*

Phylum: Basidiomycota
Class: Agaricomycetes
Order: Agaricales
Family: Fistulinaceae

Description: Commonly called "cow tongue," produces fruitbodies that are striking for their large size, tongue-shaped, with bright reddish colour (Fig. 11.3). It is found on *N. pumilio*, *N. dombeyi*, *N. obliqua*, *N. alpina* and *N. antarctica* (Toledo et al. 2016a; Rajchenberg 2006).

Lifestyle and habit: Lignicolous, causing brown rot. Found on living trees with incipient degradation.

Organoleptic characteristics: Reddish flesh, very gelatinous, fleshy texture. Soft fungal aroma and sweet taste.

11.3.4 Fistulina endoxantha *Speg*

Phylum: Basidiomycota
Class: Agaricomycetes
Order: Agaricales
Family: Fistulinaceae

Fig. 11.4 Fruitbody of *F. endoxantha* on *Nothofagus obliqua* trunk

Description: Large fruitbodies, tongue-shaped, chestnut-yellow color (Fig. 11.4). On standing *N. alpina* and *N. obliqua* (Toledo et al. 2016a, Rajchenberg 2006).

Lifestyle and habit: Lignicolous, causing brown rot. Found on living trees with incipient degradation.

Organoleptic characteristics: Chestnut-yellow flesh, firm, fleshy texture. Soft fungal aroma and sweet taste.

11.3.5 Grifola gargal *Singer*

Phylum: Basidiomycota
Class: Agaricomycetes
Order: Polyporales
Family: Meripilaceae

Description: Large, multi-layered fructifications arranged over each other, creamy white, with a dirty appearance (Fig. 11.5). It grows only on *N. obliqua* (Toledo et al. 2016a, Rajchenberg 2006).

Lifestyle and habit: Lignicolous, causing an alveolar white rot. Found on branches and stems of standing or fallen trees, in places with low tree crown cover.

Organoleptic characteristics: Very thin flesh, fleshy texture. Distinctive almond aroma and flavor.

Fig. 11.5 Fruitbody of *G. gargal* on *Nothofagus obliqua* trunk

Fig. 11.6 Fruitbodies of *R. patagonica* on *Nothofagus dombeyi* forest soil

11.3.6 **Ramaria patagonica** ***(Speg.) Corner***

Phylum: Basidiomycota
Class: Agaricomycetes
Order: Gomphales
Family: Gomphaceae

Description: Fruitbodies with branches bifurcated en the apex, in the form of a coral (Fig. 11.6), bright yellow to orange color. It grows on soil in *N. dombeyi*,

N. pumilio, *N. antarctica*, *N. obliqua* and *N. alpina* forests (Toledo et al. 2016a; Corner 1966).

Lifestyle and habit: Mycorrhizal. It is frequently observed in groups, fruiting mainly in open areas, with abundant mulch; can form rows of more than five specimens on the ground, rarely solitary.

Organoleptic characteristics: Pale orange flesh, dry velvety texture. Soft fungal aroma and sweet woody flavor.

11.3.7 Cortinarius magellanicus *Speg. Complex*

Phylum: Basidiomycota
Class: Agaricomycetes
Order: Agaricales
Family: Cortinariaceae

Description: Fruitbodies of medium size, from bright lilac to purple (Fig. 11.7), very glutinous, with argillaceous to white lamellae (Toledo et al. 2016a). A recent phylogenetic analysis showed that *C. magellanicus* represents a species complex with strong regionalism and distinct host associations, including *C. roblerauli* Salgado Salomon and Peintner, *C. magellanicus* Speg., and *C. magellanicoalbus* Salgado Salomon and Peintner in South America, where also exists other two species morphologically very similar: *Cortinarius capitellinus* E. Horak and *Cortinarius aiacapiiae* Speg. (Salgado Salomón et al. 2018). They grow on soil associated to *N. dombeyi*, *N. betuloides*, *N. pumilio*, *N. obliqua*, *N. alpina* and *N. Antarctica* forests.

Fig. 11.7 Fruitbodies of *C. magellanicus* on *Nothofagus dombeyi* forest soil

Lifestyle and habit: Mycorrhizal. It is found frequently in places covered by ferns, forming large groups.

Organoleptic characteristics: Cap with white flesh and delicate, mucilaginous texture. Strong fungal aroma and mild sweet taste.

11.3.8 Cortinarius xiphidipus *M.M. Moser and E. Horak*

Phylum: Basidiomycota
Class: Agaricomycetes
Order: Agaricales
Family: Cortinariaceae

Description: It is a pretty large mushroom, the cap usually up to 10 cm diam., pale yellowish to light chestnut color, usually covered by leaf litter due to their glutinous layer; lamellae argillaceous. Long stem (up to 15 cm.), white, solid, with remnants of veil, very diagnostic for its rooting shape (Fig. 11.8). Very abundant species, associated with *N. dombeyi*, *N. pumilio*, and *N. antarctica* (Toledo et al. 2016a; Horak 1979).

Lifestyle and habit: Mycorrhizal. It fructifies in clusters of numerous specimens, on leaf litter, usually in relatively open places.

Organoleptic characteristics: Cap with white flesh, of mucilaginous texture and firm stem. Strong fungal aroma and mild sweet taste.

Fig. 11.8 Fruitbodies of *C. xiphidipus* on *Nothofagus dombeyi* forest soil

11.3.9 Macrolepiota procera *(Scop.) Singer*

Phylum: Basidiomycota
Class: Agaricomycetes
Order: Agaricales
Family: Agaricaceae

Description: Commonly called "parasol," it is a very characteristic species in mixed forests of *N. dombeyi* and *Austrocedrus chile*nsis (D.Don) Pic.Serm.& Bizzarri. It is easily recognized by its typical large, convex cap in the form of umbrella, with pale brown colour covered by dark flattened scales (Fig. 11.9); it has a long, double ringed stem that separates easily from the cap (Toledo et al. 2016a).

Lifestyle and habit: Saprophytic. Fruits mainly in clearings of the forest and sometimes on the roadsides. It appears individually or in large groups.

Organoleptic characteristics: Cap with white flesh, light corky texture, while the stem is fibrous and inconsistent. Yeast aroma and soft fungal taste, very pleasant with nuances of nuts.

Fig. 11.9 Fruitbody of *M. procera* on *Nothofagus dombeyi* and *A. chilensis* mixed forest

Fig. 11.10 Fruitbodies of *L. nuda* on *Nothofagus dombeyi* forest soil

11.3.10 Lepista nuda *(Bull.) Cooke*

Phylum: Basidiomycota
Class: Agaricomycetes
Order: Agaricales
Family: Tricholomataceae

Description: Characterized by its blue-violet cap, same as the lamellae (Fig. 11.10). Commonly known as "blue foot" or "blewit," it is a widely distributed species in North America (Toledo et al. 2016a; Lincoff 1991). In Patagonia it has been found fruiting associated with *N. dombeyi* and *N. obliqua*.

Lifestyle and habit: Saprophytic. It appears individually or in large groups, forming witch rings on the ground.

Organoleptic characteristics: White to lilac-buff flesh, thick with tender fleshy texture, especially the cap. Fruity aroma and soft fungal taste, very pleasant. The stem is fibrous and inconsistent.

11.3.11 Lycoperdon perlatum *Pers.*

Phylum: Basidiomycota
Class: Agaricomycetes
Order: Agaricales
Family: Agaricaceae

Fig. 11.11 Fruitbodies of *L. perlatum* on *Nothofagus dombeyi* forest soil

Description: It has pear-shaped fructifications changing to globose with maturity (Fig. 11.11), up to 5 cm high and 2.5–3 cm in diameter; the surface whitish when young and brown at maturity, covered with small spines (Toledo et al. 2016a; Demoulin 1971). It has a wide global distribution; in Patagonia occurs it *N. dombeyi*, *N. antactica*, *N. pumilio*, *N. obliqua*, and *N. alpina* forests, in humid spots with shallow mulch.

Lifestyle and habit: Saprophytic. Fruits, solitary or in groups, on soils with abundant decaying wood.

Organoleptic characteristics: Fruits are consumed young, when the flesh is white and with soft texture. Aroma and taste fungal when immature, which is the only edible stage; when the gleba darkens, its qualities change and it must be discarded.

11.3.12 Cyttaria hariotii *E. Fisch*

Phylum: Ascomycota
Class: Leotiomycetes
Order: Cyttariales
Family: Cyttariaceae

Description: Commonly known as "llao-llao." Their fructifications appear as "eyes" that are outlined on globose stroma of yellow-orange color, 3–7 cm in diameter (Fig. 11.12). At maturity they present circular holes in the surface, which correspond to the mature fruiting of the fungus (Gamundi 1971; Gamundi and Horak 1993).

Fig. 11.12 Fruitbodies of *C. hariotii* on *Nothofagus antarctica* trunk

Lifestyle and habit: Obligate parasite of several *Nothofagus* species, forming tumors, developing abundantly on the branches or stems of their hosts.

Organoleptic characteristics: Thick flesh, yellowish, soft fleshy texture, more pleasant in young specimens. Soft fungal aroma and sweet fungal taste.

11.4 Ecological Aspects Associated with the Fruiting of Wild Edible Mushrooms

The study of the environmental variables associated with the fruiting of these saprophytic, mycorrhizal, and lignicolous edible species (Toledo et al. 2014; Toledo 2016) allowed to establish close associations between them. Mulch depth turned out to be positively associated with the fruiting of the mycorrhizal species *C. magellanicus* and *C. xiphidipus* (Table 11.1), in accordance with what was reported by Bergemann and Largent (2000) for the mycorrhizal species *Cantharellus formosus* Corner, with values that did not exceed 5 cm. The quality and quantity of mulch is an important factor for mycorrhizae formation in the forest soil, although many groups of mycorrhizal fungi do not fructify where there are large quantities due to inadequate soil aeration (Bergemann and Largent 2000). However, in the case of *C. xiphidipus*, all the species of this section (*xiphidipus* stirp, Moser and Horak 1975) are characterized by a long, radicant stipe, which probably constitutes an adaptation to the presence of litter layers or deep volcanic ash, frequent in the Patagonian Andes forests. On the other hand, the saprophytic species *L. perlatum*

Table 11.1 Average values of environmental variables associated with the fruiting of humicolous and mycorrhizal edible species

	C. magellanicus	*C. xiphidipus*	*R. patagonica*	*L. nuda*	*L. perlatum*	*M. procera*
% MO	19.23[a] ± (0.29)	20.53 ± (1.20)	19.17 ± (0.67)	19.36 ± (0.27)	17.21 ± (0.35)	18.75 ± (0.31)
% H	18.27 ± (0.68)	16.38 ± (0.89)	18.38 ± (0.63)	16.88 ± (0.56)	21.91 ± (1.17)	16.12 ± (0.52)
CA	84.69 ± (0.97)	80.79 ± (1.04)	84.15 ± (0.80)	77.45 ± (1.71)	76.31 ± (1.22)	82.30 ± (0.81)
Ca	19.38 ± (4.70)	7.81 ± (2.41)	12.81 ± (2.58)	8.44 ± (2.31)	13.13 ± (3.90)	18.44 ± (3.41)
Ch	28.00 ± (7.14)	5.94 ± (2.51)	13.94 ± (3.16)	3.44 ± (1.87)	12.81 ± (2.81)	11.25 ± (3.04)
Cmant	76.25 ± (3.75)	80.00 ± (4.18)	80.63 ± (3.22)	89.38 ± (2.95)	75.94 ± (2.55)	83.13 ± (1.76)
Cmd	2.50 ± (0.65)	2.81 ± (0.79)	1.25 ± (0.56)	1.25 ± (0.72)	0.31 ± (0.31)	0.31 ± (0.31)
Pm	3.64 ± (0.25)	3.93 ± (0.21)	2.98 ± (0.13)	2.98 ± (0.16)	2.62 ± (0.14)	3.04 ± (0.16)

% MO organic matter percentage, *% H* soil moisture percentage, *CA* tree crown cover, *Ca* shrub cover, *Ch* herbaceous cover, *Cmant* mulch cover, *Cmd* wood debris cover, *Pm* mulch depth

[a]Values are averages of 16 replicates ± (SD)

was negatively correlated with this variable, fruiting in sites with low mulch depth, in contrast to what was proposed by Vogt et al. (1992), who affirm that humic species develop on large amounts of organic waste.

In general, high values of organic matter were registered in soils associated to all the terricolous species (Table 11.1), coinciding with what has been reported for western- Patagonian Andes soils (La Manna et al. 2011). The range of variability of this parameter between species was narrow, indicating their intimate link with mushrooms fructification, as it has been reported for other forests (Vogt et al. 1992). The greater differences in organic matter were detected between the mycorrhizal *C. xiphidipus*, with the highest value, and *L. perlatum* (Table 11.1). The content of organic matter favors water retention capacity and nutrient availability while influencing mycorrhizal formation (Bergemann and Largent 2000). The presence of wood debris was also positively associated with the mycorrhizal species *C. magellanicus* and *C. xiphidipus*. Fallen wood shelters a large number of mycorrhizal roots (Harvey et al. 1996) and constitutes a reservoir of soil moisture that favours fruiting (Luoma et al. 2004). Soil moisture content for *C. xiphidipus* was relatively low (12.1–24.3%), probably because mulch accumulation, but associated with the presence of wood debris. On the other hand, *L. perlatum*, associated with less mulch depth and organic matter content, fructify at higher soil moisture values (14.50–27.50%).

All the humicolous and mycorrhizal species developed in sites with high tree cover, over 76%, coinciding with closed and humid sites reported by Gamundi and Horak (1993). These results match partially with those of Santos-Silva and Louro (2011) who found a strong association between mycorrhizal species and high tree cover, lower for saprophytic species, although clarifying that this last result could be conditioned by the high proportion of lignicolous and gasteroids species, the latest without representatives in our study. The coverage of herbaceous plants, however, was generally low, with a maximum of 30% for five of the six terricolous species, as compared with *Suillus luteus* fructification patches in *Pinus ponderosa* Dougl ex Laws. plantations in Patagonia, with values of 60% (Barroetaveña et al. 2008).

Fructification phenology was variable according to species. Lignicolous species showed a higher fructification abundance at the beginning of the season (fortnights 1, 2, and 3). *Aleurodiscus vitellinus* and *F. antarctica* were more abundant in the first two autumn fortnights, while *H. dusenii* was more abundant in fortnight 3 (Fig. 11.13). This could be explained by the fact that they receive moisture from the trunk or decomposing branches, which functions as reservoirs of water, and do not depend on accumulated precipitation. Terricolous species showed a more diverse fruiting pattern. *Ramaria patagonica* and *C. magellanicus* presented highest fructification abundance at the beginning of autumn, with peaks in fortnight 3 (Fig. 11.13), while *C. xiphidipus* and *L. perlatum* did so during the last 3 fortnights, quite homogeneously the former, and with a peak in fortnight 3 the latest (Toledo et al. 2014).

Therefore, a harvest plan should contemplate this variability in order to adapt harvesting and commercialization steps throughout the season. The finding frequency was also very variable between species, resulting *F. antarctica*, *C. magellanicus* complex, *R. patagonica*, and *C. hariotii* the more frequent (Fig. 11.13).

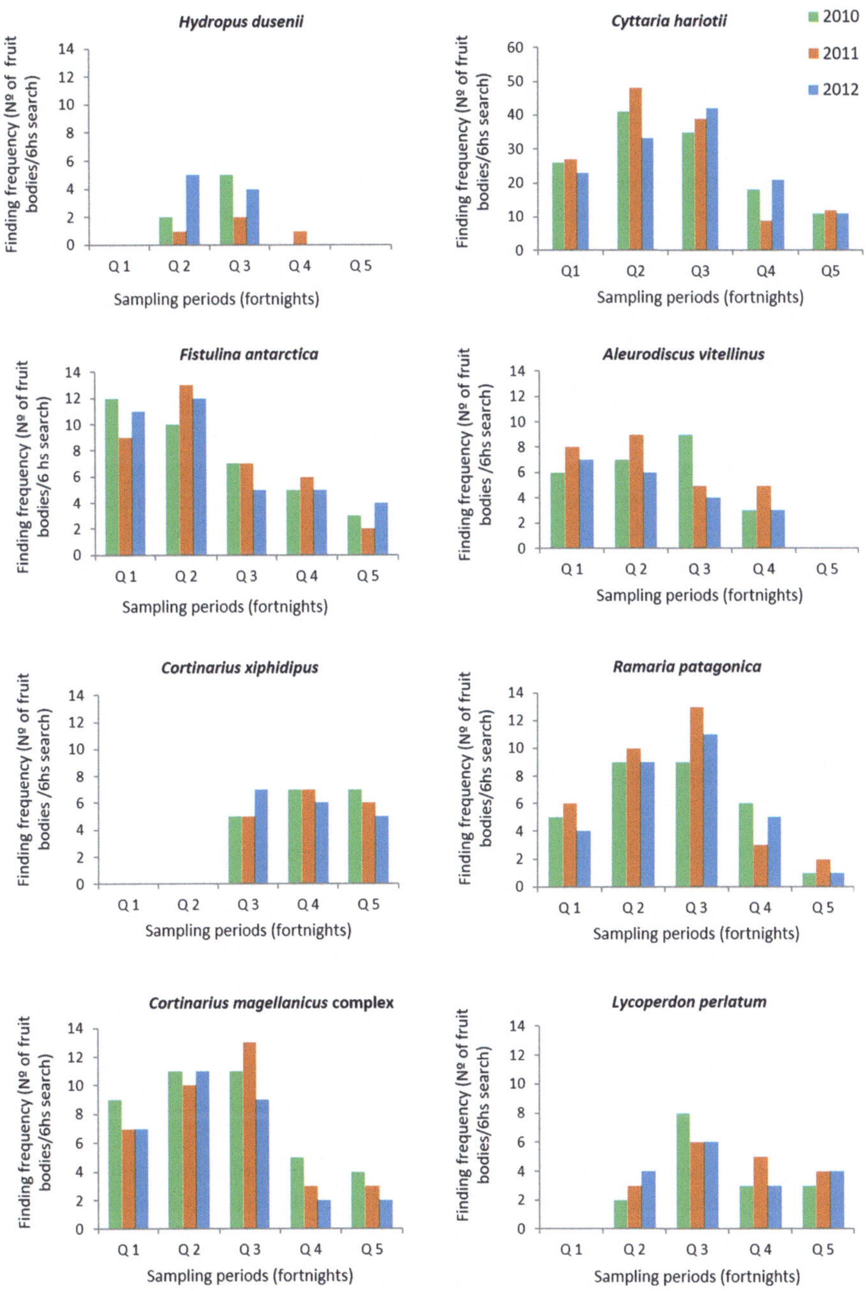

Fig. 11.13 Finding frequency per fortnight (No. fruit bodies observed/6 h search) (indicated on the axis of the ordinates), of eight wild edible mushrooms species during 2010, 2011 and 2012 fruiting seasons. *Spring season* (*C. hariotii*): F1, beginning-mid October; F2, mid-end October; F3, beginning-mid November; F4, mid- end November; and F5, beginning-mid December. *Autumn season* (*H. dusenii*, *F. antarctica*, *A. vitellinus*, *C. xiphidipus*, *R. patagonica*, *C. magellanicus* and *L. perlatum*): F1, mid-end March; F2, beginning-mid April; F3, mid-end April; F4, beginning-mid May; and F5, mid-end May

Thus, the autumn period would begin in mid-March with the offer of *F. antarctica*, followed by *A. vitellinus*, *C. magellanicus* complex, and *R. patagonica* which would extend until beginning of May. *Cortinarius xiphidipus* and *L. perlatum* could be offered from mid-April until the end of the season: *H. dusenii* in a more limited way, during April.

In relation to wood degrader species, it was observed that *F. antarctica* developed on live trunks of several *Nothofagus* species producing brown rot (Gamundi and Horak 1993; Rajchenberg 2006), associated with a high tree cover (80.23–87.56%) and an incipient state of wood degradation. *Fistulina endoxantha*, occurring on live trunks of *N. alpina* and *N. obliqua* producing brown rot (Rajchenberg 2006), was found to be mainly associated with *N. obliqua*, and a medium state of wood degradation. *Aleurodiscus vitellinus* was registered mainly growing under low tree cover (63.89–67.85%), on trunks and branches of dead *N. dombeyi*, although it has been reported associated also with *N. betuloides* and *N pumilio* branches, on living and dead wood and on bark (Gamundi and Horak 1993). *Grifola gargal*, associated with high tree cover (82.56–86.89%), appeared restricted to branches and trunks of standing or fallen *N. obliqua*, causing white alveolar rot in the heartwood as reported by Pozzi et al. (2009). *Hydropus dusenii* appeared mainly associated with *N. dombeyi*, although it is also mentioned for *N. pumilio* (Gamundi and Horak 1993), fruiting on very degraded fallen trunks and under average tree cover values (72.87–76.43%).

11.5 Conclusions and Perspectives

The environmental variables conditioned fructification with different intensity according to the substrate where fungal species growth. Terricolous species were variably conditioned by soil water availability associated with tree cover, soil organic matter and mulch characteristics, with different fruiting moments according to temperatures and accumulated precipitations. Lignicolous species use the substrate as a reservoir of moisture for their fructification, being able to fructify early in the season. Wood degradation stages also showed association with the occurrence of lignicolous species fructification, while *F. antarctica* and *F. endoxantha* developed on almost unaltered and altered wood, respectively, *H. dusenii* did so on fallen, very degraded logs.

Saprophytic and lignicolous WEM species constitute a plausible product to be cultivated (Toledo and Barroetaveña 2017). Currently, there are many studies focused on the search for optimal parameters to cultivate novel species (Lechner and Albertó 2007; Reyes et al. 2009; Zhou et al. 2015). Since they fructify with an strictly seasonal pattern, in environments with reduced and difficult accessibility, with lower frequency of finding compared with species associated with pine plantations, such as *Suillus luteus*, their increased availability would improve the chances to place them as gourmet products, or take better advantage of their medicinal properties (Morris et al. 2017; Toledo et al. 2016b).

Results obtained for these species from Patagonia show the importance of the simultaneous evaluation of multiple issues associated with WEM fructification. Some species such as *A.vitellinus* register a high number of collections, but have small fructifications that result in very little weight, while species such as *R. patagonica* or *F. antarctica* become important due to the size and fresh weight of their fruiting, despite having a lower number of collections (Toledo et al. 2014).

The narrow ranges detected for certain environmental variables associated with fruiting of humicolous and mycorrhizal edible species from Patagonian forest, such as the presence of soil wood debris, organic matter percentage, mulch depth, and coverage and tree crown cover, suggest that the fruiting of these species could be affected by small variations of these parameters. Lignicolous species, on the other hand, require different stages of wood degradation for their development. In this sense, different forest uses such as firewood extraction, logging or cattle grazing in these environments generate impacts that could affect WEM productivity, as it has been reported for northwest USA forests under silvicultural management (Colgan et al. 1999; Luoma et al. 2004).

A detailed knowledge of WEM ecology, together with forest management policies that include them as a forest product, will allow to plan their harvest and conservation. In this framework, productivity estimates as well as detailed phenology of each species should be revalidated with local studies, given that both parameters have a close relationship with the latitudinal and pluviometric variation present in the vast Andean-Patagonian region.

References

Albertó E, Curvetto N, Deschamps MRG, Lechner B (2010) Hongos silvestres y de cultivo en Argentina: historia, regiones y sistemas de producción, consumo, mercado interno y externo, legislación, oferta tecnológica e investigación y desarrollo. In: Martínez-Carrera D, Curvetto N, Sobal M et al (eds) Hacia un Desarrollo Sostenible del Sistema de Producción-Consumo de los Hongos Comestibles y Medicinales de Latinoamerica. Avances y Perspectivas en el Siglo XXI. COLPOS-UNS-CONACYT-AMC-UAEM-UPAEP-IMINAP, Puebla México, pp 333–358

Baeza R, Manuel V (1930) Nombres vulgares de las plantas silvestres de Chile y su concordancia con los nombres científicos. Imprenta El Globo, Santiago

Barroetaveña C (2006) Hongos asociados a las plantaciones forestales de la región Andino Patagónica, Manual N° 6. Centro de Investigación y Extensión Forestal Andino Patagónico, Esquel

Barroetaveña C, Toledo CV (2016a) Hongos Silvestres Comestibles Novedosos en el Bosque Nativo y en las Plantaciones de Patagonia Andina, Argentina. Cienc Invest For 22(3):73–88

Barroetaveña C, Toledo CV (2016b) The nutritional benefits of mushrooms. In: Ferreira ICFR, Morales Gómez P, Barros L (eds) Wild plants, mushrooms and nuts: functional food properties and applications. Wiley-Blackwell, West Sussex, pp 65–82

Barroetaveña C, La Manna L, Alonso V (2008) Variables affecting *Suillus luteus* fructification in ponderosa pine plantations from Patagonia (Argentina). Forest Ecol Manag 256(1):1868–1874

Barroetaveña C, Toledo CV, Rajchenberg M (2016) Hongos comestibles silvestres de plantaciones forestales y praderas de la región Andino Patagónica de Argentina, Manual N°17. Centro de Investigación y Extensión Forestal Andino Patagónico, Esquel

Barros L, Baptista P, Correia D, Casal S, Oliveira B, Ferreira ICFR (2007) Fatty acid and sugar compositions and nutritional value of five edible mushrooms from Northeast Portugal. Food Chem 105:140–145
Bergemann SE, Largent DL (2000) The site specific variables that correlate with the distribution of the Pacific Golden Chanterelle, *Cantharellus formosus*. Forest Ecol Manag 130:99–107
Boa E (2004) Los Hongos Silvestres Comestibles, Perspectiva global de su uso e importancia para la población, Productos Forestales No Madereros 17. FAO, Rome
Colgan W, Carey AB, Trappe JM, Molina R, Thysell D (1999) Diversity and productivity of hypogeous fungal sporocarps in a variably thinned Douglas-fir forest. Can J For Res 29:1259–1268
Corner E (1966) Species of *Ramaria* (Clavariaceae) withouth clamps. Trans Br Mycol Soc 49:101–113
De Michelis A, Rajchenberg M (2007) Hongos comestibles, teoría y práctica para la recolección, elaboración y conservación. INTA EEAB, Buenos Aires
Demoulin V (1971) *Lycoperdon norvegicum* Demoulin sp. nov. A new Gasteromycete with boreocontinental distribution in Europe and North America. Norweg. J Bot 18:161–167
Deschamps JR (2002) Hongos comestibles silvestres del Mercosur con valor gastronómico. Documento de trabajo N° 86. Universidad de Belgrano, Buenos Aires. http://www.ub.edu.ar/investigaciones/dt_nuevos/86_deschamps.pdf
Espinosa M (1926) Los hongos chilenos del género *Cyttaria*. Rev Chil Hist Nat 30:206–255
Gamundi IJ (1971) Las Cyttariales sudamericanas (Fungi-Ascomycetes). Darwin 16(3):461–510
Gamundi IJ, Horak E (1993) Hongos de los Bosques Andino-Patagónicos. Vazquez Mazzini, Buenos Aires
Garibay-Orijel R, Cifuentes J, Estrada-Torres A, Caballero J (2006) People using macro-fungal diversity in Oaxaca. Mexico Fungal Divers 21:41–67
Garibay-Orijel R, Caballero J, Estrada-Torres A, Cifuentes J (2007) Understanding cultural significance, the edible mushrooms case. J Ethnobiol Ethnomed 3:1–18
Garibay-Orijel R, Martínez-Ramos M, Cifuentes J (2009) Disponibilidad de esporomas de hongos comestibles en los bosques de pino-encino de Ixtlán de Juárez, Oaxaca. Rev Mex Biodivers 80:521–534
Garrido NN (1986) Survey of ectomycorrhizal fungi associated with exotic forest trees in Chile. Nova Hedwigia 43:423–442
Gorjón S, Greslebin AG, Rajchenberg M (2013) The genus *Aleurodiscus*sl (Stereaceae, Russulales) in the Patagonian Andes. Mycol Prog 12:91–108
Gunckel L (1959) Nombres indígenas relacionados con la flora chilena. Boletín de Filolología 11:191–327
Hall IR, Yun W, Amicucci A (2003) Cultivation of edible ectomycorrhizal mushrooms. Trends Biotechnol 21:433–438
Härkönen M (1998) Uses of mushrooms by Finns and Karelians. Int J Circumpolar Health 57:40–55
Harvey AE, Page-Dumroese DS, Jurgensen MF, Graham RT, Tonn JR (1996) Site preparation alters biomass, root and ectomycorrhizal development of out planted western white pine and Douglas-fir. New For 11:255–270
Horak E (1979) Fungi, Basidiomycetes, Agaricales y Gasteromycetes secotioides. Flora Criptogam Tierra del Fuego (Buenos Aires) 11:1–525
La Manna L, Buduba C, Irisarri J, Ferrari J, Cremona MV (2011) Los nutrientes del suelo en la Región Andino Patagónica: una aproximación a la interpretación de datos analíticos. Patagonia Forestal XVII:7–8
Lechner BE, Albertó E (2007) Optimal conditions for the fruit body production of natural occurring strains of *Lentinus tigrinus*. Bioresour Technol 98:1866–1869
Lincoff GH (1991) The Audubon Society field guide to North American mushrooms. Knopf, New York
Luoma DL, Eberhart JL, Molina R, Amaranthus MP (2004) Response of ectomycorrhizal fungus sporocarp production to varying levels and patterns of green-tree retention. Forest Ecol Manag 202:337–354

Martínez-Crovetto R (1968) Estudios etnobotánicos IV: Nombres de plantas y su utilidad, según los indios onas de Tierra del Fuego. Etnobiológica 3:1–20
Martins N, Morales P, Barros L, Ferreira IC (2017) The increasing demand for functional foods. In: Ferreira CFR, Morales P, Barros L (eds) Wild plants, mushrooms and nuts: functional food properties and applications. Wiley, Chichester, pp 1–13
Molares S, Toledo CV, Stecher G, Barroetaveña C (2019) Traditional mycological knowledge and processes of change in Mapuche communities from Patagonia, Argentina: A study on wild edible fungi in Nothofagaceae forests. Mycologia https://doi.org/10.1080/00275514.2019.1680219
Morris HJ, Llauradó G, Beltrán Y, Lebeque Y, Bermúdez C, García N, Gaime-Perraud I, Moukha S (2017) The use of mushrooms in the development of functional foods, drugs, and nutraceuticals. In: Ferreira CFR, Morales P, Barros L (eds) Wild plants, mushrooms and nuts: functional food properties and applications. Wiley, Chichester, pp 123–157
Moser MM, Horak E (1975) Cortinarius Fr. undnaheverwandte Gattungen in Südamerika. Beih Nova Hedwigia 52:1–628
Pildain MB, Visnovsky S, Barroetaveña C (2014) Phylogenetic diversity of true Morels (*Morchella*), the main edible non-timber product from native Patagonian forests of Argentina. Fungal Biol 118:755–763
Pilz D, Molina R (2002) Commercial harvests of edible mushrooms from the forests of the Pacific Northwest United States: issues, management, and monitoring for sustainability. Forest Ecol Manag 155:3–16
Pilz D, Molina R, Liegel L (1998) Biological productivity of Chanterelle mushrooms in and near the Olympic Peninsula Biosphere Reserve. AMBIO Special Rep 9:8–13
Pilz D, Smith J, Amaranthus M, Alexander S, Molina R, Luoma D (1999) Mushrooms and timber; managing commercial harvesting in the Oregon Cascades. J Forest 97:4–11
Pilz D, Weber NS, Carter MC, Parks CG, Molina R (2004) Productivity and diversity of morel mushrooms in healthy, burned, and insect-damaged forests of northeastern Oregon. Forest Ecol Manag 198:367–386
Pozzi C, Lorenzo L, Rajchenberg M (2009) Un hospedaje exótico del hongo comestible *Grifola gargal*. B Soc Argent Bot 44:9–10
Rajchenberg M (2006) Los Políporos (Basidiomycetes) de los Bosques Andino Patagónicos de Argentina. J Cramer, Berlin
Reyes RG, Lopez LLMA, Kumakura K, Kalaw SP, Kikukawa T, Eguchi F (2009) *Coprinus comatus*, a newly domesticated wild nutraceutical mushroom in the Philippines. IJAT 5(2):299–316
Ruan-Soto F, Mariaca R, Cifuentes J, Limón F, Pérez-Ramírez L, Sierra-Galván S (2007) Nomenclatura, clasificación y percepciones locales acerca de los hongos en dos comunidades de la Selva Lacandona, Chiapas, México. Etnobiología 5:1–20
Salgado Salomón MES, Dresch P, Horak E, Galleguillos F, Barroetaveña C, Peintner U (2018) The enigmatic *Cortinarius magellanicus* complex occurring in Nothofagaceae forests of the Southern Hemisphere. Fungal Biol 22(11):1077–1097
Santos-Silva C, Louro R (2011) Macrofungal diversity in Montado ecosystems–management and sustainability. Proceedings of Simposiso Internacional Gestión Forestal del Recurso Micológico. Instituto Europeo de Micología, Valladolid, 9–11 June 2011
Savoie JM, Largeteau ML (2011) Production of edible mushrooms in forests: trends in development of a mycosilviculture. Appl Microbiol Biotechnol 89:971–979
SAyDS (2007) Primer inventario nacional de bosques nativos: informe regional bosque andino patagónico - 1a ed- Secretaría de Ambiente y Desarrollo Sustentable de la Nación
Singer R (1969) Mycoflora Australis. Beih Nova Hedwigia 29:1–406
Singer R (1978) Keys for the identification of the Species of Agaricales II. Sydowia 31:193–237
Smith-Ramírez C (1997) Algunos usos indígenas tradicionales de la flora del bosque templado. In: Armesto JJ, Villagrán C, Arroyo MK (eds) Ecología de los bosques nativos de Chile. Editorial Universitaria, Santiago de Chile, pp 389–404
Straatsma G, Ayer F, Egli S (2001) Species richness, abundance, and phenology of fungal fruit bodies over 21 years in a Swiss forest plot. Mycol Res 105:515–523

Toledo CV (2016) Hongos comestibles silvestres de bosques nativos de *Nothofagus*en Patagonia. PhD thesis, Universidad Nacional del Comahue, Argentina

Toledo CV, Barroetaveña C (2017) Crecimiento miceliar de especies silvestres de hongos comestibles de los bosques andino-patagónicos: primeros pasos para su domesticación. Bol Soc Argent Bot 52(3):435–446

Toledo CV, Barroetaveña C, Rajchenberg M (2014) Fenología y variables ambientales que condicionan la fructificación de hongos silvestres comestibles de los bosques de *Nothofagus* spp. en Patagonia, Argentina. Rev Mex Biodiv 85:1093–1103

Toledo CV, Barroetaveña C, Rajchenberg M (2016a) Hongos comestibles silvestres de los bosques nativos de la región Andino Patagónica de Argentina. Manual N°16, Centro de Investigación y Extensión Forestal Andino Patagónico, Esquel Argentina

Toledo CV, Barroetaveña C, Fernandes A, Barros L, Ferreira I (2016b) Chemical and antioxidant properties of wild edible mushrooms from native *Nothofagus* spp. Forest, Argentina. Molecules 21(9):1201

Valenzuela-Flores E (1993) Estudio sistemático, corológico y ecológico de los Agaricales Sensu Lato de los bosques autóctonos de la Región de los Lagos en Chile. PhD thesis, Universidad de Alcalá de Henares, Madrid

Valenzuela-Flores E (2003) Hongos comestibles silvestres colectados en la X región de Chile. Bol Micol 18:1–14

Vera J (1991) Uso alimentario de recursos vegetales entre cazadores-recolectores de Bahía Laredo, cabo Negro (Magallanes). Anales Instituto Patagonia 20:156–168

Vogt K, Bloomfield J, Ammirati J, Ammirati S (1992) Sporocarp production by basidiomycetes with emphasis on forest ecosystems. In: Carrol G, Wicklow D (eds) The fungal community: its organization and role in the ecosystem. Marcel Dekker, New York, pp 563–582

Zárraga C, Vogel O (2005) Haoa Usi Mitsana Remedio de mi tierra, Pequeño libro de la medicina Yagan. Ediciones Kultrún, Valdivia

Zhou S, Tang Q, Zhang Z, Li CH, Cao H, Yang Y, Zhang JS (2015) Nutritional composition of three domesticated culinary-medicinal mushrooms: *Oudemansiella sudmusida*, *Lentinus squarrosulus*, and *Tremella aurantialba*. Int J Med Mushrooms 17(1):43–49

Part IV
Socioeconomical and Cultural Importance

Chapter 12
Truffle Cultivation in the South of France: Socioeconomic Characteristic

Pierre Sourzat

12.1 Introduction

The aim of this chapter is to present the conditions for development of truffle cultivation in its natural and agricultural environment and its socioeconomic context in the south of France. The first section presents the ancestral know-how, the truffle's place in its natural environment, the relationships between truffle fungi and agriculture, agroforestry and truffle sylviculture, agroecology, and truffle cultivation. The second section is devoted to the roles of the truffle in the territory, the sociology of truffle growers, the development of truffle cultivation in the southwest of France, and the lessons learned from truffle cultivation abroad. Finally, the third section analyzes the reasons why one becomes a truffle grower and how truffle growers become involved in the environment and contribute to its maintenance and evolution.

12.2 Truffle Cultivation from the Perspective of Agroecology and Agroforestry

12.2.1 The Ancestral Know-How

When Joseph Talon invented truffle cultivation by sowing acorns in 1810 (Chatin 1892), he initiated a method of indirect cultivation of the truffle adapted to the environmental conditions of his time. In the absence of controlled mycorrhizal plants, the abundance of black truffle spores in situ was sufficient to optimally inoculate the roots of oak seedlings. "He had noticed that truffles were generally found in dry, stony quarters at the foot of certain trees; that where they were, the ground was

P. Sourzat (✉)
Truffle Expert, Bel Air, Ussel, France

J. Pérez-Moreno et al. (eds.), *Mushrooms, Humans and Nature in a Changing World*, https://doi.org/10.1007/978-3-030-37378-8_12

generally devoid of vegetation, and placed under the direct action of the sun" (Lucas De Montigny 1862). Knowledge of the ecological requirements of the truffle was a matter of empiricism but was sufficient in a context that was clearly very favorable.

From the experience of Joseph Talon, the peasants of Provence and the south of France developed a know-how over the ensuing decades. This know-how was recorded by various authors, the best known of whom were Chatin (1892), Ferry de la Bellone (1888), De Bosredon (1887), and Pradel (1914). Taking the example of the "Manual of Trufficulture" (Pradel 1914), the titles of the most significant chapters are eloquent: considerations on truffles and truffle production; conditions favorable to truffle production; sowing in the nursery, choice of acorns, and time of seeding; planting of truffle trees, creation of natural truffle trees; formation of the "truffière," their maintenance, fertilization, reconstitution of exhausted "truffières," watering; harvesting products; evaluation of the cost of creating artificial "truffières"; some insights into the characters of taste and perfume which distinguish the best truffles; and shipping fresh truffles, culinary recipes.

In short, nearly all the extant knowledge on the truffle and known methods for its cultivation as of the beginning of the twentieth century are presented in this book. Nevertheless, production fell rapidly after the publication of the "Manual of Trufficulture" (Pradel 1914). In reflection, the Great War and its human, social, and economic consequences were important factors in the decline of truffle production. But what additional ancestral know-how was it possible to collect after Doctor Pradel's "Manual of Trufficulture"? Jean Rebière in "La Truffe du Périgord" (Rebière 1981), proposed a method of culture associated with various more modern analyses. Nevertheless, truffle production continued to decline after this publication and those of additional authors. However, the works of Grente and Delmas (1974) and Delmas (1976) have fed many hopes based in particular on the popularization of greenhouse production of seedlings inoculated under controlled conditions with truffle fungi.

A 1997 survey carried out in the Department of the Lot at the request of the Midi-Pyrénées Region (Sourzat et al. 1997) made it possible to collect and/or confirm additional know-how that contributed to the local history of the truffle as it contributed to an independent way of life. Pruning for the collection of leaves at the end of August sometimes led to the appearance of natural "truffières" in old alfalfa meadows at the edges of woods. These practices were linked to a very frugal animal husbandry. The energy for cooking food, baking bread, and heating was primarily wood-based. Peasants returned to the soil the organic matter resulting from their activities. It can be said that the cycles of carbon (energy) and nitrogen (organic matter) were perfectly regulated, without excesses that could induce unnecessary losses. Truffle cultivation was integrated into these cycles because, in the vineyards, where the truffle trees were planted, manure was brought directly to the feet of the vines where truffles were often harvested.

12.2.2 The Truffle and the Natural Environment

After the devastating phylloxera infestations of 1870–1880, the destroyed vineyards located on the hillsides were recolonized by the native oak forest (*Quercus pubescens*). They were excellent sites for development of natural "truffières" (Olivier et al. 2018). The unproductive or hard-to-reach limestone lands, the first to be abandoned due to agricultural degradation, were also suitable places for the development of natural "truffières." The maintenance of these sites was accomplished, thanks to pastoralism, which delayed and diminished the encroachment of brush and woodland. The black truffle, because of its pioneering character, had no difficulty in colonizing the abandoned limestone environments. But, in the absence of passage of the small flocks of sheep, the forest encroached on these limestone moors, and the truffle gave way to secondary stage mycorrhizal fungi.

Today, natural "truffières" are formed mainly in natural grasslands (Fig. 12.1) along the edges of woods (Sourzat 2004a). They result from the presence of an open space where the roots of the border oaks thrive in an unfertilized lawn (no additions of nitrogen in nitric or ammoniacal form), usually a *Mesobromion* grassland. The black truffle's burnt areas ("brûlé") appear about 10 m from the oak woodland edge, in the *Mesobromion*. Under the edge trees, in or near the underbrush of the natural prairie, mostly basidiomycete fruiting is observed (*Hebeloma*, *Tricholoma*, *Boletus* and *Russula*) (Sourzat 2004a). Different species of these genera are known to become important late in the successional development of the natural

Fig. 12.1 Natural truffle burnt area ("brûlé") formed by *Tuber melanosporum* in a meadow at the edge of a woodland (Le Montat, Lot, France)

environment (Sourzat 2004a). It is also sometimes possible to harvest some small *Tuber brumale* truffles in the undergrowth. When *Tuber melanosporum* develops following afforestation, it is thanks to the open space that allows the truffle to express its pioneering character.

All these elements of the ecology of the black truffle suggest that, if one could establish plantations with the agronomic and ecological characteristics of the best natural "truffières," the difficulties of truffle production would be solved. It remains to be understood, however, that many of the natural "truffières" that have been discovered these last 30–40 years have a fleeting existence (2–3 years), whereas the longevity of natural "truffières" that were found at the start of the twentieth century was often 10–20 years. Certainly, pastoralism delayed the closure of clearings and the afforestation of the moors. But this does not explain everything, especially when they appear in natural grassland maintained mainly by one or two annual mowing. The absence of summer rainfall appears sometimes to explain their decline. The possibility of a historic decline in the dominance of the truffle over other competing fungus species raises the issue of the relative virulence (Sourzat 2010).

12.2.3 Truffle and Agriculture

Truffle arboriculture, which has developed with the availability of mycorrhized plants, has not yet solved the difficulty of normalizing truffle production in traditional truffle regions. In "The truffle environment: constraints and management" (Sourzat 2010), the analysis of the distribution of production according to the size of the plots and their wooded oak environment has made it possible to identify "truffle bastions" (Fig. 12.2). These are areas within plantations where *Tuber melanosporum* is dominant over competing fungal contaminants (e.g. *Tuber brumale*) associated with the edges of surrounding oak woodlands. The pressure of fungal contamination is apparently nonexistent (or nearly so) in these truffle bastions.

If in traditional areas, truffle arboriculture finds its limits according to the size of the plantations (and in the absence of watering), on the other hand, in the cereal and wine plains of Poitou-Charentes, Center or Provence Alpes Côte d'Azur, plantations of mycorrhized trees have rapid responses with relatively simple cultivation methods (including for example tillage, watering and tree pruning). This is also the case in Western Australia where the results of truffle arboriculture are remarkable though unexplained. In New Zealand, results seem poorer than in Western Australia, raising many questions. However, yields of 50 kg per ha are now also being reached in New Zealand plantations (Guerin-Laguette, pers. comm.).

In the Lot department (in south-western France), "trufficulture" experiments (truffle silviculture) have not yielded satisfactory results. However, a site not far from Lalbenque is renowned for its exceptional success (Sourzat 2004a). The original plantation (over 50 years old) has been heavily thinned, and the residual trees have been pruned. The soil was worked during the first years. Sheep have grazed a few days in the spring. "Brûlés" have formed outside the trees' canopies, in spaces not crowded by root systems of woody plants. The calcareous lawn (*Mesobromion*)

Fig. 12.2 The high production area (pink) constitutes a Périgord black truffle bastion within a 4 ha plantation (Masquières, Lot et Garonne, France)

has been gradually nibbled by the "brûlés" that produced *Tuber melanosporum*. Species such as *Tuber brumale* and *Tuber aestivum* are rare on this site. The owner of the site never wanted to start collecting truffles until January. It is possible that early rotting truffles provided natural inoculum for enhanced production. Aerial photography of the site (Fig. 12.3) shows the impact of thinning at this site in comparison with surrounding area.

If it is desirable to talk about truffle agroforestry, why not consider it in a process of reconquest of old truffle plantations, which are not in short supply (especially in the south of the Lot department)? The essential condition will be to create sufficiently open spaces inside the oak woodlands in order to favor expression of the pioneer character of the truffle. Root systems will have to be rejuvenated by tillage that cuts roots in the first 2 years, thus allowing the profusion of young roots. The practice of inoculation, either by "truffle traps" in the form of pits or trenches, in unshaded root zones just outside the canopy dripline, should favor the appearance of new truffle mycorrhizae.

On the plateau of Valensole or in the Tricastin (Vaucluse and Drôme departments in southeastern France), the association of lavender cultivation with the planting of truffle oaks is as common as that of vines with truffles in the Lot department (Fig. 12.4). In Lascabanes (Lot, France), a truffle farmer has opted on his organic

Fig. 12.3 Contrast of tree density between the thinning plantation and its surroundings at Lalbenque (Lot)

Fig. 12.4 Truffle plantation established with lavender (Drôme, southeastern France)

farm to sow cereals (wheat, spelt) between rows of truffle trees, until the trees have reached production age. On some plantations, horses valorize the grassland resource, possibly by providing manure. In Spain, the closure of the environment and the disappearance of the natural truffle trees followed the decline of the animal husbandry which is to say the grazing of underbrush by flocks of sheep, goats, and cows (Reyna Domenech 2000). Truffle agroforestry is not lacking in examples even before the concept became popularized.

On the site of the Cahors-Le Montat Truffle Research Centre (Lot, France), some plantations which are 20–30 years old or more are useful for truffle agroforestry experiments. Experiments evaluating truffle complementation with lavender, grape, or other plants are underway. Experimentation on root and aerial system manipulation of old truffle trees can be attempted without risk to production which has already either declined or disappeared. In some plantations causes of replacement of the initial *Tuber melanosporum* mycorrhization by that of *Tuber brumale* or *Tuber aestivum* deserve to be deepened both with work on soils and trees. In general, these experiments should be preceded by molecular biological analyzes of the presence of various truffle species in the soil.

12.2.4 Agroecology and Trufficulture

The concept of agroecology is perfectly adapted to truffle cultivation because it integrates the sensitivity of the truffle ecosystem to interventions or disturbing treatments (like mechanical or chemical). The exact reasons for the exceptional production of truffles in the late nineteenth and early twentieth centuries are not known in terms of soil or environmental biology. However, it is known that the truffle was extremely dominant, to the extent that people were concerned by its virulence and/or aggressiveness toward its host trees. In fact, they were apprehensive about the shortage of firewood in northern Lot (South West) because the oaks did not grow. People thought that truffles inhibited tree growth. This dominance of *Tuber melanosporum* at the beginning of the twentieth century can be partially explained by a great sufficiency of inoculum (even in the absence of certified mycorrhizal nursery plants), resulting in the perennial production of planted and natural "truffières," but this hypothesis does not seem sufficient.

Biodiversity in the soil and on the surface of the soil appears to be a poorly known track despite the relevant arguments put forward by Callot (1999) on the importance of soil fauna in the aeration process of the soil and of fruiting bodies. Wild boars, preferentially attacking the edges of natural "brûlés" for feeding (looking for the earthworms), show that it is in the area where the truffles grow best that the earthworms are located (Sourzat 2008). There is a relationship between the presence of earthworms and those of truffles. Large animals also have a role to play (including horses, sheep and cows). We have seen how the presence of horses in calcareous moorland situations (with scattered oak trees) has favored the appearance of long-lasting and productive "brûlés" (Sourzat 2004a). It has also been seen that the disappearance of pastoralism with sheep herds has coincided with the disappearance of the truffle in many scrub woodlands in Provence and various other regions of France. The virulence of the truffle, which is illustrated by very marked and perennially productive "brûlés," is an indication of what must be sought to make the truffle dominant taking into account the biology of the environment.

The disappearance of mixed farming and the specialization of agriculture were indirect causes of the regression of the truffle. With pastoralism, in the moors and scrub woodlands where natural "truffières" had formed, the closure of the environment was delayed. Interactions between living organisms, the truffle, and its environment have not yet been explored. Some plants (e.g., grape, lavender, juniper, Jerusalem artichoke, rose hips, black spines) are considered to be beneficial (Sourzat 2004a). Before the invasion of the phylloxera root aphid (1870–1880), winemakers used to dig trenches between their vineyards and the surrounding oak woodlands in certain regions (e.g., the departments of Lot and Dordogne) so that the truffle did not damage the root systems of their grape plants. Grapes and the truffle have a long and shared history. The black truffle was introduced to new countries at the end of the twentieth century in the southern hemisphere (New Zealand, Australia, Chile) and in the northern hemisphere (USA: Oregon, California), where the vines were implanted a century earlier (in the nineteenth century).

Lavender, like grape, is widely cultivated on the Valensole plateau (Provence). Truffles are usually harvested among the roots of lavender when truffle oaks are present nearby. The development and vigor of the lavender are affected by the presence of the truffle *Tuber melanosporum*. In some cases, where juniper (*Juniperus communis*) offers the truffle shade to fruit beneath its canopy and among its roots, the shrub sometimes dies because of the intensity of the "brûlé" after several years of fruiting in its root zone. In the Pyrénées Orientales department (Pézilla de Conflent), truffle growers have noticed "brûlé" formation around *Cistus albidus* for 2 years, followed by a year or two of truffle production and then the death of the shrub.

Prunus spinosa L and *Rosa canina* L. are shrubs among the roots of which it is common to harvest truffles when they occur within "brûlés" of *Tuber melanosporum*. It even happens that the "brûlé" of natural truffières are favored by a hedge or a group of *Prunus spinosa* shrubs. Truffle production develops and the shrub is gradually annihilated. The list of plants known to be beneficial for the truffle is far from complete, although the statistical analyses of plant surveys carried out within the framework of the SYSTRUF program did not validate a positive relationship between the so-called beneficial plants and the production of the black truffle (Taschen 2015). However, molecular (Polymerase Chain Reaction) analyses of the roots of all (non-ectomycorrhizal) plants present in the truffle "brûlé" showed the presence of truffle DNA. It is not known whether the DNA corresponded to mycelium present on the root surface or inside the root tissue of these plants.

All these elements show that the concept of agroecology applied to trufficulture must necessarily integrate the concept of biodiversity into a particularly sensitive and fragile ecosystem. *Mesobromion* (calcareous lawn dominated by the grass *Bromus erectus* Huds.), which has been discussed in connection with agroforestry and truffle forestry, constitutes a very favorable environment for the formation of natural truffle trees on limestone with soft humus. Is it necessary to install it or let it grow in the plantation where it will compete for water with the tree or even the truffle?

12.3 The Socioeconomic Context of Truffle Cultivation in theTerritories

The history of the truffle is associated with that of the grape in many territories of the Occitanie region (south-central France). This history is manifested not only in cultural and agricultural habits but also in the economy of small territories. This historical and patrimonial capital is valuable to countries which do not have this prior relationship with the truffle and its culture. A better knowledge of the actors and the means used to maintain and develop this wealth should make it possible to adjust a number of choices and prepare the future in a vision that goes beyond the strict framework of truffle cultivation.

12.3.1 The Roles of Truffle and Truffle Cultivation in the Territory

The roles of the truffle and its cultivation are multiple on the territory. "The socio-economic impact of truffle cultivation on local development" was the subject of a study carried out by the Fédération Française des Trufficulteurs (FFT) with the assistance of FNADT (a French National Planning and Development Fund) and published in December 2005 (Savignac and Sourzat 2005). This study details the elements summarized below.

The black truffle is a source of economic income whose real or fancied opacity can penalize its development. It induces other activities such as gastronomic restauration, agritourism with discovery trails, rural lodgings, guest rooms in the territories where its footprint in the territorial identity is strong. It is the origin of events (truffle festivals or fairs) which attract a large public that consumes and spends money in the territory (Fig. 12.5). In France, there are a number of several truffle museums, often referred to as "Maison de la Truffe" (Home of the Truffle).

Truffle farming is an economic activity. It induces other economic activities such as those of the truffle nursery, tree planting services, the sales and installation of irrigation equipment, fencing, and "truffières" maintenance. It contributes to the preservation of traditional landscapes, prevents or limits the scrub development, and makes it possible to regain the bushy space. It may sometimes be in competition with sheep farming by creating truffle plots in vast areas of lawns or calcareous moors. Limiting shrubs is also a means of preventing forest fires.

Fig. 12.5 Festive gathering in Lalbenque (Lot, France) for a giant truffle omelette

If the black truffle *Tuber melanosporum* is at the origin of multiple activities in the territory, other species are likely to emerge in a context of free market and development of the territory. This is already the case with the white summer truffle *Tuber aestivum*, for which local markets exist during the summer (e.g. Limogne in Quercy, Lot). In Périgord, summer truffles are sold in the Sarlat market throughout the tourist season. This species also makes it possible to organize demonstrations of the search for truffles with a trained animal ("cavage" or truffle hunting with a dog or pig) outside the traditional winter season (Fig. 12.6). It appears in summer on the menu of some restaurants at the time of tourist influx. *Tuber brumale* is also a truffle with a modest market. Some truffle growers have already planted the first trees mycorrhized with the precious white Alba truffle of Italy, *Tuber magnatum*. Plantations of trees mycorrhized by the bianchetto truffle *Tuber borchii* have already been established in Lot et Garonne.

12.3.2 Sociology of Truffle Growers

Truffle growers have no specific status. The typology below is based on the status of farmers, with sub-categories that can be refined in their definition.

Farmers can be divided into several sub-categories with their own characteristics:

Fig. 12.6 Demonstration of truffle hunting with a piglet in Lalbenque (Lot, France) as an example of active agritourism

1. *Truffle growers who generate their main income from truffle production.* These few truffle growers own at least 20 ha and are rare. In order to live from truffle cultivation, they must overcome their production which is highly dependent on summer climatic conditions. In the department of the Lot, there are currently only three farmers who have fully equipped their truffle farms for irrigation. In the absence of rational irrigation, a truffle farm cannot be a viable economic entity.
2. *Truffle growers who get a secondary portion of their income from truffle production.* More numerous than the first category, they are generally nut-growers, winemakers, or sheep breeders who devote several hectares of their farm to truffles. On the economic level, they have a real desire to generate income from truffle production.
3. *Truffle growers who have one or a few plantations to perpetuate a family tradition.* They are the most numerous. They feel satisfied to get an income from their plantations when the summer climate is favorable with rainy thunderstorms. They generally believe that truffle money is good to take but do not depend on it.
4. *Truffle growers who have given up truffle farming.* Despite the presence of some old plantations on their farm that have become unproductive, they have little or no interest in truffle cultivation.

The proportion of farmers who grow truffles varies from region to region. In the traditional truffle-growing regions, analysis of areas planted and subsidized by local and regional funds shows that vocational farmers grow as much truffle acreage as do non-farmer landowners (i.e. traders, civil servants and retirees), who are much more numerous. Farmers hold most of the land that can be used for truffle cultivation.

Non-farmer truffle growers are distributed as follows.

1. *Investor truffle growers*. They plant at least 5–10 ha and have a real economic project that they do not display. They have invested substantially in equipment for their truffle project (e.g. for irrigation). They come from various primary professions: liberal professions, industrialists, and craftsmen.
2. *Passionate truffle growers*. They show their attachment to the land and to family memories with emotional motivation. However, they also have an economic incentive that encourages them to take up the challenge of truffle cultivation in earnest. They may have one or more hectares with or without irrigation.
3. *Retired truffle growers*. They value a land space received as an inheritance or bought to satisfy the pleasure of producing truffles. Economic considerations are not excluded from their project. They are often "gardeners" as truffle growers on plots limited to a few acres, sometimes more (1 acre = 4000 m^2).
4. *Small and discrete truffle growers* have only a few producing trees.
5. *Former truffle growers*. They gave up truffle farming because their plantation(s) grew old and stopped producing.

A more precise typology, locating the proportions of each type by region or department, would lead to a wider study. By way of summary, however, it can be said that the sociology of truffle growers is essentially depending in the south of France of the possibility to realize or not a cultivation integrating irrigation or not in the cultural process.

12.3.3 The Status of the Truffle Grower

There is no official professional status for truffle growers. Each truffle grower is either a farmer, a solidarity member of the "Mutualité Sociale Agricole" (Health insurance fund), a member of a socio-professional status group, or retired. This is not without problems in terms of taxation, social security coverage, the SAFER (Land Development and Rural Settlement Company), and also in cases of damage caused by large game (for example deer or wild boar), because only farmers can be compensated by the National Federation of Hunters. For the passionate truffle growers or investors, the solution remains to take the status of farmer.

12.3.4 The Development of Truffle Cultivation in the Southwest

In the Midi-Pyrénées or Occitanie region (southwestern France), this development depends on several actors. The truffle farmers' associations and unions, under the aegis of the Regional Federation of Truffle Farmers, help to develop truffle cultivation,

Fig. 12.7 Truffle plantation gardened by a retiree

maintain animation around the truffle and its culture, ensure the quality of the truffles marketed, and ensure the promotion of *Tuber melanosporum* specifically.

The Truffe Research Centre of Cahors-Le Montat (Lot, France) has a decisive role in support of truffle growers, but also with experimentation on the Montat site and with the regional network of experimenters. The Centre takes part in technical field days and training workshops on the site of the Agricultural School of Cahors Le Montat.

The Midi-Pyrénées Regional Council, the Lot and Tarn Departments, provide direct subsidies or subvention plantation establishment in general. The Department of Lot also supports the creation of "truffle gardens" (Fig. 12.7), and the renovation and uprooting of truffle woods. For several decades, there was a territorial engineer at the Truffle Research Centre of Cahors-Le Montat for the development of truffle growing on its territory. Unfortunately, this position has not been re-filled.

12.3.5 Trufficulture in France

The French Truffle Growers Federation (FFT) encourages truffle research (SYSTRUF program from 2008), experimentation (Bussereau protocol, Le Foll protocol, FranceAgriMer contribution), popularization and training of truffle growers and truffle plantation establishment, and market development. The FFT disseminates scientific, technical, and professional information through the French quarterly magazine "Le Trufficulteur" [The Truffle Grower].

12.3.6 Lessons from Truffle Farming Abroad

The natural area of the black truffle is located in the northern Mediterranean arc comprising Spain, France, and Italy. However, other countries have begun to take an interest in truffle cultivation. In the southern hemisphere, New Zealand was the first country to invest in truffle cultivation, followed by Australia and South America (Chile and Argentina). South Africa is also developing truffle cultivation with the first harvests (Miros et al. 2016).

In the northern hemisphere, the United States of America (North Carolina, Tennessee, Virginia, California, Oregon, etc.) and Canada (British Columbia) produced black truffles. If China exports the truffle *Tuber indicum*, it is also interested in the production of *Tuber melanosporum*.

Tuber uncinatum, *Tuber borchii*, and *Tuber magnatum* are species that are established or are being experimented with in various countries. The ecological plasticity of *Tuber uncinatum/Tuber aestivum* allows its successful establishment in many pedo-climatic situations, especially in Central Europe. No establishment of *Tuber indicum* is known outside its natural area of production.

The success of truffle cultivation efforts abroad depends on its ability to generate gastronomic and economic interest, especially in the countries already cited for its cultivation. The production potential of newly engaged regions, especially in the southern hemisphere, should not be underestimated. There is no water constraint insofar as watering equipment is common in agriculture. Viticulture and arboriculture are established with irrigation in climates where it does not rain for 4–6 months (Chile, Argentina, California).

12.4 Socioeconomic Motivations for Truffle Cultivation in the South of France

12.4.1 Become a Grower: Farmer, Gardener, or Future Retired Truffle Grower

The study of the socioeconomic context showed that the window of opportunity is large enough to become a truffle grower if one has land with an appropriate soil. One can be a truffle grower without being a farmer. It is enough to have a minimum of knowledge, the necessary land, the required funds, and the determination to follow through with the project. It is possible to produce truffles from the fourth or fifth year following planting if maintenance is done properly and the trees are protected from damage by big game animals.

A determining element in the success of the project is the possibility of irrigation. Climatic change (recurrent summer droughts) requires the installation of a watering system. Then, the technical itinerary must be studied in order to properly manage the constraints of the local environment (type of soil, topography, pressure of fungal contamination/competition) and the available means (labor, agricultural, or gardening equipment).

12.4.2 Planting to Harvest or Occupy Space

The motivations presented in the sociology of truffle growers point out that some people undertake truffle cultivation with serious economic objectives while others are more engaged for the pleasure of the process. In the first case, success is an important issue, while for the latter, the outcome is somewhat less urgent. However, all projects aiming at contributing to production are encouraged by growers' associations.

Occupying space with truffle trees is not sufficient reason for "truffières" establishment, despite the objective of combating desertification due to the disappearance of farms. A lack of maintenance generally results in a lack of results. Extensive truffle cultivation has become increasingly uncertain due to summer droughts and heat waves associated with global warming. Under dry and hot summer conditions, truffle development suffers, and the mycorrhizae of *Tuber melanosporum* become replaced by those of other species more adapted to such dry summer soil conditions (i.e. *Tuber aestivum*).

12.4.3 Improving Results

Where plantations on suitable soils have potential, improved outcomes are possible and should be encouraged. The potential is recognized by the presence of strong "brûlés" (burnt area extending beyond the margin of the tree's canopy) with additional space for enlargement. Analyses of mycorrhizae from root samples taken from the inner border of "brûlés" confirms the presence of *Tuber melanosporum*.

Management of the truffle's water needs is essential in the current climatic context. If the water resource does not exist naturally, it may be possible to create it by various means: water well drilling, hilly reserve, tank, basin, etc. The various systems put in place by truffle growers of the Aveyron department in the valley of the Tarn or the Dourbie are very good examples (Sourzat et al. 2014). These are masters in the art of collecting rainwater in small artificial pools. Then, by gravity or by means of a motor pump, the water is distributed through polyethylene pipes and micro-sprinklers judiciously distributed under the producing trees.

Other cultivation practices consist of adapted tillage (manual or mechanical, without root trauma), tree pruning and thinning if necessary (removing unproductive trees), and the creation of the so-called truffle traps by means of supplementary inoculation. The Short Guide to Truffle Cultivation provides key information in this domain (Sourzat 2004b).

When the plantation is young (5–10 years), if the trees have no real potential (stunted trees left in the grass, absence of *Tuber melanosporum* mycorrhizae, inappropriate soil), the recovery may be long, delicate, and expensive, so it may be preferable to tear the plants out and start again with solid fundamentals (favorable soil, well-mycorrhized plants, possibility of watering). When the orchard is older (15–25 years), in the absence of any sign of production (neither "brûlé" nor mycorrhizae of *Tuber melanosporum*) and unsuitable soil, it is better to give up. If the soil is calcareous with the presence of natural "truffières" in the immediate vicinity, it is possible to ensure a resumption of this plantation by restoration of the old plantations.

12.4.4 Restoration of Old Plantations

Reclaiming the land with truffle cultivation is not a vain project if the soil conditions are favorable and if there are (or have existed) natural "truffières" not far from the site in question. The uprooting of old truffle plantations on a large scale is a project that can only be carried out collectively. Elected representatives are appreciative of the potential benefits, but the implementation of such projects requires great efforts that do not constitute a high priority for local and regional authorities.

The natural "truffières" observed in the moors and natural grasslands alongside oak woodlands are an indication of the presence of the black truffle *Tuber melanosporum* and a favorable environment. The work of SYSTRUF showed that, in the scrub woodland of Provence, *Tuber melanosporum* persisted for many years despite the closure of the environment (Richard et al. 2005). This change in vegetation formation takes the form of coppice or a tight wood, a frequent haunt of wild boars. The presence of the black truffle was detected as DNA by molecular analyses.

In the reconquest of the land area, priority must be given to restoring old truffle plantations. These are not in short supply in traditional departments such as Lot where agricultural abandonment is obvious (Fig. 12.8). The wooded sites formed by old truffle plantations, woods of *Quercus pubescens* and *Quercus ilex* on limestone soil, are suitable for the implementation of truffières renovation or truffle silviculture (Diette and Lauriac 2005).

Different strategies of renovation and truffle forestry based on the following practices can be applied:

- Clearing,
- Thinning.
- Pruning of oak.
- Tillage (root cutting) with different tools (chisel, cultivator or discs).
- Maintenance of the soil over time according to sequences over several years (tillage, grassing, gyro grinding and pastoralism)
- Water supply (through irrigation)
- Inoculation (supplementary application of spores).

These strategies continue to be the subject of experiments and demonstration projects in the Midi-Pyrénées (Occitanie) region and also in Provence.

Fig. 12.8 Aerial view of the region of Aujols (Lot, France) showing a landscape of agricultural abandonment. Truffle plantations are more or less young

12.5 Conclusion and Perspectives

The disappearance of mixed farming and the specialization of agriculture have been indirect causes of the decline of the truffle. The traditional know-how is no longer operational in regions where there are fewer and fewer farmers. The know-how codified at the end of the nineteenth and early twentieth centuries must be adapted to environmental conditions deeply modified by new agricultural practices. After World War II, agriculture became specialized and led to the abandonment of less productive agricultural areas, favoring in turn the scrub and natural afforestation. The management of limestone spaces of truffle cultivation interest must take into account the fungal contamination pressure exerted by the old truffle plantations and the oak woods. This is true both for plantations and for truffle forestry. The truffle growers strive to protect and promote the biodiversity observed under natural conditions of production. They strive to maintain and protect a truly sensitive and fragile truffle ecosystem.

In the context of truffle development and truffle cultivation, truffle production remains an economic asset for the territories of poor limestone regions, particularly in the south of France. The direct and indirect societal impact of the precious fungus on local life is undeniable. Its worldwide reputation and its history contribute to the character and the lifeblood of the territories, to the development of a truffle tourism associated with products of gastronomy (wines, cheeses, foie gras, saffron and nuts) and the preservation of the environment. Truffle cuisine contributed to the classification of French gastronomy as a UNESCO World Heritage Site in November 2010.

The sociology of truffle growers shows that truffle production depends as much on non-farmers as on farmers. The French federation of truffle growers, however, believes by its president that it is the agricultural world that holds the key to the development of truffle cultivation because it is the farmers who own most of the land and the means of its management (Tournayre, pers. comm.). In the traditional areas where agriculture has declined and the forest has progressed, grubbing up old truffle plantations on a large scale is a project that can only be achieved in a collective framework that is difficult to implement at the moment.

Truffle cultivation remains an instrument of land reconquest for the maintenance of the quality of landscapes and the fight against fires. It is a brake on desertification when agricultural abandonment is significant. In spite of all these positive elements, there are still questions about the means to be used to multiply truffle production on its traditional area in the south of France. Control of water needs becomes necessary in a context of climate change. The first is to correct the negative effects of drought in the absence of summer storms that have become rare in the last four decades.

Acknowledgments Sincere thanks to Alexis Guerin-Laguette and Jesus Pérez-Moreno for their encouragement to propose and write this article. Heartfelt thanks to Johann Bruhn for his proofreading work, his corrections, and his very pertinent remarks.

References

Callot G (1999) La truffe, la terre, la vie [The truffle, the earth, the life]. INRA, Paris. 212 p. In French

Chatin A (1892) La Truffe [The Truffle]. Baillère, Paris. 372 p. In French

De la Bellone F (1888) La truffe: étude sur les truffes et les truffières, réédition 1999 “Rediviva” [The truffle: study on truffles and truffières, re-edition 1999]. Nîmes. 312 p. In French

De Bosredon A (1887) Manuel du Trufficulteur [Truffle Grower Manual]. Laport, Périgueux. 238 p. In French

Delmas J (1976) La truffe et sa culture [The truffle and its cultivation]. INRA, Versailles. 56 p. In French

De Montigny L (1862) Rapport de la commission chargée d’examiner les reboisements et les semis ou plantations de chênes opérés dans l’arrondissement d’Apt, en vue de la production des truffes, Société d’agriculture et d’horticulture de Vaucluse et comice agricole de l’arrondissement d’Apt, 14 et 15 septembre 1862. [Report from the commission in charge of assessing the reforestation and the sowing of oaks carried out in the district of Apt, with the view of producing truffles, Society for agriculture and horticulture of the Vaucluse and Comice Agricole (i.e. Agricultural Assembly) of the district of Apt, 14 and 15 September 1862]. Société d’agriculture et d’horticulture de Vaucluse. In French

Diette S, Lauriac A (2005) La sylviculture truffière: aperçus historiques, apports techniques et enjeux pour la région méditerranéenne [The truffle sylviculture: historical insights, technical inputs and challenges for the Mediterranean region]. forêt méditerranéenne XXVI(2):157–168. In French

Grente J, Delmas J (1974) Perspectives pour une trufficulture moderne [Perspectives for a modern truffle cultivation]. INRA, Clermont-Ferrand. 66 p. In French

Miros V, Miros P, Hall I, Zambonelli A (2016) Black truffle production in South Africa—abstract book of the 8th international workshop on edible mycorrhizal mushrooms (IWEMM8), Cahors, France, 10–13 Oct 2016. p 121
Olivier JM, Savignac JC, Sourzat P (2018) Truffe et Trufficulture [Truffle and truffle cultivation], 4th edn. Fanlac, Périgueux. 352 p. In French
Pradel L (1914) Manuel de Trufficulture [Truffle cultivation manual]. Baillère, Paris. 156 p. In French
Rebière J (1981) La Truffe du Périgord [The Périgord Truffle]. Fanlac. 204 p. In French
Reyna Domenech S (2000) Trufa, Truficultura y Selvicultura trufera [Truffle, Truffle cultivation and Truffle sylviculture]. Mundi-Prensa, Madrid. 232 p. In Spanish
Richard F, Millot S, Gardes M, Selosse MA (2005) Diversity and specificity of ectomycorrhizal fungi retrieved from an old-growth Mediterranean forest dominated by *Quercus ilex*. New Phytol 166:1011–1023
Savignac JC, Sourzat P (2005) L'impact socio-économique de la trufficulture sur le développement local—Etude réalisée par la Fédération Française des Trufficulteurs (FFT) avec le concours de la FNADT (Fonds National d'Aménagement et de Développement du Territoire). [The socio-economic impact of truffle cultivation on local development—Study carried out by the French Federation of Truffle Growers (FFT, ed) with support from the FNADT (National Fund for the Planning and Development of the Territory)]. 68 p. In French
Sourzat P (2004a) Questions d'écologie appliquées à la trufficulture—Station d'expérimentation sur la truffe [Ecological questions applied to truffle cultivation—Truffle Research Centre], Lycée professionnel agricole of Cahors-Le Montat. 64 p. In French
Sourzat P (2004b) Short guide to truffle cultivation. Station d'expérimentation sur la truffe [Truffle Research Centre], Lycée professionnel agricole of Cahors-Le Montat. 20 p. In English
Sourzat P (2008) Le principe de précaution en trufficulture—Station d'expérimentation sur la truffe [The principle of precaution in truffle cultivation—Truffle Research Centre]. Lycée professionnel agricole de Cahors-Le Montat. 72 p. In French
Sourzat P (2010) L'environnement truffier: contraintes et gestion—Station d'expérimentation sur la truffe [The truffle environment: constraints and management—Truffle Research Centre], Lycée professionnel agricole of Cahors-Le Montat. 56 p. In French
Sourzat P, Bouyssières D, Sanchez A (1997) Etude sur les conditions techniques et environnementales de la trufficulture en Midi-Pyrénées—Rapport financé par la Région Midi-Pyrénées. Truffle Research Centre], Lycée professionnel agricole of Cahors-Le Montat. 38 p. In French
Sourzat P, Genola L, Saenz W (2014) Compte rendus des actions d'expérimentation sur la truffe (selon le programme validé par FranceAgriMer et la Région Midi-Pyrénées)—[Review of experimental activities on truffle (according to the programme validated by FranceAgriMer and the Région Midi-Pyrénées)]—Truffle Research Centre, Lycée Professionnel Agricole of Cahors-Le Montat. 61 p. In French
Taschen E (2015) Biotic interactions and reproductive biology of the Black Truffle (*Tuber melanosporum* Vittad.), from spontaneous to planted truffle grounds. PhD thesis, Université de Montpellier 2. 226 p. In French with English abstract

Chapter 13
Ethnomycology in Europe: The Past, the Present, and the Future

Ornella Comandini and Andrea C. Rinaldi

13.1 Local Biology: The Mycological Side

Europe, the Old Continent, is a treasure trove of ethnobiological traditional knowledge (Anderson et al. 2011). Crafted over millennia, this knowledge was not obtained and passed between generations for the sake of spiritual satisfaction of living a peaceful life in pristine nature. Rather, awareness of the biological resources existing in the natural surroundings and the ability to use and transform them was simply necessary in order to survive and thrive. A highly interdisciplinary field, with inputs from cultural anthropology, linguistics and archeology (just to quote a few disciplines) that somehow converge into biological studies, ethnobiology is crucial for a number of reasons. Stratified over and mingled with the purely cultural pleasure of understanding how traditional societies worked and shaped their environment is the more pragmatic necessity of looking at the "the world until yesterday" to learn how human groups managed key resources and ecosystems in a sustainable way, obtaining food, materials, tools, and medicines while preserving biotic diversity and ecological integrity (Diamond 2012).

It's objectively difficult to overstate the significance and role played by fungi in basically any natural setting and in a human's everyday life. "Despite early recognition of the importance of fungi for human well-being, and archaeological evidence for human uses of fungi in food, drinks and medicines going back at least 6,000 years, historically they have remained in the shadows when compared with research on plants and animals," stressed Katherine Willis in her introduction to the recently

O. Comandini
Department of Life Sciences and the Environment, University of Cagliari, Cittadella Universitaria, Monserrato, CA, Italy

A. C. Rinaldi (✉)
Department of Biomedical Sciences, University of Cagliari, Cittadella Universitaria, Monserrato, CA, Italy
e-mail: rinaldi@unica.it

J. Pérez-Moreno et al. (eds.), *Mushrooms, Humans and Nature in a Changing World*, https://doi.org/10.1007/978-3-030-37378-8_13

released *State of the World's Fungi* (Willis 2018). In line with this general unappreciation, ethnomycology is a more recent area of research relatively to ethnobotany and ethnozoology. It focuses on the study of the interrelations between human societies and fungi, its subjects including cultural, ceremonial, and medicinal uses of mushrooms, besides their consumption as food (Wasson 1980; Singh and Aneja 1999; Boa 2004; Dugan 2011; Yamin-Pasternak 2011).

The book published in 1957 by R. Gordon Wasson and his wife Valentina Pavlovna, *Mushrooms, Russia and History*, can be safely considered as the starting point of ethnomycology as a field of study (Pavlovna Wasson and Wasson 1957). This seminal work goes deep into the connection between mushrooms and the world of man, discussing extensively the use of hallucinogenic mushrooms in Mesoamerica, particularly in Mexico, but also dissecting the attitudes of Russians, Basques, Slovaks, and other European peoples toward mushrooms. "Ours is one of the few books about mushrooms written for connoisseurs by amateurs. It is the first treatment in any language of the role played by mushrooms in the daily lives of the varied European peoples," Valentina Pavlovna (who was born in Russia and died in 1958) wrote in the preface (Pavlovna Wasson and Wasson 1957). Since the very beginning, therefore, Europe has been central to the development of ethnomycology. From this starting point, ethnomycological research in Europe has spread in several directions, not only describing mushroom use in specific cultures (e.g., Moskalenko 1987; Härkönen 1998) but also delving into the management of wild mushrooms as a resource, considering both the conservation/ecological implications of mushrooming and the economic impact on local populations (Egli et al. 1990; Boa 2004 and references therein).

This chapter reviews the ethnomycological research conducted in Europe in the last two decades. Ethnomycology studies the knowledge and practices of mushrooms that derive from a long-term engagement with the environment. As such, this type of knowledge—typically transmitted in oral form and through practical training—can understandably be associated with the past, but this would be a misleading and diminutive view. Rather, traditions evolve, as does the relationship of a people with an everchanging physical and cultural environment. Think, for example, about an Italy without tomato sauce! Well, tomatoes, together with potatoes, zucchini, corn, and many more vegetables that anyone would associate with the "traditional" Italian culinary arsenal and the Mediterranean diet did not reach Europe until mid-sixteenth century. "Much of this knowledge is traditional, that is, learned long ago and passed on with varying degrees of faithfulness for at least two or three generations. However, ethnobiological knowledge can change rapidly. Every tradition had a beginning, and was itself a new creation in its time," remarked Eugene Anderson (2011). "Ecosystems change, new plants and animals arrive, and people learn new ways of thinking; ethnobiological systems change accordingly, and are typically flexible and dynamic. Field-workers have observed new knowledge being incorporated into systems around the world". Current practices of mushrooms are not exclusively rooted in settled knowledge but they are dynamic processes. Instead, we are witnessing to an increasing fondness for mushrooming in many parts of Europe, their use as cosmetics and nutraceuticals, the raising attention to the biological and medical relevance of many species. Scientific, rather than traditional, knowledge make the fundament of these contemporary uses of mushrooms, and we should emphasize the extent to which the two systems differ, having radically dissimilar aims, methods and forms of knowledge transfer. However, despite their inherent differences, traditional and scientific forms of knowledge are not secluded in watertight compartments but tend to converge and "can learn from each other" (Mazzocchi 2006). This is particularly evident in the case of the medicinal properties of mushrooms, as described below. In the following paragraphs, we will strive to highlight these aspects as for what concerns Europeans and their mycological culture.

13.2 Recent Past and Current Trends in European Ethnomycological Research

One could rightly ask whether it has any sense to perform ethnomycological research—or any type of ethnobiological research—nowadays in such a developed part of the world such as Europe. No doubts that what we call "traditions" play a rapidly decreasing role in our modern societies, suffering significant erosion for a number of reasons (social and cultural changes, urbanization, fragmentation of rural communities, globalization, and industrialization) to the point that a large part of our biocultural heritage risks to be irremediably lost with the end of the so-called silent generation (also known as "traditionalists," born before 1946). "Ethnobiologists in Europe work to get rid of the widely held notion that ethnobiology is all about "non-Western people." European rural people are part of our professional realm. As ethnobiologists we usually study rural people's ecological knowledge in societies with high levels of self-sufficiency. We study the biocultural domains that develop in the interactions between human beings and their surrounding landscape, including perceptions of the biota, local management, and use of biological resources," is the answer by Ingvar Svanberg and colleagues in their draft on the status and directions of ethnobiological research in Europe (Svanberg et al. 2011). On top of that, despite an unquestionable wealth of mycological traditions in Europe, surprisingly few detailed studies have been performed in many areas of the continent about the gathering of mushrooms and other myco-related activities.

In the somewhat desolate scenario of European ethnomycological studies, inevitably some regions received more attention than others (this, as for all types of territory-linked endeavors, strongly depends on where scholars are located and work). Finland, for example, has been the ground for in-depth ethnomycological enquiry. Finns are generally considered to be a mycophilic bunch, but rather than being autochthonous, this positive attitude towards wild mushrooms might be the consequence of cultural influence from different traditions. Country folks living in Karelia, in the eastern part of the state, received strong inputs from their Russian neighbors that privileged *Lactarius* (e.g., *L. torminosus* (Schaeff.) Gray and *L. trivialis* (Fr.) Fr.),[1] while the more educated people of southwest Finland absorbed the French taste for *Boletus* and *Cantharellus*, permeated via the Swedish elite

[1] Names of fungal taxa follow Index Fungorum, www.indexfungorum.org

Fig. 13.1 Urban "mushroom picking." Fresh chanterelles and porcini on sale in an open-air market, central Stockholm, Sweden, summer 2018. While *Boletus* sp. are local (and advertised as such), chanterelles are generally imported from Baltic states. During 2012–2016, a mean of 89 tons of chanterelles were traded to Sweden annually, versus 1 and 5 tons of *Boletus* and *Tuber*, respectively (Widenfalk et al. 2018)

(Härkönen 1998; Yamin-Pasternak 2011; Fig. 13.1). A nationwide survey on mushroom picking was conducted in 1997–1999 and 2011 through mailed questionnaires (an interesting methodology that permits to reach a large number of participants with relatively little effort) (Turtiainen et al. 2012). With a response rate that varied from 51% to 70%, depending on the year, the results of the survey confirmed the great interest of Finns for picking wild mushrooms: from 23% to 47% of all households were engaged in picking, depending on whether the survey was conducted in an unfavorable or favorable year for mushroom appearance. The study also allowed to roughly estimate the inter-annual variation of mushroom crops, since respondents were asked to report quantities of picked mushrooms (Turtiainen et al. 2012). Commercial mushroom picking has been the subject of another investigation in the region, which documented the emergence in eastern Finland of a wild edible fungi industry based on the collection of *Boletus* spp., mainly exported to southern Europe (Cai et al. 2011). A few other studies have addressed the interesting matter of the economic importance of commercial wild mushroom gathering for rural communities, the sustainability of such an activity, and marketing of crops, namely, in Scotland (Dyke and Newton 1999), Spain (de Román and Boa 2006; Palahí et al. 2009), Italy (Sitta and Floriani 2008), and Poland (Kasper-Pakosz et al. 2016).

In addition to purely ethnomycological studies, a contribution to the field came also from more general ethnobotanical surveys considering fungi among "food plants" (Lentini and Venza 2007), sometimes with interesting outcomes. For example, local mushrooms picked for food in Castelmezzano, Lucania, southern Italy, include *Clitocybe nebularis* (Batsch) P. Kumm. (Pieroni et al. 2005), a taxon that is now generally treated as suspect since can cause gastric upset. A more systematic investigation conducted in Garfagnana, Tuscany, revealed the regular consume of at least 20 species of mushrooms, including *Grifola frondosa* (Dicks.) Gray, which is rarely reported in lists of eaten mushrooms (Pieroni 1999). "The harvesting of fungi for markets has long been economic activity in the valley. *Boletus* ssp., *Cantharellus* and, less frequently, *Amanita cesarea* [sic!] from Garfagnana are sold in regional markets, and in other parts of Tuscany," reported the author. "Dried mushrooms from the valley reached America from Genoa during the last century," (Pieroni 1999). An unusual perspective was offered by Redzic and colleagues, who have studied the mushrooms and lichens consumed by starving population and isolated guerilla fighters during the 4-years long (1992–1995) war in Bosnia and Herzegovina (Redzic et al. 2010). Using the method of "ethnobotanical interview," researchers gathered information from 51 Bosnian Muslims—some of who sought shelter for months in remote areas to escape occupation by Serb forces. Some 25 species of wild mushrooms and seven species of lichens were used during the war by local residents and soldiers, with *Agaricus campestris* L., *Lactarius piperatus* (L.) Pers., *L. delicious* (L.) Gray, *Morchella conica* (= *M. esculenta* (L.) Pers.), *Boletus edulis* Bull., and *Cantharellus cibarius* Fr. being the mushrooms mentioned more frequently. "Even though mushrooms were very unpopular in conventional nutrition in these areas due to genuine fear of poisonous mushrooms, the situation during the war was completely different," authors commented (Redzic et al. 2010).

The fascination and passion of Europeans for truffles date back centuries, although this fondness and popularity has not been constant over time. Even in this case, recent studies have observed shifting paradigms in the use of this important resource. Andrea Pieroni has discussed the climatic and social changes that are threatening the delicate "ethnoecological network among men, truffles, dogs, and the environment" that has governed until recently the gathering of valuable white truffle (*Tuber magnatum* Picco) in Piedmont, north-western Italy (Pieroni 2016). "[C]limate change, in which the summer has become a very hot and dry season; social changes, due to a more market-oriented attitude of younger gatherers; and especially environmental and macro-economic dynamics, which followed the remarkable expansion of viticulture in the study area" are the components of the mix of factors that are progressively reducing truffle harvests and menacing the transmission of a very special piece of biocultural heritage (Pieroni 2016). In Poland, truffle orchards have been established since the first decade of this century, and their demand by the local market is on the raise. As Rosa-Gruszecka and colleagues have reconstructed, this is a return, rather than a conquest of new territories; indeed, mentions of truffles in Polish literature can be traced as back as 1661, and the use of these mushrooms has a long tradition in the country, a tradition that faded away in the post-World War II period (Rosa-Gruszecka et al. 2017). Finally, a study from eastern Spain, where truffle cultivation has been promoted through autonomous community and provincial government subsidies, offered the chance to quantitatively evaluate the positive impact of such an activity, not only on local, rural economy, but also on biodiversity conservation, through oak reforestation and proper management of truffle orchards (Samils et al. 2008).

A specific range of activity of recent ethnomycological research is the documentation of practices associated with mushrooms in "marginal" (from an economic and/or geographic perspective) European territories, and on the habits of specific ethnic minorities. The Balkan peninsula, in particular, despite its vast cultural and biological diversity, has begun to attract the attention of ethnobiologists only in the last 10–15 years. Since then, several studies have addressed folk knowledge in a number of fields, in territories still primarily occupied by rural-based communities. A study recently conducted in Bulgaria, a part of the south-eastern Balkan Peninsula, has recorded the use of micromycetes in traditional Bulgarian cuisine, like the molds typical of traditional Bulgarian products, such as Green cheese, dried sausages (Lukanka, Sudjuk) and Bulgarian prosciutto (Nafpavok) (Uzunov and Stoyneva-Gärtner 2018). Authors also stressed the change in food habits of Bulgarian people, as a consequence of globalization and general change in lifestyle. Besides the ongoing interest in keeping with the traditional recipes using molds (and yeasts), researchers noted that the "processes of globalization and urbanization with enormous role of advertisements through social nets and media lead to changes in the food habits of modern Bulgarians and increased use of new healthy exotic products based on micromycetes," like those used in the production of foods of Asian origin (e.g., kefir and Miso) (Uzunov and Stoyneva-Gärtner 2018). In the Western Balkans, a survey of folk medicine and wild food plants in Albanian, Aromanian, and Macedonian villages flagged the use of dried *Bovista* spp. as externally applied for treating wounds and the consume of *Agaricus* sp. (especially by Macedonians) (Pieroni 2017).

Given the complexity of European ethnology and the ensuing "cultural mosaic" (Pan et al. 2016), it is all but surprising that minority groups have developed their own biocultural background, often eroded and threatened by assimilation in mainstream culture of local ethnic majority. Very few ethnomycological data exist in this context as for Europe (Pieroni et al. 2002). Łuczaj and colleagues recorded the traditional use of fungi (and plants) among the Ukrainian (Carpatho-Rusyns) minority in the western part of Romanian Maramureş (Łuczaj et al. 2015). Some 24 mushroom taxa were mentioned in interviews as used nowadays or until recently. While the most commonly gathered mushrooms (e.g., *B. edulis*, *C. cibarius*, *Russula virescens* (Schaeff.) Fr.) are eaten in many parts of Europe, including the Slavic area, some findings are peculiar. For example, *Entoloma clypeatum* (L.) P. Kumm. is a species frequently mentioned in the study area but not as much in other ethnomycological repertoires, probably because it can be easily confused with a large number of similar and potentially toxic species, like *Entoloma sinuatum* (Bull.) P. Kumm. (= *Entoloma lividum* Quél.). "Fungi are culturally more important than wild food plants for the studied population. Due to the large number of frequently used fungi taxa, the Ukrainians of Maramureş can definitely be described as a mycophilous community," concluded the authors (Łuczaj et al. 2015).

Although much of what reported above pertains to the traditional culinary use of mushrooms, it is fair to stress that fungi have a preeminent position in European magic and folklore as well (see Dugan 2008a, b, 2011 for a comprehensive review). It is generally believed (although in most cases there is little firm evidence to support it) that throughout Europe mushrooms were widely used in religious ceremonies and magic rituals in the pre-Christian era and that these "pagan" practices waned because of the growing opposition from the church. The use of the Red Fly Agaric, *Amanita muscaria* (L.) Lam., in the Siberian "shamanistic cultures" is particularly well known (Nyberg 1992). As ethnomycologists dig into the past, more details emerge. In Romania, for example, Constantin Drăgulescu has provided intriguing details on the traditional beliefs about fungi and their ritual uses (Drăgulescu 2004). So, "*Phallus impudicus* would be given a spell before it was used against anthrax," and some species include the word *vrăjitoare* (witch) in their Romanian names, Drăgulescu reports (Drăgulescu 2004). Also, "peasants from Muntenian villages believe that someone who eats *Laetiphorus sulphureus* in the spring will be lucky the rest of the year," and so on (Drăgulescu 2004).

13.3 Walking the Trail from Mycophoby to Mycophily

In their *Mushrooms, Russia and History*, R. Gordon Wasson and Valentina Pavlovna strived to trace the origins of the "dichotomy between mycophobic and mycophilic peoples", going well beyond the simple use of mushrooms as occasional food, but rather expanding their view on the "role of the fungal world in the culture" of a people and considering the "religious and mythological and erotic associations of the mushrooms" (Pavlovna Wasson and Wasson 1957). "As the old beliefs slowly faded away, each cultural community, no longer able to maintain alive the balanced tensions of the original involvement, clung to one face or the other of the primitive emotions, either rejecting the mushroom world or embracing the strange growths with a quasi-erotic devotion," they wrote (Pavlovna Wasson and Wasson 1957).

In Europe, it is possible to distinguish different traditions related to the consumption of edible mushrooms, rooted in culture and influenced by trade and cultural exchanges with neighboring regions, as the case of Finland reported above shows. Southwest and central Europe hosts the most mycophilic peoples, predominantly in Romance-speaking countries (e.g., Italy, Portugal, France). In Spain, Catalans and Basques are clearly mycophilic (Yamin-Pasternak 2011), while in the rest of the country, a mycophobic attitude prevails. Noteworthily, the fact that large swaths of the New World (e.g., the West Indies) were colonized either by people coming from regions of Spain were mushrooms are traditionally not eaten (like Andalucia), or from mushroom-fond French colonists, explains the current divide toward the consumption of mushrooms seen today in, say, Puerto Rico and the Domenican

Republic versus Haiti (see Nieves-Rivera 2001). Slavic-speaking populations in most eastern European countries (Belarus, Poland, Russia, Slovakia, Ukraine, countries from former-Yugoslavia) also have a longstanding mycophilic attitude. On the other hand, Germanic-speaking countries like Germany, Great Britain, and Scandinavia are predominantly mycophobic. Intriguingly, the various mycophilic or mycophobic character of populations is reflected by the existence/nonexistence and comprehensiveness of guidelines or legislation for the safe commerce of wild mushrooms (Peintner et al. 2013; Fig. 13.2) (the problem of the discrepancy between the list of mushrooms allowed for commercial use in some European countries and the species collected traditionally by communities has been recently highlighted by Kotowski 2016).

For the sake of completeness, we should note that the mycophobic vs. mycophilic divide is not universally accepted. Some researchers consider this a "disputable anthropological discourse" (Pieroni et al. 2005), while other experts believe that "a far broader and more nuanced range of cultural attitudes toward mushrooms," exist, "in which complex cultural and ecological issues appear to shape different societies' attitudes toward and usage of wild mushrooms," (Arora and Shepard 2008). For what it might matter, our personal opinion is that the mycophobic/mycophilic category is useful from an ethnomycological perspective. Notwithstanding being rooted in the cultural background of a people, the mycophobic or mycophilic attitude is not written in stone, but rather evolves as driven by social change and other factors (Greenfield 2016). Contacts with other cultures, either driven by economic or military (occupation) reasons, certainly, had a great impact on the origin of such a difference, which in most cases is probably not very ancient. As mentioned above, religion (Roman Catholic Church) most likely played a decisive role in modifying the way European peoples looked at their natural surroundings. Thus, studying the fluctuating dichotomy of mycophobic and mycophilic cultures offers a rare chance to research on how ethnobiological knowledge is shaped and develops over time, both influencing and being influenced by scientific information.

A clear example of how mycophobic and mycophilic cultures interact and influence each other is Sardinia. This large island, placed in the middle of the western Mediterranean basin, is politically part of Italy, but retains a strong cultural identity. While continental Italy can be considered one of the most mycophilic countries in Europe (Peintner et al. 2013), until recently only a handful of species were regularly picked and eaten in Sardinia, while most mushrooms were seen with suspect and prejudice (Comandini et al. 2018). *Pleurotus eryngii* (DC.) Quél., *Leccinellum corsicum* (Rolland) Bresinsky and Manfr. Binder, *Terfezia arenaria* (Moris) Trappe, *Macrolepiota procera* (Scop.) Singer, and *Agaricus* spp. are the edible mushrooms traditionally harvested in Sardinia; a few more species (*Fomes fomentarius* (L.) Fr., *Lycoperdon pyriforme* Schaeff., *L. perlatum* Pers., and *Bovista plumbea* Pers., *Hericium erinaceus* (Bull.) Pers.) were used for purposes other than food, as tinder or remedies for a number of conditions (Comandini et al. 2018). This means that the most valuable edible mushrooms present in the same area, such as *Amanita caesarea* (Scop.) Pers., porcini (*Boletus aereus* Bull., *B. edulis* and allied species), and *C. cibarius*, were simply ignored. This situation began to change dramatically about

Fig. 13.2 Mycophilic versus mycophobic attitudes in Europe. European countries with mushroom legislation or guidelines (green), without them (red), or with no information available (white). From Peintner et al. (2013). Reproduced with permission

two decades ago, when cheaper air travelling progressively opened locals to the influence of mushroom lovers coming from other Italian regions. Nowadays, more and more people enjoy foraging for wild edible mushrooms in all areas of Sardinia, mycological clubs regularly organize well-attended mycological expositions and educational events (Fig. 13.3), and a number of field guides to common Sardinian

Fig. 13.3 Mycological education in Sardinia, Italy. This informative panel has been installed in Soleminis, a forested area a few kilometers away from Cagliari, the island's main town. Each drum-like portion brings information (underneath the cover) about either edible or poisonous mushrooms common to the area, such as *Boletus aereus* and *Amanita phalloides* (Vaill. ex Fr.) Link. A famous proverb displayed in the panel reads: "All mushrooms are edible. Some only once"

mushrooms are available (Arras 2016; Comandini et al. 2018). Furthermore, both professional and amateur mycologists are tapping into the abundance of Sardinian macrofungi, describing new species and studying fungal diversity in selected habitats (Nuytinck et al. 2004; Comandini et al. 2006; Comandini and Rinaldi 2008; Brotzu and Peintner 2009; Leonardi et al. 2016, 2018; Porcu et al. 2018).

The steep increase in popularity of wild mushrooms recorded in Sardinia is not unique. Several combined reasons have brought to "an obvious change in the awareness of mushrooms by Bulgarians" starting from the 1990s (Stoyneva-Gärtner and Uzunov 2015). According to researchers, the search for nourishing food that could be freely picked in the wild following the political and economic crisis that stormed the country at the beginning of the 1990s, coupled with the appearance of Chinese restaurants offering several types of mushrooms in big cities of the country and a raised attention toward vegetarian nutrition, synergized in changing food habits of a basically mycophobic people (Stoyneva-Gärtner and Uzunov 2015).

13.4 The Renaissance of Traditional Medicine

Traditional medicine is creeping back. In Europe and other western countries, the demand for its services is in steep increase (Rinaldi and Shatty 2015). The extent of this trend is so significant that the World Health Organization in its *Traditional Medicine Strategy 2014–2023* has recently stressed the need to integrate traditional medical practices into national health care systems (World Health Organization 2013). "Over 100 million Europeans are currently T&CM [traditional and complementary medicine] users, with one fifth regularly using T&CM and the same number preferring health care which includes T&CM," noted the report, underlining that the use of T&CM as complementary therapy is common in developed countries where the health system structure is typically well developed, like many European countries and North America (World Health Organization 2013).

The medicinal value of fungi is unquestionable. The discovery by Alexander Fleming of penicillin produced by *Penicillium* mold saved millions of lives. Statins (mevastatin, lovastatin, and more) are fungal metabolites with potent hypercholesterolemic activity, widely used to reduce the risk of cardiovascular diseases. When it comes to mushrooms, these have been used in traditional medicine for thousands of years, in Asia, America, Africa, and Europe (see Box 13.1). "Nowadays, numerous preparations in the form of powders, capsules, pills and tinctures are made from medicinal mushrooms, which are available on the market as dietary supplements. These are used in the prevention and treatment of various health conditions, including modern lifestyle diseases, and are recommended for active people, athletes, and the elderly. Moreover, medicinal mushroom products have been increasingly used as feed supplements to improve health or treat diseases of companion (pets), domestic and sport animals," wrote Jure Pohleven and colleagues in a recent publication (Pohleven et al. 2016): a multi-billion-dollar market, indeed.

Some of the medicinal mushrooms most commonly used in the products mentioned above come from the traditional oriental medicine, although the same species can occur also in Europe. *Ganoderma lucidum* (Curtis) P. Karst. (reishi, lingzhi), *Cordyceps* spp. (caterpillar fungus), and *Lentinula edodes* (Berk.) Pegler (shiitake) are the main examples. Reishi (the Japanese name of *G. lucidum*, known as lingzhi in China), is probably the mushroom whose therapeutic properties have been studied in greatest detail. Two main groups of compounds extracted from *G. lucidum*, triterpenes and polysaccharides (Fig. 13.4), have been reported as having a wide

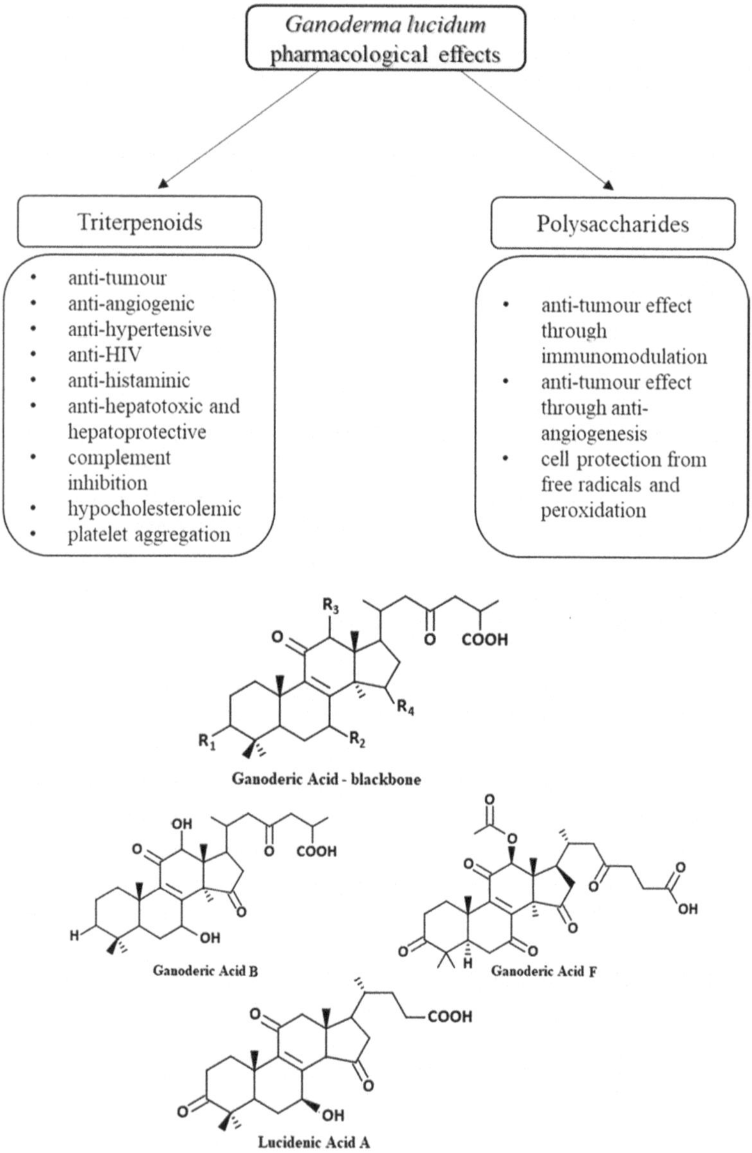

Fig. 13.4 "Medicinal mushrooms." Above, *Ganoderma lucidum* (reishi, lingzhi) pharmacological effects related to the specific group of biological compounds. Below, structural formulas of ganoderic acid and lucidenic acid, two representative triterpenoid compounds found in *G. lucidum* extracts. From Cör et al. (2018). Reproduced with permission

range of therapeutic properties, including anti-hypertensive, hypocholesterolemic, hepatoprotective, antioxidant, and antitumor activity (Cör et al. 2018). *G. lucidum* has been used for over two millennia for promoting overall health and longevity, for stimulating body's own defenses and thus preventing a wide variety of diseases. "Nowadays it is also being used in modern medicine as a supplement to cancer treatment and to fight the side-effects of chemotherapy in China and also in Western countries," wrote Darija Cör and colleagues in a recent review on reishi's biological activities (Cör et al. 2018).

In a recent review article on the use of polypores in traditional European medicine, Ulrike Grienke and colleagues focused on five species, namely, *Laetiporus sulphureus* (Bull.) Murrill, *F. fomentarius*, *Fomitopsis pinicola* (Sw.) P. Karst., *Fomitopsis betulina* (Bull.) B.K. Cui, M.L. Han and Y.C. Dai (= *Piptoporus betulinus*), and *Laricifomes officinalis* (= *Fomitopsis officinalis* (Vill.) Bondartsev & Singer) (Grienke et al. 2014). The ethnomycological tradition concerning these mushrooms is rich, especially in Central Europe, and dates back millennia (see Box 13.1). As far as medical uses are concerned, applications ranged from treatment of bladder disorders to dysmenorrhea, from rheumatisms to cancer. "Sami nomad Johan Turi wrote in the early twentieth century that a small piece of birch polypore, probably referring to *Fomes fomentarius*, was burned directly on the affected area for toothache, fractures, rheumatism, headache and pneumonia," reported Ingvar Svanberg in a recent article describing the use of bracket fungi in folk therapy among Sami and other peoples in Northern Fennoscandia (Svanberg 2018). A wide range of primary and secondary metabolites have been identified in extracts of these polypores, including polysaccharides, polysaccharide-protein complexes, organic acids, and triterpenes. Many of these substances, either as mixtures or pure compounds, have been evaluated for their biological activities, revealing, among others, promising anti-inflammatory, cytotoxic, and antimicrobial properties (Grienke et al. 2014). "Most of the pharmacological studies on bioactive compounds of polypores were conducted using crude and poorly characterized extracts. The possible mechanism of action as well as potential synergistic or antagonistic effects of multi-component mixtures derived from polypores need to be evaluated integrating pharmacological, pharmacokinetic, bioavailability-centred and physiological approaches," underlined Grienke and colleagues (2014). A comment valid for all "medicinal mushrooms," as remarked below.

However, all that glitters is not gold. Despite the longstanding tradition of using fresh and dried mushrooms or their extracts as treatments, the scientific evidence supporting the effectiveness of these preparations is overall scant, and limited to a handful of species. This is not unusual when traditional remedies come under scrutiny and are reassessed using modern methods, an obvious step before integrating traditional medicines into a conventional pharmacopeia (Rinaldi and Shatty 2015). In selected cases, clinical trials have provided promising results, demonstrating some of the medicinal properties claimed in traditional uses. Lentinan, for example, a β-glucan polysaccharide from shiitake, has been shown to benefit patients with several types of cancer when added to chemotherapy, with increased survival, reduced side effects and improved quality of life (Money 2016; Pohleven et al.

Box 13.1 The Ice Man's Fungi: Digging Up Ancient Roots

In September 1991, German hikers Helmut and Erika Simon stumbled in what later proved to be one of the most significant archeological discoveries of last century. Climbing a track in the Ötztal Alps at 3200 m of elevation, close to the Italian-Austrian border, the couple spotted a well-preserved mummy of a man, half-buried in a glacier. Extensive analyses revealed that the body belonged to a man who lived between 3350 and 3120 BC, thus at the end of Copper Age (Fig. 13.5). The mummy, nicknamed Ötzi, was found together with a full set of cloths, tools, and equipment. The study of this material opened a window of unprecedented clarity through which to observe human life in the European Neolithic (http://www.iceman.it/en/). Among the items carried by Ötzi and recovered alongside his mummy were three fungal objects that collectively make one of the clearest ethnomycological traces in Europe in a prehistoric context: a relatively large supply of tinder material prepared from the "true tinder bracket", *Fomes fomentarius* (L.) Fr. (Fig. 13.6) and two fragments of a fruiting body of the birch polypore *Fomitopsis betulina* (Bull.) B.K. Cui, M.L. Han and Y.C. Dai (= *Piptoporus betulinus*), threaded onto hide strips (Fig. 13.7) (Peintner et al. 1998). The use of *F. fomentarius* as a firestarter is well documented throughout Europe, from Neolithic until well into the twentieth century (Papp et al. 2017; Comandini et al. 2018). Ötzi had the tinder material ("amadou") stored in a "girdle bag," a belt with a sewn-on pouch, probably wrapped twice around the hips. Rather than be kept in the pouch as a piece of fruitbody in its natural state, the fungus "had been treated mechanically in order to gain a material of a wad-like consistency" (Peintner et al. 1998). Beside the tinder fungus, the belt pouch contained a scraper, a boring tool, a bone awl and a flint flake (Fig. 13.6). *F. fomentarius* is also credited with medicinal properties. Hippocrates (c. 460–c. 370 BC), the "Father of Medicine," described the use of this species for the cauterization of bleeding wounds (Pohleven et al. 2016). There are evidences that several other polypores like *Coriolopsis gallica* and *Daedalea quercina* were used as tinder in European Neolithic (Berihuete-Azorín et al. 2018). Reconstructing the possible uses of *F. betulina* by Ötzi is more complex. This species, while making a rather poor tinder material, is well known to produce a host of biologically active compounds and to have antimicrobial and styptic properties; accordingly, it has been widely used in European folk medicine until recently (see Papp et al. 2017 and references therein). Some have therefore speculated that the Ice Man "was aware of his intestinal parasites and fought them with measured doses of *Piptoporus betulinus*," (Capasso 1998), although these conclusions were later branded as unsubstantiated (Pöder and Peintner 1999; Pöder 2005). "[We] have to admit that we simply do not know the Ice Man's intentions concerning these mushrooms," concluded Reinhold Pöder, one of the researchers who have studied Ötzi's fungi (Pöder 2005).

Box 13.1 (continued)

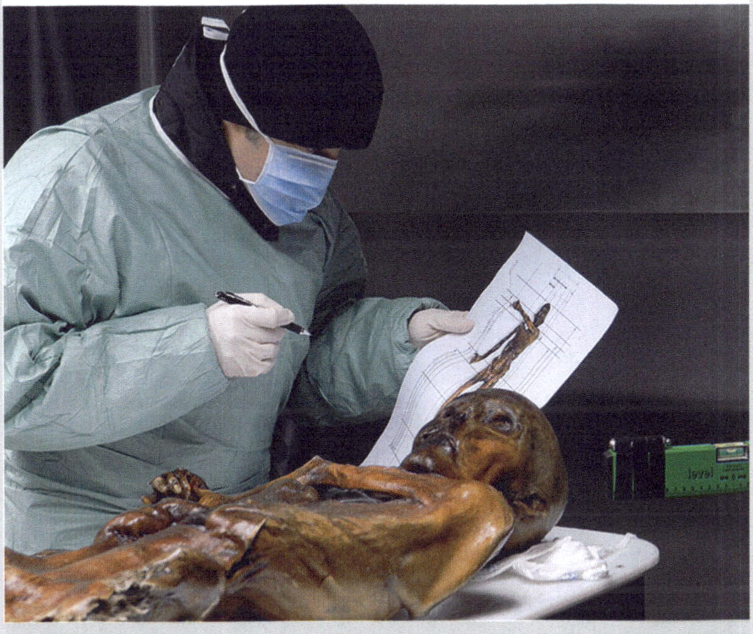

Fig. 13.5 Above, the Iceman's reconstruction by Alfons and Adrie Kennis. © South Tyrol Museum of Archaeology/Ochsenreiter. Below, scientific examination of the mummy © South Tyrol Museum of Archaeology/EURAC/Samadelli/Staschitz. Reproduced with permission

Box 13.1 (continued)

Fig. 13.6 Above, belt with sewn-on pouch found with Iceman, containing tinder fungus, a scraper, Fig. 13.6 (continued) a boring tool, a bone awl and a flint flake. © South Tyrol Museum of Archaeology/Ochsenreiter. Below, *Fomes fomentarius*. In public domain: https://it.wikipedia.org/wiki/Fomes_fomentarius#/media/File:Fomes_fomentarius.jpg

Box 13.1 (continued)

Fig. 13.7 Above, two pieces of birch polypore (birch fungus), threaded onto hide strips, were part of Ötzi's equipment. © South Tyrol Museum of Archaeology/Ochsenreiter. Below, *Fomitopsis betulina* (= *Piptoporus betulinus*). In public domain: https://commons.wikimedia.org/wiki/File:Piptoporus_betulinus_55.jpg#/media/File:Piptoporus_betulinus_55.jpg

2016). PSK and PSP, polysaccharide compounds isolated from *Trametes versicolor* (L.) Lloyd (= *Coriolus versicolor*), were shown to improve immune function in patients with certain cancers when used as adjuvants in immunochemotherapy (Pohleven et al. 2016, https://www.mskcc.org/cancer-care/integrative-medicine/

herbs/coriolus-versicolor). Some data, coming from in vitro and animal studies, or from small clinical trials, are also available for reishi, *Cordyceps* spp., *Inonotus obliquus* (Fr.) Pilát (chaga), *G. frondosa* (maitake), and *Hericium erinaceus* (Bull.) Pers. (Pohleven et al. 2016, https://www.mskcc.org/cancer-care/diagnosis-treatment/symptom-management/integrative-medicine/herbs/search). In these and many other cases, more research is needed to evaluate the real effectiveness of mushrooms products, either purified compounds or simple extracts, to determine the mechanism of action of active substances, and to assess the safety of mushrooms-based supplements. In this sense, some experts have voiced their concern. "The worst of the advertisements for medicinal mushrooms recall the era of medical quackery that tolerated patent medicines including 'Dr. Bonker's Celebrated Egyptian Oil', which was advertised as a cure for colic and cramps in humans and farm animals, and 'Dr. Solomon's Cordial Balm of Gilead' that was praised as a treatment for venereal disease and a plethora of other maladies," harshly commented Nicholas Money about the "miraculous properties of medicinal mushrooms" claimed by the plethora of products and supplements sold for the purpose of improving human health (Money 2016). Fifty or so species of mushrooms are currently known and used in various forms for their therapeutic properties; needless to say, this is just a small fraction of the species that could potentially be endowed with beneficial activities. "[With] a choice of 16 000 or more species of basidiomycetes that form mushrooms, there are lots of places to look. It is time to treat anti-aging tonics made from mushrooms as a sad phase in the history of mycology and proceed with the exploration of novel compounds with the potential to change the course of our modern plagues," pointed out Money (2016).

13.5 Magic Mirror on the Wall, Who Is the Fairest One of All?

Another twist in the recent craze about mushrooms, in Europe and in other developed parts of the world, is about their use as cosmetics. Starting from the finding that many mushrooms produce a plethora of compounds with antioxidant activity, an increasing number of studies have addressed this and related issues, and significant effort has been devoted to launching commercial skin care products based on mushrooms. Mushroom extracts and their bioactive compounds were shown to display antimicrobial, anti-inflammatory, anti-tyrosinase, anti-hyaluronidase, anti-collagenase, and anti-elastase activity (Taofiq et al. 2016a, b, 2017). These are precious properties when skin and its disorders are the target, so it's not surprising that mushroom ingredients have been heralded as very welcome new components of the arsenal of substances able to combat skin aging and at the same time reduce the severity of inflammatory skin disease and correct hyperpigmentation disorders (Fig. 13.8). The list of components in mushrooms that are beneficial to the skin is extensive. Polyphenols and phenolic compounds, including flavonoids and tannins, and saponins exert potent radical scavenging and antioxidant activity. Carotenoids,

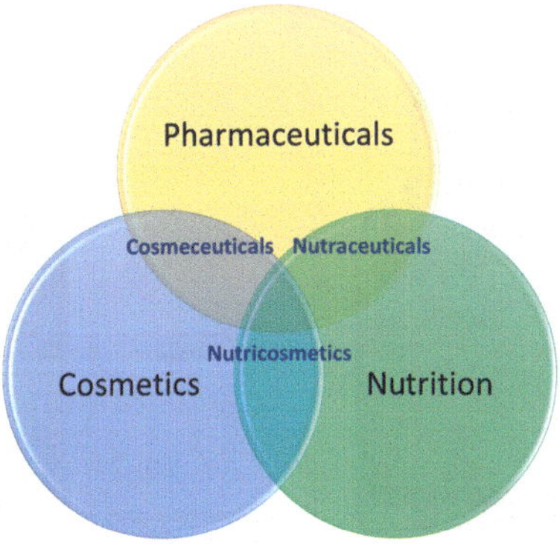

Fig. 13.8 Major applications of mushrooms in pharmaceuticals, cosmetics, and nutrition, and their interactions. Cosmeceuticals can be defined as the combination of cosmetics and pharmaceuticals and are products applied topically, such as creams, lotions, and ointments, while products with similar perceived benefits but that are ingested orally (generally as dietary supplements) are known as nutricosmetics. From Wu et al. (2016). Reproduced with permission

organic pigments found in many mushrooms, are widely used in sun lotions. Mushroom polysaccharides showed immune-modulatory activity and can help to fight skin infections when incorporated in creams; also, polysaccharides have anti-collagenase and anti-elastase activity and are thus potential ingredients to reduce skin aging (Taofiq et al. 2016a; Wu et al. 2016).

From a strictly commercial point of view, the introduction of new products containing mushrooms in the global cosmetic industry is significant, since this market is worth billions of US dollars. Species presently used—sometimes in combination—in cosmetic products are usually members of *Agaricus*, *Choiromyces*, *Cordyceps*, *Ganoderma*, *Grifola*, *Hypsizygus*, *Inonotus*, *Lentinula*, *Polyporus*, *Schizophyllum*, *Trametes*, *Tremella*, *Tricholoma*, and *Volvariella* (Hyde et al. 2010; Taofiq et al. 2016a; Wu et al. 2016). While most mushroom cosmetics are for make-up, some are indicated for revitalizing the skin, anti-aging and anti-wrinkle, as skin whiteners and as moisturizers (e.g., see Fig. 13.9). Some of these products come from very well-known brands, like Aveeno Positively Ageless by Johnson & Johnson, which contains shitake (*L. edodes*) extract, and the Mega-Mushroom line by Origins, which contains reishi (*G. lucidum*) and that promises to "help reduce visible redness, calm sensitivity and irritation" (www.origins.com/dr-weil-mega-mushroom). Great potential is also present for mycochemicals as hair cosmetics in their various formulations, including shampoos, conditioners, serums, hair styling products, and products aimed at promoting hair growth (Wu et al. 2016). In this race to the top, the last frontier of mushroom-based cosmetics is "truffle therapy." Indeed, Skin&Co, an Italian company with an operational branch in New York, uses *Tuber* extracts and other ingredients to craft an entire collection of products, including an exfoliating and regenerating bath shower, an ultra-moisturizing treatment for the body, ultra-rich body oil, facial serum and face cream (www.skinandco.it). "Over 60 years ago a woman discovered the potent enzymes contained in the Black Winter

Fig. 13.9 Cosmetics containing mushrooms or their extracts. (1) Aquamella Cream contains extracts of *Tremella fuciformis* Berk. (2) Origins Plantidote Mega-Mushroom Treatment contains *Ganoderma lucidum*. (3) Vitamega shampoo incorporates *Agaricus subrufescens* Peck. (4) Alqvimia Eternal Youth contains *Schizophyllum commune* Fr. (5) Surkran Grape Seed Lift Eye Mask contains *Tremella* polysaccharide. (6) Aveeno Positively Ageless uses *G. lucidum* and *Lentinula edodes*. (7) BeautyDiy Aqua Circulation Hydrating Gel contains *Tremella* polysaccharide. (8) La Prairie Advanced Marine Biology Night Solution contains *T. fuciformis*. (9) Sulwhasoo Hydroaid contains *S. commune* extract. (10) Yves Saint Laurent Temps Majeur Elixir De Nui contains *G. lucidum*. (11) Murad's Sleep Reform Serum contains schizophyllan, a polysaccharide from *S. commune*. (12) Kose Sekkisei Cream contains *Cordyceps sinensis* (= *Ophiocordyceps sinensis* (Berk.) G.H. Sung, J.M. Sung, Hywel-Jones & Spatafora). From Hyde et al. (2010). Reproduced with permission

Truffle found in Umbria, Italy. Today, that secret is kept in a small village in central Italy," their claim reads. "The mushrooms presently used are traditionally known to produce medicinal compounds and thus were the first to be incorporated in cosmetic applications. There are, however, numerous other mushroom species that are untested, undescribed or not yet cultivatable and that have huge potential for use in the cosmetic industry," remarked Kevin David Hyde and colleagues in an overview of the field (Hyde et al. 2010).

13.6 Conclusion

The reader that stayed with us up to this point fully realized that throughout the chapter we intended ethnomycology as the observation and recording of the interactions *sensu latu* of mushrooms and micromycetes with humans, one of the "threads of the fabric of knowledge that carry an awareness of how humans are woven into nature," (http://botanicaldimensions.org/what-is-ethnobotany/). What about the future? Is there still space for ethnomycological research in Europe? "[M]ost of Europe is now thoroughly industrial or postindustrial, hence, ethnomycology of the premodern era must often be inferred rather than directly documented from contemporary inhabitants," wrote Frank Dugan (2011). This consideration is correct, and calls for a wave of studies aimed at documenting existing traditional knowledge before this disappears irremediably, in the less developed areas of the continent and, even more urgently, in the more industrialized and urbanized ones. The intergenerational transfer of traditional mycological knowledge (Łuczaj and Nieroda 2011), and the ways of its acquisition in recently mycophilic societies, are certainly topics that must be explored, together with the uses of mushrooms among migrant communities. Examining ethnomycological classification and mushrooms nomenclature systems would reveal much of how people in Europe relate to mushrooms, making it an interesting field of research. Now that picking wild mushrooms in most European settings is a recreational activity, no longer motivated by the need to collect extra food, the sociological and anthropological reasons that drive people into the woods with this aim should be studied. This would offer the possibility of reasoning on the convergence/stratification of "traditional" and scientific mycological knowledge, mainly driven by the availability of field guides that describe edible and toxic mushrooms, and through the science divulgation activity of amatorial mycological clubs. On the other hand, picking mushroom is becoming a source of primary or secondary income for an increasing number of jobless people, also in wealthy European countries, a phenomenon that must be inspected closely, in the context of rural economy, commerce of non-timber forest products, and also of forest and mushroom conservation efforts. In brief, modern ethnomycological research in Europe has still a lot to discover, following the thread connecting traditional uses of mushrooms to the role these fantastic organisms currently play in our societies, up to the sustainable contribution that fungi are bound to provide to human well-being in the near future.

References

Anderson EN (2011) Ethnobiology: overview of a growing field. In: Anderson EN, Pearsall D, Hunn E, Turner N (eds) Ethnobiology. Wiley, Hoboken, pp 1–14

Anderson EN, Pearsall D, Hunn E, Turner N (eds) (2011) Ethnobiology. Wiley, Hoboken

Arora D, Shepard GH Jr (2008) Mushrooms and economic botany. Econ Bot 62:207–212

Arras L (2016) Funghi in Sardegna. Ilisso Edizioni, Nuoro

Berihuete-Azorín M, Girbal J, PiqueÂ R et al (2018) Punk's not dead. Fungi for tinder at the Neolithic site of La Draga (NE Iberia). PLoS One 13:e0195846

Boa E (2004) Wild edible fungi. A global overview of their use and importance to people. FAO, Rome

Brotzu R, Peintner U (2009) *Cortinarius cedretorum var. halimiorum*, a new variety of a *Phlegmacium* associated with *Halimium halimifolium* (Cistaceae) in Mediterranean costal sand dunes. Bresadoliana 1:25–44

Cai M, Pettenella D, Vidale E (2011) Income generation from wild mushrooms in marginal rural areas. Forest Policy Econ 13:221–226

Capasso L (1998) 5300 years ago, the Ice Man used natural laxatives and antibiotics. Lancet 352:1864

Comandini O, Contu M, Rinaldi AC (2006) An overview of *Cistus* ectomycorrhizal fungi. Mycorrhiza 16:381–395

Comandini O, Rinaldi AC (2008) *Lactarius cistophilus* Bon & Trimbach + *Cistus* sp. Descr Ectomycorrhizae 11(12):83–88

Comandini O, Paulis S, Rinaldi AC (2018) Sardinia: mycovisions from a charming land. Curr Res Environ Appl Mycol 8:474–491

Cör D, Knez Ž, Knez Hrnčič M (2018) Antitumour, antimicrobial, antioxidant and antiacetylcholinesterase effect of *Ganoderma lucidum* terpenoids and polysaccharides: a review. Molecules 23:649

Diamond J (2012) The world until yesterday: what can we learn from traditional societies? Viking Press, New York

Drăgulescu C (2004) Mushrooms in Romanian ethnobotany. Moeszia 2:66–70

Dugan FM (2008a) Fungi in the ancient world: how mushrooms, mildews, molds, and yeast shaped the early civilizations of Europe, the Mediterranean, and the Near East. APS Press, St. Paul

Dugan FM (2008b) Fungi, folkways and fairy tales: mushrooms & mildews in stories, remedies & rituals, from Oberon to the Internet. North American Fungi 3:23–72

Dugan FM (2011) Conspectus of world ethnomycology. APS Press, St. Paul

Dyke AJ, Newton AC (1999) Commercial harvesting of wild mushrooms in Scottish forests: is it sustainable? Scott For 53:77–85

Egli S, Ayer E, Chatelain E (1990) Die Beschreibung der Diversitat von Makromyzeten. Erfahrungen aus pilzokologischen Langzeitstudien im Pilzreservat La Chanéaz, FR. Mycol Helv 9:19–32

Greenfield PM (2016) Social change, cultural evolution, and human development. Curr Opin Psychol 8:84–92

Grienke U, Zöll M, Peintner U, Rollinger J (2014) European medicinal polypores—a modern view on traditional uses. J Ethnopharmacol 154:564–583

Härkönen M (1998) Uses of mushrooms by Finns and Karelians. Int J Circumpolar Health 40:40–55

Hyde KD, Bahkali AH, Moslem MA (2010) Fungi—an unusual source for cosmetics. Fungal Divers 43:1–9

Kasper-Pakosz R, Pietras M, Łuczaj L (2016) Wild and native plants and mushrooms sold in the open-air markets of southeastern Poland. J Ethnobiol Ethnomed 12:45

Kotowski M (2016) Differences between European regulations on wild mushroom commerce and actual trends in wild mushroom picking. Slovak Ethnol 64:169–178

Lentini F, Venza F (2007) Wild food plants of popular use in Sicily. J Ethnobiol Ethnomed 3:15

Leonardi M, Comandini O, Rinaldi AC (2016) Peering into the Mediterranean black box: *Lactifluus rugatus* ectomycorrhizae on *Cistus*. IMA Fungus 7:275–284

Leonardi M, Neves MA, Comandini O, Rinaldi AC (2018) *Scleroderma meridionale* ectomycorrhizae on *Halimium halimifolium*: expanding the Mediterranean symbiotic repertoire. Symbiosis 76:199–208

Łuczaj L, Nieroda Z (2011) Collecting and learning to identify edible fungi in southeastern Poland: age and gender differences. Ecol Food Nutr 50:319–336

Łuczaj L, Kinga Stawarczyk K, Kosiek T et al (2015) Wild food plants and fungi used by Ukrainians in the western part of the Maramureş region in Romania. Acta Soc Bot Pol 84:339–346

Mazzocchi F (2006) Western science and traditional knowledge: despite their variations, different forms of knowledge can learn from each other. EMBO Rep 7:463–466

Money NP (2016) Are mushrooms medicinal? Fungal Biol 120:449–453

Moskalenko SA (1987) Slavic ethnomedicine in the Soviet Far East. Part I: herbal remedies among Russians/Ukrainians in the Sukhodol Valley, Primorye. J Ethnopharmacol 21:231–251

Nieves-Rivera AM (2001) Origin of mycophagy in the West Indies. Inoculum 52:1–3

Nuytinck J, Verbeken A, Leonardi M et al (2004) Characterization of *Lactarius tesquorum* ectomycorrhizae on *Cistus* sp., and molecular phylogeny of related European *Lactarius* taxa. Mycologia 96:272–282

Nyberg H (1992) Religious use of hallucinogenic fungi: a comparison between Siberian and Mesoamerican cultures. Karstenia 32:71–80

Palahí M, Pukkala T, Bonet JA et al (2009) Effect of the inclusion of mushroom values on the optimal management of even-aged pine stands of Catalonia. Forest Sci 55:503–511

Pan C, Pfeil BS, Videsott P (2016) Die Volksgruppen in Europa. Handbuch der europäischen Volksgruppen Band 1. Verlag Österreich GmbH, Wien

Papp N, Rudolf K, Bencsik T, Czégényi D (2017) Ethnomycological use of *Fomes fomentarius* (L.) Fr. and *Piptoporus betulinus* (Bull.) P. Karst. in Transylvania, Romania. Genet Resour Crop Evol 64:101–111

Pavlovna Wasson V, Wasson RG (1957) Mushrooms, Russia and history. Pantheon Books, New York

Peintner U, Pöder R, Pümpel T (1998) The iceman's fungi. Mycol Res 102:1153–1162

Peintner U, Schwarz S, Mešić A et al (2013) Mycophilic or mycophobic? Legislation and guidelines on wild mushroom commerce reveal different consumption behaviour in European countries. PLoS One 8:e63926

Pieroni A (1999) Gathered wild food plants in the Upper Valley of the Serchio River (Garfagnana), Central Italy. Econ Bot 53:327–341

Pieroni A (2016) The changing ethnoecological cobweb of white truffle (*Tuber magnatum* Pico) gatherers in South Piedmont, NW Italy. J Ethnobiol Ethnomed 12:18

Pieroni A (2017) Traditional uses of wild food plants, medicinal plants, and domestic remedies in Albanian, Aromanian and Macedonian villages in South-Eastern Albania. J Herb Med 9:81–90

Pieroni A, Nebel S, Quave C et al (2002) Ethnopharmacology of Liakra: traditional weedy vegetables of the Arbëreshë of the Vulture Area in Southern Italy. J Ethnopharmacol 81:165–185

Pieroni A, Nebel S, Santoro RF, Heinrich M (2005) Food for two seasons: culinary uses of non-cultivated local vegetables and mushrooms in a south Italian village. Int J Food Sci Nutr 56:245–272

Pöder R, Peintner U (1999) Laxatives and the Ice Man. Lancet 353:926

Pöder R (2005) The Ice Man's fungi: facts and mysteries. Int J Med Mushrooms 7:357–359

Pohleven J, Korošec T, Gregori A (2016) Medicinal mushrooms. MycoMedica, Kranjska, Gora. http://www.goba.eu/en/library/medicinal-mushrooms-book. Accessed 15 Jan 2019

Porcu G, Mua A, Casula M, Sanna M (2018) *Russula purpureovirescens* sp. nov. una nuova specie del sottogenere *Heterophyllidia* raccolta in Sardegna. Micol Vegetazione Mediterr 33:3–16

Redzic S, Barudanovic S, Pilipovic S (2010) Wild mushrooms and lichens used as human food for survival in war conditions; Podrinje-Zepa Region (Bosnia and Herzegovina, W. Balkan). Hum Ecol Rev 17:175–187

Rinaldi A, Shatty P (2015) Traditional medicine for modern times: facts and figures. http://www.scidev.net/global/medicine/feature/traditional-medicine-modern-times-facts-figures.html. Accessed 10 Jan 2019

de Román M, Boa E (2006) The marketing of *Lactarius deliciosus* in Northern Spain. Econ Bot 60:284–290

Rosa-Gruszecka A, Hilszczańska D, Gil W, Kosel B (2017) Truffle renaissance in Poland—history, present and prospects. J Ethnobiol Ethnomed 13:36

Samils N, Olivera A, Danell E et al (2008) The socioeconomic impact of truffle cultivation in rural Spain. Econ Bot 62:331–340

Singh J, Aneja KR (eds) (1999) From ethnomycology to fungal biotechnology: exploiting fungi from natural resources for novel products. Kluwer Academic, New York

Sitta N, Floriani M (2008) Nationalization and globalization trends in the wild mushroom commerce of Italy with emphasis on porcini (*Boletus edulis* and allied species). Econ Bot 62:307–322

Stoyneva-Gärtner MP, Uzunov BA (2015) An ethnobiological glance on globalization impact on the traditional use of algae and fungi as food in Bulgaria. J Nutr Food Sci 5:413

Svanberg I, Łuczaj L, Pardo-De-Santayana M, Pieroni A (2011) History and current trends of ethnobiological research in Europe. In: Anderson EN, Pearsall D, Hunn E, Turner N (eds) Ethnobiology. Wiley, Hoboken, pp 189–212

Svanberg I (2018) Ethnomycological notes on *Haploporus odorus* and other polypores in Northern Fennoscandia. J North Stud 12:73–91

Taofiq O, González-Paramás AM, Martins A, Barreiro MF, Ferreira ICFR (2016a) Mushrooms extracts and compounds in cosmetics, cosmeceuticals and nutricosmetics—a review. Ind Crops Prod 90:38–48

Taofiq O, Heleno SA, Calhelha RC et al (2016b) Development of mushroom-based cosmeceutical formulations with anti-inflammatory, anti-tyrosinase, antioxidant, and antibacterial properties. Molecules 21:1372

Taofiq O, Heleno SA, Calhelha RC et al (2017) The potential of *Ganoderma lucidum* extracts as bioactive ingredients in topical formulations, beyond its nutritional benefits. Food Chem Toxicol 108:139–147

Turtiainen M, Saastamoinen O, Kangas K, Vaara M (2012) Picking of wild edible mushrooms in Finland in 1997–1999 and 2011. Silva Fenn 46:569–581

Uzunov BA, Stoyneva-Gärtner MP (2018) Micromycetes in traditional and modern Bulgarian cuisine. Food Nutr Curr Res 1:19–25

Wasson RG (1980) The wondrous mushroom. Mycolatry in Mesoamerica. Mc Graw-Hill, New York

Widenfalk O, Jakobsson M, Hammarström A, Widenfalk L (2018) Trade and production of plants and plant products in Sweden—a knowledge base for pest risk analysis. Greensway AB, Uppsala. https://www.slu.se/globalassets/ew/org/centrb/riskv/pub/trade-and-production-of-plants-and-plant-products-in-sweden.pdf. Accessed 20 Jan 2019

Willis KJ (ed) (2018) State of the World's fungi 2018. Report. Royal Botanic Gardens, Kew, Richmond. https://stateoftheworldsfungi.org/2018/reports/SOTWFungi_2018_Full_Report.pdf. Accessed 15 Jan 2019

World Health Organization (2013) WHO traditional medicine strategy 2014-2023. WHO, Geneva. http://www.who.int/medicines/publications/traditional/trm_strategy14_23/en/. Accessed 5 Feb 2019

Wu Y, Choi M-H, Li J et al (2016) Mushroom cosmetics: the present and future. Cosmetics 3:22

Yamin-Pasternak S (2011) Ethnomycology: fungi and mushrooms in cultural entanglements. In: Anderson EN, Pearsall D, Hunn E, Turner N (eds) Ethnobiology. Wiley, Hoboken, pp 213–230

Part V
Ecology with Emphasis on Wild Edible Fungi

Chapter 14
Interactions Between Soil Mesofauna and Edible Ectomycorrhizal Mushrooms

Faustino Hernández-Santiago, Irma Díaz-Aguilar, Jesús Pérez-Moreno, and Jorge L. Tovar-Salinas

14.1 Introduction

Biological diversity, or biodiversity, according to the Convention of Biological Diversity 1992, is the observed variability of living organisms inhabiting the polar regions, forests, savannas, deserts, from the urban zones and agricultural areas to the marine and aquatic ecosystems, as well as that of the ecological complexes of which they are part of; the latter encompass the diversity within species, among species and that of the ecosystems, equally (Magurran 2010). Biodiversity is a key component in nature because it directly influences the sustainability and resilience of ecosystems. However, microorganisms and soil fauna are rarely considered in biodiversity conservation programmes, despite their richness in species, functional importance and the supplied services to the ecosystem. Therefore, an estimation of appropriate scales, quantification, and environmental factors, relevant to their evaluation, is essential (Fragoso et al. 2001; Griffith 2012; Okabe 2013).

The soil is a living ecosystem, complex and dynamic, in which vital processes are developed for the function of terrestrial ecosystems. The storage of carbon, the decomposition of organic matter, and the nutrient cycle, all of which are controlled, take place mainly by biological activity (Orgiazzi et al. 2016). The decomposition of organic matter mostly occurs by the enzymatic activity of fungi and bacteria. However, the fractioning of plant and animal residues, the transformation of vegetal

F. Hernández-Santiago · I. Díaz-Aguilar (✉) · J. L. Tovar-Salinas
Ecología de Suelos, Edafología, Campus Montecillo, Colegio de Postgraduados, Montecillo, Texcoco, Estado de México, México
e-mail: idiaz@colpos.mx; idiaz@ualberta.ca

J. Pérez-Moreno
Microbiología, Edafología, Campus Montecillo, Colegio de Postgraduados, Texcoco, México

J. Pérez-Moreno et al. (eds.), *Mushrooms, Humans and Nature in a Changing World*, https://doi.org/10.1007/978-3-030-37378-8_14

residues into humic compounds, and the dispersal of microbial propagules are performed mostly by microarthropods such as mites and springtails; hence, they are denominated as 'decomposers or litterfall transformers' (Swift et al. 1979, 2012). On the other hand, fungi increase the quantity and efficiency of nutrient absorption by plants through the formation of symbiotic associations called mycorrhizas (Smith and Read 2008; Swift et al. 2012; Orgiazzi et al. 2016).

Thus far, the significant effects of fungi and mesofauna, mites and springtails, have been evaluated separately regarding the functions of the soil. However, the mutualist interactions, the diversity patterns, and the function of mesofauna in the fruiting bodies of macromycetes, as a temporal mesofauna habitat, and the biological associations between the mesofauna and this fungal habitat are practically unknown. Komonen (2003) notes that superior fungi, or macromycetes, emerge as islands rich in species along the forest soil landscape and that in addition to performing important ecological functions as a reproductive structure, these fungi also provide the soil invertebrates with a habitat and food. Therefore, the macromycete sporomes should also be considered an important reservoir of soil fauna diversity and a network of biological activity.

The most frequent insect–fungus relation in nature is the use of fungi as a food source (Palacios-Vargas and Gómez 1991). An important part of the diet of the more relevant groups in the edaphic fauna, among them springtails and mites, is composed of hyphae and spores of a wide diversity of inferior and superior fungi (Christiansen 1964). The mesofauna present in macromycetes fulfils functions such as spore dispersal, decomposition of the fruiting body, and/or regulation of the cohabitating populations (Okabe 1999). Likewise, the mites and springtails inhabiting the sporome can be indicators of the soil and ecosystem quality, since they are sensitive to any disturbance because of their unique biological characteristics (Greenslade 2007; Gulvik 2007; Walter and Proctor 2013).

On the other hand, mite and springtail diversity in the soil is reflected in the relations between these mesofauna with other organisms, of which phoresis is of particular interest. Phoresy is a migratory movement favouring dispersal by insects, and it is the means by which mites and springtails are transported from one site to the next (Binns 1982; Houck and Oconnor 1991; Walter and Proctor 2013; Hofstetter and Moser 2014). This behaviour has been documented in fossil records, e.g., amber from the Eocene and Miocene (Penney et al. 2012; Grünemaier 2016), which suggests that phoresis is an ancestral means of dispersion. The free-living mite fauna, such as edaphic mites, tend to exit this environment, either in a phoretic manner or through vertical movements between the layers of litterfall and soil, and climb the rocks, tree trunks, and plant branches, temporally establishing themselves in sites that provide them with favourable climatic conditions, security, and a source of food (Hoffmann and Riverón 1992). OConnor (2003) notes that astigmatid mites, which naturally inhabit patchy habitats such as decomposing wood, manure, carrion, fungal fruiting bodies, and treeholes, are dispersed by phoresy atop any insect frequenting the habitat.

Mite–insect–fungi relations occur in a wide spectrum of shapes and forms ranging from parasitic to mutualistic (Hofstetter and Moser 2014), and since insect and mites often share habitats and resources, a phoretic relation with fungivorous insects could facilitate fungus–microarthropod associations in the fruiting bodies, despite the extreme spatio-temporal heterogeneity of the habitat (O'Connell and Bolger 1997a). Phoresis can influence the multi-trophic relations among the soil organisms (Athias-Binche 1994), and the fruiting bodies of macromycetes should be no exception; diptera are generally related to the transport of phoretic mites in fungal cultures (Fletcher and Gaze 2008).

14.2 Classification and General Characteristics of Soil Mesofauna

Classification of the edaphic mesofauna. Soil mesofauna, according to its trophic specialization, is classified under the following criteria: adaptation and soil preference, body size and feeding regimen, morphological aspects such as shape and body pigmentation, motility in the different soil layers, chemical and mechanical sensitivity, light sensitivity (photophobia), resistance to humidity, and desiccation (Christiansen 1964).

Adaptation and soil preference. Based on the classification by Krausee (1928/29), Christiansen (1964), and Stebaeva (1970), soil mesofauna is grouped into the following: (1) the epedaphic species, which are those living on the surface of the soil and litterfall; (2) the hemiedaphic species, which are those living in the organic soil; (3) the euedaphic species, which are those found in mineral soil and pores; (4) troglomorphs, which are those inhabiting subterranean environments, such as caves and grottoes; and (5) synecomorphs, which live freely in the nests of social insects, such as ant and termite colonies. Stebaeva (1970) includes atmobiotic species, which are microarthropods inhabiting microphytic plants such as herbs, shrubs, and tree trunks and branches, although they can also be found on the surface of the litterfall.

Body size. Wallwork (1970), based on body length, classifies the edaphic fauna into (1) microfauna, which is composed of organisms smaller than 0.2 mm; (2) mesofauna, which includes species with a body length ranging from 0.2 to 10 mm; and (3) macrofauna, which includes species larger than 10 mm. On the other hand, because the environment of a soil organism is a function of the organism's size and because the texture of the soil influences the available space in the habitable pores, Swift et al. (1979) created a classification according to body width. In this classification, the microfauna (1) comprises organisms smaller than 0.1 mm; this group includes nematodes, rotifers, and tardigrades. The mesofauna (2) are organisms whose body width is within a range of 0.1–2 mm; this category includes mites, springtails, diplurans, proturans, enchytraeids, and pseudoscorpions. The macrofauna (3) are those organisms larger than 2 mm and includes ants, termites, arachnids, beetles, millipedes, centipedes, woodlice, and earthworms. Since this classification is conventional, and several arthropods may be considered microfauna during their early development stages and meso- or even macrofauna when they are adults, the use of the term microarthropods is convenient when referring to diminutive aptera arthropods in the soil (Palacios-Vargas 2003).

Feeding regimen. According to their feeding preferences (Krantz and Lindquist 1979; Krantz 2009; Walter and Proctor 2013), soil fauna is classified as follows: (1) Macrophytophages are organisms feeding on the remains of superior plants or their decomposing tissues; saprophages (feeding on vegetal or animal material in decomposition), detritivores (ingesting pieces of organic detritus), and xylophages (feeding from decomposing wood) are included in this category. (2) Microphytophages are organisms feeding on microbiota, among which the bacteriophages (bacteria), mycophages (fungi), and algivores (algae) are included, along with bryophages (mosses), and pollinivores (pollen). (3) Predators are those that hunt their prey using hunting strategies (stalking and ambush or pursuit). (4) Omnivores are saprophages and predators. (5) Necrophages are those feeding on carrion. (5) Coprophages are those feeding on faecal matter from diverse animals.

14.2.1 Characteristics of Subclass Acari

Mites are the most diverse group of the phylum Arthropoda, a monophyletic taxon of terrestrial and marine invertebrates (Weygoldt 1998), subphylum Chelicerata, class Arachnida, and subclass Acari (Lindquist et al. 2009a; Zhang 2011). These are some of the oldest terrestrial microarthropods, with known fossils dating as far back as the early Devonian (Norton et al. 1988). Their small size (100–500 μm) and evolutive plasticity have allowed them to colonize most terrestrial and aquatic habitats due to their great capacity to exploit the available trophic resources in each environment and their different and diverse feeding habits, which have allowed them to establish widely diverse trophic interactions (Vacante 2016). Mites are megadiverse; approximately one million species may exist (Walter and Proctor 2013), of which 53,657 Acari species have been recorded worldwide (Zhang 2011), with 2892 species in Mexico alone (Palacios-Vargas et al. 2014).

Taxonomy. Mites belong in the subclass Acari, and according to Lindquist et al. (2009a), they are divided into two superorders (Parasitiformes and Acariformes), both of which include several orders. Parasitiformes includes the orders Opilioacarida, Holothyrida, Ixodida, and Mesostigmata. Acariformes includes the orders Trombidiformes, which is divided into two suborders—Sphaerolichida and Prostigmata—and Sarcoptiformes, which contains the suborders Endeostigmata and Oribatida, the cohort Astigmata being derived from the latter (Norton 1998). Within the group of edaphic mesofauna, species from the order Mesostigmata, suborder Prostigmata, suborder Oribatida, and cohort Astigmata (Palacios-Vargas et al. 2014) are frequently found, with a lower frequency of species being identified from the suborders Endeostigmata and Sphaerolichida.

14.2.1.1 Superorder Parasitiformes

Order Mesostigmata. This is the order with highest diversity within the Superorder Parasitiformes, and it exhibits a wide range of habits and associations. The order Mesostigmata is divided into several suborders: Sejida, Trigynaspida, and Monogynaspida. The cohorts Uropodina and Gamasina are derived from the latter and their feature morphology varies from the other Acari in the order. Their size ranges between 0.2 and 4.5 mm. They are characterized by a three-segmented chelicerae, a tritosternum (fluid transporter), a couple of respiratory spiracles in the medial part of the idiosoma, and dorsal and ventral shields, and the coxa in their legs are free and motile (Lindquist et al. 2009b). Most species are free-living predators inhabiting litterfall, decomposing wood, compost, nests, and carrion. These species are not overly abundant, and their main function is to regulate the populations of microarthropods in the soil. Other species are parasites or symbionts of mammals, birds, reptiles, or other arthropods (Walter and Proctor 2013). Few species are fungivores (family Ameroseiidae) or saprophagous (Krantz and Lindquist 1979). Generally, they are predators of nematodes (Walter and Ikonen 1989) and collembolans, enchytraeids, other mites (Prostigmata, soft bodied Astigmata and hard bodied Oribatida), and of nymphs, larvae, and eggs from insects, proturans, pauropods, and symphylans (Walter et al. 1988). A total of 109 families, 878 genera and 11,424 species have been recorded worldwide (Zhang 2011). However, these records have increased to 110 families, 916 genera, and 12,017 species in the space of 2 years (Walter and Proctor 2013). Pérez et al. (2014) report a record of 50 families, 158 genera, and 507 species in Mexico.

14.2.1.2 Superorder Acariformes

(a) Order Trombidiformes

Suborder Sphaerolichida. These are small edaphic mites that are rarely found in large quantities. They have globular bodies with ornamented cuticles and being weakly sclerotized, possess prodorsum with two pairs of trichobothrial setae, and do not have rutella. Known species have been found in moss, soil, and litterfall, and, presumably, they are mycophages or ambush predators (Walter 1988; Walter et al. 2009). Only 21 species in 2 monotypic families (Sphaerolichidae and Lordalycidae) have been recorded worldwide (Zhang et al. 2011). However, Walter and Proctor (2013) report that this record has recently increased to 25 species. In Mexico, only one genus and one species in the family Lordalycidae has been recorded (Pérez et al. 2014).

Suborder Prostigmata. This suborder is a very diverse and heterogeneous group, both in morphology and biology, as well as in behaviour. The body size of an adult specimen can be in the range of 0.1–10 mm. In general, these organisms are little sclerotized; have globular, pyriform, rhomboid, or vermiform shapes; and feature varied colouration. Their mouthparts (chelicerae, palps, and subcapitulum) show widely varied modifications, especially in the phytophagous species (Lindquist 1998); they feature simple eyes, the coxa in their legs are fused to the idiosoma, and the tracheal system usually shows a couple of spiracles near the base of the chelicerae or the anterior part of the prodorsum. Species that parasitize birds and mammals, including humans (Hoffmann 1990), are found within this group. In the soil, they are predaceous on other microarthropods and nematodes and contribute to soil fertility by incorporating their liquid faeces as well as the remains of their prey, which are easily degraded by microorganisms (Walter and Proctor 2013). Prostigmata mites are the most diverse group within the order Trombidiformes, with 25,800 species described worldwide (Zhang et al. 2011). In Mexico, 1764 species have been recorded (Palacios-Vargas et al. 2014).

(b)Order Sarcoptiformes

Suborder Endeostigmata. Most genera have cosmopolitan distributions inhabiting in a wide range of different soils and are frequently found in extreme habitats (high temperature, especially in deserts, sand, and deep soil). These mites are very small, have globular or extremely elongated shapes, have branched setae along their bodies, feature rutella, and have one to two pairs of trichobothrial setae on the prodorsum; furthermore, some species are characterized by their ability to jump. They feed on fungi, algae, soft-bodied microinvertebrates, such as rotifers and tardigrades, and some species are nematode predators, such as those from genera *Alycus* and *Alicorhagia* (Walter 1988). However, species from families Nanorchestidae and Nematalycidae apparently feed on fluids (Walter and Proctor 2013). Ten families, 27 genera and 110 species have been recorded worldwide (Walter et al. 2011; Walter and Proctor 2013). In Mexico, six families, 11 genera, and 15 species have been recorded (Pérez et al. 2014).

Suborder Oribatida. This suborder plays an important role in the decomposition of organic matter since they fragment the vegetal material and make it more accessible to fungi and bacteria; furthermore, their 'pellets' (faeces) contribute to soil structure. Their interaction with fungi is considered one of their main functions because they are involved in spore dispersal and control fungal population growth by making fungi one of their main sources of feeding (Martínez 2009). Likewise, some omnivorous species consume any type of vegetal and animal material (Siepel and de Ruiter-Dijkman 1993). Oribatid mites are divided into six major groups, the early-derivative Palaeosomata and Enarthronota, the small group Parhyposomata, the middle-derivative Mixonomata and Desmononata, and the highly derived Brachypylina (Schaefer et al. 2010). In general, their body sizes range from 0.15 to 2 mm; they possess chelate-dentated chelicerae and rutella; their distinctive sensory organs are located in the prodorsum (bothridial setae); they feature highly sclerotized bodies; range in colour from light brown to black; and possess large anal and genital apertures that are closed by paired plates, with the latter possessing papillae. A total of 162 families, 1289 genera and subgenera, and 11,107 species and subspecies have been recorded worldwide (Subías 2019). In Mexico, 105 families, 251 genera and 435 species have been recorded (Pérez et al. 2014).

Cohort Astigmata. This group is widely diverse in shape and behaviour. The species inhabiting the soil are specialized to live in the most varied and ephemeral habitats, and they are very successful in symbiotic associations, e.g., dispersal through phoresy and parasitism (Norton 1998; Houck and OConnor 1991; Martínez 2009). They are small, ranging from 0.2 to 1.8 mm in length, have little sclerotization, lack segmentation, feature cuticular respiration, lack trichobothria, and have leg coxa fused to the ventral section of the podosoma, with males characterized by aedeagus (copulating organ) and adanal suckers. In general, they move slowly and show paedomorphosis, which is the retention of juvenile structures into adulthood. Soil Astigmata feature chelate-dentated chelicerae adapted to fragment vegetal material, fungi, and algae (Martínez 2009). Moreover, they also ingest liquids and other materials resulting from carrion decomposition (OConnor 2009). The cohort Astigmata comprises 77 families, 1128 genera, and 6150 species (Schatz et al. 2011). However, these records increased to 80 families, 1133 genera, and 6220 species in 2013 (Walter and Proctor 2013). Their diversity in Mexico is represented by 351 species in 43 families and 140 genera (Pérez et al. 2014).

14.2.2 Characteristics of Class Collembola

Collembolans, or springtails, are in the level of the class Insecta within the superclass Hexapoda. Hexapods represent a monophyletic lineage that, along with crustaceans, is found within the Pancrustacea clade, in Phylum Arthropoda. The lineage includes insects and a number of apterygota groups such as Collembola, Diplura, Protura, and Archaeognatha (Sasaki et al. 2013; Misof et al. 2014; Dell'Ampio et al. 2014; Faddeeva et al. 2015; Derst et al. 2016). Recent phylogenetic analyses suggest that Hexapoda is derived from a single terrestrialization event, within the Ecdysozoa, and from an originally crustacean lineage that evolved into the terrestrial surface (von Reumont et al. 2012; Rota-Stabelli et al. 2013). However, Dell'Ampio et al. (2014) consider that these phylogenetic relations are not yet well defined and thus are still considered unresolved. Regardless, the fossil records suggest that collembola are indeed a very old lineage present since the Devonian period, approximately 400 million years ago (Hirst and Maulik 1926).

Springtails are characterized by a body segmented into three regions: (1) head with antennae and entognatha mouthparts, (2) a three-segmented thorax, and (3) an abdomen with six segments. They possess three unique characteristics that define them as a group: (1) the furca, which is an organ used to jump and is situated ventrally in the fourth abdominal segment; (2) the ventral tube or collophore, which is a ventral projection in the first abdominal segment and related with hydric regulation and may, in addition, function as an adhesive organ; and (3) the retinaculum in the third abdominal segment, whose function is to hold the furca (Eisenbeis 1982; Hopkin 1997; Bellinger et al. 1996–2018).

Springtails are microarthropods that constitute, along with mites, a very important component of the soil mesofauna in every terrestrial ecosystem. The species within this class range between 0.12 and 17 mm in length (Bellinger et al. 1996–2018), are common and abundant in the different soil layers, and often show density records of >100,000 individuals/m^2 in the soil (Hopkin 1997). Some species exhibit a gregarious behaviour due to the production of pheromones that are secreted under stressful conditions, such as when the species is at risk of dehydration (Verhoef and Prast 1989) or in the presence of predators (Negri 2004). They have a short life span and are parthenogenetic (Hopkin 1997).

Collembolans can be found from the sea level to up to 7000 masl, practically inhabiting every existing biotype: soils, litterfall, moss and hepatics, fungi (including macromycetes), tree bark, decomposing tree trunks, bat guano, and other cave biotypes (Christiansen 1964; Palacios-Vargas 2014), as well as in rivers, lakes, and the seashore (Deharveng et al. 2008). Likewise, even if they have not colonized the open sea, some utilize the sea currents for their dispersal. Additionally, they inhabit the snow layers, glaciers (Hågvar 2010a, b), and regions covered by enormous ice sheets, such as the Arctic and Antarctica (Ávila-Jiménez and Coulson 2011), and they are frequently found in the nests of social insects, birds, and mammals (Christiansen 1964; Cutz-Pool et al. 2007).

Taxonomy. Traditionally, two divisions have been considered in the Class: Arthropleona and Symphypleona sensu *lato*, according to their morphology, (e.g., cylindrical or globular body). Currently, four orders are considered within the Class Collembola: Poduromorpha, Entomobryomorpha, Neelipleona, and Symphypleona (Deharveng 2004; Janssens and Christiansen 2011; Palacios-Vargas 2014; Baquero and Jordana 2015; Bellinger et al. 1996–2018). Approximately 8990 species have been recorded worldwide (Bellinger et al. 1996–2018), and in Mexico there are currently records of 1058 species (Cruz-Leal et al. 2016).

14.2.2.1 Order Poduromorpha

This order is characterized by an elongated body divided into thorax and abdomen, with well-separated segments (or tergites); the prothorax is clearly developed and possesses setae; dorsally visible, post-antennal organ with numerous vesicles; dorsoventrally depressed body; and antennae with four segments with a sense organ in the third antennal segment (Palacios-Vargas 2014). The furca is occasionally reduced or even absent, and it is distinctive the presence or absence of pseudocelli on the head and body, of morphology, and disposition characteristic of each species (Baquero and Jordana 2015). Worldwide, 11 families, 341 genera, and 3410 species have been recorded (Bellinger et al. 1996–2018).

14.2.2.2 Order Entomobryomorpha

The members in this order are characterized by a reduced first thoracic segment, lacking setae or scales, in contrast with Poduromorpha, whose first thoracic segment is well developed. These species have elongated bodies with the posterior segments of the abdomen sometimes fused to each other. The body presents setae and/or scales of different shapes in some genera, the species has long antennae with four to six segments, and furca is always present. The post-antennal organ is generally formed by one or a few vesicles, with trichobothria over head and abdomen (Jordana 2012; Palacios-Vargas 2014; Baquero and Jordana 2015). This order constitutes one of the most diverse Collembola taxons, with a worldwide record of 12 families, 263 genera, and 4267 species (Bellinger et al. 1996–2018).

14.2.2.3 Order Neelipleona

This order features small (0.3–1 mm) organisms, each with a globular body, divided in two parts: head and thorax fused with the abdominal segments presenting sensory fields placed in small depressions, absent eyes, and antennae shorter than the head and with four segments (Baquero and Jordana 2015; Palacios-Vargas 2014). They are mainly a euedaphic group inhabiting the soil. However, several described species inhabit caves with a high content of organic material (Kováč and Papáč 2010; Papáč and Palacios-Vargas 2016). This is one of the less diverse taxons in the order Collembola, with a worldwide record of 1 family, 5 genera, and 33 species (Janssens and Christiansen 2011). In 2016, these records were increased to 6 genera (Papáč and Palacios-Vargas 2016) and 59 species (Bellinger et al. 1996–2018).

14.2.2.4 Order Symphypleona

Organisms from this order feature a globular body as a result of the semi-fusion of the thorax and the first abdominal segments; the body is divided into three parts: the head, large abdomen, and small abdomen. In general, the size of the body ranges from 0.12 to 4 mm; the antennae are longer than the head, with four segments and sometimes with rings; the eyes are generally present; the furca and retinaculum are well developed, and the body shows diverse pigmentation patterns, with organisms being striped, dotted, patterned, or lacking either pattern in the case of edaphic or cave species (Palacios-Vargas 2014). Bellinger et al. (1996–2018) recorded 11 families, 126 genera, and 1254 species worldwide.

14.3 Origin and Evolution of Acari, Collembola, and Ectomycorrhizal Fungi

Mites and springtails are among the oldest terrestrial microarthropods, with known fossil records dating back to the Devonian period. At the end of the Palaeozoic era, life had invaded the land and produced a boom in diversity first by plants, soon followed by arthropods and finally by animals. The first hexapods appeared (or at least the first ones that were found preserved as fossils) during the Devonian period, approximately 410 million years ago (Grimaldi and Engel 2005). The ancestral evidence of collembolans was found in the paleontological deposit of Rhynie chert, in Scotland, where the *Rhyniella praecursor* fossil was found (Hirst and Maulik 1926), with an estimated age dating back to the early Devonian, probably Lochkovian-Pragian (maximal age), 396–407 million years ago (Whalley and Jarzembowski 1981; Rice et al. 1995; Engel and Grimaldi 2004). The palaeoenvironment from Rhynie was very likely that of a marsh (Tasch 1957 referenced by Grimaldi and Engel 2005).

In a fossil deposit north of New York, close to the city of Gilboa, the earliest evidence of oribatid mites was found, *Protochthonius gilboa* and *Devonacarus sellnicki* (Norton et al. 1988), which date back to approximately 390 million years ago, thus corresponding to the mid-Devonian (Grimaldi and Engel 2005; Sidorchuk 2018). Other important findings in Rhynie chert were some Acari fossils reported by Hirst (1923). However, these have been re-identified and placed within the suborders Endeostigmata and Prostigmata by Dubinin (1991), although a controversy concerning their taxonomy is ongoing (Sidorchuk 2018). The Gilboa fossils are younger than those from Rhynie; however, both represent the first evidence of the existence, from the Palaeozoic era, of terrestrial apterygota hexapods (Collembola) and chelicerates (Acari) (Fig. 14.1). Primitive terrestrial plants appeared in the Ordovician period (490–443 million years ago), and possibly, the first terrestrial mites and collembolans appeared and emigrated towards those early plant communities during the Upper Silurian (443–414 million years ago). However, the Silurian fauna is unbeknownst, and the only available knowledge is that it consisted of primitive arachnids and myriapods (Jeram et al. 1990).

On the other hand, numerous palaeobotanic, morphoanatomical, and phylogenetic studies based on molecular techniques revealed that the co-evolution maintained between the ectomycorrizal fungi and plant roots date back to the Palaeozoic era, during the Ordovician, with the origin of the first terrestrial plants (Remy et al. 1994; Wellman and Gray 2000; Honrubia 2009; Veneault-Fourrey and Martin 2013); this evolutionary history suggests that the symbiosis with mycorrhizas facilitated the colonization of the land by the first plants (Hibbett et al. 2000; Heckman et al. 2001; Field et al. 2015). It is therefore likely that the first fungal mycorrhizas colonized land before plants did (Veneault-Fourrey and Martin 2013). The fossil evidence of mycorrhizal fungi was observed for the first time in the most recent lineages of arbuscular mycorrhiza, Glomeromycota, 480–460 million years ago (Heckman et al. 2001; Parniske 2008) (Fig. 14.1). On the other hand, Heckman et al. (2001) noted that the results of their study on protein sequence analysis of

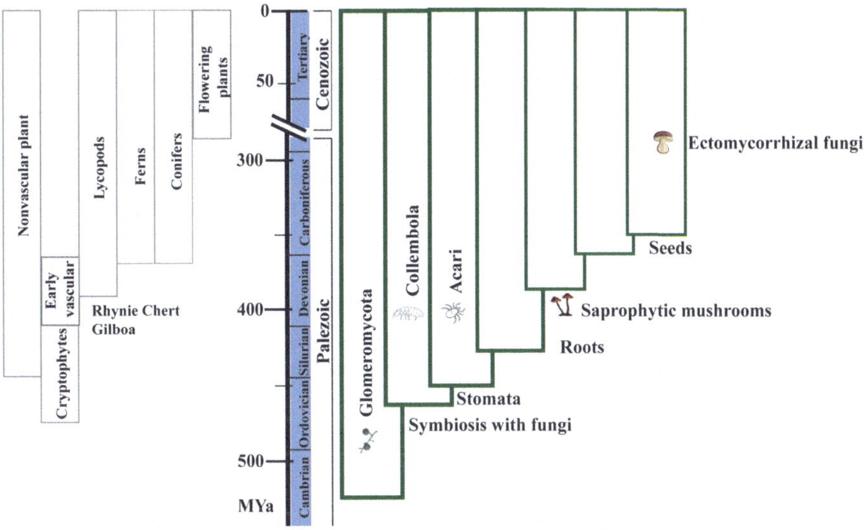

Fig. 14.1 Approximate ages in millions of years of collembolans, mites, and ectomycorrhizal fungi according to the fossil record and molecular phylogenies

orthologous fungi groups suggest that the main fungal lineages were already present 1 billion years ago.

Ectomycorrhizal fungi originated independently from arbuscular mycorrhiza and likely emerged repeatedly, in time and space, as an adaptive diversification alternative to the several lineages of arboreal taxa and the new ecological and environmental demands (Hibbett et al. 2000; Tedersoo and Smith 2013), e.g., in response to desiccation and nutrient deficiency threats associated with terrestrial existence (Pirozynski and Malloch 1975). Despite the first well-conserved fossils of ectomycorrhizal associations, pine-*Rhizopogon* and pine-*Suillus* dated back to only 50 million years during the Middle-Eocene (LePage et al. 1997), the origin of these fungi could have occurred much before, with the apparition of hosting plants with roots (Honrubia 2009). Bruns and Shefferson (2004) indicate two geological times, during the Cretaceous (130 million years ago, Fig. 14.1), when the fossil record for the Pinaceae and many angiosperm lineages first became common, and the Eocene–Oligocene transition, when the climate cooled and Northern Hemisphere forests became dominated by the latter plant groups such as Rosid clade. The discovery of Agaricomycetes from the Cretaceous period, with well-developed fruiting bodies and a complete and intact pileus with lamellae and peduncle, hints at the evolutionary stasis of the sporome at approximately 99 million years ago (Cai et al. 2017). Therefore, many of the morphological traits of fruiting bodies are highly adaptive and have evolved in response to their dispersal and new niches (Nagy et al. 2011).

According to the fossil records and feeding habits, mycophage or fungivore, of several species of Acari and Collembola, the association with fungal species (ectomycorrhizas, saprophytes, arbuscular, among others) could have been established more than 400 million years ago (Fig. 14.1). However, paleontological evidence of the mesofauna–sporome association is rather scarce, since no fruiting bodies of Basidiomycetes occurring before the early Cretaceous have been observed. A recent amber fossil find suggests that macromycetes have been used as a habitat for fungivore insects since the Albian stage, at approximately 100 million years ago (Schmidt et al. 2010). Furthermore, the use of the sporome as a food source is hinted at by the feeding preference of Staphylinidae beetles for fungi in the order Agaricales, as revealed by mid-Cretaceous Burmese amber (Cai et al. 2017). Hibbett et al. (1997) mentioned the presence of edaphic fauna—one mite and diverse insects—along with the fruiting body of a fossilized fungus in a piece of amber from the Cretaceous, whose species was *Archaeomarasmius leggetti* , a saprophytic fungus. However, they do not suggest any feeding preferences, and instead only mention that the edaphic fauna was used as an indicator of the potential habitat where the fungus was found. Therefore, intact sporomes preserved in amber shed light on the hypothesis that higher fungi, including the major groups, had already diversified by the Early Cretaceous (Cai et al. 2017).

Particulate mycophagia is an old feeding habit, and very likely the ancestral feeding manner of the first oribatid mites in the Devonian (Norton 1985). In contrast, Lindquist (1965, 1975) mentions that mycophagic species of mesostigmatan mites seem to have emerged repeatedly and independently from their predatory ancestors. For example, the species *Hoploseius tenuis* (Blattisociidae), which only lives in shelf fungi, modified its body by elongating it, thus allowing it to inhabit the pores in the hymenium, as well the apical section of the fixed chelicerae, which likely assists in feeding by enabling the mite to scrape the walls of the pores in the fungus to feed.

14.4 Ecology of the Interaction Between Mesofauna (Acari and Collembola) and Macromycetes

The sporomes, or fruiting bodies of macromycetes, of diverse sizes and consistencies (fleshy, leathery, corky, or woody) develop on the forest ground of temperate and tropical ecosystems in symbiosis with forest species (ectomycorrhizas), on wood, in decomposing organic matter (saprophytes), or as parasites in living tree trunks. The macromycetes include species of Basidiomycetes and Ascomycetes, with visible fruiting body structures (e.g. basidiomata); these fungi persist only for short periods or with longer life spans, such as that of the shelf or bracket fungi. On the other hand, the formation of erect fruiting bodies is associated with efficient spore dispersal, and according to their life cycle, they can be inhabited in a temporal or ephemeral manner by a vast diversity of micro-, meso-, and macrofauna.

The analysis of mesofauna–sporome interactions implies a link between the ecology of mites and springtails; on the other hand, the fruiting body is viewed as a

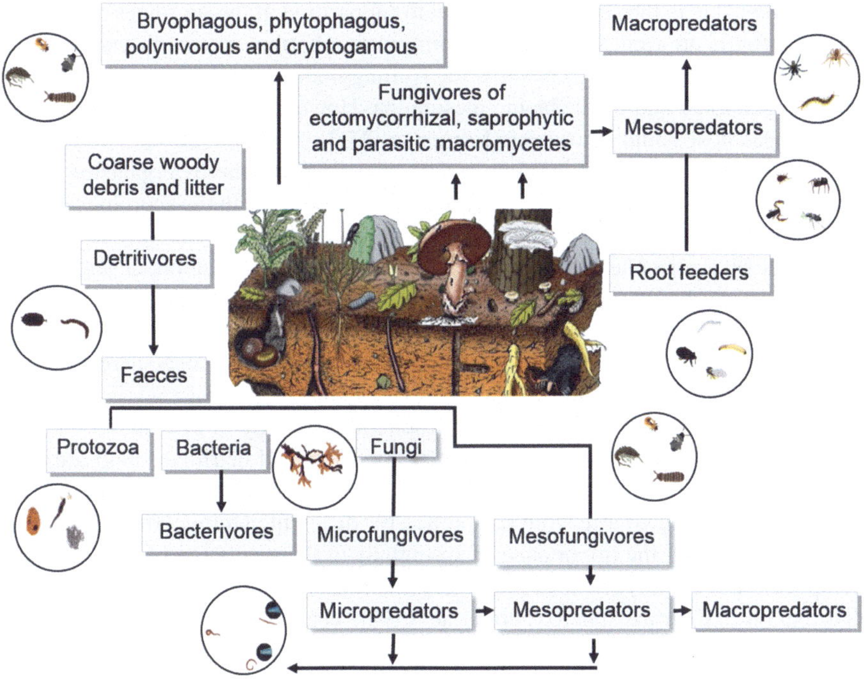

Fig. 14.2 Ecology of the edaphic fauna and their interaction with fungal sporomes: ectomycorrhizal, saprophytic, and parasitic macromycetes. Soil fauna: micro and mesofungivores (tylenchid nematodes, enchytraeids, collembolans, mites, beetles); micro and mesopredators (nematodes, mites, pseudoscorpions, staphylinid beetles, spiders); macro predators (carabid beetles, centipedes, ants, spiders); root feeders (tylenchid nematodes, symphylans, scarab beetles); detritivores (mites, collembolans, millipedes); protozoa (ciliates, amoebas); bacteria (nitrifying and phosphate-solubilizing bacteria); and bacterivores (ciliates, nematodes). On the sporome, in addition to mycophagy, interspecific relationships of predation, necrophagy, cannibalism, and parasitism may occur (based on the functional groups of Shaw et al. 1991)

reservoir of diversity and network of biological activity. In addition, the interactions between the sporome and mesofauna should not be viewed as an isolated process since they involve different edaphic organisms. Micro-, meso-, and macrofauna spend all or part of their life cycle in the soil, in both organic and mineral horizons, forming functional groups (Fig. 14.2). Due to their feeding preferences, the epedaphic and hemiedaphic fauna close to the fruiting bodies become grouped into micro- and mesofungivores, micro- and mesopredators, macropredators, rhizophages, and detritivores (Shaw et al. 1991); these groupings have a simplified the way of describing the different positions within the trophic chain that exists in the soil. However, these positions not only represent the links within the trophic chain but also reveal a complex network of services, collaboration, and competence, enabling the function of the ecosystem.

The study of edaphic fauna–sporome interactions has shown that some organisms such as those belonging to the micro-, meso-, and macrofauna do not spend all of their life cycle in the soil, but rather that the apparition of the sporomes during the rainy season is an event favouring the emigration of some species in the different functional groups (Fig. 14.2) towards the fruiting bodies of ectomycorrhizal fungi, mainly epedaphic and hemiedaphic mesofauna. However, the preference of a specific fungal host could also exist (Komonen 2003). This colonization can occur through the wind, vertical migration, and phoresy, in particular by organisms of the order Coleoptera (beetles) and Diptera (flies) that reproduce in the sporomes. Fungi with short life cycles, among which are included the edible ectomycorrhizal mushrooms (EEMs), are ephemeral resources; therefore, the colonization, development, and emigration should occur within a limited lapse of time, as happens in temperate climates during the warmest months with higher rain precipitation (Walter and Proctor 2013). Likewise, the fruiting bodies of a given fungal species share a similar edaphic fauna in a geographical area (Komonen 2003).

Jørgensen et al. (2003) mention that despite the diversity of available food (e.g. nematodes, faecal pellets of other animals, fine roots and detritus), fungi are presumably the main source of food for most Collembola species. Furthermore, collembolans have shown a feeding preference for different ectomycorrhizal species, thus demonstrating the importance of springtails in the regulation of diversity and composition of ectomycorrhizal fungal communities (Kanters et al. 2015) and their distribution in the rhizosphere (Hiol et al. 1994). The foraging effect on fungi and detritus influences essential processes in the ecosystems, such as the cycles of nitrogen and carbon (Filser 2002; Staaden et al. 2011).

Sitta and Süss (2012) mentioned that among the diverse interactions between fungi and edaphic fauna, mycophagy represents the highest impact to the trade and consumption of wild EEMs because certain insect orders and arthropod classes feed on the sporome of fresh or decomposing fungi, and these insects also spend all or part of their biological cycles on these sporomes (Hackman and Meinander 1979; Bruns 1984; Hanski 1989; Krivosheina 2008). The soil fauna associated with the fungal sporome is diverse, and it is mainly related with the morphological and organoleptic characteristics of EEMs and with the biological cycle of each macromycetes species, which highlights the importance of fruiting bodies in maintaining the biodiversity of the edaphic fauna (Komonen 2003).

14.5 Relation Between Acari and Collembola with the Sporomes of EEMs

The mycophagic habits of diverse species of mites and springtails have been reported by several authors. The first reports were of collembolans affecting cultures of fungi, e.g. Folsom (1933) mentioned five taxa considered as plague in the culture of fungi, and Thomas (1939) reported several species present in cultured and wild edible fungi. In temperate forests, many Acari and Collembola species can be found in shelf or bracket fungi (Graves 1960; Pielou and Verma 1968; Matthewman and Pielou 1971; Lindquist 1975, 1995; O'Connell and Bolger 1997a, b) and in the sporomes of ectomycorrhizal fungi (Palacios-Vargas and Gómez-Anaya 1994; O'Connell and Bolger 1997a; Nakamori and Suzuki 2005). However, several aspects are still unknown concerning the ecological relation between mites and springtails with the sporomes of EEMs, and these aspects require further research. This chapter presents recent findings in the mesofauna–sporome association and the feeding preferences that mites and springtails have for wild EEMs from forest zones with temperate and cold climates in the central region of the Neovolcanic Axis in Mexico.

14.5.1 Areas of Study

14.5.1.1 Forest Zones in the Volcanoes Iztaccíhuatl and Popocatépetl

The collection of EEMs was carried out in the area of influenceof the Iztaccíhuatl-Popocatépetl National Park, in September 2016. The volcanoes are located in the central part of the Sierra Nevada, between coordinates 18°59′ and 19°16′25″ N and 98°34′54″ and 98°16′25″ W, at elevations of 5500 masl for Popocatépetl and 5220 masl for Iztaccíhuatl. The local market of Ozumba is near this area, located in the municipality of Ozumba de Alzate, Estado de Mexico, where a great variety of wild EEMs from the forest zones of the volcanoes Iztaccíhuatl and Popocatépetl are traded (Pérez-Moreno et al. 2010). The obtained mushrooms were collected from different forest zones located in the foothills of the Iztaccíhuatl volcano, Paso de Cortés and San Pedro Nexapa, with the support of local wild mushroom traders. The type of forests from which the sporomes were collected is the temperate forests of pine-oak (ocotero and zacatonal, as they are known by the local mushroom traders).

14.5.1.2 Forest Zones in Mount Tláloc, San Pablo Ixayoc

Another collection of EEMs was done in the Mount Tláloc in the municipality of San Pablo Ixayoc, Texcoco, Estado de México, which is located within the Sierra Nevada mountain range. Mount Tláloc is situated in the western reach of the municipality of Texcoco, at an altitude of 4125 masl between parallels 19°23′43″ and 19°28′37″ N and meridians 98°42′51″ and 98°48′12″W. The ecosystems found in Mount Tláloc along the altitudinal gradient are oak (*Quercus rugosa* and *Quercus laurina*), fir (*Abies religiosa*), and pine (*Pinus hartwegii*) forests. The collections were mainly done in oak and pine forests.

14.5.2 Collection and Identification of Sporomes and Mesofauna

Fresh and completely developed sporomes of EEMs were collected in each forest area, and each specimen was photographed for later morphological identification (Lodge et al. 2004). The sporomes were sectioned at the base of the stipe and immediately deposited in a hermetic plastic bag; 96% alcohol was later added to preserve both the fungus and the mesofauna present in it. In the laboratory, the mesofauna was sifted from the hymenium using three different opening sizes, 0.850, 0.500, and 0.150 mm; the mesofauna was retained in the latter. Mites and springtails were simultaneously separated, identified to order and suborder levels, and their abundances estimated under a Zeizz Stemi 2000-C stereoscope. Some specimens were mounted for their correct identification, with observations being made of ectomycorrhizal spores in the gut contents and faecal pellets; previous to mounting, the springtails and mites were clarified with 85% lactic acid or lactophenol. The specimens were mounted on slides using Hoyer's medium before being dried in an oven at 45 °C (Palacios-Vargas and Mejía-Recamier 2007; Walter and Krantz 2009). Another mounting media used for mites (Oribatida and Mesostigmata) and springtails (Entomobryomorpha and Symphypleona) was the water-soluble synthetic resin DMHF (dimethyl hydantoin formaldehyde). The taxonomic identification of the fungal material was done through the techniques proposed by Largent et al. (1977), and the works of Lincoff (1981), Singer et al. (1990, 1991, 1992), Kirk et al. (2001), and Tulloss (1994), among others, were also consulted. The scientific nomenclature of the fungi was based on the Index Fungorum (www.indexfungorum.org). Once the fungi were identified, their abundance was estimated (number of sporomes per species).

14.5.3 Collembola–Ectomycorrhizal Sporome Interactions

Springtails are the most common visitors and, numerically, the biggest group that can be found in the lamellae of macromycetes, especially on the surface of the hymenium (Yamashita and Hijii 2003), where they feed on the hyphae and/or spores in an opportunistic manner (Mateos et al. 1996; Sawahata et al. 2000; Greenslade et al. 2002; Nakamori and Suzuki 2005). The main mycophagic springtails found were of the family Hypogastruridae (Greenslade et al. 2002; Yamashita and Hijii 2003; Castaño-Meneses et al. 2004; Nakamori and Suzuki 2005), which often can be found in almost all EEMs and represent 60–99% of all collembolans found in the analysed sporomes (Palacios-Vargas and Gómez-Anaya 1994; Mateos et al. 1996; Takahashi et al. 2005). Likewise, during a collection of different EEMs in the forest zones of the volcanoes Iztaccíhuatl and Popocatépetl, 7061 mites and springtails were found, of which the latter were the most abundant and pertained to the order Poduromorpha (93.8%); among these Poduromorpha, springtails of the family

Hypogastruridae, were predominant. Collembolans of the orders Entomobryomorpha and Symphypleona were found in low abundance, 0.93% and 0.47%, respectively.

Sawahata et al. (2002) note that rain precipitation stimulates springtails to emerge from the soil in search of food and to migrate vertically, which explains why they are more frequent and numerous in fruiting bodies during the rainy season. Palacios-Vargas and Gómez (1991) observed dense formations of springtails travelling in the direction of nearby fungi, climbing the stipe until reaching the hymenium, perhaps attracted by some bioactive compound or by the production of spores. Moreover, they noted that the number of springtails present in fungi is independent of the pileus length. In species of the genus *Boletus*, springtails penetrate from the base of the stipe or directly from the hymenophore. When many springtails present (hundreds of individuals), they can cause grave damage due to the construction of a great quantity of branched tunnels, which may reduce the fruiting body to a spongy pulp (Palumbo and Sitta 2007 referenced by Sitta and Süss 2012).

Unlike what was found in the forest zones of the volcanoes, in Mount Tláloc, we quantified a total of 434 specimens in the collected macromycetes; the highest collembolan abundance in the sporomes was shown for the order Entomobryomorpha (31.6%), whereas the lowest percentages were shown for the orders Symphypleona (13.1%) and Poduromorpha (11.8%). The difference between the number of individuals collected in the forest zones of the volcanoes Iztaccíhuatl and Popocatépetl and Mount Tláloc is possibly due to the end of the rainy season at the time of collection in the latter. However, the rainy period extended until late September in the forest zones of the volcanoes, which favoured the collection of a higher number of sporomes and the mesofauna present in them.

Mateos et al. (1996) found 5575 Collembolan specimens belonging to nine species in diverse wild Basidiomycetes mushrooms of Mediterranean climate, including EEMs. The species *Ceratophysella tergilobata* (Hypogastruridae) represented 99.8% of the total number specimens and was found in 94.8% of the examined basidiomata; therefore, this species is a commonplace inhabitant of the sporomes in the region. In Mexico, Palacios-Vargas and Gómez-Anaya (1994) presented a list of the mycetophilic collembolans identified in the different sporomes of macromycetes in the country, including some species of EEMs. In this regard, they mention that the species *Ceratophysella gibbosa* (Hypogastruridae) showed the highest abundance, and the macromycetes showing the highest abundance were *Polyporus* sp. (38.04%), *Russula* sp. (15.75%), and *Boletus* sp. (8.5%), whereas *Leccinum* sp., *Cortinarius* sp. and *Amanita* sp. showed the lowest abundance recorded (2.41% in each one of them). On the other hand, Sawahata (2006a) notes that springtails can consume from 1 to 92% of the hymenium area in Agaricales fungi, which is correlated with the density of collembolans in the fruiting body and the fungal species.

We quantified the abundance of springtails, by order, in 158 sporomes of EEMs, comprising 18 species, from the forest zones of the volcanoes Iztaccíhuatl and Popocatépetl. The highest mean abundances of Poduromorpha individuals per sporome was 143 in *Ramaria* aff. *flava*, 122 in *Laccaria bicolor*, 116 in *Ramaria rubrievanescens*, and 113 in *Lactarius deliciosus* (Fig. 14.3). The sporomes of species in the genera *Amanita*, *Morchella*, *Russula*, and *Hebeloma* showed the lowest

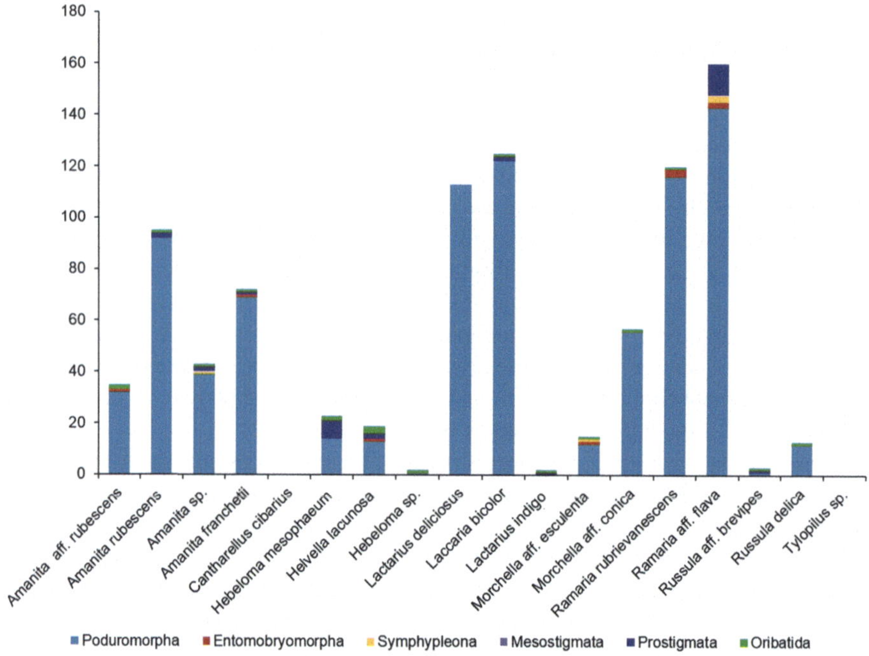

Fig. 14.3 Average abundance of Acari and Collembola, at the level of order and suborder, on the sporomes of edible ectomycorrhizal mushrooms collected in the forested areas of the Iztaccíhuatl and Popocatépetl volcanoes

abundance of Poduromorpha springtails, and in the fruiting bodies of *Cantharellus cibarius* and *Tylopilus* sp., no mesofauna was found.

In the collembolans showing a preference for feeding on fungi, Palacios-Vargas and Gómez (1991) suggested the following types of mouthparts: (1) chewing mouthparts, whose mandibles possess numerous molar surfaces, as observed in *Ceratophysella*; (2) no mandibles but showing thick and squared maxillaries used for breaking spores, such as in *Brachystomella*; and (3) mouthparts modified as stilettos within a buccal cone, used in the suction of liquids from fungal tissues (hyphae), such as in *Odontella*. However, most of the identified springtails belonged to the type of chewing mouthparts.

Some Collembola specimens, when observed under a microscope, showed gut contents composed by fungal material such as gill trama and spores, hyphae and spores, or only spores in different proportions (Fig. 14.4), or no gut contents were observed, which indicates that collembolans have preferences for different EEMs. However, the proportion of spores and hyphae in the gut content varies with the species of springtail (Greenslade et al. 2002) and fungus (Sawahata et al. 2000) in question. Further, Sawahata et al. (2002) mentioned that the spores of some fungi species are physically broken in the intestinal tract of certain springtail species, whereas the spores of other fungi species retain their original shape. In the case of the spores of

Fig. 14.4 Springtails feeding on spores of edible ectomycorrhizal mushrooms. (**a**) Poduromorpha, Hypogastruridae found on *Hebeloma mesophaeum*, (**b**) spores of *H. mesophaeum* in the gut contents, (**c**) *Russula delica*, (**d**) Poduromorpha found on *R. delica*, (**e**) spores of *R. delica* in the gut contents, (**f**) *Morchella* aff. *conica*, (**g**) Entomobryomorpha found on *M.* aff. *conica*, and (**h**) hyphae and spores of *M.* aff. *conica* in the gut contents

EEMs found in the gut contents or faecal pellets of the springtails observed in this study, they did not show any damage such as rupture or deformation (Fig. 14.4), which provides evidence of the potential that these collembolans have for spore dispersal (Lilleskov and Bruns 2005; Boddy and Jones 2008; Sitta and Süss 2012; Halbwachs and Bässler 2015).

14.5.4 Interaction of Acari with Sporomes of EEMs

Fungi with short life cycles, among these, the EEMs, are ephemeral resources; thus, the colonization and emigration of soil Acari must occur within a limited time. Therefore, the mite populations present in the sporome in forest zones with temperate and cold climates tend to be dominated by species with short generation times, such as Astigmata and Mesostigmata. Most Mesostigmata species found in shelf fungi are mycetophilic, whereas those found in macromycetes are predators; namely, they feed on other mites and nematodes inhabiting the sporomes. However, some species, such as *Epicroseius walteri*, *Ameroseius* sp. and *Asperolaelaps rotundus* in the family Ameroseiidae, are fungivores (Walter and Proctor 2013). In the forest zones of the volcanoes Iztaccíhuatl and Popocatépetl, we observed that the most abundant prostigmatan mites, especially from the species *Linopodes* sp. (Eupodidae, Prostigmata), had some affinity for particular fungi species or growth forms, such as the sporomes of *Helvella lacunosa*, since they feature ribbed stipes and cavernous holes through which they ran and hid at great speed. Similarly, Okabe and Amano (1993) mention that Prostigmata of the family Eupodidae were found in most sporomes of different fungal species in the central area of Kanto, Japan.

In temperate and boreal forests, the acarofauna present in shelf fungi is dominated by oribatid mites (Pielou and Verma 1968; Matthewman and Pielou 1971; O'Connell and Bolger 1997a, b). Shelf fungi, both annual and perennial, possess a more diverse fauna of oribatid mites and relatively long life cycles, longer than Agaricales and Boletales (O'Connell and Bolger 1997b). On the contrary, the interaction between oribatid mites and EEMs has been mainly observed in feeding preference experiments in the laboratory (Schneider et al. 2005), but not in the field (Sawahata 2006a). The presence of oribatid mites has been reported only on *Sarcodon scabrosus* and *Phellodon melaleucus* (Telepholaceae) fruit bodies in a beech forest (Sawahata 2006b).

In the forests of Mount Tláloc, a total of 96 edible EEMs were collected, belonging to 22 species, and these contained 434 mesofauna specimens. In contrast with those mites previously found in the forests zones of volcanoes Iztaccíhuatl and Popocatépetl, mites were found in higher proportion, with the most abundant being Oribatida (33%), whereas Prostigmata (9.90%) and Mesostigmata (0.69%) were less abundant. Overall, the abundance of mesofauna present in the sporomes of the EEMs collected in Mount Tláloc was very low, and it was only in the sporomes of *Morchella* sp., *Russula delica*, and *Morchella* aff. *conica* that diverse individuals of different mite and springtail orders and suborders were observed (Fig. 14.5). As previously mentioned, the end of the rainy season could have been an influencing factor on the abundance of mesofauna. Therefore, several collection trips during the rainy season would be adequate to detect any variations in abundance and diversity of mesofauna present in the sporomes of wild EEMs.

Oribatid mites display different feeding preferences among ectomycorrhizal species (Schneider et al. 2005), and many of these oribatid mites can be 'choosy generalists', which ordinarily possess a wide feeding spectrum but feed selectively instead

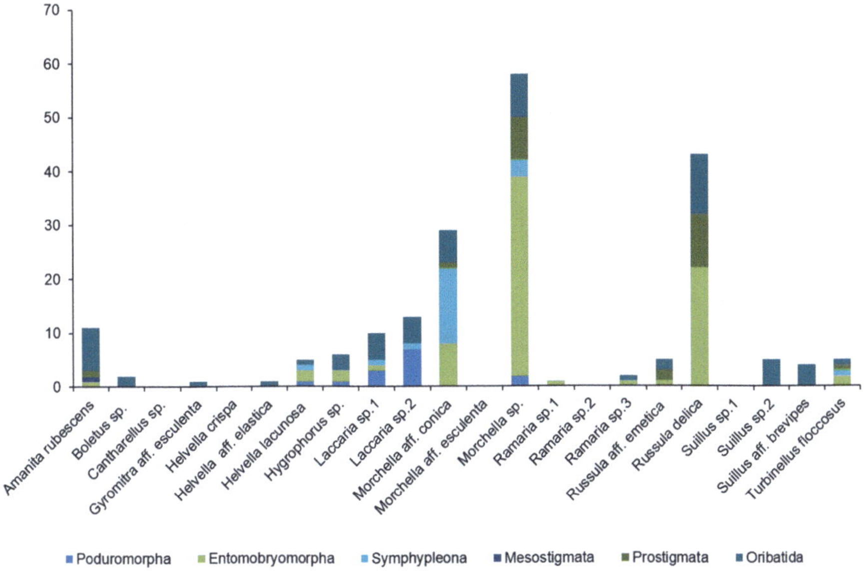

Fig. 14.5 Average abundance of Acari and Collembola, at the level of order and suborder, on the sporomes of edible ectomycorrhizal mushrooms collected in the forested areas of the Mount Tláloc, San Pablo Ixayoc

when high-quality food becomes available (Schneider and Maraun 2005). On the other hand, the presence of spores and trama found in the pellets of some oribatid mites could suggest feeding preferences of EEMs; for example, spores from *Ramaria rubrievanescens* and *Helvella lacunosa* were found in the pellets of oribatid mites (Fig. 14.6). Notably, many oribatid and prostigmata mites (*Linopodes* sp., Eupodidae) were observed carrying many eggs, thus suggesting that having high nutritional quality sporomes as a feeding source enable the good development of the offspring (Fig. 14.7). However, the nutritional value of the spores for fungivorous mites also needs to be addressed (Okabe 1999). In addition, there were several mite morphospecies, of which immature stages were collected from the EEMs, suggesting that mites permanently inhabit the sporomes (Okabe 2013).

The diverse groups of springtails and mites observed include species that regularly display mycophagic or fungivore behaviour, as suggested by their feeding on trama, spores, and hyphae of EEMs, which is consistent with the reports of Hopkin (1997), Brand and Dunn (1998), and Rusek (1998). The mesofauna present in the sporomes of edible EEMs fulfils functions as spore dispersers, fruiting body decomposers, and/or regulators of populations of other microarthropods cohabitating in this microhabitat. However, we cannot exclude the possibility that the presence of soil mites and mainly springtails in the sporomes is due only to the shelter found while they were being hosted in the lamellae of the hymenium. It is still unknown how a large number of springtails can coexist with no clear and obvious evidence of

Fig. 14.6 Oribatid mites (Acari, Oribatida) feeding on spores of edible ectomycorrhizal mushrooms. (**a**) *Ramaria rubrievanescens*, (**b**) Oribatida found on *R. rubrievanescens*, (**c**) spores of *R. rubrievanescens* in the pellet, (**d**) *Helvella lacunosa*, (**e**) Oribatida found on *H. lacunosa*, and (**f**) spores of *H. lacunosa* in the pellet

feeding preferences because spores or trama were not observed in the gut contents of many collembolans on examined sporomes.

14.5.5 Spore Dispersal Mechanisms

The spore dispersal pattern of fungi can affect the genetic flux, population structure, and fungal communities. Several studies have demonstrated the potential of fungivore invertebrates, including oribatids, collembolans, dipterans, beetles, diplopods, and their respective predators to disperse spores of macromycetes (Lilleskov and Bruns 2005; Boddy and Jones 2008; Halbwachs and Bässler 2015) through

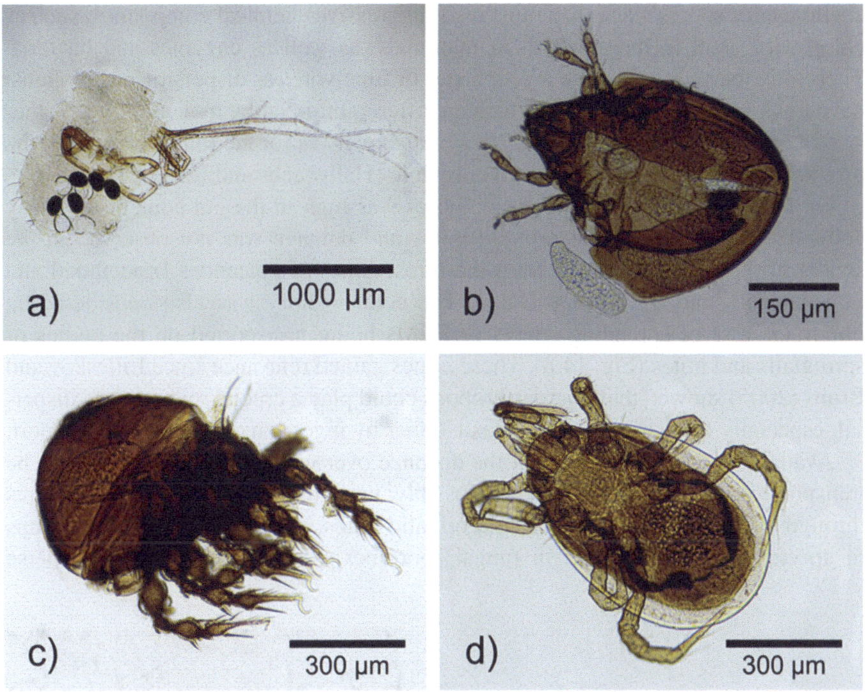

Fig. 14.7 Eggs contained in mite bodies inhabiting on different edible ectomycorrhizal mushrooms. (**a**) Prostigmatan mite, Eupodidae, on *Helvella lacunosa*, (**b**) Oribatid mite on *Ramaria* sp., (**c**) Oribatid mite on *Helvella lacunosa*, and (**d**) Mesostigmatan mite on *Russula* sp.

endozoochory (dispersal by consumption) and/or ectozoochory (dispersal through body adhesion). Mycophagy can increase spore dispersal during the reproductive stage of sporomes (Lilleskov and Bruns 2005; Boddy and Jones 2008; Sitta and Süss 2012). However, the mycophagic habit of soil microarthropods can also damage young sporomes and, consequently, result in a diminished spore production (Sitta and Süss 2012).

Diverse basidiomycetes produce resupinate sporocarps, lacking stipe and developing homogenously upon the substrate, and though spore dispersal occurs actively, spores often do not seem well positioned for aerial dispersion. In this regard, Lilleskov and Bruns (2005), in a study with the ectomycorrhizal species *Tomentella sublilacina*, stained the faeces of fungivores and their predators with DAPI (4′,6-diamidino-2-phenylindole), observing that 7–73% of the spores featured intact nuclei in fungivores, whereas in predators, this percentage was lower, thus suggesting the spore dispersal potential of *T. sublilacina* through microarthropods and their predators in the soil trophic chain. Maraun et al. (1998) note that several soil microarthropods ingest, but do not digest, the spores of EEMs, thus representing an important dispersal mechanism in forest ecosystems after any disturbance. On the contrary, the passage through the gut contents of vertebrates is a tremendous challenge for spores since they must resist aggressive chemical compounds, such as chlorhydric acid in the stomach of mammals, as well as enzymes and bacteria. Therefore, the spores that use mycophages or fungivores as dispersion vectors must have thick walls for protection, which are often reinforced by melanin. The synthesis of melanin is complex (Solano 2014) and expensive for the fungus; however, the provided protection benefits seem worthwhile (Halbwachs and Bässler 2015).

On the other hand, the amount of spores was high in the gut contents of many collembolans and pellets of oribatid mites and damage was not observed in the spores of the EEMs collected from the forest zones of volcanoes Iztaccíhuatl and Popocatépetl, and the Mount Tláloc. The ectozoochory was also considered an important way of spreading spores of EEMs being transported on the bodies of springtails and mites (Fig. 14.8). These issues gained relevance since Lilleskov and Bruns (2005) showed that microarthropods could play a crucial role in spore dispersal, especially for ectomycorrhizal taxa, either by ingestion or by passive transport.

Available evidence is scarce of the distance over which spores of EEMs can be transported by microarthropods, but possibly, these distances reach tens of metres through wind or phoresy. On the soil, oribatid mites, one of the less mobile groups of species commonly found in fungal sporomes, have the potential to disperse

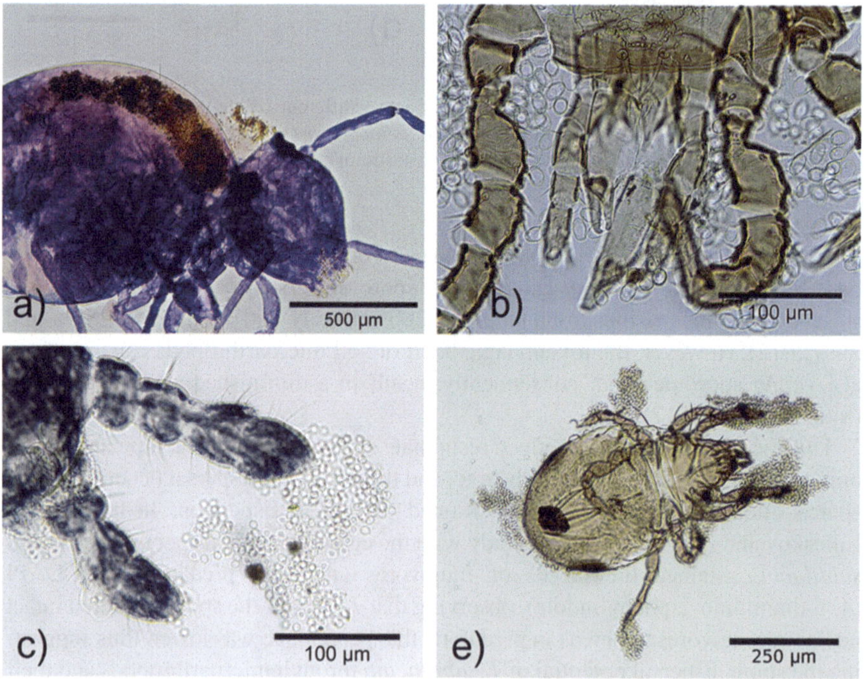

Fig. 14.8 Dispersion of spores of edible ectomycorrhizal mushrooms by ingestion: (**a**) Symphypleona, Collembola, with spores of *Cortinarius* sp. in the gut contents, and by adhesion to the body, (**b**) Zerconidae mite, Mesostigmata, with spores of *Helvella lacunosa*, (**c**) Poduromorpha, Collembola, with spores of *Russula delica*, and (**d**) Oribatid mite with spores of *Russula delica*

spores for a greater distance than expected (Lilleskov and Bruns 2005). Behan and Hill (1978) mentioned that the estimated linear distance for oribatid mites is of <20.5 cm day^{-1}, and lower rates ($\leq$20 cm week^{-1}). On the contrary, the highest potential rate for springtails is of 20 cm week^{-1} and the poor dispersants only 0.5 and 1.2 cm week^{-1} (Ojala and Huhta 2001). For passive dispersers, spore ornamentation is greatly useful for dispersal since these adhere to the setae or exoskeleton of mites or springtails, and spore transport at some distance from the source is significant even if it is small or through vast distances, in the soil. Additionally, Lehmitz et al. (2011) have mentioned that oribatid mites can be dispersed by wind up to 160 m above ground level and that 90% of the wind-dispersed oribatid mites generally live in forest habitats, which is considered an important factor in the colonization of habitats by these species (Karasawa et al. 2005; Lindo and Winchester 2008; Lehmitz et al. 2011), including colonization of the sporomes of EEMs and spore dispersal. However, Calhim et al. (2018) indicated that the challenges affecting success in spore establishment and the chance of reaching a suitable microhabitat for germination may be more important for successfull dispersal that challenges related to spore transportation.

14.5.6 Presence of Bioactive Compounds in Sporomes of EEMs in Response to Mesofauna Invasion

The insect–fungus relation is based on chemical or olfactory signals, which is why mycophagic insects and spore dispersers prefer those sporomes containing substances with appealing smell and taste, whereas those containing insecticides and toxic repellents are less prone to be attacked by arthropods (Martin 1979; Mier et al. 1996; Wang et al. 2002; Staaden et al. 2011). The nature of prey–predator interactions, namely, the consumption of an organism by another, is essential in understanding the dynamics of the populations, community structure and its diversity (Miner et al. 2005). Herbivores form a vast functional class of predators that consume but do not kill their prey, since they only eliminate parts of it, which is rarely lethal in the short term. Despite the relevance of plant–herbivore interactions in the ecosystems, a wide diversity of vertebrates, molluscs, nematodes, arthropods, and microarthropods extract essential nutrients from fungal organisms by facultative or obligatory foraging (Döll et al. 2013).

In a manner similar to plants, fungi are non-motile organisms unable to escape this type of attack by predators. As a response, plants have adapted to attack by herbivores by developing different chemical defences, such as secondary metabolites (Mithöfer and Boland 2012). The toxic secondary metabolites produced by fungi could also mediate resistance against fungivores (Spiteller 2008; Rohlfs and Churchill 2011). However, no evidence has been discovered on whether fungi are capable of changing the composition of their secondary metabolites in response to grazing by fungivorous microarthropods (Döll et al. 2013). A common response for how eukaryotic organisms become injured is the activation of hydrolases and lipoxygenases that drive the lipid peroxidation producing a variety of degradation products. After a lesion occurs in fungi, linoleic acid becomes oxidized by a typical mushroom lipoxygenase and degrades through the intermediate acid (8E,12Z,10*S*)-10-hydroperoxi-8,12-octadecadienoic into (3R)-1-octen-3-ol and (8E)-10-oxo-8-decenoic acid. The (3R)-1-octen-3-ol acts as a chemical defence against parasites and predators due to odour intensification, and it is present in most fungi (Spiteller 2008).

Fungi are a potential source of bioactive compounds for phytoprotection that may be constituted by organic compounds with insecticidal properties or genes encoding toxic proteins. Several fungi species, such as those in genera *Lepista* or *Cantharellus*, are not consumed by insects, and when the sporome is attacked, only a few species are involved (Bruns 1984; Wang et al. 2002). This corroborates the observation in the samples of *Cantharellus cibarius* s.l. collected in the forest zones of volcanoes Iztaccíhuatl and Popocatépetl, and *Cantharellu*s sp. in the Mount Tláloc, where no mesofauna was found in the sporomes of the EEMs (Fig. 14.3). Danell (1994) note that *C. cibarius* is not attractive to 'predators and parasites', and mentioned that selective forces probably favour the formation of repellent compounds. Because the abundance of mites and springtails in the fruiting bodies was null, this can be interpreted as a true absence (non-existent in the sporome). Importantly, even other insects, especially fly larvae, which are commonly found (Krivosheina 2008), were absent in the sporome of *C. cibarius*. Likewise, the fresh sporome of *C. cibarius* was rarely attacked by microarthropods.

The defence mechanism of *Cantharellus* has not been completely proven yet; however, the potential natural repellent or insecticide effects may be attributed to acetylenic lipids or to oxylipins (Blacklock et al. 2010). Several authors mention that fresh *C. cibarius* sporomes can produce cibaric acid (9Z,13Z,15E)-14,18-dihydroxi-12-keto-9,13,15octadecatrienoic acid) in quantities as high as 100 mg kg^{-1} of fresh weight, particularly in response to injury. Unlike the potent defensive sesquiterpenes formed in injured fruit bodies of *Lactarius* species, cibaric acid possesses very weak antimicrobial and cytotoxic activity (Pang and Sterner 1991; Pang et al. 1992; Anke et al. 1996; Gry and Andersson 2014), being not toxic to human consumption but not especially tasty to springtails and mites (Fig. 14.9).

Guevara and Dirzo (1999) note the probable existence of a natural hierarchy in the preference or palatability of certain fungi species. Martin (1979) suggested that the fungus–insect interaction is based on chemical signals; therefore, mycophagic and spore disperser arthropods prefer good-tasting sporomes, whereas those that contain insecticidal substances and toxic repellents are less vulnerable to attack by arthropods (Mier et al. 1996; Wang et al. 2002). Staaden et al. (2011) mentioned that springtails are capable of perceiving olfactory signals from different fungi species, including EEMs, which allows them to turn away from highly toxic fungi. Moreover, they are also capable of detecting and responding with less preference for fungi that have been previously foraged by other springtails (Döll et al. 2013).

Mier et al. (1996) elaborated extracts from dry sporomes of edible and toxic fungi to identify possible bioactive compounds with insecticidal properties. Of the

Fig. 14.9 Ectomycorrhizal mushrooms with different bioactive compounds. (**a**) *Cantharellus cibarius* contains cibaric acid (9Z,13Z,15E)-14,18-dihydroxy-12-keto-9,13,15octadecatrienoic acid), non-toxic to humans, (**b**) *Cortinarius* sp. contains orellanine, (2,2′-bipyridine)-3,3′,4,4′-tetrol-1,1′-dioxide, toxic to man, (**c**) Entomobryomorpha, and (**e**) Symphypleona with spores of *Cortinarius* sp. in their gut contents (**d** and **f**)

175 Ascomycetes and Basidiomycetes species evaluated, 79 inhibited insect development and some of the more toxic extracts originated from EEMs, such as *Boletus badius*, *B. luridus*, *B. edulis*, *B. erythropus*, *Cantharellus tubaeformis*, *Clavulina cinerea*, *Cortinarius purpurascens*, *Hygrophorus chrysodon*, *H. niveus*, *Suillus bovinus*, *S. subtomentosus*, *Tricholoma equestre*, *T. saponaceum*, *T. sejuctum*, *T. sulphureum*, and *Xerocomellus chrysodon*. Later, Wang et al. (2002) suggested that

proteins, lectins, and haemolysins are responsible for most insecticidal activity in fungal sporomes. Most likely, the more known insecticidal properties of wild ectomycorrhizal mushrooms are those from *Amanita muscaria*, where the 'ibotenic acid' is the bioactive compound attributed with insecticidal properties (Takemoto et al. 1964). Likewise, the ether extract of *Ramaria eryuanensis*, identified as ergosta-7,22-dien-3beta, 5a,6 beta-triol, showed insecticidal activity against diamondback moth larvae (*Plutella xylostella*) (Dahao et al. 2010). However, this does not imply that the compounds are toxic or unsavoury for all mycophagic arthropods.

Although wild ectomycorrhizal mushrooms of the genus *Cortinarius* are mostly considered toxic and orellanine, (2,2′-bipyridine)-3,3′, 4,4′-tetrol-1,1′-dioxyde, is the toxin present in several species of Cortinariaceae (Oubrahim et al. 1997), the collembolans of orders Poduromorpha, Entomobryomorpha, and Symphypleona do not seem to find it unpleasant and feed on these fungi. In the forest zone of Mount Tláloc, 11 sporomes of the genus *Cortinarius* sp. were collected and evaluated, with many spores being found in the gut contents of springtails, thus suggesting that the noted mushrooms are a highly nutritious food source (Figs. 14.8a and 14.9) despite possessing toxic compounds or micromorphological structures used as a defence against microarthropod foraging. On the other hand, Nakamori and Suzuki (2005) evaluated the preference of three collembolans species in the family Hypogastruridae for sporomes of *Cortinarius salor*, *Lactarius quietus*, and *Russula emetica*, observing that each collembolan species had a preference for certain fungal species. The latter could probably be due to the chemical (bioactive compounds) and/or micro- and macromorphological characteristics of EEMs. However, this suggestion does not mean that these characteristics can be applied to all mycophagic arthropods and that the perception of highly nutritious food sources is important in edaphic microarthropods, since these resources are scarce and, for a limited time, are heterogeneously distributed through the soil landscape (Bengtsson et al. 1991).

14.5.7 Impact of Collembola and Acari in the Trade of Wild EEMs

The collection of wild edible fungi has great economic potential that, if handled correctly, could represent a real alternative in the sustainable management of forests and generate important environmental and social benefits. Furthermore, if the commercialization of these species is to proceed to the international market, a series of issues, among them, the presence of edaphic fauna in the sporome, must first be considered. Worldwide, great demand exists for the consumption of wild fungi in a fresh, dry, or pickled form. However, along with the fungi, larvae from fungivores, springtails, mites, and other dead arthropods are also consumed, and though there is no evidence of harmful side effects, it could create the sensation of disgust in the consumers (Yen 2009).

Due to the existence of a 'palatable ranking' of the different fungi species by fungivore insects, including microarthropods, some species of international economic importance, such as *Boletus edulis* s.l., *Suillus luteus* s.l., *Tricholoma matsutake* s.l., *Amanita*sect. *Caesarea Lactarius* sect. *Dapetes*, *Russula* spp., and *Tuber* spp., are predisposed to be colonized by soil microarthropods to some degree (Sitta and Süss 2012). In addition, the diverse and previously mentioned EEMs have shown great economic importance for their demand as food or nutraceutics in the international market (Pilz and Molina 1996).

In the forest zones of temperate and cold climates in the central region of the Neovolcanic Axis in Mexico, EEMs exist with the potential to be used as a non-timber forest product that can be exported into the international market, which could be the most valuable product for sustainable forests. The international trade of wild edible fungi is valued in billions of dollars annually. One of the reasons for such an elevated cost is that most of these fungi cannot be cultured (Yun and Hall 2004) and are of great interest in the gourmet cuisine of several European and North-American countries (Karwa et al. 2011). Among the species with economic potential are *Amanita* sect. *Caesarea* and *Cantharellus cibarius* s.l., the latter has an estimated annual value of $1.67 billion USD, according to the retail market. The main countries currently demanding these species are Canada, France, Italy, Spain, the United States, China, and Germany (Watling 1997; Hall et al. 2003; Arora and Dunham 2008).

The fresh sporomes of *C. cibarius* and kindred species are rarely attacked by microarthropods, which represents an advantage in their commercialization. This could be due to the presence of insecticidal or repellent compounds in the sporome. Danell (1994) mentioned that, despite the slow development of the sporome, less than 1% of *Cantharellus* is infested by Diptera larvae in comparison with most Agaricales and Boletales species, which show an infestation rate of 40–80%. The mycophages reported in the literature are mainly polyphagic; these include dipteran larvae in the family Limoniidae, followed with less frequency by some species of *Suillia* and *Drosophila* (Krivosheina 2008).

Similarly, in the study region, some other species were observed with less frequency, possibly because the rainy season had ended; however, these species may have the potential to be exploited, as is the case of with Boletaceae. Collembolans are often found in the sporomes of the genus *Boletus*, occasionally in great number, although not as many as in fungi with lamellar hymenium. Furthermore, the percentage of sporomes attacked varies considerably among the *Boletus* species, possibly due to an environmental effect and the growing season; for example, *B. edulis* is less frequently attacked than *B. reticulatus* (Sitta and Süss 2012).

The aptitude of wild EEMs for human consumption will depend on the number of arthropods incorporated in a standard weight sample. The Food and Drug Administration (FDA 1995) performs a 'waste or dirt' test in the tolerance of certain species of 'maggots' (no more than 20 individuals or not more than 5 when measuring more than 2 mm in length) and for 'mites' (no more than 75 individuals) in processed or dehydrated samples. The impact of insects smaller than 1–2 mm (e.g., mites, springtails, Diptera eggs, and first instar larvae) is irrelevant (Sitta and Süss

2012). Therefore, the sporomes lacking edaphic mesofauna, such as *Cantharellus cibarius* s.l. and *Amanita* sect. *Caesarea*, are the species with highest potential for use and commercialization, because they could meet the exportation sanitary norms.

14.6 Conclusions

The presence of the mesofauna in sporomes of wild EEMs hints at ecological interactions occurring in the soil that are usually barely known. The fruiting bodies have tremendous importance in the activity of the edaphic fauna, since they show the trophic and habitat preferences of mites and springtails. Therefore, the role that macromycetes play on the forest soil during the rainy season is that of a great quality food source and of a habitat used in the completion of their life cycle. Mesofauna–sporome interactions are a consequence of the biological diversity present in a determined time and space due to the short life period of the sporome. The importance and complexity of these ecological relations is awakening increasing interest among mycologists and soil ecologists because the ecology of the soil is a research field of extreme importance for the future. Therefore, a coordinated global effort is needed to address the more important issues affecting soil sustainability, as well as filling the empty spaces in the knowledge of soil ecology. Despite their functional relevance, the interactions between soil organisms are barely known in a country with such macromycetes megadiversity as Mexico; therefore, it is necessary to know the mesofauna–sporome interactions to understand the multiple ecosystem services existent in the forest zones of the Neovolcanic Axis in Mexico.

Acknowledgements Technical support from the students Marbella Beltrán Rentería y Lizbeth Álvarez Ramos from the Instituto Tecnológico de Ciudad Altamirano, Mexico and from MSc Jorge M. Valdés Carrasco from Colegio de Postgraduados, especially for valuable suggestions and improvements in microscopy, as well as Mexico is deeply acknowledged. JPM acknowledges the financial support from the CONACyT 2018-000007-01EXTV and COMECyT for the development of an international stay at Kunming Institute of Botany, Chinese Academy of Sciences in Kunming, China. Funding from the CONACyT project "Biotechnologies of highly-prized or bioculturally important edible ectomycorrhizal mushrooms for reforestation with native Mexican trees using frontier scientific methodologies" is also acknowledged.

References

Anke H, Morales P, Sterner O (1996) Assays of the biological activities of two fatty acid derivatives formed in the edible mushrooms *Cantharellus cibarius* and *C. tubaeformis* as a response to injury. Planta Med 62(2):181–183. https://doi.org/10.1055/s-2006-957850

Arora D, Dunham SM (2008) A new, commercially valuable chanterelle species, *Cantharellus californicus* sp. nov., associated with live oak in California, USA. Econ Bot 62(3):376–391. https://doi.org/10.1007/s12231-008-9042-7

Athias-Binche F (1994) La phorésie chez les acariens – aspects adaptatifs et évolutifs [Phoresy in acarina – adaptive and evolutionary aspects]. Editions du Castillet, Perpignan

Ávila-Jiménez ML, Coulson SJ (2011) A holarctic biogeographical analysis of the Collembola (Arthropoda, Hexapoda) unravels recent post-glacial colonization patterns. Insects 2(3):273–296. https://doi.org/10.3390/insects2030273

Baquero E, Jordana R (2015) Clase collembola. órdenes poduromorpha, entomobryomorpha, neelipleona y symphypleona. Revista IDE@–SEA 36:1–11

Behan VM, Hill SB (1978) Feeding habits and spore dispersal of oribatid mites in the North American Arctic. Rev Ecol Biol Sol 15(4):497–516

Bellinger PF, Christiansen KA, Janssens F (1996–2018) Checklist of the Collembola of the World. http://www.collembola.org. Accessed December 2018

Bengtsson G, Hedlund K, Rundgren S (1991) Selective odor perception in the soil Collembola *Onychiurus armatus*. J Chem Ecol 17(11):2113–2125. https://doi.org/10.1007/BF00987995

Binns ES (1982) Phoresy as migration–some functional aspects of phoresy in mites. Biol Rev 57(4):571–620. https://doi.org/10.1111/j.1469-185X.1982.tb00374.x

Blacklock BJ, Scheffler BE, Shepard MR, Jayasuriya N, Minto RE (2010) Functional diversity in fungal fatty acid synthesis. The first acetylenase from the Pacific golden chanterelle, *Cantharellus formosus*. J Biol Chem 285(37):28442–28449. https://doi.org/10.1074/jbc.M110.151498

Boddy L, Jones TH (2008) Interactions between Basidiomycota and invertebrates. In: Boddy L, Frankland JC, van West P (eds) Ecology of saprotrophic basidiomycetes. Academic, London, pp 155–179

Brand RH, Dunn CP (1998) Diversity and abundance of springtails (Insecta: Collembola) in native and restored tallgrass prairies. Am Midl Nat 139(2):235–242. https://doi.org/10.1674/0003-0031(1998)139[0235:DAAOSI]2.0.CO;2

Bruns TD (1984) Insect mycophagy in the Boletales: fungivore diversity and the mushroom habitat. In: Wheeler Q, Blackwell M (eds) Fungus-insect relationships: perspectives in ecology and evolution. Columbia University Press, New York, pp 91–129

Bruns TD, Shefferson RP (2004) Evolutionary studies of ectomycorrhizal fungi: recent advances and future directions. Can J Bot 82(8):1122–1132. https://doi.org/10.1139/b04-021

Cai C, Leschen RAB, Hibbett DS, Xia F, Huang D (2017) Mycophagous rove beetles highlight diverse mushrooms in the cretaceous. Nat Commun. https://doi.org/10.1038/ncomms14894

Calhim S, Halme P, Petersen JH, Læssøe T, Bässler C, Heilmann-Clausen J (2018) Fungal spore diversity reflects substrate-specific deposition challenges. Sci Rep. https://doi.org/10.1038/s41598-018-23292-8

Castaño-Meneses G, Palacios-Vargas JG, Cutz-Pool LQ (2004) Feeding habits of Collembola and their ecological niche. An Inst Biol Ser Zool 75(1):135–142

Christiansen K (1964) Bionomics of Collembola. Annu Rev Entomol 9(1):147–178. https://doi.org/10.1146/annurev.en.09.010164.001051

Cruz-Leal JI, Palacios-Vargas JG, Montejo-Cruz M (2016) Importancia nacional e internacional de la colección de Collembola (Hexapoda) de la Facultad de Ciencias, UNAM. Bol Soc Mex Ento (ns) 2:75–80

Cutz-Pool LQ, Villegas-Guzmán GA, Palacios-Vargas JG (2007) Colémbolos (Hexapoda: Collembola) asociados a nidos de *Neotoma palatina* (Rodentia: Muridae) en Zacatecas. México Brenesia 67:83–86

Dahao W, Jinming G, Juntao F, Xing Z (2010) Insecticidal activity of chemical components from *Ramaria eryuanensis*. Plant Prot 36(5):110–112

Danell E (1994) *Cantharellus cibarius*: mycorrhiza formation and ecology. Dissertation, University of Uppsala

Deharveng L (2004) Recent advances in Collembola systematics. Pedobiologia 48(5–6):415–433. https://doi.org/10.1016/j.pedobi.2004.08.001

Deharveng L, D'Haese CA, Bedos A (2008) Global diversity of springtails (Collembola; Hexapoda) in freshwater. Hydrobiologia 595:329–338. https://doi.org/10.1007/978-1-4020-8259-7_36

Dell'Ampio E, Meusemann K, Szucsich NU, Peters RS, Meyer B, Borner J et al (2014) Decisive data sets in phylogenomics: lessons from studies on the phylogenetic relationships of primarily wingless insects. Mol Biol Evol 31(1):239–249. https://doi.org/10.1093/molbev/mst196

Derst C, Dircksen H, Meusemann K, Zhou X, Liu S, Predel R (2016) Evolution of neuropeptides in non-pterygote hexapods. BMC Evol Biol. https://doi.org/10.1186/s12862-016-0621-4

Döll K, Chatterjee S, Scheu S, Karlovsky P, Rohlfs M (2013) Fungal metabolic plasticity and sexual development mediate induced resistance to arthropod fungivory. Proc R Soc B 280(1771):20131219. https://doi.org/10.1098/rspb.2013.1219

Dubinin VB (1991) Class acaromorpha. Ticks, mites or gnathosomous chelicerates. In: Rohdendorf BB, Davis DR (eds) Fundamentals of palaeontology. Arthropoda, tracheata, chelicerata, vol 9. Smithsonian Institution Libraries, The National Science Foundation Washington, DC, Amerind Publishing, New Delhi, pp 681–722

Eisenbeis G (1982) Physiological absorption of liquid water by Collembola: absorption by the ventral tube at different salinities. J Insect Physiol 28(1):11–20. https://doi.org/10.1016/0022-1910(82)90017-8

Engel MS, Grimaldi DA (2004) New light shed on the oldest insect. Nature 427(6975):627–630. https://doi.org/10.1038/nature02291

Faddeeva A, Studer RA, Kraaijeveld K, Sie D, Ylstra B, Mariën J et al (2015) Collembolan transcriptomes highlight molecular evolution of hexapods and provide clues on the adaptation to terrestrial life. PLoS One 10(6):e0130600. https://doi.org/10.1371/journal.pone.0130600

FDA (1995) Defect levels handbook. The food defect action levels. Levels of natural or unavoidable defects in foods that present no health hazards for humans. http://www.fda.gov/Food/GuidanceRegulation/GuidanceDocumentsRegulatoryInformation/SanitationTransportation/ucm056174.htm. Accessed 28 Nov 2017

Field KJ, Pressel S, Duckett JG, Rimington WR, Bidartondo MI (2015) Symbiotic options for the conquest of land. Trends Ecol Evol 30(8):477–486. https://doi.org/10.1016/j.tree.2015.05.007

Filser J (2002) The role of Collembola in carbon and nitrogen cycling in soil. Pedobiologia 46(3–4):234–245. https://doi.org/10.1078/0031-4056-00130

Fletcher JT, Gaze RH (2008) Mushroom pest and disease control. A colour handbook. Manson Publishing, London

Folsom JW (1933) The economic importance of Collembola. J Econ Entomol 26(5):934–939. https://doi.org/10.1093/jee/26.5.934

Fragoso C, Reyes-Castillo P, Rojas P (2001) La importancia de la biota edáfica en México. Acta Zool Mex (ns) 1:1–10

Graves RC (1960) Ecological observations on the insects and other inhabitants of woody shelf fungi (Basidiomycetes: Polyporaceae) in the Chicago area. Ann Entomol Soc Am 53(1):61–78. https://doi.org/10.1093/aesa/53.1.61

Greenslade P (2007) The potential of Collembola to act as indicators of landscape stress in Australia. Aust J Exp Agri 47(4):424–434. https://doi.org/10.1071/EA05264

Greenslade P, Simpson JA, Grgurinovic CA (2002) Collembola associated with fungal fruit-bodies in Australia. Pedobiologia 46(3/4):345–352. https://doi.org/10.1078/0031-4056-00142

Griffith GW (2012) Do we need a global strategy for microbial conservation? Trends Ecol Evol 27(1):1–2. https://doi.org/10.1016/j.tree.2011.10.002

Grimaldi D, Engel MS (2005) Evolution of the insects. Cambridge University Press, Cambridge

Grünemaier M (2016) Phoretic springtail (Collembola: Sminthuridae) on a false blister beetle (Coleoptera: Oedemeridae) in Eocene Baltic amber. Palaeodiversity 9(1):9–13. https://doi.org/10.18476/pale.v9.a2

Gry J, Andersson C (2014) Mushrooms traded as food. Vol II sec 2. Nordic risk assessments and background on edible mushrooms, suitable for commercial marketing and background lists for industry, trade and food inspection. Risk assessments of mushrooms on the four guidance lists. Nordic Council of Ministers, Copenhagen. https://doi.org/10.6027/TN2014-507. Accessed 15 Jan 2017

Guevara R, Dirzo R (1999) Consumption of macro-fungi by invertebrates in a Mexican tropical cloud forest: do fruit body characteristics matter? J Trop Ecol 15(5):603–617
Gulvik ME (2007) Mites (Acari) as indicators of soil biodiversity and land use monitoring: a review. Pol J Ecol 55(3):415–440
Hackman W, Meinander M (1979) Diptera feeding as larvae on macrofungi in Finland. Ann Zool Fenn 16:50–83
Hågvar S (2010a) A review of Fennoscandian arthropods living on and in snow. Eur J Entomol 107(3):281–298. https://doi.org/10.14411/eje.2010.037
Hågvar S (2010b) Primary succession of springtails (Collembola) in a Norwegian glacier foreland. Arctic Antarct Alp Res 42(4):422–429. https://doi.org/10.1657/1938-4246-42.4.422
Halbwachs H, Bässler C (2015) Gone with the wind – a review on basidiospores of lamellate agarics. Mycosphere 6(1):78–112. https://doi.org/10.5943/mycosphere/6/1/10
Hall IR, Yun W, Amicucci A (2003) Cultivation of edible ectomycorrhizal mushrooms. Trends Biotechnol 21(10):433–438. https://doi.org/10.1016/S0167-7799(03)00204-X
Hanski I (1989) Fungivory: fungi, insects and ecology. In: Wilding N, Collins NM, Hammond PM, Webber JF (eds) Insect—fungus interactions. Academic, London, pp 25–68
Heckman DS, Geiser DM, Eidell BR, Stauffer RL, Kardos NL, Hedges SB (2001) Molecular evidence for the early colonization of land by fungi and plants. Science 293(5532):1129–1133. https://doi.org/10.1126/science.1061457
Hibbett DS, Grimaldi D, Donoghue MJ (1997) Fossil mushrooms from Miocene and cretaceous ambers and the evolution of homobasidiomycetes. Am J Bot 84(7):981–991. https://doi.org/10.2307/2446289
Hibbett DS, Gilbert LB, Donoghue MJ (2000) Evolutionary instability of ectomycorrhizal symbioses in basidiomycetes. Nature 407(6803):506–508. https://doi.org/10.1038/35035065
Hiol FH, Dixon RK, Curl EA (1994) The feeding preference of mycophagous Collembola varies with the ectomycorrhizal symbiont. Mycorrhiza 5(2):99–103. https://doi.org/10.1007/BF00202340
Hirst S (1923) XLVI.–on some arachnid remains from the old red sandstone (rhynie chert bed, Aberdeenshire). Ann Mag Nat Hist 12(70):455–474. https://doi.org/10.1080/00222932308632963
Hirst S, Maulik S (1926) On some arthropod remains from the Rhynie chert (old red sandstone). Geol Mag 63(2):69–71. https://doi.org/10.1017/S0016756800083692
Hoffmann A (1990) Los trombicúlidos de México (Acarida: Trombiculidae). Parte taxonómica. Publicaciones especiales del Instituto de Biología 2, UNAM, México
Hoffmann A, Riverón R (1992) Biorrelaciones entre los musgos y su acarofauna en México. Trop Bryol 6:105–110. https://doi.org/10.11646/bde.6.1.12
Hofstetter RW, Moser JC (2014) The role of mites in insect-fungus associations. Annu Rev Entomol 59:537–557. https://doi.org/10.1146/annurev-ento-011613-162039
Honrubia M (2009) Las micorrizas: una relación planta-hongo que dura más de 400 millones de años. Anales Jard Bot Madrid 66(1):133–144. https://doi.org/10.3989/ajbm.2226
Hopkin SP (1997) Biology of the springtails (Insecta: Collembola). Oxford University Press, Oxford
Houck MA, OConnor BM (1991) Ecological and evolutionary significance of phoresy in the Astigmata. Annu Rev Entomol 36:611–636. https://doi.org/10.1146/annurev.en.36.010191.003143
Janssens F, Christiansen KA (2011) Class Collembola Lubbock, 1870. In: Zhang ZQ (ed) animal biodiversity: an outline of higher-level classification and survey of taxonomic richness. Zootaxa 3148:192–194. https://doi.org/10.11646/zootaxa.3148.1.34
Jeram AJ, Selden PA, Edwards D (1990) Land animals in the Silurian: arachnids and myriapods from Shropshire, England. Science 250(4981):658–661. https://doi.org/10.1126/science.250.4981.658
Jordana R (2012) Synopses on Palaearctic Collembola. Volume 7/1. Capbryinae & Entomobryini. Soil Organisms 84(1):1–390

Jørgensen HB, Elmholt S, Petersen H (2003) Collembolan dietary specialisation on soil grown fungi. Biol Fertil Soils 39(1):9–15. https://doi.org/10.1007/s00374-003-0674-6

Kanters C, Anderson IC, Johnson D (2015) Chewing up the wood-wide web: selective grazing on ectomycorrhizal fungi by Collembola. Forests 6(8):2560–2570. https://doi.org/10.3390/f6082560

Karasawa S, Gotoh K, Sasaki T, Hijii N (2005) Wind-based dispersal of oribatid mites (Acari: Oribatida) in a subtropical forest in Japan. J Acarol Soc Jpn 14(2):117–122. https://doi.org/10.2300/acari.14.117

Karwa A, Varma A, Rai M (2011) Edible ectomycorrhizal fungi: cultivation, conservation and challenges. In: Rai M, Varma A (eds) Diversity and biotechnology of ectomycorrhizae. Soil biology, vol 25. Springer, Heidelberg, pp 429–453

Kirk PM, Cannon PF, David JC, Stalpers JA (2001) Ainsworth & Bisby's Dicitionary of the fungi, 9th edn. CAB International, Wallingford

Komonen A (2003) Hotspots of insect diversity in boreal forests. Conserv Biol 17(4):976–981. https://doi.org/10.1046/j.1523-1739.2003.02076.x

Kováč L, Papáč V (2010) Revision of the genus *Neelus* Folsom, 1896 (Collembola, Neelida) with the description of two new troglobiotic species from Europe. Zootaxa 2663:36–52. https://doi.org/10.11646/zootaxa.2663.1.2

Krantz GW (2009) Habits and habitats. In: Krantz GW, Walter DE (eds) A manual of acarology, 3rd edn. Texas Tech University Press, Lubbock, pp 64–82

Krantz GW, Lindquist EE (1979) Evolution of phytophagous mites (Acari). Annu Rev Entomol 24:121–158. https://doi.org/10.1146/annurev.en.24.010179.001005

Krausee A (1928/29) Zur terminologie der edaphischen biozönosen. Int Ent Z 22:110–111

Krivosheina NP (2008) Macromycete fruit bodies as a habitat for dipterans (Insecta, Diptera). Entomol Rev 88(7):778–792. https://doi.org/10.1134/S0013873808070038

Largent DL, Johnson D, Watling R (1977) How to identify mushrooms to genus III: microscopic features. Mad River Press, Eureka

Lehmitz R, Russell D, Hohberg K, Christian A, Xylander WER (2011) Wind dispersal of oribatid mites as a mode of migration. Pedobiologia 54(3):201–207. https://doi.org/10.1016/j.pedobi.2011.01.002

LePage BA, Currah RS, Stockey RA, Rothwell GW (1997) Fossil ectomycorrhizae from the middle Eocene. A J Bot 84(3):410–412. https://doi.org/10.2307/2446014

Lilleskov EA, Bruns TD (2005) Spore dispersal of a resupinate ectomycorrhizal fungus, *Tomentella sublilacina*, via soil food webs. Mycologia 97(4):762–769. https://doi.org/10.3852/mycologia.97.4.762

Lincoff GH (1981) The Audubon society field guide to north American mushrooms. Alfred A. Knopf, New York

Lindo Z, Winchester NN (2008) Scale dependent diversity patterns in arboreal and terrestrial oribatid mite (Acari: Oribatida) communities. Ecography 31(1):53–60. https://doi.org/10.1111/j.2007.0906-7590.05320.x

Lindquist EE (1965) An unusual new species of *Hoploseius* Berlese (Acarina: Blattisociidae) from Mexico. Can Entomol 97(11):1121–1131. https://doi.org/10.4039/Ent971121-11

Lindquist EE (1975) Associations between mites and other arthropods in forest floor habitats. Can Entomol 107(4):425–437. https://doi.org/10.4039/Ent107425-4

Lindquist EE (1995) Remarkable convergence between two taxa of ascid mites (Acari: Mesostigmata) adapted to living in pore tubes of bracket fungi in North America, with description of *Mycolaelaps* new genus. Can J Zool 73(1):104–128. https://doi.org/10.1139/z95-013

Lindquist EE (1998) Evolution of phytophagy in trombidiform mites. Exp Appl Acarol 22(2):81–100. https://doi.org/10.1023/A:1006041609774

Lindquist EE, Krantz GW, Walter DE (2009a) Classification. In: Krantz GW, Walter DE (eds) A manual of acarology, 3rd edn. Texas Tech University Press, Lubbock, pp 97–103

Lindquist EE, Krantz GW, Walter DE (2009b) Order Mesostigmata. In: Krantz GW, Walter DE (eds) A manual of acarology, 3rd edn. Texas Tech University Press, Lubbock, pp 124–232

Lodge DJ, Ammirati JF, O'Dell TE, Mueller GM, Huhndorf SM, Wang CJ et al (2004) Terrestrial and lignicolous macrofungi. In: Mueller GM, Bills GF, Foster MS (eds) Biodiversity of fungi: inventory and monitoring methods. Academic, New York, pp 127–158

MagurranAE(2010)Q&a:whatisbiodiversity?BMCBiol.https://doi.org/10.1186/1741-7007-8-145

Maraun M, Visser S, Scheu S (1998) Oribatid mites enhance the recovery of the microbial community after a strong disturbance. Appl Soil Ecol 9(1–3):175–181. https://doi.org/10.1016/S0929-1393(98)00072-9

Martin MM (1979) Biochemical implications of insect mycophagy. Biol Rev 54(1):1–21. https://doi.org/10.1111/j.1469-185X.1979.tb00865.x

Martínez PA (2009) Oribatida. In: Momo FR, Falco LB (eds) Biología y ecología de la fauna del suelo. Imago Mundi. UNGS, Buenos Aires, pp 79–102

Mateos E, López R, Barranco T, Hoyo P, Llimona X (1996) Colémbolos (Hexapoda, Collembola) asociados con carpóforos de basidiomicetes recolectados en el SW de Cataluña. Rev Catalana Micol 19:99–107

Matthewman WG, Pielou DP (1971) Arthropods inhabiting the sporophores of *Fomes fomentarius* (Polyporaceae) in Gatineau Park, Quebec. Can Entomol 103(6):775–847. https://doi.org/10.4039/Ent103775-6

Mier N, Canete S, Klaebe A, Chavant L, Fournier D (1996) Insecticidal properties of mushroom and toadstool carpophores. Phytochemistry 41(5):1293–1299. https://doi.org/10.1016/0031-9422(95)00773-3

Miner BG, Sultan SE, Morgan SG, Padilla DK, Relyea RA (2005) Ecological consequences of phenotypic plasticity. Trends Ecol Evol 20(12):685–692. https://doi.org/10.1016/j.tree.2005.08.002

Misof B, Liu S, Meusemann K, Peters RS, Donath A, Mayer C et al (2014) Phylogenomics resolves the timing and pattern of insect evolution. Science 346(6210):763–767. https://doi.org/10.1126/science.1257570

Mithöfer A, Boland W (2012) Plant defense against herbivores: chemical aspects. Annu Rev Plant Biol 63:431–450. https://doi.org/10.1146/annurev-arplant-042110-103854

Nagy LG, Walther G, Házi J, Vágvölgyi C, Papp T (2011) Understanding the evolutionary processes of fungal fruiting bodies: correlated evolution and divergence times in the Psathyrellaceae. Syst Biol 60(3):303–317. https://doi.org/10.1093/sysbio/syr005

Nakamori T, Suzuki A (2005) Preference of three collembolan species for fruit-bodies of three species of basidiomycete fungi. Pedobiologia 49(2):119–125. https://doi.org/10.1016/j.pedobi.2004.09.003

Negri I (2004) Spatial distribution of Collembola in presence and absence of a predator. Pedobiologia 48(5–6):585–588. https://doi.org/10.1016/j.pedobi.2004.07.004

Norton RA (1985) Aspects of the biology and systematics of soil arachnids, particularly saprophagous and mycophagous mites. Quaest Entomol 21:523–541

Norton RA (1998) Morphological evidence for the evolutionary origin of Astigmata (Acari: Acariformes). Exp Appl Acarol 22(10):559–594. https://doi.org/10.1023/A:1006135509248

Norton RA, Bonamo PM, Grierson JD, Shear WA (1988) Oribatid mite fossils from a terrestrial Devonian deposit near Gilboa, New York. J Paleontol 62(2):259–269. https://doi.org/10.1017/S0022336000029905

O'Connell T, Bolger T (1997a) Fungal fruiting bodies and the structure of fungus-micro-arthropod assemblages. Biol Environ 97(3):249–262

O'Connell T, Bolger T (1997b) Stability, ephemerality and dispersal ability: microarthropod assemblages on fungal sporophores. Biol J Linn Soc 62(1):111–131. https://doi.org/10.1006/bijl.1997.0144

OConnor BM (2003) Mites. In: Resh VH, Cardé RT (eds) Encyclopedia of insects. Academic, San Diego, pp 726–733

OConnor BM (2009) Cohort Astigmatina. In: Krantz GW, Walter DE (eds) A manual of acarology, 3rd edn. Texas Tech University Press, Lubbock, pp 565–657

Ojala R, Huhta V (2001) Dispersal of microarthropods in forest soil. Pedobiologia 45(5):443–450. https://doi.org/10.1078/0031-4056-00098

Okabe K (1999) Vectoring of *Hypocrea nigricans* (Hypocreales: Hypocreaceae) by three fungivorous mite species (Acari: acaridae). Exp Appl Acarol 23(8):653–658. https://doi.org/10.1023/A:1006218919274

Okabe K (2013) Influence of spatio-temporal resource availability on mushroom mite diversity. Exp Appl Acarol 61(3):299–310. https://doi.org/10.1007/s10493-013-9696-4

Okabe K, Amano H (1993) Mite species collected from field mushrooms (II): Mesostigmata, Prostigmata and Astigmata. J Acarol Soc Jpn 2(1):19–28. https://doi.org/10.2300/acari.2.19

Orgiazzi A, Bardgett RD, Barrios E, Behan-Pelletier V, Briones MJI, Chotte JL et al (2016) Global soil biodiversity atlas. European Commission, Publications Office of the European Union, Luxembourg

Oubrahim H, Richard JM, Cantin-Esnault D, Seigle-Murandi F, Trécourt F (1997) Novel methods for identification and quantification of the mushroom nephrotoxin orellanine. Thin-layer chromatography and electrophoresis screening of mushrooms with electron spin resonance determination of the toxin. J Chromatogr A 758(1):145–157. https://doi.org/10.1016/S0021-9673(96)00695-4

Palacios-Vargas JG (2003) Los microartrópodos (Collembola) de la selva tropical húmeda. In: Álvarez-Sánchez J, Naranjo-García E (eds) Ecología del suelo en la selva tropical húmeda de México. Instituto de Biología, Facultad de Ciencias, UNAM, México, DF, Instituto de Ecología AC, Xalapa, pp 217–225

Palacios-Vargas JG (2014) Biodiversidad de Collembola (Hexapoda: Entognatha) en México. Rev Mex Biodiv 85:220–231. https://doi.org/10.7550/rmb.32713

Palacios-Vargas JG, Gómez JA (1991) Los colémbolos y su relación con los hongos. In: Navarrete-Heredia JL, Quiroz-Rocha GA (eds) Memorias del 1er Simposio Nacional sobre la Interacción Insecto-Hongo. XXVI Congreso Nacional de Entomología, Veracruz, pp 99–114

Palacios-Vargas JG, Gómez-Anaya JA (1994) Lista actualizada de colémbolos micetófilos de México (Hexapoda: Entognatha). Folia Entomol Mex 92:21–30

Palacios-Vargas JG, Mejía-Recamier BE (2007) Técnicas de colecta, montaje y preservación de microartrópodos edáficos. Las Prensas de Ciencias, Facultad de Ciencias, UNAM, México

Palacios-Vargas JG, Mejía-Recamier BE, de Oyarzabal A (2014) Guía ilustrada para los artrópodos edáficos. Las Prensas de Ciencias, Facultad de Ciencias, UNAM, México

Pang Z, Sterner O (1991) Cibaric acid, a new fatty acid derivative formed enzymatically in damaged fruit bodies of *Cantharellus cibarius* (chanterelle). J Org Chem 56(3):1233–1235. https://doi.org/10.1021/jo00003a054

Pang Z, Sterner O, Anke H (1992) (8E)-10-Hydroxydec-8-enoic acid: its isolation from injured fruit bodies of *Cantharellus tubaeformis* and synthetic preparation. Acta Chem Scand 46:301–303

Papáč V, Palacios-Vargas JG (2016) A new genus of Neelidae (Collembola) from Mexican caves. ZooKeys 569:37–51. https://doi.org/10.3897/zookeys.569.5984

Parniske M (2008) Arbuscular mycorrhiza: the mother of plant root endosymbioses. Nat Rev Microbiol 6(10):763–775. https://doi.org/10.1038/nrmicro1987

Penney D, McNeil A, Green DI, Bradley RS, Jepson JE, Withers PJ, Preziosi RF (2012) Ancient Ephemeroptera–Collembola symbiosis fossilized in amber predicts contemporary phoretic associations. PLoS One. https://doi.org/10.1371/journal.pone.0047651

Pérez TM, Guzmán-Cornejo C, Montiel-Parra G, Paredes-León R, Rivas G (2014) Biodiversidad de ácaros en México. Rev Mex Biodiv 85:399–407. https://doi.org/10.7550/rmb.36160

Pérez-Moreno J, Lorenzana-Fernández A, Carrasco-Hernández V, Yescas-Pérez A (2010) Los hongos comestibles silvestres del Parque Nacional Izta-Popo, Zoquiapan y Anexos. Colegio de Postgraduados, SEMARNAT, CONACyT, México

Pielou DP, Verma AN (1968) The arthropod fauna associated with the birch bracket fungus, *Polyporus betulinus*, in eastern Canada. Can Entomol 100(11):1179–1199. https://doi.org/10.4039/Ent1001179-11

Pilz D, Molina R (eds) (1996) Managing forest ecosystems to conserve fungus diversity and sustain wild mushroom harvests. Gen. Tech. Rep. PNW-GTR-371. US Department of Agriculture, Forest Service, Pacific Northwest Research Station, Portland, OR. https://doi.org/10.2737/PNW-GTR-371

Pirozynski KA, Malloch DW (1975) The origin of land plants: a matter of mycotrophism. Biosystems 6(3):153–164. https://doi.org/10.1016/0303-2647(75)90023-4

Remy W, Taylor TN, Hass H, Kerp H (1994) Four hundred-million-year-old vesicular arbuscular mycorrhizae. Proc Natl Acad Sci U S A 91(25):11841–11843. https://doi.org/10.1073/pnas.91.25.11841

Rice CM, Ashcroft WA, Batten DJ, Boyce AJ, Caulfield JBD, Fallick AE, Hole MJ, Jones E, Pearson MJ, Rogers G, Saxton JM, Stuart FM, Trewin NH, Turner G (1995) A Devonian auriferous hot spring system, Rhynie, Scotland. J Geol Soc Lond 152(2):229–250. https://doi.org/10.1144/gsjgs.152.2.0229

Rohlfs M, Churchill ACL (2011) Fungal secondary metabolites as modulators of interactions with insects and other arthropods. Fungal Genet Biol 48(1):23–34. https://doi.org/10.1016/j.fgb.2010.08.008

Rota-Stabelli O, Daley AC, Pisani D (2013) Molecular timetrees reveal a Cambrian colonization of land and a new scenario for ecdysozoan evolution. Curr Biol 23(5):392–398. https://doi.org/10.1016/j.cub.2013.01.026

Rusek J (1998) Biodiversity of Collembola and their functional role in the ecosystem. Biodivers Conserv 7(9):1207–1219. https://doi.org/10.1023/A:1008887817883

Sasaki G, Ishiwata K, Machida R, Miyata T, Su ZH (2013) Molecular phylogenetic analyses support the monophyly of Hexapoda and suggest the paraphyly of Entognatha. BMC Evol Biol. https://doi.org/10.1186/1471-2148-13-236

Sawahata T (2006a) Hymenial area of agaric fruit bodies consumed by Collembola. Mycoscience 47(2):91–93. https://doi.org/10.1007/s10267-005-0272-6

Sawahata T (2006b) Oribatid mites collected from fruit bodies and hyphal mats of ectomycorrhizal fungi in a Japanese beech forest. Edaphologia 79:23–26

Sawahata T, Soma K, Ohmasa M (2000) Number and food habit of springtails on wild mushrooms of three species of Agaricales. Edaphologia 66:21–33

Sawahata T, Soma K, Ohmasa M (2002) The seasonal change in abundance of *Hypogastrura denisana* Yosii on agaric mushrooms in relation to its life cycle. Edaphologia 69:35–45

Schaefer I, Norton RA, Scheu S, Maraun M (2010) Arthropod colonization of land – linking molecules and fossils in oribatid mites (Acari, Oribatida). Mol Phylogenet Evol 57(1):113–121. https://doi.org/10.1016/j.ympev.2010.04.015

Schatz H, Behan-Pelletier VM, OConnor BM, Norton RA (2011) Suborder Oribatida van der Hammen, 1968. In: Zhang ZQ (ed) animal biodiversity: an outline of higher-level classification and survey of taxonomic richness. Zootaxa 3148:141–148. https://doi.org/10.11646/zootaxa.3148.1.26

Schmidt AR, Dörfelt H, Struwe S, Perrichot V (2010) Evidence for fungivory in cretaceous amber forests from Gondwana and Laurasia. Palaeontographica 283(4–6):157–173. https://doi.org/10.1127/palb/283/2010/157

Schneider K, Maraun M (2005) Feeding preferences among dark pigmented fungal taxa ("Dematiacea") indicate limited trophic niche differentiation of oribatid mites (Oribatida, Acari). Pedobiologia 49(1):61–67. https://doi.org/10.1016/j.pedobi.2004.07.010

Schneider K, Renker C, Maraun M (2005) Oribatid mite (Acari, Oribatida) feeding on ectomycorrhizal fungi. Mycorrhiza 16(1):67–72. https://doi.org/10.1007/s00572-005-0015-8

Shaw CH, Lundkvist H, Moldenke A, Boyle JR (1991) The relationships of soil fauna to long-term forest productivity in temperate and boreal ecosystems: processes and research strategies. In: Dyck WJ, Mess CA (eds) Long-term field trials to assess environmental impacts of harvesting. Proceedings, IEA/BE T6/A6 Workshop, Florida, USA IEA/BE T6/A6 Report No.5. Forest Research Institute, Rotorua, New Zealand, FRI Bulletin No. 161, pp 39–77

Sidorchuk EA (2018) Mites as fossils: forever small? Int J Acarol 44(8):349–359. https://doi.org/10.1080/01647954.2018.1497085

Siepel H, de Ruiter-Dijkman EM (1993) Feeding guilds of oribatid mites based on their carbohydrase activities. Soil Biol Biochem 25(11):1491–1497. https://doi.org/10.1016/0038-0717(93)90004-U

Singer R, García J, Gómez LD (1990) The Boletineae of Mexico and Central America I&II. Nova Hedwigia 98:1–70

Singer R, García J, Gómez LD (1991) The Boletineae of Mexico and Central America III. Nova Hedwigia 102:1–99

Singer R, García J, Gómez LD (1992) The Boletineae of Mexico and Central America IV. Nova Hedwigia 105:1–62

Sitta N, Süss L (2012) Insects parasitizing edible ectomycorrhizal mushrooms. In: Zambonelli A, Bonito GM (eds) Edible ectomycorrhizal mushrooms: current knowledge and future prospects. Soil biology, vol 34. Springer, Heidelberg, pp 335–353

Smith SE, Read DJ (2008) Mycorrhizal symbiosis, 3rd edn. Academic, London

Solano F (2014) Melanins: skin pigments and much more—types, structural models, biological functions, and formation routes. New J Sci. https://doi.org/10.1155/2014/498276

Spiteller P (2008) Chemical defence strategies of higher fungi. Chem Eur J 14(30):9100–9110. https://doi.org/10.1002/chem.200800292

Staaden S, Milcu A, Rohlfs M, Scheu S (2011) Olfactory cues associated with fungal grazing intensity and secondary metabolite pathway modulate Collembola foraging behaviour. Soil Biol Biochem 43(7):1411–1416. https://doi.org/10.1016/j.soilbio.2010.10.002

Stebaeva SK (1970) The life forms of springtails (Collembola). Zool Zh 49(10):1437–1455

Subías LS (2019) Actualización online de marzo: Listado sistemático, sinonímico y biogeográfico de los ácaros oribátidos (Acariformes: Oribatida) del mundo (excepto fósiles). http:/bba.bioucm.es/cont/docs/RO_1.pdf. Accessed 29 Mar 2019

Swift MJ, Heal OW, Anderson JM (1979) Decomposition in terrestrial ecosystems. Studies in ecology, vol 5. University of California Press, Berkeley

Swift MJ, Bignell DE, Moreira FMS, Huising EJ (2012) El inventario de la biodiversidad biológica del suelo: conceptos y guía general. In: FMS M, Huising EJ, Bignell DE (eds) Manual de biología de suelos tropicales. Muestreo y caracterización de la biodiversidad bajo suelo. Secretaria de Medio Ambiente y Recursos Naturales (SEMARNAT), Instituto Nacional de Ecología (INE), México, pp 29–52

Takahashi KH, Tuno N, Kagaya T (2005) Abundance of mycophagous arthropods present on different species of fungi in relation to resource abundance at different spatial scales. Eur J Entomol 102(1):39–46. https://doi.org/10.14411/eje.2005.006

Takemoto T, Nakajima T, Sakuma R (1964) Isolation of a flycidal constituent" ibotenic acid" from *Amanita muscaria* and *A. pantherina*. J Pharm Soc Jpn 84:1233–1234

Tedersoo L, Smith ME (2013) Lineages of ectomycorrhizal fungi revisited: foraging strategies and novel lineages revealed by sequences from belowground. Fungal Biol Rev 27(3–4):83–99. https://doi.org/10.1016/j.fbr.2013.09.001

Thomas CA (1939) The animals associated with edible fungi. J N Y Entomol Soc 47(1):11–37

Tulloss RE (1994) Type studies in *Amanita* section *Vaginatae*. I: some taxa described in this century (studies 1-23) with notes on description of spores and refractive hyphae in *Amanita*. Mycotaxon 52(1):305–396

Vacante V (2016) The handbook of mites of economic plants: identification, bio-ecology and control. CAB International, Wallingford

Veneault-Fourrey C, Martin F (2013) New insights into ectomycorrhizal symbiosis evolution and function. In: Kempken F (ed) The Mycota: a comprehensive treatise on fungi as experimental systems for basic and applied research. XI agricultural applications, 2nd edn. Springer, Heidelberg, pp 273–293

Verhoef HA, Prast JE (1989) Effects of dehydration on osmotic and ionic regulation in *Orchesella cincta* (L.) and *Tomocerus minor* (Lubbock) (Collembola) and the role of the coelomoduct kidneys. Comp Biochem Physiol A 93(4):691–694. https://doi.org/10.1016/0300-9629(89)90485-4
von Reumont BM, Jenner RA, Wills MA, Dell'Ampio E, Pass G, Ebersberger I et al (2012) Pancrustacean phylogeny in the light of new phylogenomic data: support for Remipedia as the possible sister group of Hexapoda. Mol Biol Evol 29(3):1031–1045. https://doi.org/10.1093/molbev/msr270
Wallwork JA (1970) Ecology of soil animals. McGraw-Hill, London
Walter DE (1988) Predation and mycophagy by endeostigmatid mites (Acariformes: Prostigmata). Exp Appl Acarol 4(2):159–166. https://doi.org/10.1007/BF01193873
Walter DE, Ikonen EK (1989) Species, guilds, and functional groups: taxonomy and behavior in nematophagous arthropods. J Nematol 21(3):315–327
Walter DE, Krantz GW (2009) Collecting, rearing, and preparing specimens. In: Krantz GW, Walter DE (eds) A manual of acarology, 3rd edn. Texas Tech University Press, Lubbock, pp 83–96
Walter DE, Proctor HC (2013) Mites: ecology, evolution and behavior. Life at a microscale, 2nd edn. Springer, Dordrecht
Walter DE, Hunt HW, Elliott ET (1988) Guilds or functional groups? An analysis of predatory arthropods from a shortgrass steppe soil. Pedobiologia 31(3/4):247–260
Walter DE, Lindquist EE, Smith IM, Cook DR, Krantz GW (2009) Order trombidiformes. In: Krantz GW, Walter DE (eds) A manual of acarology, 3rd edn. Texas Tech University Press, Lubbock, pp 233–420
Walter DE, Bolton S, Uusitalo M, Zhang ZQ (2011) Suborder Endeostigmata Reuter, 1909. In: Zhang ZQ (ed) animal biodiversity: an outline of higher-level classification and survey of taxonomic richness. Zootaxa 3148:139–140. https://doi.org/10.11646/zootaxa.3148.1.25
Wang M, Triguéros V, Paquereau L, Chavant L, Fournier D (2002) Proteins as active compounds involved in insecticidal activity of mushroom fruitbodies. J Econ Entomol 95(3):603–607. https://doi.org/10.1603/0022-0493-95.3.603
Watling R (1997) The business of fructification. Nature 385(6614):299–300. https://doi.org/10.1038/385299a0
Wellman CH, Gray J (2000) The microfossil record of early land plants. Phil Trans R Soc Lond B 355(1398):717–732. https://doi.org/10.1098/rstb.2000.0612
Weygoldt P (1998) Evolution and systematics of the Chelicerata. Exp Appl Acarol 22(2):63–79
Whalley P, Jarzembowski EA (1981) A new assessment of *Rhyniella*, the earliest known insect, from the Devonian of Rhynie, Scotland. Nature 291(5813):317. https://doi.org/10.1038/291317a0
Yamashita S, Hijii N (2003) Effects of mushroom size on the structure of a mycophagous arthropod community: comparison between infracommunities with different types of resource utilization. Ecol Res 18(2):131–143. https://doi.org/10.1046/j.1440-1703.2003.00541.x
Yen AL (2009) Edible insects: traditional knowledge or western phobia? Entomol Res 39(5):289–298. https://doi.org/10.1111/j.1748-5967.2009.00239.x
Yun W, Hall IR (2004) Edible ectomycorrhizal mushrooms: challenges and achievements. Can J Bot 82(8):1063–1073. https://doi.org/10.1139/b04-051
Zhang ZQ (2011) Phylum Arthropoda von Siebold, 1848. In: Zhang, ZQ (ed) Animal biodiversity: an outline of higher-level classification and survey of taxonomic richness. Zootaxa 3148:99–103. https://doi.org/10.11646/zootaxa.3148.1.14
Zhang ZQ, Fan QH, Pesic V, Smit H, Bochkov AV, Khaustov AA et al (2011) Order Trombidiformes Reuter, 1909. In: Zhang ZQ (ed) Animal biodiversity: an outline of higher-level classification and survey of taxonomic richness. Zootaxa 3148:129–138. https://doi.org/10.11646/zootaxa.3148.1.24

Chapter 15
Diversity and Importance of Edible Mushrooms in Ectomycorrhizal Communities in Mexican Neotropics

Roberto Garibay-Orijel, Andrés Argüelles-Moyao, Julieta Álvarez-Manjarrez, Rodolfo Enrique Ángeles-Argáiz, Olimpia Mariana García-Guzmán, and Haydée Hernández-Yáñez

15.1 Introduction

Fungi carry out multiple ecosystem functions such as decomposing organic material, recycling nutrients, optimizing nutrient availability, and improving soil structure and stability. Due to their essential role in ecosystems, mycorrhizal fungi are one of the most important functional groups. Approximately 95% of terrestrial plants develop mycorrhizae and depend on this relationship to establish, survive, and develop (Finlay and Read 1986; Van der Heijden et al. 2015). Mycorrhizae are basically formed by three major fungi groups: Glomeromycota, which forms arbuscular mycorrhizae; Ascomycota and Basidiomycota, which form ectomycorrhizae, ectendomycorrhizae, arbutoid, and orchidiod mycorrhizae, among others. The ectomycorrhizal association is established between more than 5000 fungal species and plants in the families Betulaceae, Casuarinaceae, Dipterocarpaceae, Ericaceae, Fabaceae, Fagaceae, Myrtaceae, Pinaceae, Salicaceae, etc. Although the number of plant species involved is not high, most of these are dominant shrubs and trees of temperate and boreal ecosystems and therefore of great ecological importance (Brundrett 2009).

Traditionally, the study of ectomycorrhizal fungi (EMF) has been based on their sexual reproductive structures or fruit bodies. However, the knowledge of their diversity is biased by this method because most of these fungi do not have sexual

R. Garibay-Orijel (✉) · A. Argüelles-Moyao · J. Álvarez-Manjarrez · R. E. Ángeles-Argáiz
Laboratorio de Sistemática y Ecología de Ectomicorrizas, Instituto de Biología, Universidad Nacional Autónoma de México, Ciudad de México, Mexico
e-mail: rgaribay@ib.unam.mx

O. M. García-Guzmán
Laboratorio de Ecología Bacteriana, Instituto de Ecología, Universidad Nacional Autónoma de México, Ciudad de México, Mexico

H. Hernández-Yáñez
GIS Laboratory, Smithsonian Conservation Biology Institute, Front Royal, VA, USA

J. Pérez-Moreno et al. (eds.), *Mushrooms, Humans and Nature in a Changing World*, https://doi.org/10.1007/978-3-030-37378-8_15

reproduction or either, produce hypogeous fruit bodies, or are inconspicuous. Their ecological knowledge is also influenced by this bias, because it is the vegetative state of these fungi that has the most impact in the ecosystem, not their reproductive state. The production of fruit bodies only reflects the energy invested in sexual reproduction and is not an extension of their vegetative state. Therefore, the development of molecular techniques to study mycorrhizae directly from the roots as well as to study the mycelium in the soil has revolutionized our understanding of these fungi. The use of these techniques has shed light on numerous unknown lineages; the fact that many fungi thought to be saprobes are in fact mycorrhizal; and the discrepancy between the diversity and structure of fungi communities when studied by their mycorrhizae than when studied by the fruiting bodies found (Horton and Bruns 2001).

With almost 400 species of edible mushrooms, Mexico possess the second largest edible mushroom biocultural heritage worldwide (Garibay-Orijel and Ruan-Soto 2014), just after China. There, more than 15 million indigenous people and 105 million mestizos have incorporated wild edible mushrooms into their traditions and diet. The genera *Pinus* and *Quercus* have one of their diversification centers in Mexico. The diversity and endemism of both genera in Mexico is, therefore, high (Gernandt and Pérez-de la Rosa 2014; Aguilar-Romero et al. 2016). Both pines and oaks depend on their relationships with EMF for their establishment, growth, and survival. In order to develop a sustainable management of edible ectomycorrhizal mushrooms we must understand their role in ectomycorrhizal communities as these sustain the production of their fruit bodies. The study of EMF as components of mycorrhizal symbiosis provides information not only on their ecology, taxonomy, and systematics but also on their relationship with their hosts and their role in the forest. Therefore, it is now possible to know which species of fungi are associated with certain plant species. This information is essential because it allows us to propose reforestation and restoration plans that are adequate for the plant species and the soil conditions in specific regions. Additionally, this information is useful in forestry plantations, where selecting for the correct fungi would give trees greater growth and higher survival rates. Industrialized countries are already strengthening their forestry sector with the use of this technology. However, in Mexico, the use of this technology is just beginning and it will not continue to develop if the basic information on the fungi diversity and ecology is not produced.

In this chapter we explore the diversity and structure of ectomycorrhizal fungi communities associated with the main forest ecosystems of the Mexican Neotropics, with particular emphasis on edible species.

15.2 Materials and Methods

The convergence of two major ecoregions, the Neartic and the Neotropic, makes Mexico a biodiverse country. Most studies on the diversity and molecular ecology of ectomycorrhizal fungi have taken place in the Holarctic (specifically in USA,

Canada and Europe). Unfortunately, fungi of the temperate forests of the Mexican Neotropics have been poorly studied. Identifying the EMF in the Neotropics is fundamental for our understanding of their biological history, evolution and distribution. This study provides information on the diversity of EMF of the Mexican Neotropics.

We integrated data from fruit body sampling from 2009 to 2015 with data from ectomycorrhizal sampling to evaluate the diversity of EMF. For this purpose, we selected study sites that represented the main vegetation types that host ectomycorrhizal fungi: *Abies religiosa* forests in el Zarco, Estado de Mexico (Argüelles-Moyao et al. 2016); *Alnus* forests in La Malinche, Tlaxcala and in Acatlán, Veracruz (Kennedy et al. 2011); *Pinus* spp. forests in Cuitzeo, Michoacán (Garibay-Orijel 2008); *Quercus* spp. forests in Cuitzeo, Michoacán (García-Guzmán et al. 2017); and tropical dry forests in Chamela, Jalisco (Álvarez-Manjarrez et al. 2018).

15.2.1 Sampling of Fruit Bodies

EMF fruit bodies were collected extensively during the rainy season of the year (June to November) from 2009 to 2015. Most of the sampling took place in Amanalco, "el Zarco," "Nevado de Toluca" and Temascaltepec in "Estado de Mexico," and Cuitzeo in Michoacán. The objective was to collect all EMF species encountered, although emphasis was placed in groups of ectomycorrhizal fungi that had been previously overlooked in mycological studies in Mexico. Therefore, sampling focused on those fungi that produced hypogeous fruit bodies (truffles and false truffles) or resupinate fruit bodies. Collected specimens were described, photographed, and dehydrated according to the techniques suggested by Cifuentes et al. (1986) and Halling (1996). Samples were never dried in temperatures over 60 °C to preserve the DNA, and in some cases small fragments of hymenium (0.5 cm^2) were placed in 96% ethanol. Specimens were identified by observation and measurement of their microscopic characteristics following Largent et al. (1984). Specimens were then deposited in the fungi collection of the "Herbario Nacional" (MEXU) at UNAM. Fungi identification was done using dichotomous keys and specialized monographies. Additionally, we contrasted conventional taxonomic identification with identification with DNA nucleotide similarity sequences based on the ITS region.

15.2.2 Sampling of Mycorrhizae

In each sampling site we made 10 parallel transects where mycorrhizae were sampled. Each transect was 60 m long and separated by 20 m. We took soil cores of 5 cm in diameter by 30 cm long. In each transect we took three soil cores separated by 20 m. From each sample, we haphazardly obtained 15–20 mycorrhizae. In total, we sampled 500–1000 mycorrhizae per site.

The study of *Abies religiosa* EMF was done at the locality of el Zarco, in "la Sierra de las Cruces, Estado de Mexico" (19° 17′ 36″ N, 99° 21' 18″ W). This locality has a monodominant forest of *Abies religiosa*. Soils in this area are humic Andosols and ocric Andosols, both of medium texture. The area has a mean annual temperature of 10.2 °C with an average rainfall of 1241.7 mm (INEGI 2011). In 2013, we sampled seven plots with similar tree characteristics (diameter at breast height, average distance between trees, data not shown). Details of the study site and sampling design can be found in Argüelles-Moyao et al. (2016).

The study of EMF associated with *Alnus* spp. was done in two sites in Tlaxcala and two sites in Veracruz. The sites in Tlaxcala are located in "La Malinche National Park" (19° 16′ 04″ N, 98° 02′ 07″ W, 3283 m asl and 19° 11′ 17″ N, 97° 58′ 58″ W, 2929 m asl). Soils in these sites are Andosols, climate is cold subtropical (Cw2) with a mean annual temperature that fluctuates between 12 °C and 18 °C, and a mean average rainfall of 620–928 mm. Forests of *Alnus jorullensis* and *Pinus montezumae* are found within these sites. The other two sites are found in the Acatlán Volcano, Naolinco, Veracruz (19° 40′ 29″ N, 96° 51′ 07″ W, 1816 m asl and 19° 40′ 58″ N, 96° 51′ 28″ W, 1880 m asl). These sites have Andosol soils, climate is C(wa) humid subtropical with a mean annual temperature that fluctuates between 15.6 °C and 16.0 °C, and a mean average rainfall of 2400 mm. In these sites the only mycorrhizal host was *A. acuminata.* The four sites were sampled in 2010; details on sampling design and molecular procedures can be found in Kennedy et al. (2011).

The study of EMF associated with oak-pine forests and oak forests were done at Cuitzeo basin. The basin has an area of approximately 4000 km^2 and is located north of Michoacán and south of Guanajuato. Soils in the area are Vertisols, Luvisols, Andosols, and Acrisols. Climate is subtropical highland (Cwb) with summer and winter rains. Mean annual temperature is 15 °C and has a year average of 1000 mm rainfall (Mendoza et al. 2011). Dominant vegetation types are scrubs, *Quercus*, *Pinus-Quercus*, and *Abies* in a fragmented landscape with croplands in mosaic pattern. Sampling of these oak-pine forests was done in 2007. We located eight samplings sites with a dominant vegetation of *Pinus* spp. and with a different proportion of *Quercus* spp. Details on sampling design can be found in Garibay-Orijel (2008).

At Cuitzeo basin four more sampling sites within oak forest were chosen in 2010. In those sites, *Quercus deserticola*, *Q. castanea*, *Q. obtusata*, *Q. magnolifolia*, *Q. rugosa*, and *Q. laeta* are the only hosts for ectomycorrhizal fungi. Sampling details for this site can be found in García-Guzmán et al. (2017).

The study of EMF associated with tropical dry forest was done at Chamela Biological Station, Jalisco (19° 30′ N, 105° 03′ W) during the years of 2012–2014. Soils in this area are heterogeneous: eutric and chromic Cambisol, chromic and haplic Lixisol, haplic Ferrasol, and rendzic, litic Leptosols. Climate is warm subhumid (Aw0) in summer and dry in winter (Bshw). Mean annual temperature is 26.9 °C with an average rainfall of 1394 mm. The dominant plant family in this area is Fabaceae. Sampling details can be found in Alvarez-Manjarrez et al. (2018).

15.2.3 Molecular Techniques

Taxonomic identification of ectomycorrhizae and sporocarps was based on the Internal Transcribed ribosomal Spacer (ITS1 and ITS2) of the nuclear DNA region. This region provides sufficient resolution to differentiate samples at the species level and it is one of the most represented in the genetic databases (Horton and Bruns 2001), mainly because it is used as the barcode of life for fungi (Schoch et al. 2012).

The process of extraction, amplification, and sequencing followed the protocol by Izzo et al. (2005) with modifications. DNA was extracted with the kit XNAP REDExtract-N-Amp (Sigmal-Aldrich, St. Louis, Mo, EUA). The advantage of this method is that it allows the DNA of both the fungus and the plant to be extracted from the mycorrhizae. Therefore, it is possible to identify the fungi that are associated with each tree without ambiguity. Although the DNA of the fungi and the plant are mixed in a solution, it is possible to isolate them by using selective amplification with PCR. For this reason, the fungi ITS (ITS1–5.8S-ITS2) region was amplified with the primers ITS1F, ITS4, or ITS4B (Gardes and Bruns 1993; White et al. 1990). When necessary, the chloroplast *trnL* region was amplified for the plants with the primers proposed by Taberlet et al. (1991). PCR products were cleaned with ExoSAP-IT (USB Corporation, Cleveland, Ohio, EUA) and were sequenced with Big Dye Terminator Kit (Applied Biosystems, Foster City, CA, EUA) under the terms specified by the manufacturers. Finally, DNA sequences were obtained with an ABI 3000 automatic sequencer from the "Laboratorio de Secuenciación de la Biodiversidad y la Salud" of the Biology Institute of the "Universidad Nacional Autónoma de México."

DNA sequences were clustered according to a 97% nucleotide similarity in OTUs (Operational Taxonomic Units). This percentage is accepted as standard in EMF molecular ecology studies (Peay et al. 2008). Sequences of each OTU were later compared with the database in GenBank (NCBI) through BLAST (Altschul et al. 1997) and the assigned taxonomy followed the criteria from García-Guzmán et al. (2017). OTUs were considered species only when they were assigned to a taxon.

Possible degree of endemism was determined following two criteria: (a) distribution in Mexico, if the species had a restricted distribution between study sites; and (b) world distribution, if the species had already been registered in another part of the world. This was determined by the presence of at least one ITS sequence in GenBank. Those species restricted to the Mexican Neotropics and not previously sequenced in another part of the world were considered potentially endemic.

15.3 Results and Discussion

15.3.1 Diversity of Edible Ectomycorrhizal Fungi in the Mexican Neotropics

DNA sequences from the ITS region were obtained for approximately 1300 sporocarps and 2683 ectomycorrhizae, most of these acquired from the "Eje Neovolcánico Transversal" sampling sites. Approximately 4000 samples have been sequenced. These sequences, grouped at 97% of nucleotide similarity, represent 804 fungal species. Out of the 804 species, 693 are ectomycorrhizal. The remaining 111 species are parasitic fungi, saprobes, or endophytes present in sampled mycorrhizae or from fruit bodies of saprobe species. In Mexico, 371 species of wild edible mushrooms are consumed traditionally by indigenous and mestizo people and 229 of these are ectomycorrhizal (Garibay-Orijel and Ruan-Soto 2014). So these 229 edible mushrooms represent 33% of the known ectomycorrhizal fungi from Mexican Neotropics.

The 693 species of the EMF collected belong to 85 genera. The genera with the highest species richness were *Tomentella* (77 species), *Inocybe* (57), *Russula* (54), *Sebacina* (47), *Ramaria* (30), *Amanita* (23), *Thelephora* (21), *Lactarius* (19), and *Clavulina* (16) (Fig. 15.1). These numbers are not estimates, but effectively collected species either as fruit bodies or mycorrhizae. Owing to this sampling effort, one of the most important contributions of this work was the collection of more than 100 fruit bodies of the genus *Tomentella*, out of which only four species were known to Mexico. It is now obvious that this genus is more diverse than what was previously thought. Furthermore, most of these species are new to science (Alvarez-Manjarrez et al. 2016). We were also able to collect more than 50 specimens of the family *Sebacinaceae*, out of which only three species were previously known to Mexico.

Information was generated on the geographic distribution of all species. Species with the widest distribution in the temperate forests were: *Tuber separans* s.l., *Rhizopogon* group *ellenae* sp. 1, *Laccaria vinaceobrunnea*, *Sistotrema confluens*, *Laccaria trichodermophora*, *Amanita rubescens* s.l., *Rhizopogon* group *ellenae* sp. 2, *Suillus pseudobrevipes*, *Helvella* cf. *lacunosa*, and *Laccaria laccata* s.l. Out of these species, *A. rubescens* s.l., *H.* cf. *lacunosa*, *L. laccata* s.l., *L. trichodermophora*, *L. vinaceobrunnea*, and *S. pseudobrevipes* are considered edible fungi and traditionally part of the Mexican cuisine (Garibay-Orijel and Ruan-Soto 2014).

When we include all of the species collected, either by fruit bodies or mycorrhizae, into an accumulation curve with 50 random permutations (Fig. 15.2), it is evident that we have not achieved to reduce the steep slope of species diversity. The diversity of EMF is much greater than what our sampling has managed to capture. Moreover, the richness estimator, Chao 1, determined 2611 potential EMF species for the Mexican Neotropics. Most of our sampling was centered in the temperate zone of the Neotropics; therefore, we do not have sufficient data of the tropical ecosystems in Mexico. Including more sampling sites from the tropics would surely increase Chao 1's estimation. Another reason why the species accumulation curve

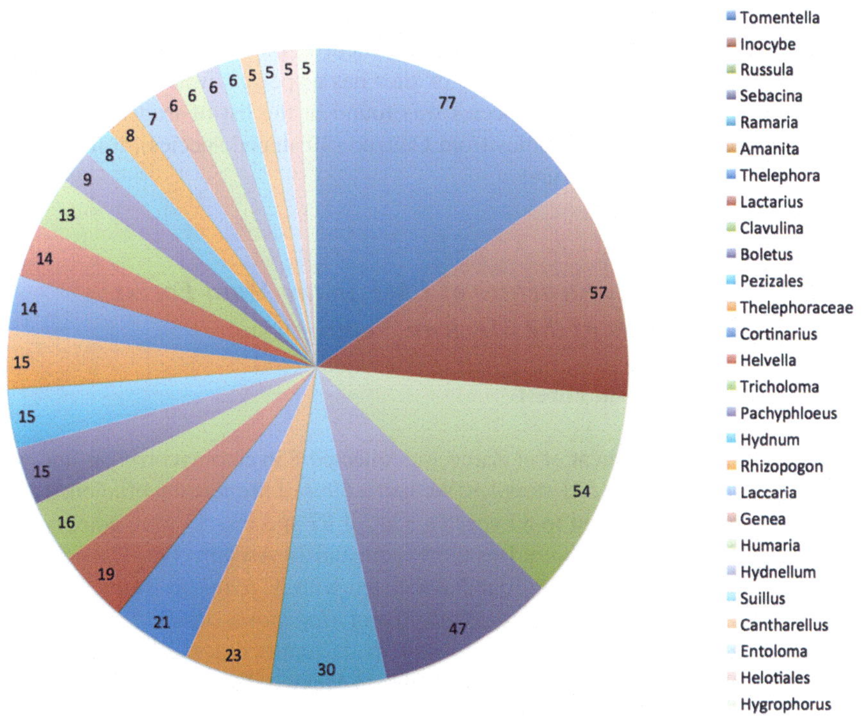

Fig. 15.1 Species richness of ectomycorrhizal fungi by genera. Only genera with the highest richness are shown

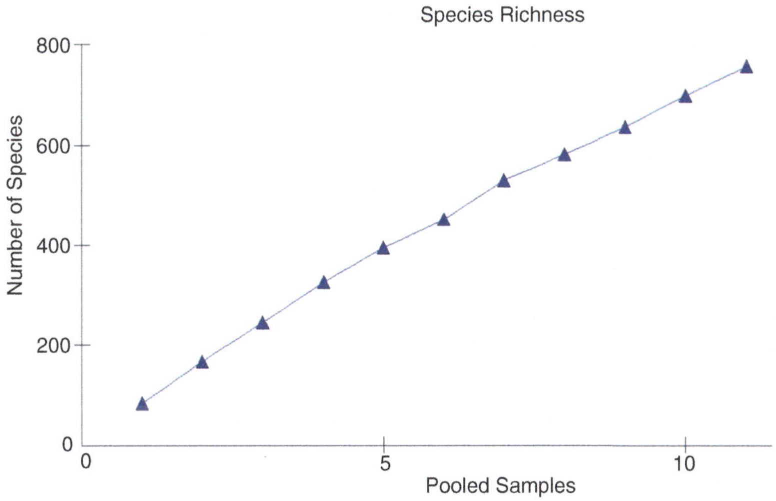

Fig. 15.2 Ectomycorrhizal fungal species accumulation curve of the Mexican Neotropics

has not reached a plateau is that there are very few shared species between study sites: a high level of beta diversity. This is evident in the results of the similarity analysis of the sampling sites. No pair of sites has a species overlap of over 25%. The greatest number of shared species was found in the list of fruit bodies from Temascaltepec and the fruit bodies from Cuitzeo (30 shared species), followed by "Nevado de Toluca" and Amanalco.

15.3.2 Community Structure of EMF in the Main Forest Ecosystems of the Mexican Neotropics

15.3.2.1 *Abies religiosa* Forest

In the *Abies religiosa* forest of el Zarco we collected 856 ectomycorrhizas and we obtained sequences from 591 samples. We had a 69% of sequencing efficiency and the sequences corresponded to 87 species. Out of 87 species, 83 were ectomycorrhizal fungi from which 69 were Basidiomycota and 14 Ascomycota. The ectomycorrhizal species of edible fungi found associated with *A. religiosa* were *Clavulina* cf. *cinerea*, *Clavulina* aff. *rugosa*, *Helvella* cf. *lacunosa*, *Russula americana*, *Russula* cf. *chloroides*, and *Tremelloscypha* sp. 1 (reported as *T. dichroa*). These species have been previously reported as edible in Mexico (Franco-Mass et al. 2012; Burrola et al. 2013; Bandala et al. 2014).

The families with the highest species richness were Inocybaceae (21 species), Thelephoraceae (15 species), Russulaceae, and Sebacinaceae (9 species). The genera with the highest richness were *Inocybe* (21 species), *Tomentella* (10 species), and *Russula* (8 species).

The families with the greatest abundance of mycorrhizae were Clavulinaceae (146 mycorrhizae), Thelephoraceae (115), Inocybaceae (98), and Russulaceae (74). The genera with the greatest abundance of mycorrhizae were *Clavulina* (82 mycorrhizae), *Membranomyces* (62), and *Russula* (22). The species with the highest abundance of mycorrhizae were *Clavulina* cf. *cinerea* (14.39%), *Membranomyces* sp. 1 (10.88%), and *Russula* sp. 1 (3.86%) (Fig. 15.3). Simpson's diversity index for this forest was 0.95 and the Chao 1 richness estimator was 145.3. The 32.2% of species found had values of nucleotide similarity inferior to 97% with their closest sequence in GenBank.

15.3.2.2 *Alnus* spp. Forests

In the *Alnus* forests we collected 562 ectomycorrhizas and obtained 416 sequences, these represent a sequencing efficiency of 74%. We found a total of 23 species in the four sampling sites. Out of the EMF associated with *Alnus*, only the genera *Lactarius* and *Clavulina* have been reported having edible species. Nevertheless, the ITS sequence of *Lactarius* sp. 1 had a 99% similarity to *L. omphaliformis*, *Lactarius* sp.

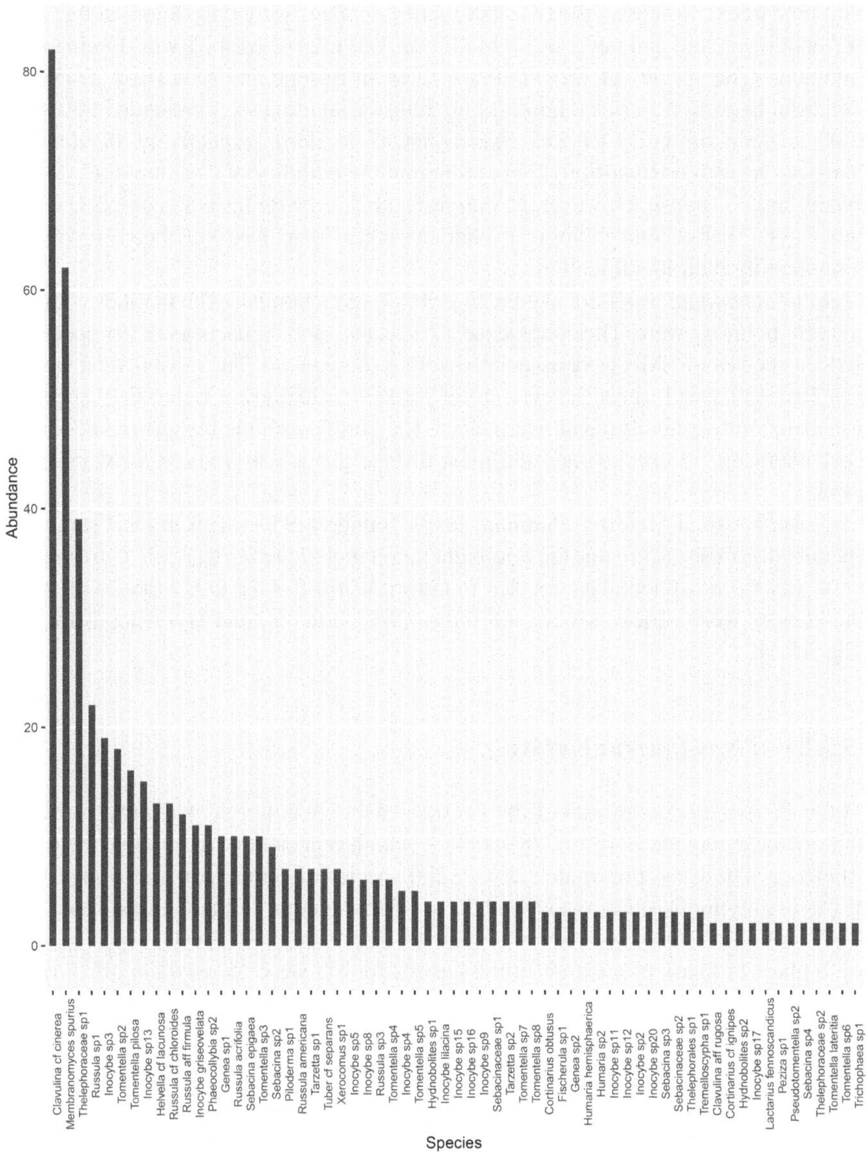

Fig. 15.3 Community structure of ectomycorrhizal fungi associated with *Abies religiosa* in el Zarco, Estado de México

4 had a 99% genetic similarity with *L. cyathuliformis*, both European nonedible species. Montoya et al. (2004) recorded the edible wild mushroom species of La Malinche (including *Alnus* forest). But the species recorded do not match our sampled species. A possibility is that edible mushrooms of this biome are mainly symbionts with pine trees and not *Alnus*, because the forests of La Malinche are mixed

with pine trees. Moreover, the two species of *Lactarius* were only found in the sites of Acatlán volcano and never at La Malinche. Yet, *Clavulina* sp. 1 was dominant in both sites. The species of *Clavulina* are eaten in Europe and Asia (Boa 2005), as well as in tropical America (Henkel et al. 2004). Clavulina sp. 1 presented 99% and 100% genetic similarity with *C. cinerea* and *C. cristata*, respectively. In Mexico, Garibay-Orijel and Ruan-Soto (2014) reported the edibility of *C. cinerea*, *C. coralloides*, and *C. rugosa*. However, *C. cinerea* and *C. cristata* form a taxonomic complex with shared morphological characteristics that has not been resolved, complicating their identification.

Alnus acuminata was associated with EMF of seven families. The families with the greatest richness were Thelephoraceae (7 species) and Sebacinaceae (6 species). Sixteen species of EMF were associated with *A. acuminata.* The most diverse genus was *Tomentella* with seven species. *Alnus jorullensis* had lower richness (9 species) distributed in five families and five genera. The family with the highest richness was Thelephoraceae (3 species) and the most diverse genus was *Tomentella* with three species.

Clavulina sp. 1 was the dominant taxon found in 35% of examined samples. Specifically, this species was the dominant taxon in *A. acuminata*, while *Cortinarius* sp. 1 was the most abundant species found in *A. jorullensis*. Other taxa with high abundances were various species of *Tomentella*, *Alnicola*, and Hymenogastraceae (Fig. 15.4).

15.3.2.3 *Pinus-Quercus* Forests

At the *Pinus-Quercus* forests of the Cuitzeo basin we collected 1600 mycorrhizae and obtained sequences from 978 samples, which represent 61.12% of sequencing efficiency. These samples matched with 206 species of ectomycorrhizal fungi. Out of these, 149 are Basidiomycota and 57 are Ascomycota. In the Cuitzeo basin, 76 species of edible fungi were reported (Reyes-García et al. 2009). The species with the highest biomass production in the basin's forests were *Lyophyllum* aff. *loricatum*, *Lactarius indigo* var. *indigo*, *Laccaria* sp., *Boletus frostii*, and *Amanita arkansana* (Torres-Gómez et al. 2018). Some of these species, like *Lactarius indigo* var. *indigo* and *Laccaria* sp., were found to form mycorrhizas with pine trees but in low frequencies.

The EMF species belonged to 19 families: Thelephoraceae (40 species), Russulaceae and Sebacinaceae (26), and Cortinariaceae (16). The genera with the highest richness were *Tomentella* (26 species), *Russula* (22), *Inocybe* (10), and *Sebacina* (8).

The most abundant species of EMF were Pezizomycetes sp.1, *Phialocephala fortinii*, *Russula* grp. *delica*, *Cenococcum geophilum*, *Rhizoscyphus* sp., *Lactarius* sect. *piperites*, *Russula* aff. *cyanoxantha*, *Wilcoxinia rehmii*, Sebacinaceae sp.1, Ascomycota sp.1, Atheliaceae sp.1, *Russula* grp. *pectinatoides*, *Lactarius* sect. *deliciosi*, Leotiomycetes sp.1, and *Inocybe* sp. Together, these species represent 26.33% of examined samples (Fig. 15.5). However, the most abundant EMF of the area

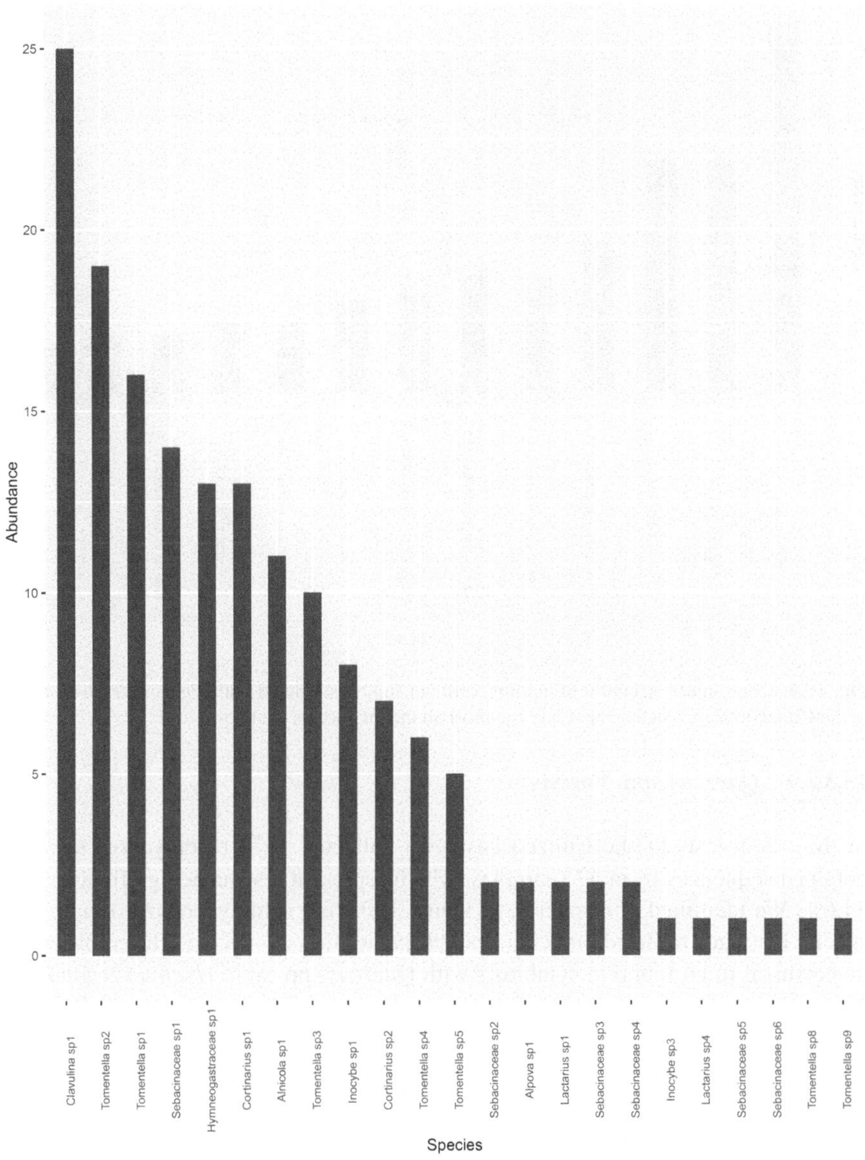

Fig. 15.4 Community structure of ectomycorrhizal fungi associated with *Alnus* forests

(Pezizomycetes sp.1) was only identifiable to the class level. These and many of the other studied species are potentially new taxa for science. The importance of anamorphic (i.e. without evident fruiting bodies) fungal species is clear in these ecosystems; within the most common EMF, anamorphic fungi constituted 60% of the richness and 65.82% of the abundance. We should also highlight the Russulaceae family with 5 species with evident fruiting bodies that in total colonized 27.03% of the roots.

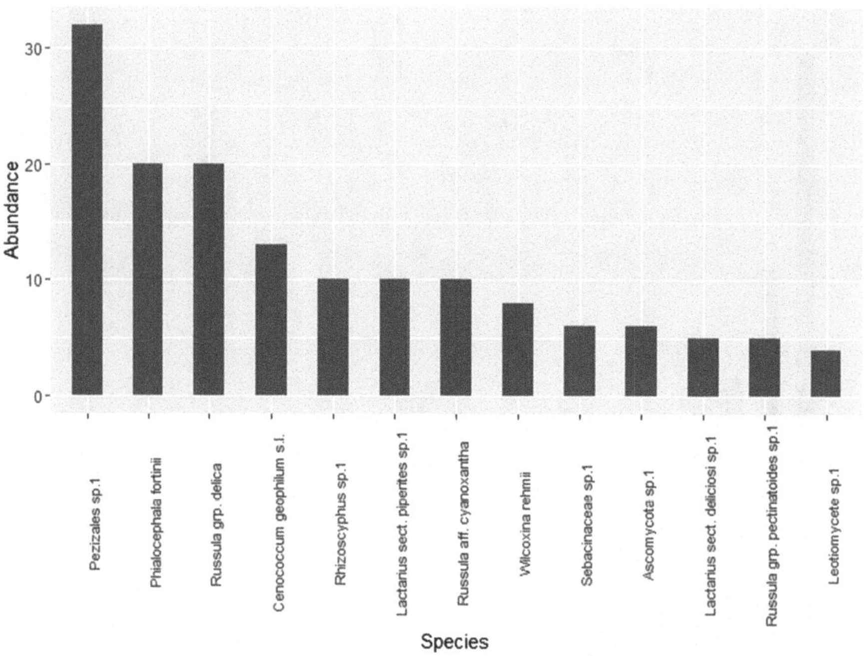

Fig. 15.5 Community structure of ectomycorrhizal fungi associated with *Pinus-Quercus* forests in the Cuitzeo basin, Michoacan. Only the most abundant species are shown

15.3.2.4 *Quercus* spp. Forests

In the oak forests of the Cuitzeo basin we collected 1671 ectomycorrhizas and obtained sequences from 573 samples, which represent a sequencing efficiency of 34.6%. We identified 158 species, of which 140 were ectomycorrhizal fungi, 112 species belonged to Basidiomycota, and 28 species to Ascomycota. The edible ectomycorrhizal fungi found in symbiosis with *Quercus* spp. were *Hydnum repandum*, *Lactarius yazooensis*, *Russula* aff. *brevipes*, *Russula* aff. *cyanoxanta*, and *Russula* aff. *olivacea.*

The community had 22 families; the best represented in terms of richness were Thelephoraceae (31 species), Russulaceae (24), Sebacinaceae (20), and Inocybaceae (13). The species belonged to 34 genera. *Tomentella* (25 species), *Russula* (22), *Sebacina* (19), *Inocybe* (13), and *Pachyphlodes* (5) were the most diverse in this community.

The families with the highest abundant mycorrhizae were Russulaceae (22.79%), Sebacinaceae (25.53%), Thelephoraceae (19.12%), Inocybaceae (6.62%), and Pyronemataceae (5.88%). The genera with higher abundance were *Sebacina* (25.53%), *Russula* (20.77%), *Tomentella* (16.91%), *Inocybe* (6.62%), *Humaria* (3.68%), *Pachyphlodes* (3.31%), and *Clavulina* (2.39%). These seven genera represented 79.21% of the mycorrhizae. The most abundant species in the community were *Sebacina* sp. 1 with 6.2% of the abundance, followed by *Sebacina* aff. *epigaea* with 4.1% and *Russula* aff. *curtipes* with 3.4% (Fig. 15.6). The site Aguila 1 showed the greatest richness (72 species), the greatest diversity (H′ = 3.99), less dominance (D = 0.019), and greater evenness (1/D = 53.78). In contrast, the site Aguila 2 showed a lower level of richness (38 species), the lowest diversity (H′ = 2.67), the

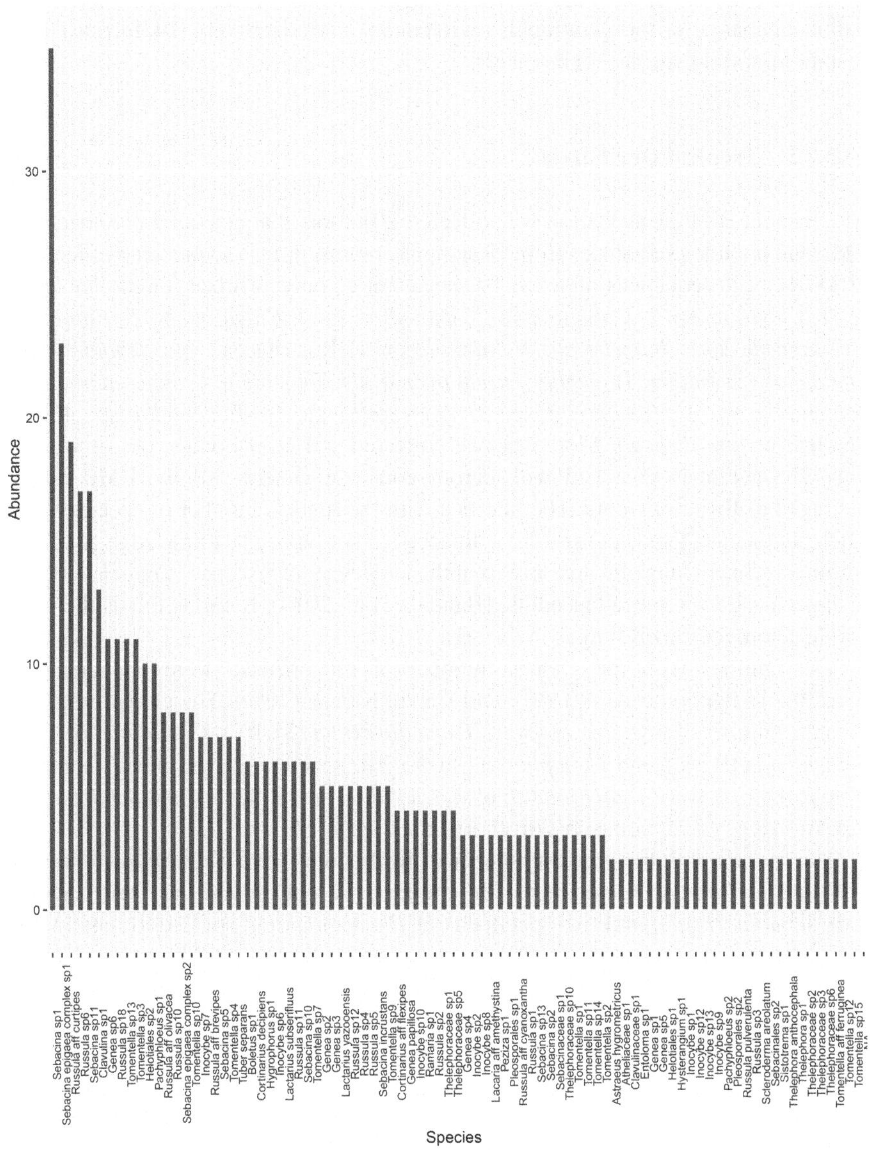

Fig. 15.6 Community structure of ectomycorrhizal fungi associated with *Quercus* forests in the Cuitzeo basin, Michoacán

greatest dominance (D = 0.118), and the lowest evenness (1/D = 8.45). Although these sites are geographically close, they shared few species. This scenario suggests a high species turnover within this forest (García-Guzmán et al. 2017). Nevertheless, both sites are very similar in diversity, dominance and evenness. The Chao 1 estimated a potential richness of 208 species including all sampling sites. The 35.71% of the species had a nucleotide similarity of less than 97% when compared to their best match in GenBank. These results reveal the lack of data and information in the public databases for this geographic region and suggest that these oak forests could sustain important species' endemism.

15.3.2.5 Tropical Dry Forest

At Chamela Biological Station we collected 201 roots that appeared colonized by EMF and obtained sequences from 98 samples, representing a sequencing efficiency of 48.7%. The sequences matched 19 species of ectomycorrhizal fungi: *Clavulina* sp. 1, *Clavulina* sp. 2, *Clavulina* sp. 3, *Clavulina* sp. 4, *Clavulina* sp. 5, *Entoloma* sp., *Inocybe* sp.1, *Inocybe* sp. 2, *Russula* sp. 1, Russulaceae sp., *Sebacina* sp., *Thelephora versatilis*, *Thelephora pseudoversatilis*, *Tomentella brunneoincrustata*, *Tomentella* sp. 1, *Tomentella* sp. 2, *Tomentella* sp. 3, *Tomentella* sp. 4 and *Tremelloscypha dichroa*. Moreover, we collected 101 fruit bodies, out of which, only 28 specimens matched with ectomycorrhizal genera. All the EMF found belonged to Basidiomycota. The families with the highest number of species were Thelephoraceae (8 species) and Clavulinaceae (5 species). The knowledge of edible ectomycorrhizal fungi in this area is null. However, in Chiapas, *Tremelloscypha dichroa* has been reported as edible (Bandala et al. 2014). It seems this is the only edible ectomycorrhizal fungus in the area.

The families with the greatest number of mycorrhizae were Sebacinaceae (41.2%), Thelephoraceae (33.9%), and Clavulinaceae (20%). The genera with the greatest number of mycorrhizae were *Tremelloscypha* (31.1%), *Clavulina* (20.1%), *Tomentella* (19.2%), and *Thelephora* (14.6%). The species with the greatest number of mycorrhizae were *T. dichroa* (26.6%), *Clavulina* sp.1 (14.6%), and *Sebacina* sp. (14.6%) (Fig. 15.7). Simpson's diversity index was 0.85.

The species with the greatest number of fruit bodies were *Thelephora versatilis* (18) and *T. dichroa* (6 specimens). In contrast, we only found one fruit body of *T. brunneoincrustata* and two specimens of *Phaeoclavulina* sp. Unfortunately, for the latter species, we did not find its ectomycorrhizal form. Moreover, for the remaining EMF, we did not find fruit bodies. Interestingly, *T. dichroa* had a high production of fruit bodies, as well as high ectomycorrhizal colonization. None of the mycorrhizae sequences found in this tropical dry forest had nucleotide similarities of 97% or more when compared with the sequences in GenBank. But they were 100% identical to the specimens used recently to describe them as new species (Ramírez-López et al. 2015; Álvarez-Manjarrez et al. 2016).

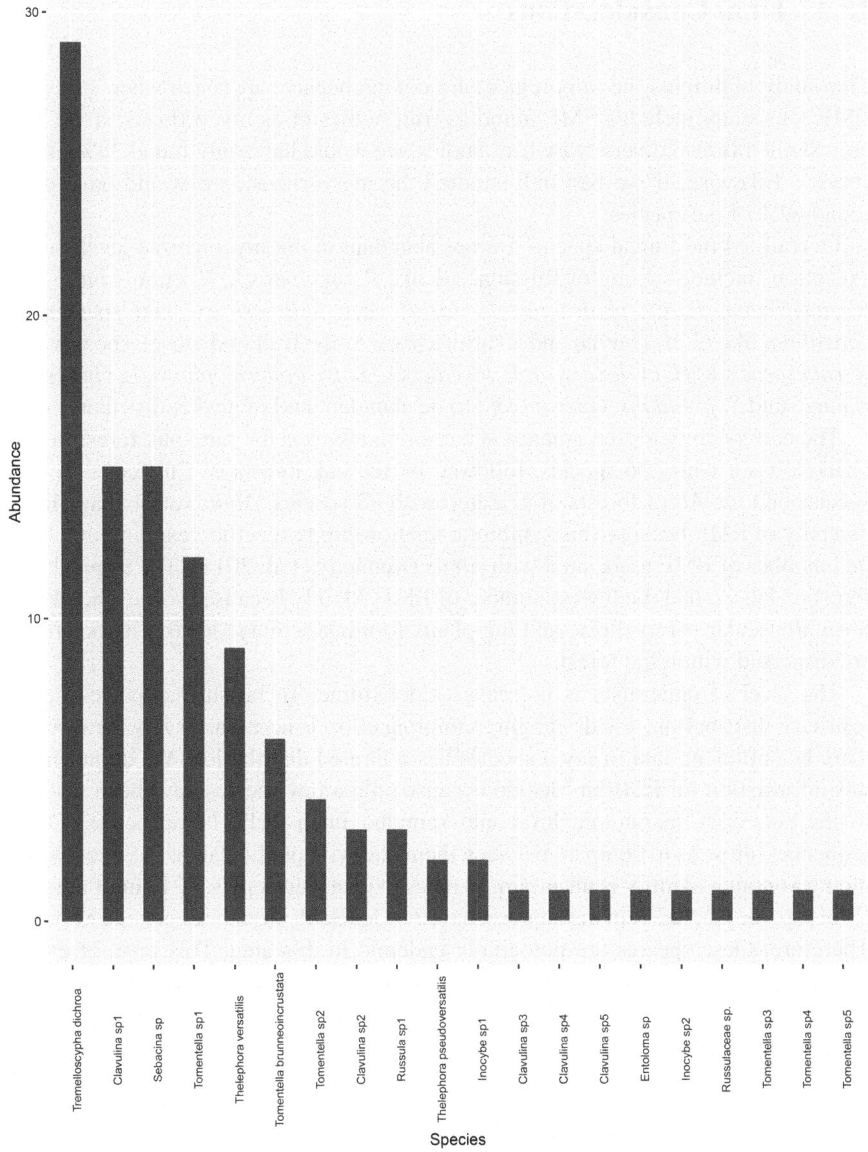

Fig. 15.7 Community structure of ectomycorrhizal fungi associated with the tropical dry forest of Chamela, Jalisco

15.4 Final Considerations

This study highlights the importance of a comprehensive approach when studying EMF. Our study includes EMF found as fruit bodies or as mycorrhizas. If we had only studied fungal diversity by fruit bodies, we would have only found 38% of the species. Likewise, if we had only studied the mycorrhizae, we would have only found 69% of the species.

Overall, edible fungal species are not abundant at the mycorrhizae level, some exceptions include species of Russulaceae like *R. brevipes* s.l., *R. cyanoxantha* s.l., *R. americana*, *R.* cf. *chloroides*, *Lactarius* sect. *deliciosi* sp., and species of *Clavulina*, like *C.* cf. *cinerea* and *C.* aff. *rugosa*. At the fruit body level, species like *A. rubescens* s.l., *H.* cf. *lacunosa*, *L. laccata* s.l., *L. trichodermophora*, *L. vinaceobrunnea*, and *S. pseudobrevipes* proved to be abundant and of a wide distribution.

The ecosystem with the greatest species richness was the pine-oak forest of the Cuitzeo basin with 206 species, followed by the oak forests of Cuitzeo with 153 species and the *Abies* forests of el Zarco with 83 species. *Alnus* forests have little diversity of EMF because this symbiotic relationship is specific, resulting in a limited number of EMF associated with *Alnus* (Kennedy et al. 2011). The tropical dry forest of Jalisco had the lowest number of EMF. Mainly because most of the plants form arbuscular mycorrhizae and the plants forming ectomycorrhizal associations are dispersed within the forest.

The level of endemism is not easy to determine. To be sure a species has a restricted distribution, a wide-ranging sampling effort is necessary. Only then would there be sufficient data to say a species has a limited distribution. We do not have this information for EMF in Mexico because only a few species have been studied to the necessary taxonomic level and sampled intensively. Nevertheless, DNA sequences allow us to compare between them and with public databases, even when their taxonomic identity is uncertain. In the Mexican Neotropics, 421 out of the 690 (71%) species included in the analysis have no GenBank sequences outside Mexico. Therefore, these species are potentially endemic to this area. This level of endemism may seem exaggerated, but it is comparable with the phanerogamic flora of the MegaMexico bioregion, which according to Rzedowski (1991) can hold 72% endemism. Locally, the endemism of EMF ranged between 32.2% in the *A. religiosa* forests up to a 100% in the tropical dry forest.

Acknowledgments This research was funded by the projects UNAM PAPIIT-IN218210, UNAM PAPIIT IV201015, CONACYT-239266 y CONACyT-Michoacan MICH-2009-C05-112966.

References

Aguilar-Romero R, García-Oliva F, Pineda-García F, Torres I, Peña-Vega E, Ghilardi A, Oyama H (2016) Patterns of distribution of nine *Quercus* species along an environmental gradient in a fragmented landscape in Central Mexico. Bot Sci 94:471–482

Altschul SF, Madden TL, Schaffer AA, Zhang JH, Zhang Z, Miller W, Lipman DJ (1997) Gapped BLAST and PSI-BLAST—a new generation of protein database search programs. Nucleic Acids Res 25:3389–3402

Alvarez-Manjarrez J, Garibay-Orijel R, Smith ME (2018) Caryophyllales are the main hosts of a unique set of ectomycorrhizal fungi in a Neotropical dry forest. Mycorrhiza 28:103–115

Alvarez-Manjarrez J, Villegas-Ríos M, Garibay-Orijel R, Contreras-Pacheco M, Kõljalg U (2016) Tomentella brunneoincrustata, the first described species of the Pisonieae-associated Neotropical Tomentella clade, and phylogenetic analysis of the genus in Mexico. Mycol Prog 15:10. https://doi.org/10.1007/s11557-015-1152-x

Argüelles-Moyao A, Garibay-Orijel R, Márquez-Valdelamar LM, Arellano-Torres E (2016) *Clavulina-Membranomyces* is the most important lineage within the highly diverse ectomycorrhizal fungal community of *Abies religiosa*. Mycorrhiza 27(1):53–65

Bandala VM, Montoya L, Villegas R, Cabrera T, Gutiérrez MJ, Acero T (2014) "Nangañaña" (*Tremelloscypha gelatinosa*, Sebacinales) hongo silvestre comestible del bosque tropical deciduo en la depresión central de Chiapas, México. Acta Bot Mex 106:149–159

Boa E (2005) Los hongos silvestres comestibles: Perspectiva global de su uso e importancia para la población 17, Productos Forestales no Maderables. FAO, Rome

Brundrett MC (2009) Mycorrhizal associations and other means of nutrition of vascular plants: understanding the global diversity of host plants by resolving conflicting information and developing reliable means of diagnosis. Plant Soil 320:37–77

Burrola-Aguilar C, Garibay-Orijel R, Argüelles-Moyao A (2013) *Abies religiosa* forests harbor the highest species density and sporocarp productivity of wild edible mushrooms among five different vegetation types in a neotropical temperate forest region. Agric Syst 87:1101–1115

Cifuentes J, Villegas M, Pérez-Ramírez L (1986) Hongos. In: Lot A, Chiang F (eds) Manual de Herbario. Consejo Nacional de la Flora de México AC, Mexico, pp 55–64

Finlay RD, Read DJ (1986) The structure and function of the vegetative mycelium of ectomycorrhizal plants II. The uptake and distribution of phosphorus by mycelial strands interconnecting host plants. New Phytol 103:157–165

Franco-Mass S, Burrola-Aguilar C, Arana-Gabriel Y (2012) Hongos silvestres comestibles: un recurso forestal no maderable del Nevado de Toluca. Universidad Autónoma del Estado de México, Mexico

García-Guzmán OM, Garibay-Orijel R, Hernández E, Arellano-Torres E, Oyama K (2017) Word-wide meta-analysis of Quercus forests ectomycorrhizal fungal diversity reveals southwestern Mexico as a hotspot. Mycorrhiza 27:811–822

Gardes M, Bruns TD (1993) ITS primers with enhanced specificity for basidiomycetes—application to the identification of mycorrhizae and rusts. Mol Ecol 2:113–118

Garibay-Orijel R, Ruan-Soto F (2014) Listado de los hongos silvestres consumidos como alimento tradicional en México. In: Moreno-Fuentes A, Garibay-Orijel R (eds) La Etnomicología en México, Estado del arte. Red de Etnoecología y Patrimonio Biocultural (CONACYT), UAEH, Instituto de Biologia, UNAM, Sociedad Mexicana de Micología, Asociación Etnobiológica Mexicana, GIDEM, Sociedad Latinoamericana de Etnobiología, Mexico, pp 95–115

Garibay-Orijel R (2008) Recuadro 13.1 Importancia funcional de los hongos ectomicorrizógenos. In: Soberón J, Halffter G, Llorente-Bousquets J (eds) Capital natural de México, Conocimiento actual de la biodiversidad, vol I. CONABIO, Maxico, pp 373–375

Gernandt D, Pérez-de la Rosa JA (2014) Biodiversidad de Pinophyta (coníferas) en México. Rev Mex Biodiv 85:126–133

Halling RE (1996) Recommendations for collecting mushrooms. In: Alexiades MN (ed) Selected guidelines for ethnobotanical research: a field manual. New York Botanical Garden, New York, pp 135–141

Henkel T, Aime MC, Chin M, Andrew C (2004) Edible mushrooms from Guyana. Mycologist 18:104–111

Horton TR, Bruns TD (2001) The molecular revolution in ectomycorrhizal ecology: peeking into the black-box. Mol Ecol 10:1855–1871

INEGI (2011) Anuario estadístico del Estado de México. http://inegi.mx/est/contenidos/espanol/sistemas/aee11/mex/mapas.pdf. Accessed 14 May 2013

Izzo DA, Meyer M, Trappe MC, North M, Bruns TD (2005) Hypogeous ectomycorrhizal fungal species on roots and in small mammal diet in a mixed-conifer forest. For Sci 51:243–254

Kennedy P, Garibay-Orijel R, Higgins L, Ángeles-Argáiz R (2011) Ectomycorrhizal fungi in Mexican *Alnus forests* support the host co-migration hypothesis and continental-scale patterns in phylogeography. Mycorrhiza 21:559–568

Largent D, Johnson D, Watling R (1984) How to identify mushrooms to genus III: microscopic features. Mad River Press, Eureka, CA

Mendoza ME, Granados EL, Geneletti D, Pérez-Salicrup EL, Salinas V (2011) Analyzing land cover and land use change processes at watershed level: a multitemporal study in the Lake Cuitzeo watershed, Mexico (1975–2003). Appl Geogr 31:237–250

Montoya A, Kong A, Estrada-Torres A, Cifuentes J, Caballero J (2004) Useful wild fungi of La Malinche National Park, México. Fung Div 17:115–143

Peay K, Kennedy P, Bruns TD (2008) Fungal community ecology: a hybrid beast with a molecular master. Bioscience 58:799–781

Ramírez-López I, Villegas-Ríos M, Salas-Lizana R, Garibay-Orijel R, Alvarez-Manjarrez J (2015) *Thelephora versatilis* and *Thelephora pseudoversatilis*: two new cryptic species with polymorphic basidiomes inhabiting tropical deciduous and sub-perennial forests of the Mexican Pacific coast. Mycologia 107(2):346–358

Reyes-García MG, Gómez-Peralta M, Zamora-Equihua V (2009) Guía de Hongos de los alrededores de Morelia. Universidad Michoacana de San Nicolás de Hidalgo, Morelia

Rzedowski J (1991) Diversidad y orígenes de la flora fanerogámica de México. Acta Bot Mex 14:3–21

Schoch CL, Seifert KA, Huhndorf S, Robert V, Spouge JL, Levesque CA, Chen W, Fungal Barcoding Consortium (2012) Nuclear ribosomal internal transcribed spacer (ITS) region as a universal DNA barcode marker for fungi. PNAS 109:6241–6246

Taberlet P, Gielly L, Pautou G, Bouvet J (1991) Universal primers for amplification of three non-coding regions of chloroplast DNA. Plant Mol Biol 17:1105–1109

Torres-Gómez M, Garibay-Orijel R, Casas A, Pérez-Salicrup DR (2018) Scattered ectomycorrhizal trees intermingled within *Cupressus lusitanica* plantations sustain the diversity and availability of edible mushrooms. Agric Syst 92:575–588

van der Heijden M, Martin FM, Selosse MA, Sanders IR (2015) Mycorrhizal ecology and evolution: the past, the present, and the future. New Phytol 205:1406–1423

White TJ, Bruns TD, Lee SB, Taylor JW (1990) Amplification and direct sequencing of fungal ribosomal RNA genes for phylogenetics. In: Innis MA, Gelfand DH, Sninsky JJ, White TJ (eds) PCR protocols-a guide to methods and applications. Academic, New York, pp 315–322

Chapter 16
A Checklist of Ectomycorrhizal Mushrooms Associated with *Quercus humboldtii* in Colombia

Natalia Vargas and Silvia Restrepo

16.1 Introduction

The native *Quercus humboldtii* is a dominant species in the Colombian Andean mountains, expanding from the Darien in Panama (8° N) to the southern montane cordilleras in Colombia (1° N) (Pulido et al. 2006; Cárdenas and Salinas 2006; Orwa et al. 2009), within a wide altitudinal range from 750 to 3450 m asl (Fundación Natura 2007; Avella and Cárdenas 2010). These ecosystems occupy small continuous and discontinuous relicts in the Colombian departments of Antioquia, Boyacá, Caldas, Cauca, Cundinamarca, Chocó, Huila, Nariño, Quindio, Santander, Valle del Cauca, and Tolima (Myers and Lynch 1997; Pulido et al. 2006; Fundación Natura 2007).

Oak forests establish symbiotic relationships with ectomycorrhizal mushrooms (ECM). This beneficial ecological interaction is established between plant roots and fungal mycelium, playing an essential role in the dynamics of forest ecosystems. The association allows the exchange of nutrients, especially phosphorus and nitrogen, from the fungi to the host plant and carbohydrates from the host to the fungi, and it constitutes an overall communication system among several trees by translocating nutrients (Read 1998; Pérez-Moreno and Read 2004).

Most studies related to macrofungi in Colombia have focused on oak forests (*Q. humboldtii*), because it is estimated that fungal diversity in these ecosystems is high (Franco-Molano et al. 2000). Nearly 99% of the local fungal diversity has been reported in the Andean Mountains, including saprotrophic, pathogenic, entomopathogenic, and ectomycorrhizal fungi (Vasco-Palacio and Franco-Molano 2013). Moreover, native oak forests are the habitat for many ectomycorrhizal species with a potential high value, because of their nutritive and medicinal characteristics and their importance to local people as an economic income (Boa 2004, Pérez-Moreno

N. Vargas (✉) · S. Restrepo
Laboratory of Mycology and Plant Pathology, Universidad de Los Andes, Bogotá, Colombia
e-mail: n.vargas20@uniandes.edu.co; srestrep@uniandes.edu.co

J. Pérez-Moreno et al. (eds.), *Mushrooms, Humans and Nature in a Changing World*, https://doi.org/10.1007/978-3-030-37378-8_16

2012). This fact raises the attention on managing good practices of the ecosystem products.

In this chapter, we compile information on ECM species and their distribution in Colombia, based on national reports of specimens collected in *Q. humboldtii* forests. Regarding the use of mushrooms, we provide references in which the edibility of some ectomycorrhizal mushrooms was reported. We include 2 national and 16 local species records from the departments of Boyacá, Cundinamarca, and Santander. We suggest that more studies aiming to assess the regional fungal diversity are key components to move forward in the knowledge of the Colombian fungi and their conservation.

16.2 Methods

16.2.1 Sampling

Fruiting bodies of ECM associated with *Quercus humboldtii* were collected in the forests of Boyacá and Santander. The departments distributed along the Andean mountains present two highest precipitation seasons during the year (Guzmán et al. 2014): April–June and October–December, which favour the fructification of fungal species. We collected specimens during the rainy season of April–May 2014 and 2015 in the departments of Boyacá (Municipio de Villa de Leyva, Vereda Capilla 05°39′ 26.78″ N, 73°30′ 46.41″ O; Municipio de Arcabuco, Vereda Piedras Blancas, 05°48.546″ N, 73°28.751″ O; Municipio de Arcabuco, 5°45′ 35.38″ N, 73°26′ 47.10″ O) and Santander (Vereda San José de La Montaña 06°02′ 29.82″ N, 73°00′ 02.8″ O). We included records from Cundinamarca based on previously collected specimens stored in the ANDES Herbarium (Universidad de los Andes, Bogotá) and registered in the SPECIFY database version 6.6.02 software (www.specifysoftware.org).

16.2.2 Descriptions

Macroscopic and microscopic features were analyzed for each collected sample. Fruit bodies were dried and packaged in plastic bags, stored in the ANDES Herbarium (Universidad de los Andes, Bogotá), and registered in the SPECIFY database. We used taxonomic keys and guides by Arora (1986), Halling (1989), Halling and Mueller (1999), Franco-Molano et al. (2000), Tulloss (2000, 2002, 2005), Mata et al. (2003), Halling and Mueller (2005), and Phillips (2005).

16.2.3 Search for ECM Reports in Colombia, the Mycorrhizal Trophic Status and the Edibility of Wild Mushrooms

We made an extensive literature search on fungal diversity lists for the country, with the earliest record by Hooker and Kunth in 1822 and until 2018. We included reports from published papers, field guides, and books. Articles showing the ectomycorrhizal trophic status of fungal genera included in our ECM checklist were searched in the ISI Web of Knowledge (www.webofknowledge.com). The family and author of each species followed Index Fungorum (http://www.indexfungorum.org/names/names.asp). Finally, we provide references in which the edibility of some ectomycorrhizal species is reported (Smith 1964, Arnolds 1995, Polese and Lamaison 1999, Boa 2004, Phillips 2005, Pérez-Moreno et al. 2010, Burrola-Aguilar et al. 2012, Eyssartier et al. 2011, Smith and Bonito 2012).

16.3 Results and Discussion

A total of 120 ECM species were compiled from reports indicating their association to *Q. humboldtii* forests (Table 16.1). Figure 16.1 shows a map of Colombia with the number of ECM species per family and per department. About 11 out of 14 departments located in the Andean mountain system have records of ECM species, except Risaralda, Chocó, and Norte de Santander. Caldas, Quindio, and Tolima are among the departments with the lowest number of ECM reports. The department with the highest number of repords is Antioquia, followed by Boyacá and Cundinamarca. The fungal families with the major number of ECM species per department were Boletaceae (in five departments), Amanitaceae (in two departments), and Russulaceae (in two departments).

Information on the ectomycorrhizal trophic status of fungal genera was confirmed by literature regarding anatomical, chemical, and/or molecular analyses (Table 16.2).

A total of 18 species are new records: 2 national and 16 local records for the departments of Boyacá, Cundinamarca, and Santander (Table 16.3, Fig. 16.2)

16.3.1 Morphological Description of the Two National Records

Inocybe tahquamenonensis D.E. Stuntz 1954. Material studied NVE 303 ANDES_F802 Fig. 2A, A1, A2—Colombia, Boyacá, Municipio de Arcabuco, Vereda Peñas Blancas, 20 May 2012, in *Q. humboldtii.* This species occurs in north temperate regions in eastern North America (Phillips 2005; Matheny and Moreau 2009). *Pileus*: 1.5–4 cm cm wide, convex to plano-convex to decurved when mature, dark purplish-brown to reddish or blackish-brown, with pronounced scales concolorus

Table 16.1 Ectomycorrhizal species reported in *Q. humboldtii* in Colombia. Species reported as edible are indicated

Species	Department	References in which the species is reported in Colombia; the references reporting the species associated with *Q. humboldtii* are in bold. Species reported as edible, without confirmed consumption, are indicated with the word "Edible" (Franco-Molano et al. 2000), followed by the references in brackets [].
Basidiomycota		
Agaricales		
Amanitaceae		
Amanita		
A. advena (Tulloss et al. 1992)	ANT	**Tulloss et al. (1992)**, Franco-Molano and Uribe-Calle (2000), Vasco-Palacio and Franco-Molano (2013)
A. arocheae (Tulloss et al. 1992)	ANT	**Tulloss et al. (1992)**, Franco-Molano and Uribe-Calle (2000), **Halling and Mueller (2005)**, Vasco-Palacio and Franco-Molano (2013)
A. aureomonile (Tulloss and Franco-Mol. 1992)	VAL	**Tulloss et al. (1992)**, Franco-Molano and Uribe-Calle (2000), Vasco-Palacio and Franco-Molano (2013)
A. brunneolocularis (Tulloss et al. 1992)	ANT, BOY, VAL	Saldarriaga et al. (1988), **Tulloss et al. (1992)**, **Franco-Molano and Uribe-Calle (2000)**, Franco-Molano et al. (2000), Halling and Mueller (2005), Vasco-Palacio and Franco-Molano (2013), Soto-Medina and Bolaños-Rojas (2013).
A. citrina (Pers. 1797)	BOY	**Vargas et al. (2017)**.
A. colombiana (Tulloss et al. 1992)	ANT, BOY, SAN	**Tulloss et al. (1992)**, **Franco-Molano et al. (2000)**, Franco-Molano and Uribe-Calle (2000), Vasco-Palacio and Franco-Molano (2013), **Vargas et al. (2017)**, **Vargas and Restrepo (2019)**.
A. flavoconia (G.F. Atk. 1902)	ANT, CUN, BOY, SAN	Saldarriaga et al. (1988), **Tulloss et al. (1992)**, Franco-Molano and Uribe-Calle (2000), **Franco-Molano et al. (2000)**, **Halling and Mueller (2005)**, Cepero de García et al. (2012), Vasco-Palacio and Franco-Molano (2013), **Vargas et al. (2017)**.
A. fuligineodisca (Tulloss et al. 1992)	ANT, BOY, **CUN**, NAR, SAN	Saldarriaga et al. (1988), **Tulloss et al. (1992)**, Franco-Molano and Uribe-Calle (2000), **Franco-Molano et al. (2000)**, **Halling and Mueller (2005)**, Vasco-Palacio and Franco-Molano (2013), **Vargas et al. (2017)**, **this study**.
A. gemmata (Fr.) Bertill. 1866	CUN	Nasi (1977), **Guzmán and Varela (1978)**, Tulloss et al. (1992), Franco-Molano and Uribe-Calle (2000), Vasco-Palacio and Franco-Molano (2013). Edible [Boa 2004, Hall et al. 2011, Burrola-Aguilar et al. 2012]
A. humboldtii (Singer 1963)	CUN, NAR	**Singer (1963)**, Denis (1970), **Guzmán and Varela (1978)**, Pulido (1983), Tulloss et al. (1992), Wu et al. (1997), Vasco-Palacio and Franco-Molano (2013)
A. inaurata Secr. 1833	BOY	**Singer (1963)**, Denis (1970), Pulido (1983), Tullos et al. (1992), Vasco-Palacio and Franco-Molano (2013)

(continued)

Table 16.1 (continued)

Species	Department	References in which the species is reported in Colombia; the references reporting the species associated with *Q. humboldtii* are in bold. Species reported as edible, without confirmed consumption, are indicated with the word "Edible" (Franco-Molano et al. 2000), followed by the references in brackets [].
A. muscaria (L.) Lam. 1783	SAN	**Vargas et al. (2019)**
A. picea (Tulloss et al. 1992)	BOY	**Tulloss et al. (1992)**, Franco-Molano and Uribe-Calle (2000), Vasco-Palacio and Franco-Molano (2013)
A. sepultipes (Vargas and Restrepo 2019)	BOY, SAN	**Vargas and Restrepo (2019)**
A. sororcula (Tulloss et al. 1992)	ANT, BOY, SAN	**Tulloss et al. (1992)**, **Franco-Molano et al. (2000)**, Franco-Molano and Uribe-Calle (2000), Vasco-Palacio and Franco-Molano (2013), **Vargas et al. (2017)**, **Vargas and Restrepo (2019)**
A. virosa	BOY	**Vargas et al. (2017)**.
A. xylinivolva (Tulloss et al. 1992)	ANT, CAU, CUN, NAR, BOY, SAN	Saldarriaga et al. (1988), **Tulloss et al. (1992)**, Franco-Molano and Uribe-Calle (2000), **Franco-Molano et al. (2000)**, Vasco-Palacio and Franco-Molano (2013), **Vargas et al. (2017)**.
Cortinariaceae		
Cortinarius		
C. aurantiobrunneus (Ammirati et al. 2007)	SAN	**Vargas and Restrepo (2019)**
C. boyacensis Singer	BOY	**Singer (1963)**, Denis (1970), Wu et al. (1997), Franco-Molano and Uribe-Calle (2000), Vasco-Palacio and Franco-Molano (2013)
C. harrisonii (Ammirati et al.)	SAN	**Vargas and Restrepo (2019)**
C. iodes (Berk. and M.A. Curtis 1853)	ANT, **CUN**, NAR, BOY, **SAN**	**Franco-Molano et al.** (**2000**, 2010), López-Quintero et al. (2007), Cepero de García et al. (2012), Vasco-Palacio and Franco-Molano (2013), **this study**.
C. violaceus (L.) (Gray 1821)	ANT, **BOY**	**Franco-Molano et al.** (**2000**, 2010), Vasco-Palacio and Franco-Molano (2013). Edible [Smith 1964, Pérez-Moreno et al. 2010, Eyssartier et al. 2011, Burrola-Aguilar et al. 2012]. Not edible according to Phillips (2005)
Rozites colombiana (Halling and Ovrebo 1987)	ANT	Halling and Ovrebo (1987), Saldarriaga et al. (1988), Franco-Molano and Uribe-Calle (2000), **Franco-Molano et al. (2000)**, **Halling and Mueller (2005)**, Vasco-Palacio and Franco-Molano (2013)

(continued)

Table 16.1 (continued)

Species	Department	References in which the species is reported in Colombia; the references reporting the species associated with *Q. humboldtii* are in bold. Species reported as edible, without confirmed consumption, are indicated with the word "Edible" (Franco-Molano et al. 2000), followed by the references in brackets [].
Hydnangiaceae		
Laccaria		
L. amethystina Cooke	ANT, CUN	**Guzmán and Varela (1978)**, Mueller (1996), **Franco-Molano et al. (2000)**, Halling and Mueller (2005), Vasco-Palacio and Franco-Molano (2013). Edible [Boa 2004, Garibay-Orijel et al. 2007, Pérez-Moreno et al. 2010, Eyssartier et al. 2011, Burrola-Aguilar et al. 2012]
L. gomezii (Singer and Mueller 1988)	ANT, HUI	**Mueller and Singer (1988)**, Mueller (1996), Franco-Molano and Uribe-Calle (2000), Franco-Molano et al. (2000), Halling and Mueller (2005), Vasco-Palacio and Franco-Molano (2013)
L. laccata (Scop.) (Cooke 1884)	ANT, BOY, CAL, CUN, QUI	**Pulido (1983)**, Saldarriaga et al. (1988), Mueller (1996), Nieves-Rivera et al. (1997), Franco-Molano and Uribe-Calle (2000), **Franco-Molano et al. (2000)**, **Halling and Mueller (2005)**, Montoya et al. (2005), Betancur et al. (2007), López-Quintero et al. (2007), Vasco-Palacio and Franco-Molano (2013). Edible [Smith 1964, Polese and Lamaison 1999, Boa 2004, Phillips 2005, Pérez-Moreno et al. 2010, Eyssartier et al. 2011, Hall et al. 2011, Burrola-Aguilar et al. 2012]
L. ohiensis (Mont.) (Singer 1947)	VAL	**Mueller (1996)**, Soto-Medina and Bolaños-Rojas (2013), Vasco-Palacio and Franco-Molano (2013)
L. proxima (Boud.) Pat. 1887		**Mueller (1996)**, Vasco-Palacio and Franco-Molano (2013). Edible [Phillips 2005, Pérez-Moreno et al. 2010, Burrola-Aguilar et al. 2012].
Hygrophoraceae		
Hygrophorus		
H. hondurensis (Murrill) (Murrill 1912)	PNN	**Boekout and Pulido (1989)**, Franco-Molano et al. (2010), Vasco-Palacio and Franco-Molano (2013)
H. obconicus (Peck 1909)	PNN	**Boekout and Pulido (1989)**, Franco-Molano et al. (2010), Vasco-Palacio and Franco-Molano (2013)
H. quercuum (Singer 1973)	BOY	**Singer (1973)**, Wu et al. (1997), Franco-Molano and Uribe-Calle (2000)
Inocybaceae		
Inocybe		
I. calamistrata (Fr.) (Gillet 1876)	ANT	Franco-Molano et al. (2010)
I. fastigiata (Schaeff.) (Quél. 1872)	CUN	**Guzmán and Varela (1978)**

(continued)

Table 16.1 (continued)

Species	Department	References in which the species is reported in Colombia; the references reporting the species associated with *Q. humboldtii* are in bold. Species reported as edible, without confirmed consumption, are indicated with the word "Edible" (Franco-Molano et al. 2000), followed by the references in brackets [].
I. hystrix (Fr.) (P. Karst. 1879)	ANT	**López-Quintero et al. (2007)**, Vasco-Palacio and Franco-Molano (2013)
I. jalopensis (Murrill) (Singer 1958)	CUN	**Singer (1963)**, Denis (1970), **Guzmán and Varela (1978)**, Franco-Molano and Uribe-Calle (2000), Vasco-Palacio and Franco-Molano (2013)
I. rimosa (Bull.) P. (Kumm. 1871)	CUN	Guzmán and Varela (1978), Franco-Molano and Uribe-Calle (2000), Vasco-Palacio and Franco-Molano (2013)
I. tequendamae (Singer 1963)	CUN	**Singer (1963)**, Denis (1970), Guzmán and Varela (1978), Wu et al. (1997), Franco-Molano and Uribe-Calle (2000), Vasco-Palacio and Franco-Molano (2013)
I. tahquamenonensis (Stuntz 1954)	**BOY**	**This study**
Tricholomataceae		
Tricholoma		
T. cystidiosum (A.H. Sm. 1941)	ANT	**Saldarriaga et al. (1988)**, Franco-Molano and Uribe-Calle (2000), Vasco-Palacio and Franco-Molano (2013)
T. caligatum (Viv.) Ricken 1914	**BOY**	Cepero de García et al. (2012), **this study**. Edible [Boa 2004, Phillips 2005]
Tricholomopsis humboldtii Singer, Ovrebo and Halling 1990	ANT	Singer et al. (1990), Franco-Molina et al. (2000)
Boletales		
Boletaceae		
Aureoboletus auriporus (Peck) (Pouzar 1957)	ANT	**Franco-Molano et al. (2000)**, Franco-Molano and Uribe-Calle (2000), Vasco-Palacio and Franco-Molano (2013)
Austroboletus subvirens (Hongo) (Wolfe 1980)	ANT, HUI	**Halling (1989)**, **Franco-Molano et al. (2000)**, Franco-Molano and Uribe-Calle (2000), Vasco-Palacio and Franco-Molano (2013)
Boletellus		
B. ananas (M.A. Curtis) (Murrill 1909)	ANT, VAL	Halling (1989, 1996), **Franco-Molano et al. (2000)**, Franco-Molano and Uribe-Calle (2000), Vasco-Palacio and Franco-Molano (2013). Edible [Burrola-Aguilar et al. 2012]
B. russellii (Frost) Gilbert	CAU	**Halling (1989)**, Franco-Molano and Uribe-Calle (2000), Vasco-Palacio and Franco-Molano (2013). Edible [Smith 1964, Boa 2004, Burrola-Aguilar et al. 2012, Kuo and Methven 2014]

(continued)

Table 16.1 (continued)

Species	Department	References in which the species is reported in Colombia; the references reporting the species associated with *Q. humboldtii* are in bold. Species reported as edible, without confirmed consumption, are indicated with the word "Edible" (Franco-Molano et al. 2000), followed by the references in brackets [].
Boletus		
B. atkinsonianus (Murrill) Sacc. and (Trotter 1912)	ANT	**Halling (1989**, 1996), Vasco-Palacio and Franco-Molano (2013)
B. fuligineotomentosus (Singer 1973)	VAL	**Singer (1973)**, Halling (1989), Wu et al. (1997), Franco-Molano and Uribe-Calle (2000), Vasco-Palacio and Franco-Molano (2013)
B. neoregius (Halling and Muell. 1999)	ANT, CUN, **BOY**	**Franco-Molano et al. (2000**, 2010); **Halling and Mueller (2005)**; Vasco-Palacio and Franco-Molano (2013), **This study.**
B. orquidianus (Halling 1989)	ANT	Halling (1989), **Franco-Molano et al. (2000)**, Franco-Molano and Uribe-Calle (2000), Vasco-Palacio and Franco-Molano (2013)
B. pseudorubinellus (A.H. Sm. and Thiers 1971)	ANT, CAU	**Halling (1989)**, **Franco-Molano et al. (2000)**, Franco-Molano and Uribe-Calle (2000), Vasco-Palacio and Franco-Molano (2013)
B. pulverulentus (Opat. 1836)	CUN	**Halling (1989)**, **Franco-Molano et al. (2000)**, Franco-Molano and Uribe-Calle (2000), Vasco-Palacio and Franco-Molano (2013). Edible [Phillips 2005, Hall et al. 2011]
B. pyrrhosceles (Halling 1992)	ANT, NAR	Halling (1989), **Franco-Molano et al. (2000)**, Franco-Molano and Uribe-Calle (2000), Vasco-Palacio and Franco-Molano (2013)
B. reticulatus (Schaeff. 1763)	CAU	Hooker and Kunth (1822), NA. Edible [Boa 2004, Kuo and Methven 2014]
B. subtomentosus L. 1753	ANT, BOY	**Halling (1989)**, Franco-Molano and Uribe-Calle (2000), Vasco-Palacio and Franco-Molano (2013). Edible [Arnolds 1995, Phillips 2005]
B. truncatus (Singer, Snell and E.A. Dick) (Pouzar 1966)	ANT, BOY	**Halling (1989)**, Franco-Molano and Uribe-Calle (2000), Cepero de García et al. (2012), Vasco-Palacio and Franco-Molano (2013). Edible [Boa 2004, Hall et al. 2011]
Xerocomellus chrysenteron (Bull.) (Sutara 2008)	ANT, NAR, **BOY**	Saldarriaga et al. (1988), Franco-Molano et al. (2010), Vasco-Palacio and Franco-Molano (2013), **this study**. Edible [Arnolds 1995, Boa 2004, Phillips 2005, Hall et al. 2011, Burrola-Aguilar et al. 2012]. Sinonimo: *Boletus chrysenteron* Bull 1791
Leccinum		
L. andinum Halling	ANT	**Halling (1989)**, **Franco-Molano et al. (2000)**, Franco-Molano and Uribe-Calle (2000), Halling and Mueller (2005), Cepero de García et al. (2012), Vasco-Palacio and Franco-Molano (2013).

(continued)

Table 16.1 (continued)

Species	Department	References in which the species is reported in Colombia; the references reporting the species associated with *Q. humboldtii* are in bold. Species reported as edible, without confirmed consumption, are indicated with the word "Edible" (Franco-Molano et al. 2000), followed by the references in brackets [].
L. rugosiceps (Peck) Singer	ANT, CAU, TOL	Halling (1996), **Franco-Molano et al.** (**2000**, 2010), Halling and Mueller (2005), Vasco-Palacio and Franco-Molano (2013). Edible [Boa 2004, Phillips 2005, Hall et al. 2011]
L. talamancae Halling, L.D. Gómez and Lannoy	ANT	**Franco-Molano et al.** (**2000**, 2010), Halling and Mueller (2005), López-Quintero et al. (2007), Vasco-Palacio and Franco-Molano (2013)
Phylloporus		
P. phaeoxanthus Singer and L.D Gómez	ANT	**Franco-Molano et al.** (**2000**, 2010), Halling and Mueller (2005), Vasco-Palacio and Franco-Molano (2013)
P. fibulatus Singer, Ovrebo and Halling	ANT, NAR	**Singer et al. (1990)**, Halling et al. (1999), Wu et al. (1997), **Franco-Molano et al. (2000)**, Franco-Molano and Uribe-Calle (2000), Vasco-Palacio and Franco-Molano (2013)
P. centroamericanus (Singer and Gómez 1984)	ANT, **BOY**	**Franco-Molano et al.** (**2000**, 2010), Halling and Mueller (2005), Vasco-Palacio and Franco-Molano (2013)
P. purpurellus Singer	CAU	**Singer (1973)**, Wu et al. (1997), Halling et al. (1999), Franco-Molano and Uribe-Calle (2000), Vasco-Palacio and Franco-Molano (2013)
Pulveroboletus ravenelii (Berk. and M.A. Curtis) (Murrill 1909)	ANT, VAL	Boekout and Pulido (1989), **Franco-Molano et al.** (**2000**, 2010), Soto-Medina and Bolaños-Rojas (2013)
Strobilomyces confusus (Singer 1945)	HUI	**Halling (1989)**, Franco-Molano and Uribe-Calle (2000), Halling and Mueller (2005), Cepero de García et al. (2012), Vasco-Palacio and Franco-Molano (2013). Edible [Boa 2004, Phillips 2005, Hall et al. 2011, Burrola-Aguilar et al. 2012].
Tylopilus		
T. bulbosus (Halling and Muell. 2001)	ANT	**Sierra et al. (2011)**
T. indecisus (Peck) (Murrill 1909)	BOY	**Peña-Cañón and Henao-Mejía (2014)**. Edible [Peña-Cañón and Henao-Mejía 2014]
T. obscurus (Halling 1989)	ANT, **HUI**, SAN	**Halling (1989)**, **Franco-Molano et al.** (**2000**, 2010), **Halling and Mueller (2005)**, Vasco-Palacio and Franco-Molano (2013), **Vargas and Restrepo (2019)**, **this study**.
T. umbrosus (G.F. Atk) A.H. Sm. and Thiers	NAR	Franco-Molano et al. (2010), NA

(continued)

Table 16.1 (continued)

Species	Department	References in which the species is reported in Colombia; the references reporting the species associated with *Q. humboldtii* are in bold. Species reported as edible, without confirmed consumption, are indicated with the word "Edible" (Franco-Molano et al. 2000), followed by the references in brackets [].
Calostoma cinnabarinum Desv. 1809	ANT, HUI, CAL, **BOY**	Saldarriaga et al. (1988), Dumont and Umaña (1978), **López-Quintero et al. (2007)**, Betancur et al. (2007), Cepero de García et al. (2012), Vasco-Palacio and Franco-Molano (2013), **this study**.
Xanthoconium separans (Peck) halling and Both.	ANT, NAR	**Franco-Molano et al.** (**2000**, 2010). Edible [Boa 2004]
Gyroporaceae		
Gyroporus castaneus (Bull.) Quél. 1886	CAU	**Halling and Mueller (2005)**, Franco-Molano et al. (2010). Edible [Smith 1964, Polese and Lamaison 1999, Boa 2004, Phillips 2005, Hall et al. 2011]
Sclerodermataceae		
Scleroderma		
S. albidum Pat. and Trab.	CUN	**Guzmán and Varela (1978)**
Cantharellales		
Cantharellaceae		
Cantharellus		
C. cibarius Fr. 1821	ANT, CUN	**Guzmán and Varela (1978)**, Franco-Molano and Uribe-Calle (2000), **López-Quintero et al. (2007)**, Vasco-Palacio and Franco-Molano (2013). Edible [Smith 1964, Arnolds 1995, Polese and Lamaison 1999, Boa 2004, Phillips 2005, Garibay-Orijel et al. 2007, Pérez-Moreno et al. 2010, Eyssartier et al. 2011, Hall et al. 2011, Burrola-Aguilar et al. 2012, Kuo and Methven 2014]
C. cinnabarinus (Schwein.) Schwein. 1832	CUN	**Guzmán and Varela (1978)**, Franco-Molano and Uribe-Calle (2000), Vasco-Palacio and Franco-Molano (2013). Edible [Smith 1964, Boa 2004, Phillips 2005, Garibay-Orijel et al. 2007,
C. lateritius (Berk.) (Singer 1951)	**CUN**, NAR	**Petersen and Mueller (1992)**, Franco-Molano and Uribe-Calle (2000), Cepero de García et al. (2012), Vasco-Palacio and Franco-Molano (2013), Henkel et al. (2014), **this study**. Edible [Phillips 2005]
Craterellus		
C. boyacensis Singer 1963	ANT, BOY, HUI, SAN	**Singer (1963)**, Denis (1970), Wu et al. (1997), Franco-Molano and Uribe-Calle (2000), Halling and Mueller (2005), Vasco-Palacio and Franco-Molano (2013), Henkel et al. (2014), **Vargas and Restrepo (2019)**.
C. fallax A.H. Sm. 1968	CUN, **BOY**	Wu and Mueller (1995), Franco-Molano and Uribe-Calle (2000), Vasco-Palacio and Franco-Molano (2013), Henkel et al. (2014), **this study**. Edible [Boa 2004, Phillips 2005, Hall et al. 2011]

(continued)

Table 16.1 (continued)

Species	Department	References in which the species is reported in Colombia; the references reporting the species associated with *Q. humboldtii* are in bold. Species reported as edible, without confirmed consumption, are indicated with the word "Edible" (Franco-Molano et al. 2000), followed by the references in brackets [].
Pseudocraterellus sinuosus (Fr.) (Corner 1958)	ANT	**Wu and Mueller (1995)**, Franco-Molano and Uribe-Calle (2000), Vasco-Palacio and Franco-Molano (2013), Henkel et al. (2014).
Hydnaceae		
Hydnum repandum L. 1753	ANT	Henao (1989), **López-Quintero et al. (2007)**, Cepero de García et al. (2012), Vasco-Palacio and Franco-Molano (2013). Edible [Arnolds 1995, Polese and Lamaison 1999, Boa 2004, Phillips 2005, Garibay-Orijel et al. 2007, Pérez-Moreno et al. 2010, Eyssartier et al. 2011, Burrola-Aguilar et al. 2012]
Gomphales		
Gomphaceae		
Ramaria		
R. botrytis (Pers.) (Ricken 1918)	BOY	**Peña-Cañón and Henao-Mejía (2014)**. Edible [Polese and Lamaison 1999, Boa 2004, Pérez-Moreno et al. 2010, Eyssartier et al. 2011, Hall et al. 2011, Burrola-Aguilar et al. 2012, Peña-Cañón and Henao-Mejía 2014]
R. cyaneigranosa (Marr and Stuntz 1974)	BOY	**Peña-Cañón and Henao-Mejía (2014)**. Edible [Peña-Cañón and Henao-Mejía 2014]
R. flava (Schaeff.) Quél. 1888	BOY	**Peña-Cañón and Henao-Mejía (2014)**. Edible [Boa 2004, Garibay-Orijel et al. 2007, Pérez-Moreno et al. 2010, Burrola-Aguilar et al. 2012, Peña-Cañón and Henao-Mejia 2014]
R. formosa (Pers.) Quél. 1888	CAL	**Betancur et al. (2007)**, Vasco-Palacio and Franco-Molano (2013). Edible [Boa 2004, Hall et al. 2011, Burrola-Aguilar et al. 2012, Peña-Cañón and Henao-Mejía 2014]. Eyssartier et al. (2011) report the species as toxic.
R. secunda (Berk.) Corner	BOY	**Ruíz and Henao-Mejía (2006)**. Edible [Hall et al. 2011, Peña-Cañón and Henao-Mejía 2014]
Hymenochaetales		
Hymenochaetaceae		
Coltricia		
C. cinnamomea (Jacq.) (Murrill 1904)	ANT	**Henao (1989)**, Vasco-Palacio and Franco-Molano (2013)
C. focicola (Berk. and M.A. Curtis) (Murrill 1908)	CUN	**Guzmán and Varela (1978)**, Vasco-Palacio and Franco-Molano (2013)
C. perennis (L.) Murrill 1903	ANT	**Henao (1989)**, Vasco-Palacio and Franco-Molano (2013)
Russulales		

(continued)

Table 16.1 (continued)

Species	Department	References in which the species is reported in Colombia; the references reporting the species associated with *Q. humboldtii* are in bold. Species reported as edible, without confirmed consumption, are indicated with the word "Edible" (Franco-Molano et al. 2000), followed by the references in brackets [].
Russulaceae		
Lactarius		
L. atroviridis Peck	ANT, BOY	**Franco-Molano et al.** (**2000**, 2010), Halling and Mueller (2005), Vasco-Palacio and Franco-Molano (2013)
L. caucae Singer	CAU	**Singer (1973)**, Wu et al. (1997), Franco-Molano and Uribe-Calle (2000), Vasco-Palacio and Franco-Molano (2013)
L. costaricensis Singer	NAR	Franco-Molano et al. (2010), **Halling and Mueller (2005)**, Vasco-Palacio and Franco-Molano (2013)
L. chrysorrheus Fr. 1838	ANT, CUN, **BOY**, SAN	**Guzmán and Varela (1978)**, **Franco-Molano et al. (2000)**, Franco-Molano and Uribe-Calle (2000), Vasco-Palacio and Franco-Molano (2013), **Vargas and Restrepo (2019)**, **this study**.
L. fragilis (Burl.) Hersler and A.H Sm.	ANT, SAN	**Franco-Molano et al.** (**2000**, 2010), Vasco-Palacio and Franco-Molano (2013), **Vargas and Restrepo (2019)**.
L. gerardii Peck	ANT	**Franco-Molano et al. (2000)**, Vasco-Palacio and Franco-Molano (2013). Edible [Wang et al. 2004, Phillips 2005, Stubbe et al. 2010, Hall et al. 2011]
L. lignyotus (Fr.) (Kuntze 1891)	SAN	**Vargas and Restrepo (2019)**
L. quercuum Singer	BOY	**Singer (1963)**, Denis (1970), Wu et al. (1997), Vasco-Palacio and Franco-Molano (2013).
L. rimosellus (Peck 1906)	ANT, SAN	**Franco-Molano et al.** (**2000**, 2010), **Vargas and Restrepo (2019)**.
Lactifluus		
Lf. deceptivus Peck	ANT, BOY, **SAN**	Franco-Molano et al. (2000, 2010), **Halling and Mueller (2005)**, Vasco-Palacio and Franco-Molano (2013), **this study**.
Lf. indigo (Schwein)	ANT, BOY, CUN, NAR	**Franco-Molano et al.** (**2000**, 2010); Halling and Mueller (2005); Cepero de García et al. (2012), Vasco-Palacio and Franco-Molano (2013). Edible [Smith 1964, Boa 2004, Phillips 2005, Pérez-Moreno et al. 2010, Hall et al. 2011, Burrola-Aguilar et al. 2012]
Russula		
R. boyacensis Singer	BOY	**Singer (1963)**, Denis (1970), Wu et al. (1997), Franco-Molano and Uribe-Calle (2000), Vasco-Palacio and Franco-Molano (2013)
R. brevipes Peck	CAL, CUN	**Guzmán and Varela (1978)**, Montoya et al. (2005), Vasco-Palacio and Franco-Molano (2013). Edible [Boa 2004]

(continued)

Table 16.1 (continued)

Species	Department	References in which the species is reported in Colombia; the references reporting the species associated with *Q. humboldtii* are in bold. Species reported as edible, without confirmed consumption, are indicated with the word "Edible" (Franco-Molano et al. 2000), followed by the references in brackets [].
R. caucaensis (Singer 1989)	CAU	**Singer (1989)**, Franco-Molano and Uribe-Calle (2000), Wu et al. (1997), Vasco-Palacio and Franco-Molano (2013).
R. columbiana Singer	CUN	**Singer (1963)**, Wu et al. (1997), Franco-Molano and Uribe-Calle (2000), Vasco-Palacio and Franco-Molano (2013)
R. compacta Frost	ANT	Franco-Molano et al. (2010), **Halling and Mueller (2005)**, Vasco-Palacio and Franco-Molano (2013)
R. cremoricolor (Earle 1902)	SAN	**Vargas and Restrepo (2019)**
R. cyanoxantha (Schaeff) Fr.	CUN, **BOY**, SAN	**Guzmán and Varela (1978)**, Franco-Molano and Uribe-Calle (2000), Vasco-Palacio and Franco-Molano (2013), **Vargas and Restrepo (2019)**. Edible [Boa 2004, Hall et al. 2011]
R. emetica (Schaeff) Fr.	ANT, BOY	**Singer (1963)**, Saldarriaga et al. (1988), Franco-Molano and Uribe-Calle (2000), Sierra et al. (2011), Vasco-Palacio and Franco-Molano (2013)
R. humboldtii Singer	CUN	**Singer (1963)**, Denis (1970), Wu et al. (1997), Franco-Molano and Uribe-Calle (2000), Vasco-Palacio and Franco-Molano (2013)
R. idroboi Singer	CUN	**Singer (1963)**, Wu et al. (1997), Franco-Molano and Uribe-Calle (2000), Vasco-Palacio and Franco-Molano (2013)
R. peckii Singer	ANT	Franco-Molano et al. (2010), Vasco-Palacio and Franco-Molano (2013), NA
R. puiggarii (Speg.) Sing.	ANT	**López-Quintero et al. (2007)**, Vasco-Palacio and Franco-Molano (2013)
R. sardonia Fr. 1838	**BOY**	**This study**. Edible [Boa 2004, Hall et al. 2011]. Edibility suspected [Phillips 2005]
R. semililacea Singer	CUN	**Singer (1989)**, Wu et al. (1997), Franco-Molano and Uribe-Calle (2000), Vasco-Palacio and Franco-Molano (2013)
R. silvestris (Singer) Reumaux	ANT	**López-Quintero et al. (2007)**, Vasco-Palacio and Franco-Molano (2013)
R. virescens (Schaeff) Fr.	ANT	**Franco-Molano et al. (2000**, 2010), Vasco-Palacio and Franco-Molano (2013). Edible [Smith 1964, Boa 2004, Phillips 2005, Sitta and Davoli 2012, Hall et al. 2011]
Thelephorales		
Thelephoraceae		
Thelephora		
T. cervicornis (Corner 1968)	QUI	Vasco-Palacio and Franco-Molano (2013), NA

(continued)

Table 16.1 (continued)

Species	Department	References in which the species is reported in Colombia; the references reporting the species associated with *Q. humboldtii* are in bold. Species reported as edible, without confirmed consumption, are indicated with the word "Edible" (Franco-Molano et al. 2000), followed by the references in brackets [].
T. palmata (Scop.) Fr. 1821	ANT, SAN	**Henao (1989)**, Vasco-Palacio and Franco-Molano (2013), **Vargas and Restrepo (2019)**.
Ascomycota		
Eurotiales		
Elaphomycetaceae		
Elaphomyces muricatus Fr. 1829	CUN	**Guzmán and Varela (1978)**, Vasco-Palacio and Franco-Molano (2013).
Pezizales		
Helvellaceae		
Helvella		
H. lacunosa (Afzel. 1783)	ANT	**Tobón (1991)**, Vasco-Palacio and Franco-Molano (2013). Edible [Polese and Lamaison 1999, Boa 2004, Pérez-Moreno et al. 2010, Burrola-Aguilar et al. 2012]. Not recommended according to Smith (1964),
H. macropus (Pers.) P. Karst	ANT	**Tobón (1991)**, Vasco-Palacio and Franco-Molano (2013)

[a]Abbreviations for the departments: Antioquia (ANT), Boyacá (BOY), Caldas (CAL), Cauca (CAU), Cundinamarca (CUN), Huila (HUI), Nariño (NAR), Quindio (QUI), Santander (SAN), Tolima (TOL), and Valle del Cauca (VAL). The new reports found in this study with distributions in the departments of Boyacá, Cundinamarca, and Santander are in bold letters (*BOY*, *CUN*, and *SAN*). *NA* no available information on the host species

with the pileus surface; margin even with scales. *Context* reddish-purple, 4 mm wide. *Lamellae* adnexed to adnated, concolorous with the pileus surface, close to slightly distant. *Spore print* brown. *Stipe* 3–6 × 0.4–0.7 cm, concolorus with the pileus surface, cylindrical, with abrupt scales. *Basidiospores*: 6–9 × 5–7.5 μm, cruciform. *Basidia*: 24–31 × 13–16 μm. *Hymenophoral trama* interwoven hyphae.

Russula sardonia Fr. 1838. Material studied NVE 633 ANDES_F650 Fig. 2C, C1—Colombia, Boyacá, Municipio de Arcabuco, km 5 via Arcabuco-Gachantiva. 15 Dec. 2013, in *Q. humboldtii.* This species occurs in north temperate regions, found in Europe and western North America (Phillips 2005).

Pileus: 4–6.5 cm wide, convex, to flat in mature specimens and with a depression, violet, purplish or brownish-red, greenish or ochre to yellowish, hard, glabrous. *Context* white, 1–2 cm wide. *Lamellae* adnexed to slightly decurrent, at first cream to pale golden yellow, narrow. *Stipe*: 3.0–8.0 cm long × 1.0–1.5 cm wide, whitish to very pale lilac upper half, to greyish dark lilac in the lower half, uniform; surface fibrillose to slightly pruinose. *Spore print* cream. *Basidiospores:* 7–9 × 6–8 μm, ovoid with warts up to 0.5 μm high, joined into ridges forming fine irregular lines or rugose ornamentation. *Basidia:* 50–60 × 10–14 μm. *Hymenophoral trama*: ovoid cells up to 30 μm long, regularly arranged. *Pleurocystidia* spindle-shaped or cylindrical, without septa.

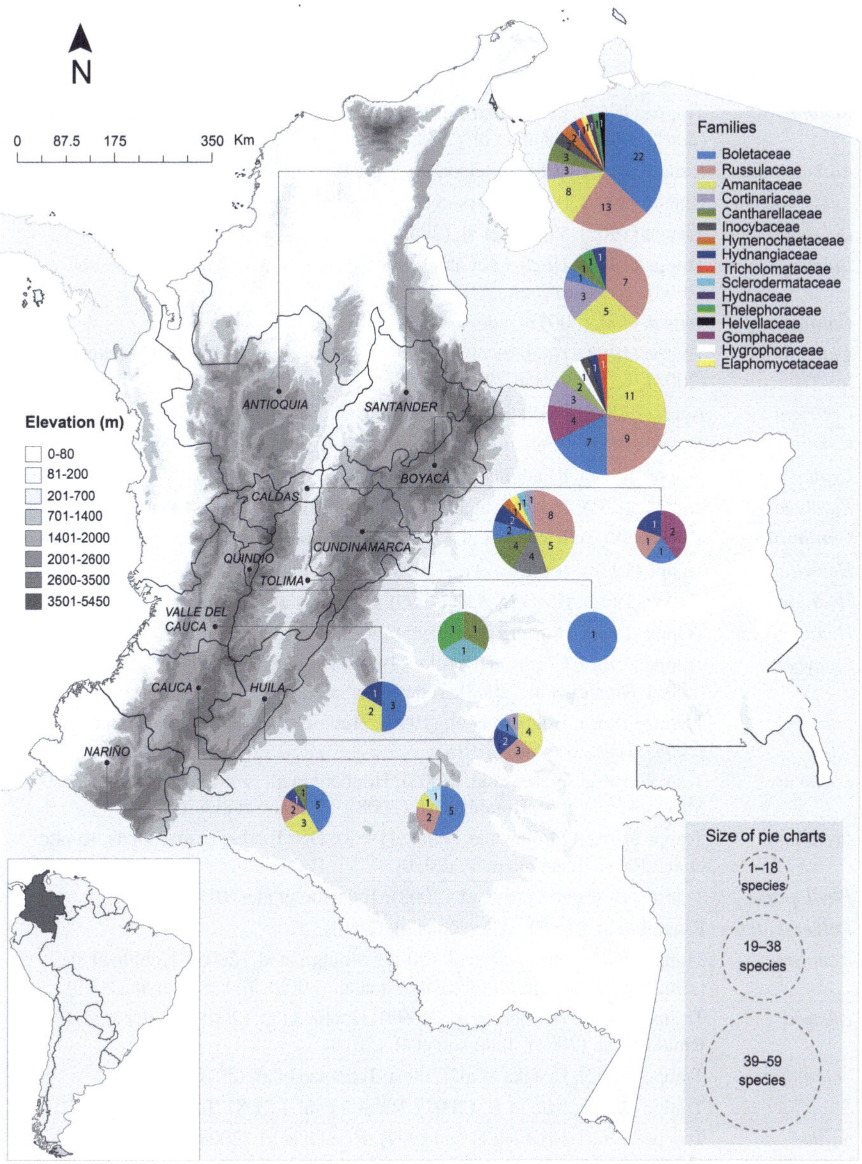

Fig. 16.1 The departments of Colombia where ECM species have been reported with *Quercus humboldtii*. The number of ECM species per family is indicated inside each pie chart. The delineated departments in the map encompass the Andean cordillera and are a part of the total departments in Colombia

Table 16.2 References reporting the ectomycorrhizal characteristic of fungal genera included in this study

Species	References
Amanita	Trappe (1962), Cripps and Miller (1995), Högberg et al. (1999), Hobbie et al. (2001, 2002), Rinaldi et al. (2008), Tedersoo et al. (2010), Wolfe et al. (2012).
Aureoboletus	Rinaldi et al. (2008), Tedersoo et al. (2010).
Austroboletus	Rinaldi et al. (2008), Tedersoo et al. (2010).
Boletellus	Trappe (1962), Rinaldi et al. (2008), Tedersoo et al. (2010).
Boletus	Trappe (1962), Högberg et al. (1999), Hobbie et al. (2001), Rinaldi et al. (2008), Tedersoo et al. (2010).
Calostoma	Wilson et al. (2007), Tedersoo et al. (2010).
Cantharellus	Trappe (1962), Högberg et al. (1999), Hobbie et al. (2001, 2002), Rinaldi et al. (2008), Tedersoo et al. (2010).
Coltricia	Agerer (2006), Tedersoo et al. (2007), Rinaldi et al. (2008), Tedersoo et al. (2010).
Cortinarius	Trappe (1962), Högberg et al. (1999), Rinaldi et al. (2008), Tedersoo et al. (2010).
Craterellus	Trappe (1962), Rinaldi et al. (2008), Tedersoo et al. (2010).
Elaphomyces	Trappe (1962), Tedersoo et al. (2003), Rinaldi et al. (2008), Tedersoo et al. (2010).
Gyroporus	Trappe (1962), Rinaldi et al. (2008), Tedersoo et al. (2010).
Helvella	Trappe (1962), Tedersoo et al. (2006), Rinaldi et al. (2008), Tedersoo et al. (2010).
Hydnum	Trappe (1962), Högberg et al. (1999), Rinaldi et al. (2008), Tedersoo et al. (2010).
Hygrophorus	Trappe (1962), Högberg et al. (1999), Rinaldi et al. (2008), Tedersoo et al. (2010).
Inocybe	Trappe (1962), Cripps and Miller (1995), Högberg et al. (1999), Hobbie et al. (2001), Rinaldi et al. (2008), Tedersoo et al. (2010).
Laccaria	Trappe (1962), Högberg et al. (1999), Hobbie et al. (2001), Rinaldi et al. (2008), Tedersoo et al. (2010).
Lactarius	Trappe (1962), Flores et al. (2005), Högberg et al. (1999), Hobbie et al. (2001), Miller et al. (2006), Rinaldi et al. (2008), Tedersoo et al. (2010).
Leccinum	Trappe (1962), Molina and Trappe (1982), Den Bakker et al. (2004), Rinaldi et al. (2008), Tedersoo et al. (2010).
Phylloporus	Trappe (1962), Rinaldi et al. (2008), Tedersoo et al. (2010).
Pulveroboletus	Rinaldi et al. (2008), Tedersoo et al. (2010).
Ramaria	Trappe (1962), Humpert et al. (2001), Nouhra et al. (2005), Hobbie et al. (2002), Hosaka et al. (2006), Rinaldi et al. (2008), Tedersoo et al. (2010).
Russula	Trappe (1962), Högberg et al. (1999), Hobbie et al. (2001), Miller et al. (2006), Rinaldi et al. (2008), Tedersoo et al. (2010).
Scleroderma	Trappe (1962), Hosaka et al. (2006), Tedersoo et al. (2010).
Strobilomyces	Trappe (1962), Sato et al. (2007), Rinaldi et al. (2008), Tedersoo et al. (2010).
Suillus	Trappe (1962), Högberg et al. (1999), Rinaldi et al. (2008), Tedersoo et al. (2010).
Thelephora	Trappe (1962), Agerer and Weiss (1989), Rinaldi et al. (2008), Tedersoo et al. (2010).
Tricholoma	Trappe (1962), Högberg et al. (1999), Hobbie et al. (2001), Rinaldi et al. (2008), Tedersoo et al. (2010).
Tylopilus	Trappe (1962), Rinaldi et al. (2008), Tedersoo et al. (2010).
Xanthoconium	Rinaldi et al. (2008), Tedersoo et al. (2010).
Xerocomellus	Trappe (1962)

(continued)

Table 16.3 New records of ECM fungi collected in *Q. humboldtii* forests, in the departments of Boyacá, Cundinamarca, and Santander

Species	Collection(s)/Herbarium Catalog number[a]	Department[b]	Locality
Amanita fuligineodisca	Observation	CUN	Vereda Santa Barbara, km 19 via Mosquera-La Mesa
Boletus neoregius	NVE474/ANDES_F974	BOY	Vereda Capilla, Municipio de Villa de Leyva
Calostoma cinnabarinum	NVE315/ANDES_F814, NVE462/ANDES_F962	BOY	Km 0.7, via Arcabuco-Gachantiva, Municipio de Arcabuco; Vereda Capilla, Municipio de Villa de Leyva
Cantharellus lateritius	ANDES_F71–72	CUN	Parque Nacional natural Chicaque, Municipio de San Antonio del Tequendama
Cortinarius iodes	NVE233–235/ANDES_F732–734. ANDES_F523, ANDES_F529. NVE482–483/ANDES_F982–983	BOY, CUN, SAN	Vereda Peñas Blancas, Municipio de Arcabuco; Vereda Santa Barbara, Municipio de Bojacá; Vereda San José de la Montaña
Cortinarius violaceus	NVE405/ANDES_F905	BOY	Vereda Peñas Blancas, Municipio de Arcabuco
Craterellus falax	NVE307/ANDES_F806	BOY	Vereda Peñas Blancas, Municipio de Arcabuco
Inocybe tahquamenonensis	NVE303/ANDES_F802	BOY	Vereda Peñas Blancas, Municipio de Arcabuco
Laccaria laccata	NVE291/ANDES_F791	SAN	Vereda San José de la Montaña
Lactifluus chrysorrheus	NVE358/ANDES_F857,	BOY	Vereda Peñas Blancas, Municipio de Arcabuco
Lactifluus deceptivus	NVE508/ANDES_F2008	SAN	Vereda San José de la Montaña
Phylloporus centroamericanus	NVE429/ANDES_F929	BOY	Vereda Capilla, Municipio de Villa de Leyva
Russula cyanoxantha	NVE244/ANDES_F743, NVE460/ANDES_F960	BOY	Vereda Peñas Blancas, Municipio de Arcabuco
Russula sardonia	NVE 633/ANDES_F650	BOY	Km 5, via Arcabuco-Gachantiva, Municipio de Arcabuco
Tricholoma caligatum	Observation	BOY	Vereda Capilla, Municipio de Villa de Leyva
Xerocomellus chrysenteron	NVE 449/ANDES_F949	BOY	Km 2, via Gachantiva-Arcabuco

[a]Collections made by Natalia Vargas Estupiñán (NVE)
[b]Departments: Boyacá (BOY), Cundinamarca (CUN), and Santander (SAN)

Fig. 16.2 Basidiome and microscopy of new national and some local records. (**a**) Basidiomes of *Inocybe tahquamenonensis*, (**a1**) Spores of *I. tahquamenonensis*, (**a2**) Basidia of *I. tahquamenonensis*; (**b**) *Cortinarius violaceus*; (**c**) Basidiome of *Russula sardonia*, (**c1**) Spores, basidia and pleurocystidia of *R. sardonia*; (**d**) *Phylloporus centroamericanus;* (**e**) *Boletus neoregius*; (**f**) *Calostoma cinnabarinum*; (**g**) *Xerocomellus chrysenteron*; (**h**) *Craterellus fallax*; (**i**) *Amanita fuligineodisca.* White scale bars correspond to 1 cm. Black scale bars correspond to 10 μm

16.3.2 Notes on Some Taxa Not Included in the Checklist

16.3.2.1 Basidiomycota

Amanitaceae: All the species in Colombia, with the exception of *A. savannae* (described from a wet savanna by Tulloss and Franco-Molano, 2008), belong to a clade of symbiotic species (Subgenera *Lepidella* and *Amanita*) (Wolfe et al. 2012). We did not include the species *A. ceciliae* in the list, because the Colombian and Mesoamerican species for "*A. ceciliae*" probably is *A. sororcula* Tulloss, Ovrebo & Halling (http://amanitaceae.org/?Amanita%20ceciliae). The species *A. muscaria* has been widely reported in exotic pine plantations (Pulido 1983; Franco-Molano et al. 2000; Franco-Molano and Uribe-Calle 2000; Montoya et al. 2005), but Vargas et al. (2019) reported it in association with *Q. humboldtii* in Santander.

Cortinariaceae: The genus *Cortinarius* is one of the most diverse genera containing over 2000 spp. (Kirk et al. 2008), and a large number of taxa occurring in Colombia have not been determined yet to species level.

Entolomataceae: The species *E. ferrugineogranulatum* reported by Soto-Medina and Bolaños-Rojas (2013) and Horak (1977) was not included in the checklist, since it was reported in open lands (*potreros*) and on rotten wood in rain forests. The same occurs for the species *E. lyophylliforme* reported by Horak (1977) in a tropical rain forest near Buenaventura at 180 m asl. The species *E. venezuelanum* (Dennis) E. Horak 1978 was collected in forests dominated by *Colombobalanus excelsa* (Soto-Medina and Bolaños-Rojas 2013) and was not included in the checklist. The genus *Entoloma* is reported to have ectomycorrhizal species (Rinaldi et al. 2008; Tedersoo et al. 2010); however, the subgenera *Nolanea* and *Leptonia* are nonmycorrhizal (Tedersoo et al. 2010).

Hydnangiaceae: The species *Laccaria ohiensis* has been collected in forests dominated by the native oak species *Colombobalanus excelsa* in the department Valle del Cauca (Soto-Medina and Bolaños-Rojas 2013). A report of *Laccaria laccata* was made by Sánchez (2006) in Norte de Santander; however, there is no specification on its host.

Boletaceae: the species *Boletus orquidianus* (=*Xerocomus orquidianus*) was reported for the country but not associated with Fagaceae (Halling 1989). However, it was later collected in an oak forest in Antioquia (Franco-Molano et al. 2000). The species *Boletus pavonius* and *B. purpurascens* are reported from Santander (Hooker and Kunth 1822; Vasco-Palacio and Franco-Molano 2013) on the banks of the Magdalena river growing on decomposed wood (Hooker and Kunth 1822), and was not included in this checklist. A specimen identified as *B. reticulatus* was collected in a temperate region between Popayán and Almaguer (Hooker and Kunth 1822), probably in oak forests; however, no vegetation information for this specimen was reported.

We did not include species in the genus *Phlebopus*, since the ecology of the genus is ambiguous showing some species cultivated as saprotrophs (Thoen and Ducousso 1990; Wilson et al. 2012) or engaging in multipartite symbiotic interactions (Zhang et al. 2015).

The genus *Chalciporus* has been reported as ectomycorrhizal by Rinaldi et al. (2008); however, Tedersoo et al. (2010) concluded that there is not enough evidence to show its mycorrhizal habit. For this reason, we did not include in the checklist two species reported in Colombia: *Chalciporus piperatus* (Bull.) Bataille and *Chalciporus caribaeus* Pegler reported by López-Quintero et al. (2007) and Franco-Molano et al. (2010), respectively.

Suillus luteus NVE425 Andes_F925, was collected in *Q. humboldtii* in the department of Boyacá. However, species in this genus are restricted to Pinaceae, and therefore *S. luteus* was not included in the present ECM checklist. Anatomical, chemical, and molecular analyses must be performed to confirm the association with *Q. humboldtii*. Previous studies have reported this species in Colombia associated with conifers (Franco-Molano et al. 2000) and introduced with *Pinus* spp. (Guzmán and Varela 1978) in Antioquia, Caldas, and Cundinamarca.

Cantharellaceae*:* the species *Cantharellus cinereus* was previously reported by Vasco-Palacio and Franco-Molano (2013), but was not included in this checklist, since the study by Guzmán and Varela (1978) did not report it.

Gomphales: Species in the genus *Ramaria* are reported both as ectomycorrhizal and saprotroph (Humpert et al. 2001; Tedersoo et al. 2010). We include in the checklist the species *R. cyaneigranosa* whose mycorrhizal status was confirmed by Nouhra et al. (2005); it belongs to the subgenus *Laeticolora* characterized by a terricolous habit (Humpert et al. 2001). Other species included in the checklist (Table 16.1) belong to the subgenera *Ramaria* (*R. botrytis*, *R. secunda*, *R. flava*) and the subgenus *Laeticolora* (*R. formosa*), have terricoluos habit, and are suggested to be mycorrhizal (Humpert et al. 2001; Hobbie et al. 2002; Smith and Read 2008). In contrast, the saprotrophic species *Ramaria stricta* grows on wood debris (Hosaka et al. 2006), belongs to a phylogenetic clade of lignicolous taxa (Humpert et al. 2001), and is not included in the checklist.

16.3.2.2 Ascomycota

Pezizales: Within this order several species are expected to form ectomycorrhizal symbiosis including species in the genera *Humaria, Genea, Trichophaea, Geopora, Helvella, Hydnotria, Peziza*, and *Sarcosphaera* (Tedersoo et al. 2006). The species *Peziza patena* has been reported in Colombia by Vasco-Palacios and Franco-Molano (2013), but there is no specific data on its distribution or host.

16.3.3 Notes on Other Orders

Geastrales: According to the studies referenced by Rinaldi et al. (2008) and the phylogenetic analysis by Hosaka et al. (2006), we did not included *Geastrum* as ECM.

Trechisporales: Dunham et al. (2007) observed some traits that characterize ECM formation in species from the genus *Trechispora*. In Colombia, eight species of this genus are reported, but they were not included here; further analyses on the trophic status must be made for this particular genus.

16.3.4 Edible Ectomycorrhizal Fungi in Native Oak Forests

Regarding our search on previous reports on edibility of wild fungi, 37 ectomycorrhizal species associated with *Q. humboldtii* forests in the Andean mountains in Colombia are potentially edible (Table 16.1). Among them, some are locally reported as edible for the department of Boyacá: *Ramaria secunda* and *Laciflluus indigo* by Ruíz and Henao-Mejía (2006), *Tylopilus indecisus, Ramaria flava, R. cyaneigranosa, R. botrytis*, and the other five species of *Ramaria* by Peña-Cañon and Henao-Mejía (2014), *Lactifluus deceptivus* (pers. comm. Yeina Niño Fernández, March 2014) and *Russula cyanoxantha* (pers. comm. Angélica Ruíz, May 2012).

Given that the native oak forests provide environmental benefits and services, their maintainance and protection are vital (Chaves et al. 2007). Encouraging public awareness on the importance of fungal diversity and its conservation, in localities where its edibility is traditionally known, might allow a sustainable production. In countries such as the United States, and some countries in Europe, conservation activities have been proposed based on the impact of intense harvesting of edible mushrooms over long time periods (Pilz and Molina 2002).

16.4 Concluding Remarks

The conservation of ectomycorrhizal fungi associated to oak ecosystems is a priority, regarding that oak forests suffered habitat loss, rare fungal species are associated with this host, and a potential use of ECM species is known. Moreover, ectomycorrhizal fungi play a key role in sequestering carbon in soils (Soudzilovskaia et al. 2019), hence their conservation and restoration might be one strategy to act on when considering actions for climate change mitigation.

Following the evaluation criteria provided by the *Micheli Guide to Fungal Conservation* (http://www.fungal-conservation.org/micheli.htm), in this report we accomplished the following citeria: mentioning ECM fungi in a conservation context, listing departments in the Colombian Andes where ECM fungal species occur with native *Q. humboldtii*, and showing different regions in the Andean cordillera where there is a lack of knowledge on fungal diversity and where exploration of ECM diversity should be enhanced.

Acknowledgements We would like to thank the Faculty of Science for funding the field trips. We are grateful also to Orlando Vargas, Martha Estupiñán, and Julián Mosquera for their collaboration in the field trips.

References

Agerer R, Weiss M (1989) Studies on ectomycorrhizae. XX. Mycorrhizae formed by *Thelephora terrestris* on Norway spruce. Mycologia 81:444–453

Agerer R (2006) Fungal relationships and structural identity of their ectomycorrhizae. Mycol Prog 5:67–107

Ammirati J, Garnica S, Halling RE, Mata M, Mueller GM, Carranza J (2007) New species associated with and in Costa Rica. Can J Bot 85(9):794–812

Arnolds E (1995) Conservation and management of natural populations of edible fungi. Can J Bot 73:987–998

Arora D (1986) Mushrooms demystified. Ten Speed Press, Berkeley, CA

Avella A, Cárdenas LM (2010) Conservación y uso sostenible de los bosques de roble en el corredor de conservación Guantiva-La Rusia-Iguaque, departamentos de Santander y Boyacá, Colombia. Colombia For 13:5–25

Betancur M, Calderón H, Betancourt G, Gallego A (2007) Hongos macromycetes en dos relictos de bosque húmedo tropical montano bajo de la vereda la Cuchilla, Marmato, Caldas. Boletín Científico Centro de Museos Museo de Historia Natural 11:19–31

Boa E (2004) Wild edible fungi: a global overview of their use and importance to people. In: Nonwood forest products, vol 17. FAO, Rome, p 148

Boekout T, Pulido MM (1989) The occurrence of macrofungi and their habitats in vegetations along the Parque Los Nevados transect. In: Van Der Hammen, T., S. Diaz-Piedrahita, and V. J. Alvarez. (eds). La Cordillera Central Colombiana Transecto Parque Los Nevados (segunda parte). Estudios de Ecosistemas Tropandinos: The Sierra Nevada de Santa Marta. pp 485–505

Burrola-Aguilar C, Garibay-Orijel R, Hernández Téllez M. 2012. Los hongos comestibles silvestres del estado de México: Propuesta para su aprovechamiento. En: Sánchez J and Mata G (eds). Hongos comestibles y medicinales en Iberoamerica, investigación y desarrollo en un entorno multicultural. El Colegio de la Frontera Sur, Chiapas, pp 39–49

Cárdenas D, Salinas NR (2006) Libro rojo de plantas de Colombia. Especies maderables amenazadas I parte. Versión preliminar. Instituto Amazónico de Investigaciones Científicas (SINCHI), Ministerio de Ambiente, Vivienda y Desarrollo Sostenible (MAVDT), Bogotá, p 169

Cepero de García M, Restrepo R, Franco-Molano AE, Cardenas M, Vargas N (2012) Biología de Hongos. Ediciones Uniandes. Universidad de Los Andes, Bogotá, p 520

Chaves ME, Santamaría M, Sánchez E (2007) Alterativas para la conservación y uso sostenible de la biodiversidad en los Andes Colombianos. Resultados 2001–2007. Instituto de Investigaciones de Recursos Biológicos Alexander van Humboldt, Bogotá, Colombia. 276 pp.

Cripps CL, Miller OK (1995) Ectomycorrhizae formed *in vitro* by quaking aspen: including *Inocybe lacera* and *Amanita pantherina*. Mycorrhiza 5:357–370

Den Bakker HC, Zuccarello GC, Kuyper TW, Noordeloos ME (2004) Evolution and host specificity in the ectomycorrhizal genus *Leccinum*. New Phytol 163:201–215

Denis RW (1970) Fungus Flora of Venezuela and adjacent countries. Kew Bull Add Ser 3:584

Dumond KP, Umaña MI (1978) Los Hongos de Colombia VII. *Laterna* y *Calostoma cinnabarina* en Colombia. Caldasia 12:349–352

Dunham SM, Larsson KH, Spatafora JW (2007) Species richness and community composition of mat-forming ectomycorrhizal fungi in old- and second-growth Douglas-fir forests of the HJ Andrews experimental Forest, Oregon, USA. Mycorrhiza 17:633–645

Eyssartier G, Masson-Deblaize I, Joly P (2011) 1001 setas. Servilibro ediciones SA, Madrid, p 264

Flores R, Díaz G, Honrubia M (2005) Mycorrhizal synthesis of *Lactarius indigo* with five Neotropical pine species. Mycorrhiza 15:563–570

Franco-Molano AE, Aldana R, Halling R (2000) Setas de Colombia. Universidad de Antioquía, Colombia, Guía de campo, p 156

Franco-Molano AE, Uribe-Calle E (2000) Hongos Agaricales y Boletales de Colombia. Biota Colomb 1:25–43

Franco-Molano AE, Corrales A, Vasco-Palacios AM (2010) Macrohongos de Colombia II. Listado de especies de los ordenes Agaricales, Boletalles, Cantharelalles y Russulales (Agaricomycetes, Basidiocmycota). Actual Biol 32:89–113

Fundación Natura (2007) Elementos conceptuales para la conservación y uso sostenible de los bosques de roble negro (*Colombobalanus excelsa*) y roble común (*Quercus humboldtii*), en jurisdicción de CAS y CORPOBOYACÁ. Fundación Natura, Bogotá, p 16

Garibay-Orijel R, Caballero J, Estrada-Torres A, Cifuentes J (2007) Understanding cultural significance, the edible mushrooms case. J Ethnobiol Ethnomed 3:1. https://ethnobiomed.biomedcentral.com/articles/10.1186/1746-4269-3-4

Guzmán G, Varela L (1978) Hongos de Colombia. III. Observaciones sobre los hongos, liquenes y mixomicetos de Colombia. Caldasia 12:309–338

Guzmán D, Ruíz JF, Cadena M (2014) Regionalización de Colombia según la estacionalidad de la precipitación media mensual, a través Análisis de Componentes Principales (ACP). Grupo de Modelamiento de Tiempo, Clima y Escenarios de Cambio Climático Subdirección de Meteorología-IDEAM. http://modelos.ideam.gov.co/media/dynamic/clima/colombia/regionalizacion-colombia.pdf

Hall IR, Lyon T, Wang Y, Buchanan P (2011) A list of putative edible or medicinal ectomycorrhizal mushrooms. Truffles and Mushrooms, Consulting Ltd., Dunedin, p 44

Halling R (1989) A synopsis of Colombian Boletes. Mycotaxon 34:93–113

Halling R, Mueller G (1999) New Boletes from Costa Rica. Mycologia 91:893–899. https://doi.org/10.2307/3761543

Halling R, Mueller G (2005) Common mushrooms of the Talamanca Mountains, Costa Rica. The New York Botancial Garden Press, New York, p 195

Halling R, Ovrebo CL (1987) A new species of Rozites from oak forests of Colombia, with notes on biogeography. Mycologia 79:674–678

Henao LG (1989) Notas sobre afiloforales colombianos (Basidiomycetes: Aphyllophoralles). Caldasia 16:1–9

Henkel TW, Wilson AW, Aime MC, Dierks J, Uehling JK, Roy M, Schimann H, Wartchow F, Mueller GM (2014) Cantharellaceae of Guyana II: new species of *Craterellus*, new south American distribution records for *Cantharellus guyanensis* and *Craterellus excelsus*, and a key to the Neotropical taxa. Mycologia 106:307–324

Hobbie EA, Weber NS, Trappe JM (2001) Mycorrhizal vs. saprotrophic status of fungi: the isotopic evidence. New Phytol 150:601–610

Hobbie EA, Weber NS, Trappe JM, Van Klinken GJ (2002) Using radiocarbon to determine the mycorrhizal status of fungi. New Phytol 156:129–136

Högberg P, Plamboeck AH, Taylor AF, Fransson PM (1999) Natural 13C abundance reveals trophic status of fungi and host-origin of carbon in mycorrhizal fungi in mixed forests. Proc Natl Acad Sci U S A 96:8534–8539

Hooker W, Kunth CS (1822) Synopsis plantarum aequinoctialium orbis novae. Vol 1

Horak E (1977) *Entoloma* in South America I. Sydowia 30:40–111

Hosaka K, Bates ST, Beever RE, Castellano MA, Colgan W, Domínguez LS, Nouhra E, Geml J, Giachini AJ, Kenney S, Simpson N, Spatafora J, Trappe J (2006) Molecular phylogenetics of the gomphoid-phalloid fungi with an establishment of the new subclass Phallomycetidae and two new orders. Mycologia 98:949–959

Humpert AJ, Muench EL, Giachini AJ, Castellano MA, Spatafora JW (2001) Molecular phylogenetics of *ramaria* and related genera: evidence from nuclear large subunit and mitochondrial small subunit rDNA sequences. Mycologia 93:465–477

Kirk PM, Cannon PF, Minter DW, Stalpers JA (2008) Dictionary of the fungi, 10th edn. CABI Publishing, Wallingford, p 771

Kuo M, Methven AS (2014) Mushrooms of the Midwest. University of Illinois Press, Urbana. http://search.ebscohost.com.ezproxy.uniandes.edu.co:8080/login.aspx?direct=true&db=nlebk&AN=816582&lang=es&site=ehost-live

López-Quintero C, Vasco-Palacios A, Franco-Molano AE (2007) Capítulo 3: Macrohongos de un bosque de roble, *Quercus humboldtii* Bonpl., en la vereda Contrafuerte, municipio de Andes (Colombia). In: Reserva Natural Regional Cuchilla Jardín Támesis Antioquia: una mirada a su biodiversidad. Red de Investigación en Biodiversidad y Conservación –Redbio–, Corporación Autónoma Regional del Centro de Antioquía, Medellín. pp 21–34

Mata M, Halling RE, Mueller GM (2003) Macrohongos de Costa Rica, vol 2. Instituto Nacional de Biodiversidad –INBio, Santo Domingo de Heredia, Costa Rica, p 240

Matheny PB, Moreau PA (2009) A rare and unusual lignicolous species of *Inocybe* (Agaricales) from eastern North America. Brittonia 61:163–171

Miller SL, Larsson E, Larsson KH, Verbeken A, Nuytinck J (2006) Perspectives in the new Russulales. Mycologia 98:960–970

Molina R, Trappe JM (1982) Lack of mycorrhizal specificity by the ericaceous hosts *Arbutus menziesii* and *Arctostaphylos uva-ursi*. New Phytol 90:495–509

Montoya F, Arias D, Betancur-Agudelo M (2005) Contribución al conocimiento de los hongos Macromicetos del Resguardo Indígena Nuestra Señora de la Candelaria de la Montaña Riosucio, Caldas. Boletín Científico (Centro de Museos) Museo de Historia Natural 9:19–30

Mueller G (1996) Distribution and species composition of *Laccaria* in tropical and subtropical America. Rev Biol Trop 44:131–135

Mueller GM, Singer R (1988) *Laccaria gomezii*, a new agaric species from querceta of Colombia and Costa Rica. Mycotaxon 33:223–227

Myers C, Lynch J (1997) *Eleutherodactylus laticorpus*, a peculiar new frog from the Cerro Tacarcuna area, Panamanian-Colombian frontier. Am Mus Novit 3196:1–12

Nasi M (1977) Los hongos superiores de la Sabana de Bogotá y alrededores: descripción botánica, consideraciones ecológicas y bioquímicas, métodos de recolección e identificación, posibilidades de aprovechamiento en Colombia. Tésis de Maestría. Universidad de los Andes, Colombia

Nieves-Rivera A, Santos C, Betancourt C (1997) Notas sobre los Agaricales del Páramo de Guasca, Departamento de Cundinamarca, Colombia. Caldasia 19:349–351

Nouhra ER, Horton TR, Cazares E, Castellano M (2005) Morphological and molecular characterization of selected *Ramaria* mycorrhizae. Mycorrhiza 15:55–59

Orwa C, Mutua A, Kindt R, Jamnadass R, Simons A (2009) Agroforestree Database: a tree reference and selection guide version 4.0. http://www.worldagroforestry.org/af/treedb

Peña-Cañón ER, Henao-Mejía LG (2014) Conocimiento y uso tradicional de hongos silvestres de las comunidades campesinas asociadas a bosques de roble (*Quercus humboldtii*) en la zona de influencia de la Laguna de Fúquene, Andes Nororientales. Etnobiología 12:28–40

Pérez-Moreno J, Read DJ (2004) Los hongos ectomicorrízicos, lazos vivientes que conectan y nutren a los árboles en la naturaleza. Interciencia 29:239–247

Pérez-Moreno J, Lorenzana Fernández A, Carrasco Hernández V, Yescas-Pérez A (2010) Los hongos comestibles silvestres del Parque Nacional Izta-Popo, Zoquiapan y Anexos. Colegio de Postgraduados, SEMARNAT, CONACyT. Montecillo, Texcoco, Estado de México. p 167

Pérez-Moreno J (2012) Los hongos comestibles ectomicorrícicos y su biotecnología. En: Sánchez J and Mata G (eds) Hongos comestibles y medicinales en Iberoamerica, investigación y desarrollo en un entorno multicultural. Chiapas: El Colegio de la Frontera Sur. pp. 19–28

Petersen RH, Mueller G (1992) New south American taxa of *Cantharellus, C. nothofagorum, C. xanthoscyphus* and *C. lateritius* var. *colombianus*. Boletín de la sociedad Argentina de Botánica 28:195–200

Phillips R (2005) Mushrooms and other fungi of North America. Firefly Books, Buffalo, NY, p 319

Pilz D, Molina R (2002) Commercial harvests of edible mushrooms from the forests of the Pacific Northwest United States: issues, management, and monitoring for sustainability. For Eco Manage 155(1–3):3–16

Polese JM, Lamaison JL (1999) The great encyclopedia of mushrooms. Könemann Verlagsgesellschaft, Cologne, p 240

Pulido M (1983) Estudios en Agaricales Colombianos: Los hongos de Colombia IX. Instituto de Ciencias Naturales, Museo de Historia Natural, Universidad Nacional de Colombia, p 143

Pulido M, Cavelier J, Cortés SP (2006) Structure and composition of Colombian montane oak forests. In: Kapelle M (ed) Ecology and conservation of neotropical montane oak forests. Springer-Verlag, Berlin Heidelberg, pp 141–151

Read D (1998) Mycorrhizal status of pines. In: Richardson D (ed) Ecology and biogeography of *Pinus*. Cambridge University Press, Cambridge, pp 324–340

Rinaldi AC, Comandini O, Kuyper TW (2008) Ectomycorrhizal fungal diversity: separating the wheat from the chaff. Fungal Divers 33:1–45

Ruíz A, Henao LG (2006) Hongos comestibles de Iguaque, Serie Especies Colombianas: 14. Instituto de Investigación de Recursos Biológicos Alexander von Humboldt, Bogotá, p 14

Saldarriaga Y, Pineda F, García G, Velásquez L, Guzmán G (1988) Nuevos Registros de Agaricales en Colombia. Rev Mex Micol 4:333–342

Sánchez N (2006) Hongos Basidiomycetes: una contribución al conocimiento de 14 géneros en Norte de Santander. Respuestas 2:14–26

Sato H, Yumoto T, Murakami N (2007) Criptic species and host specificity in the ectomycorrhizal genus *Strobilomyces* (Strobilomycetaceae). Am J Bot 94:1630–1641

Sierra JD, Arias J, Sánchez M (2011) Registro Preliminar de Macrohongos (Ascomycetes y Basidiomycetes) en el Bosque Húmedo Montano del Alto El Romeral (Municipio de Angelópolis, Departamento de Antioquia-Colombia). Rev Fac Nal Agr Medellín 64:6159–6174

Singer R (1963) Oak mycorrhiza fungi in Colombia. Mycopathologia 20:239–252

Singer R (1973) Diagnoses Fungorum Novorum Agaricalium III. Sydowia 7:1–106

Singer R (1989) New taxa and new combinations of Agaricales (diagnoses Fungorum Novorum Agaricalium IV). Fieldiana Bot 21:1–133

Singer R, Ovrebo C, Halling R (1990) A new species of *Phylloporus* and a new species of *Tricholomopsis* from Colombia with notes on *Phylloporus boletinoides*. Mycologia 82:452–459

Sitta N, Davoli P (2012) Edible ectomycorrhizal mushrooms: international markets and regulations. In: Zambonelli A, Bonito G (eds) Edible Ectomycorrhizal mushrooms, current knowledge and future prospects. Springer-Verlag, Berlin Heidelberg, Germnay, pp 355–380

Smith A (1964) The mushroom hunter's field guide, revised and enlarged. The University of Michigan Press, Ann Arbor, p 264

Smith SE, Read DJ (2008) Mycorrhizal symbiosis, 2nd edn. Academic, London, p 800

Smith M, Bonito G (2012) Systematics and ecology of edible ectomycorrhizal mushrooms. In: Zambonelli A, Bonito G (eds) Edible ectomycorrhizal mushrooms, current knowledge and future prospects. Springer-Verlag, Berlin Heidelberg, Germnay, pp 17–39

Soto-Medina E, Bolaños-Rojas AC (2013) Hongos macroscópicos en un bosque de niebla intervenido, vereda Chicoral, Valle del Cauca, Colombia. Biota Colombiana 14:1–12

Soudzilovskaia N, van Bodegom P, Terrer C, van't Zelfde M, McCallum I, McCormack L, Fisher J, Brundrett M, César de Sá N, Tedersoo L (2019) Global mycorrhizal plant distribution linked to terrestrial carbon stocks. Nat Commun 10:5077

Stubbe D, Nuytinck J, Verbeken A (2010) Critical assessment of the Lactarius gerardii species complex (Russulales). Fungal Biology 114(2–3):271–283

Tedersoo L, May TW, Smith ME (2010) Ectomycorrhizal lifestyle in fungi: global diversity, distribution, and evolution of phylogenetic lineages. Mycorrhiza 20:217–263. https://doi.org/10.1007/s00572-009-0274-x

Tedersoo L, Suvi T, Beaver K, Saar I (2007) Ectomycorrhizas of *Coltricia* and *Coltriciella* (Hymenochaetales, Basidiomycota) on Caesalpiniaceae, Dipterocarpaceae and Myrtaceae in Seychelles. Mycol Prog 6:101–107

Tedersoo L, Hansen K, Perry BA, Kjoller R (2006) Molecular and morphological diversity of pezizalean ectomycorrhiza. New Phytol 170:581–596

Tedersoo L, Kõljalg U, Hallenberg N, Larsson KH (2003) Fine scale distribution of ectomycorrhizal fungi and roots across substrate layers including coarse woody debris in a mixed forest. New Phytol 159:153–165

Thoen D, Ducousso M (1990) Mycorrhizal habit and sclerogenesis of *Phlebopus sudanicus* (Gyrodontaceae) in Senegal. Agric Ecosyst Environ 28:519–523

Tobón LE (1991) Ascomycetes de Colombia: Discomicetes del departamento de Antioquia. Caldasia 16:327–336

Trappe JM (1962) Fungus associates of ectotrophic mycorrhizae. Bot Rev 28:538–606

Tulloss RE, Ovrebo C, Halling R (1992) Studies on *Amanita* (Amanitaceae) from Andean Colombia. Mem NY Bot Gar 66:1–46

Tulloss RE (2000) Nomenclatural changes in *Amanita*. Mycotaxon 75:329–332

Tulloss RE (2002) Tabular key to rubescent species of *Amanita* Section *Validae*. http://eticomm.net/~ret/amanita/key.dir/valirube.pdf

Tulloss RE (2005) Appendix A6: draft key to species of *Amanita* occurring in the Northeastern USA and Eastern Canada. http://eticomm.net/~ret/amanita/key.dir/nekey.pdf
Tulloss RE, Franco-Molano AE (2008) Studies in *Amanita* subsection Vittadiniae I—a new species from Colombian savanna. Mycotaxon 105:317–323
Vargas N, Pardo de la Hoz C, Franco-Molano AE, Jiménez P, Restrepo S, Grajales A (2017) Defining the phylogenetic position of *Amanita* species in Colombia. Mycologia 109:261–276
Vargas N, Restrepo S (2019) Macrohongos (Basidiomycota y Ascomycota) de los ecosistemas exocársticos de El Peñón (Andes), Santander, Colombia. In: Lasso CA, Barriga J, Fernánde-Auderset J (eds) Biodiversidad subterránea y epigea de los sistemas cársticos de El Peñón (Andes), vol VII. Serie Fauna Silvestre Neotropical, Santander, Boadilla del Monte
Vargas N, Goncalves S, Franco-Molano AE, Restrepo S, Pringle A (2019) In Colombia the Eurasian fungus *Amanita muscaria* is expanding its range into tropical *Quercus humboldtii* forests. Mycologia. https://doi.org/10.1080/00275514.2019.1636608
Vasco-Palacios A, Franco-Molano AE (2013) Diversity of Colombian macrofungi (Ascomycota - Basidiomycota). Mycotaxon 121:499. http://mycotaxon.com/resources/checklists/VascoPalacios-v121-checklist.pdf
Wilson AW, Hobbie EA, Hibbett DS (2007) The ectomycorrhizal status of *Calostoma cinnabarinum* determined using isotopic, molecular, and morphological methods. Can J Bot 85:385–393
Wilson AW, Binder M, Hibbett DS (2012) Diversity and evolution of ectomycorrhizal host associations in the Sclerodermatineae (Boletales, Basidiomycota). New Phytol 194:1079–1095
Wolfe B, Tulloss RE, Pringle A (2012) The irreversible loss of a decomposition pathway marks the single origin of an ectomycorrhizal symbiosis. PLoS One 7:e39597
Wu Q, Mueller G (1995) The genus *Craterellus* (Basidiomycetes, Aphyllophorales) in Costa Rica and Colombia. Doc Mycol XXV:487–496
Wu Q, Mueller G, Obrevo C (1997) An index to genera, species, and infraspecfic taxa of Basidiomycete fungi described by Rolf Singer. In: Mueller G, Wu Q (eds) Mycological contributions of Rolf Singer: field itinerary, index to new taxa, and list of publications. Fieldiana, botany, vol 38. Field Museum of Natural History, Chicago, pp 13–99

Chapter 17
Modifications of Community Structure in Ectomycorrhizal Arctic Fungi as a Consequence of Global Warming

Luis N. Morgado and József Geml

17.1 Introduction

Evidence derived from tree rings, sediment layers and ice cores revealed that temperatures during the past few decades exceeded those of the past four millennia (Mann et al. 1999; Mann and Jones 2003; Salzer et al. 2014). Since the second half of the nineteenth century, the Earth's surface warmed an average of 0.68 °C and accelerated from 1970 onwards with the 10 warmest years on record (131 years) occurring on the twenty-first century (Post 2013). This ongoing climate change is expected to be a major threat to biodiversity and will greatly alter ecosystem dynamics in the near future (Schöroter et al. 2005; Pimm 2009; Montoya and Raffaelli 2010).

Although there is an average global climate warming, specific regions are experiencing a more intense climate change. For example, the Arctic has some of the highest rates of warming on the planet, varying between 0.06 and 0.1 °C in air temperature increase per year in the past five decades. Because the permafrost regions at high latitudes are estimated to hold approximately 50% of Earth's reactive carbon (Tarnocai et al. 2009), a large fraction of this C may become increasingly vulnerable to mobilization due to warming-induced melting coupled with higher microbial decomposition rates (Anisimov et al. 2007; Hansen et al. 2010; Comiso and Hall

L. N. Morgado (✉)
Section for Genetics and Evolutionary Biology, Department of Biosciences, University of Oslo, Oslo, Norway

Biodiversity Dynamics Research Group, Naturalis Biodiversity Center, Leiden, The Netherlands
e-mail: l.n.morgado@ibv.uio.no

J. Geml
Biodiversity Dynamics Research Group, Naturalis Biodiversity Center, Leiden, The Netherlands

MTA-EKE Lendület Environmental Microbiome Research Group, Eszterházy Károly University, Eger, Hungary

J. Pérez-Moreno et al. (eds.), *Mushrooms, Humans and Nature in a Changing World*, https://doi.org/10.1007/978-3-030-37378-8_17

2014), which in turn might induce a positive feedback to the ongoing climate warming. Other climate feedbacks include changes in sea ice cover, length of ice-free periods (Arrigo and van Dijken 2011; Post 2013), greening of the surrounding land surface and tree line advancement (Kharuk et al. 2013; Zhang et al. 2013). All of these reduce the surface albedo, resulting in positive feedbacks to warming (Chapin III et al. 2005; Post et al. 2009). For example, a greening of the Arctic, driven by increases in shrub density (Sturm et al. 2005; Loranty and Goetz 2012; Tape et al. 2006, 2012), could potentially result in increased C sequestration (Welker et al. 1997; Sistla et al. 2013; Pattison and Welker 2014). Increases in shrub density and canopy growth can further alter the tundra by local snow-trapping in winter, increasing soil insulation, causing higher winter and spring-time soil temperatures and altering the rates of N and C turnover.

There is mounting evidence that arctic tundra plant communities are already responding to increased summer warming, both at local and circumpolar scales (Sturm et al. 2001; Wahren et al. 2005; Walker et al. 2006; Pattison and Welker 2014). The general trends include increases in plant leaf litter, graminoid plants and shrub coverage and decreases in lichens and bryophytes coverage, as well as leaf C:N ratio (Sturm et al. 2001; Wahren et al. 2005; Welker et al. 2005; Walker et al. 2006; Mercado-Díaz 2011; Pattison and Welker 2014). These trends vary with tundra type, plant functional groups and species. The aboveground vegetation changes are likely accompanied by changes belowground such as soil moisture, soil nutrient pools, fine-root abundance and turnover dynamics, all of which interplay with the microbial community dynamics (e.g. Read et al. 2004; Dickie and Reich 2005; Dickie et al. 2005; Strand et al. 2008; Toljander et al. 2006; Twieg et al. 2009; Peay et al. 2011). In the arctic tundra, fungi are the major component of the soil microbial biomass and play a critical role in ecosystem functioning (Callaghan et al. 2005). However, despite their recognized importance and the recent advances regarding belowground processes related with fungal dynamics such as C and N cycling (e.g., Schimel and Bennet 2004; Borner et al. 2008; Schaeffer et al. 2013), as well as microbial community responses to environmental changes (e.g., Clemmensen et al. 2006; Campbell et al. 2010; Deslippe et al. 2011; Deslippe et al. 2012), our understanding about the compositional and functional changes of arctic fungal communities in response to climate change remains rudimentary at best.

Almost all arctic plants are highly dependent on mutualistic relationships with mycorrhizal fungi for survival in these nutrient-limited environments (Gardes and Dahlberg 1996; Hobbie et al. 2009). Such associations include ectomycorrhizal (ECM), arbuscular mycorrhizal, ericoid and arbutoid mycorrhizal fungi (Väre et al. 1992; Newsham et al. 2009). It has been estimated that between 61 and 86% of N in Arctic tundra plants is obtained through mycorrhizal fungi (Hobbie and Hobbie 2006). Currently, our ability to predict the response of fungal communities to climate change factors is hampered both by the few detailed descriptions of the members of these communities and by our limited understanding on the ecology of many fungal species.

Globally, approximately 100,000 species of fungi have been described, but their true diversity may be as high as 6 million species (Blackwell 2011; Taylor et al.

2014). The Arctic in particular has been an understudied region, as the first works for molecular fungal diversity assessments in selected arctic sites started in the last decade (Bjorbækmo et al. 2010; Geml et al. 2012). Traditionally, fungal biodiversity studies have been based almost entirely on collection and taxonomic study of sporocarps. These studies assess only a fraction of the diversity of the fungal community because of their cryptic life style and the sporadic nature of the fructification process. However, in recent years an increasing number of molecular studies have been devoted to arctic fungi. The vast majority of these focused on root-associated fungi, particularly ECM, amassing valuable information on their diversity, biogeographic patterns (Bjorbaekmo et al. 2010; Blaalid et al. 2012; Geml et al. 2012; Timling et al. 2012) and their responses to experimental warming (Clemmensen et al. 2006; Deslippe et al. 2011; Morgado et al. 2015, 2016). ECM species are amongst the most ecologically important taxa, and seemingly represent one of the most diverse fungal guilds. Despite these important advances, the effects of long-term climate changes on soil fungal communities' structure and dynamics remain largely unknown in terms of potential changes in ecological functions as well as in taxonomic diversity. Similarly, there is currently very limited information on edible fungal species' response to climate change. Kauserud et al. (2012a, b) studied fruiting patterns throughout 1970 and 2008 and reported a shift in mushroom fruiting phenology through time. Fruiting seasons of most species are starting later in the autumn and their annual fruiting period is also ending later. Karavani et al. (2018) found that soil moisture and the minimum autumn temperatures were positively correlated with the yield of edible mushroom in a Mediterranean ecosystem; and that according to the current climate change scenarios predicted for that biome, fructifications of edible fungal species may be enhanced by climate change.

In community ecology, functional traits can be defined as biological features that play a role in the ecology of the community (Díaz and Cabido 2001). Therefore, community composition is intrinsically linked with organismal functional traits. These traits are influenced by environment and biotic interactions, and they determine the suitability of the organism in a habitat and in a community. In turn, these traits can influence ecosystem functions. Traits that influence the organism's response to the environment are considered 'response traits' while those that influence ecosystem function are known as 'effect traits' (Lavorel and Garnier 2002). Importantly, these traits may function simultaneously as response and effect traits (Koide et al. 2014). Two selected examples, (1) melanized fungi and (2) ECM extramatrical exploration types, are briefly introduced in the text that follows.

Melanins are dark macromolecules composed of various types of indolic and phenolic monomers, usually complexed with proteins and/or carbohydrates (Butler and Day 1998). When present, they are located in the cell wall or extracellular matrix of fungi, and constitute a considerable portion of total fungal biomass and likely require a considerable energetic investment (Rast and Hollenstein 1977; Butler and Day 1998). This feature has been extensively argued, and was recently shown in physiological experiments (Fernández and Koide 2013) to increase individual tolerance to several environmental stressors, such as freezing (Robinson 2001) and hydric stress (Fernández and Koide 2013). Indeed, the fungal communities of arid and seasonally water-stressed environments, as well as communities with extreme environments such as Antarctic, have a high proportion of melanized fungi (Onofri et al. 2007; Querejeta et al. 2009; Sterflinger et al. 2012). In turn, melanins are resistant to decomposition and usually considered recalcitrant. Because fungi are an important component of total soil biomass, the abundance of melanized mycelia in the habitat are likely to be an important component of C soil pools (Malik and Haider 1982; Butler et al. 2005).

Belowground, ECM fungal mycelium morphology can be divided in two parts: (1) the ectomycorrhizae, a morphological structure composed of fungal hyphae and plant roots; and (2) the extramatrical mycelium (EMM), i.e. the mycelium external to the ectomycorrhizae that grows into the surrounding soil, with the crucial functions of foraging the different soil layers for nutrients, as well as seeking new roots for colonization (Martin et al. 2001; Anderson and Cairney 2007). The EMM may form an intricate hyphal network interconnecting plant roots that pave the way for inter-plant C and nutrients exchange (Selosse et al. 2006). EMM of different taxa are known to have distinct anatomical and physiological features that are attributable to various foraging strategies (Colpaert et al. 1992; Agerer 2001; Hobbie and Agerer 2010). The main characteristics to classify EMM are the mycelium exploration types (ET), presence/absence of rhizomorphs (vessel-like structures) and hyphae hydrophobicity (Agerer 2001; Hobbie and Agerer 2010; Peay et al. 2011; Lilleskov et al. 2011; Cairney 2012). Several studies linked the EMM characteristics with the type of N pools they explore in the soil, and with their roles in soil–plant interaction, taking into account cost–benefit energy ratio for both fungi and host plant (e.g., Agerer 2001; Lilleskov et al. 2002; Hobbie and Agerer 2010; Lilleskov et al. 2011; Cairney 2012). Additionally, plant species with abundant EMM generally showed stronger potential to produce extracellular enzymes than species with scarce EMM (Tedersoo et al. 2012), an essential feature to acquire organically bounded N. It has been hypothesized that species with EMM of the medium-distance fringe, and long-distance exploration types might have a higher potential to explore recalcitrant nutrient-pools through extracellular enzyme activity, and that species with contact, short and medium-distance smooth exploration types might be associated with labile nutrient soil-pools (e.g. Lilleskov et al. 2002; Hobbie and Agerer 2010; Lilleskov et al. 2011). Therefore, the hyphal exploration type strategy may be interconnected with soil N turnover ratio, plant mineral nutrition and inter-plant nutrient transfer. In exchange, plants may allocate more or less C-derived photosynthates to the symbiotic fungi. The fate of the allocated C will greatly depend on the life span and turnover ratio associated with the EMM.

The main goal of this chapter is to understand how the arctic ECM fungal community responds to changes in climatic conditions. The focus is on the long-term effects of summer warming on belowground ECM fungal community composition, diversity and functional traits. The increase in temperature was passively achieved using open top chambers (OTC). It has been repeatedly shown that OTCs have no significant experimental bias and provide a reasonable approximation to predicted climatic changes in the Arctic (e.g. Marion et al. 1997; Sharkhuu et al. 2013; Bokhorst et al. 2013).

17.2 Methods

17.2.1 Study Site

The sampling site is part of the International Tundra Experiment (ITEX) (Henry and Molau 1997; Welker et al. 1997). The area is located at the Arctic Long Term Ecological Research site in the Toolik Lake region, northern foothills of the Brooks Range, Alaska, USA (68°37′ N, 149°32′ W; 760 m a.s.l.). The region lies within the bioclimatic subzone E, which covers approximately 36% of the Arctic's dry land surface. Mean air annual temperature is −7 °C and annual precipitation ranges between 200 and 400 mm, with approximately 50% falling as snow and average snow depth of 50 cm (DeMarco et al. 2011). Mean temperatures in July range from 9 to 12 °C (Walker et al. 2005). The distribution of vegetation depends on edaphic factors determined by topography and geological history. The oldest soils were developed on glacial till from the Sagavanirktok glacial advance (>300,000 years ago), the next oldest soils on till from the Itkillik I advance (ca. 60,000 years ago) and the youngest soils on till from the Itkillik II advance (ca. 10,000 years ago) (Hobbie and Kling 2014). The availability of N limits the primary productivity, and the net ecosystem productivity is approximately 10–20 g C m^{-2} $yr.^{-1}$ (McGuire et al. 2000).

The sampling focused on two tundra types, the dry heath and the moist acidic tundra. The dry tundra has soils with low water content, occurs on exposed, rocky interfluve areas with thin organic soil layer (0–10 cm) and retain little snow cover during winter due to wind exposure. The soils are well drained and have a deeply thawed mineral layer. The vegetation is dominated by *Dryas octopetala*, *Salix polaris*, *Vaccinium* spp. and fruticose-lichens. The moist tundra soils have high water content, occur on rolling topography with silt to gravely soils and have a thicker organic layer (0–30 cm). Thaw depth is between 30 and 50 cm, and in many locations the soil does not thaw to the mineral layer. The vegetation is dominated by *Betula nana*, *Salix pulchra*, and the sedge *Eriophorum vaginatum*. Detailed descriptions of the plant communities can be found in Walker et al. (1999) and Kade et al. (2005), and their detailed response to the increased summer temperatures in Walker et al. (1999), Wahren et al. (2005), Welker et al. (2005) and Mercado-Díaz (2011).

17.2.2 Experimental Design

In the end of July 2012, we sampled soil from 20 plots across the dry and the moist tussock tundra. In each tundra type, we sampled five plots (replicates) that were subjected to 18 years of passively increased summer air temperature by hexagonal open top chambers (OTCs, see also Fig. 17.1), subsequently referred to as 'warmed plots', and five adjacent areas with unaltered conditions ('control plots'). The sampling was performed with a soil corer with a diameter of 2 cm and depth of 20 cm.

Fig. 17.1 Open top chambers in the dry tundra, Toolik Lake, Alaska

In each of the 20 plots, five soil cores were taken, thoroughly mixed (pooled) and kept frozen until lyophilization.

The OTCs are 1 m^2 and 0.4 m high and constructed of translucent fiberglass (Marion et al. 1997; Walker et al. 1999). Within the OTCs, the summer air and upper soil temperature increased by a mean daily average of 1.5 °C (Walker et al. 1999; Welker et al. 2000; Walker et al. 2006). Every year, since 1994, the OTCs were set up as soon as 50% of the plot ground area was snow-free (usually early June) and were removed at the end of August or early September, following the International Tundra Experiment (ITEX) protocol (Welker et al. 1997).

17.2.3 DNA Metabarcoding

The DNA extraction, PCR protocol, Ion Torrent sequencing and data clean-up procedures were described in detail in Geml et al. (2014a). For each sample, 1 g of lyophilized soil was used to carry out two independent DNA extractions that were pooled to optimize extraction homogenization. In the PCR, we targeted the ITS2 region of the nuclear ribosomal internal transcribed spacer, which is currently accepted as the universal barcode marker for fungi (Schoch et al. 2012). We used primers fITS7 (Ihrmark et al. 2012) and ITS4 (White et al. 1990). The ITS4 primer was labelled with sample-specific Multiplex Identification DNA-tags (MIDs). The amplicon library was sequenced using an Ion 318TM Chip by an Ion Torrent

Personal Genome Machine (PGM; Life Technologies, Guilford, CT, U.S.A.) at Naturalis Biodiversity Center. For the initial bioinformatics steps, the online platform Galaxy (Afgan et al. 2016: https://main.g2.bx.psu.edu/root) was used to sort samples according to MIDs/samples. Primers and adapters were removed. We used MOTHUR v. 1.32.1 (Schloss et al. 2009) for subsequent sequence analyses. Sequences shorter than 150 bp and longer than 400 bp were removed because they are usually of low quality. We then used a threshold of average Phred score of at least 20 in sliding window of 50 bp to discard low-quality sequences. The quality-filtered sequences were normalized following Gihring et al. (2012), by random sub-sampling so that each sample contained equal number of sequences. We then clustered the sequences into operational taxonomic units (OTUs) using OTUpipe (Edgar 2010), with the simultaneous removal of putatively chimeric sequences using de novo and reference-based filtering (the curated dataset of fungal ITS sequences of Nilsson et al. (2011), with the default settings). We used a 97% sequence similarity clustering threshold following many other fungal ecology studies (e.g. O'Brien et al. 2005; Higgins et al. 2007; Geml et al. 2008, 2009; Amend et al. 2010; Tedersoo et al. 2010; Geml et al. 2012; Kauserud et al. 2012a, b; Brown et al. 2013; Blaalid et al. 2013; Geml et al. 2014b; Davey et al. 2015). Global singletons were discarded from further analysis. The reference database published by Kõljalg et al. (2013) was used to determine the taxonomic affinity of the OTUs using USEARCH v7 (Edgar 2010). OTUs with less than 80% similarity to any identified fungal sequence were excluded from the final analysis due to unreliable classification, and/ or uncertainty regarding ecological role.

17.2.4 Statistical Analyses

OTU was considered present in a given sample if it had at least 5 sequences, following the suggestion of Lindahl et al. (2013) to minimize false positives (e.g., OTUs that are common in one sample, but may be low-abundant contaminants in others). We determined ECM basidiomycete relative sequence abundance on a *per plot* basis. To avoid overestimation of OTU richness and to focus on the more distributed species, we removed OTUs that were present in only one of the plots. However, we considered the identity of these OTUs and joined the sequence counts at the genus level. We used the R Vegan package (Oksanen et al. 2012; R core team 2017) to run general non-metric multidimensional scaling (GNMDS) on a primary matrix of experimental plots by relative abundance of OTUs. The dataset was subjected to 500 iterations per run using the Bray-Curtis similarity index and a random starting number. We then used the ENVFIT function of the R Vegan package to correlate a secondary matrix of experimental plots by taxa richness, taxa relative sequence abundance and hyphal exploration types. For each variable, we calculated the Kendal's Tau correlation value of the secondary matrix with axes 1 and 2 of the GNMDS. Variables with significant correlation values ($\alpha = 0.05$) were overlaid in the GNMDS ordinations. We tested whether ECM basidiomycetes community composition was statistically different between the treatments, using a multi-response permutation procedure (MRPP) with R Vegan package, and determined any preferences of individual OTUs for either control and warmed plots in moist tussock and dry tundra using Indicator Species Analyses (Dufrêne and Legendre 1997) with R Labdsv package (Roberts 2016). We also tested for significant differences in OTU richness regarding genera and EMM characteristics using Students *t*-test.

17.2.5 ECM Fungi and EMM Determination

We followed the publication of Tedersoo and Smith (2013) to select the basidiomycete ECM taxa. For most OTUs we used at least 90% sequence similarity to determine genera. Because Sebacinales have a diverse ecology, we selected ECM OTUs based on their supported phylogenetic placement using Maximum Likelihood (ML) and Bayesian Inference (BI), with at least 70% bootstrap (ML) and/or 0.95 posterior probability (BI) amongst sequences of taxa that were morphologically confirmed as ECM by Glen et al. (2002), Urban et al. (2003), Ryberg et al. (2009) and Tedersoo and Smith (2013).

To determine the EMM characteristics, we followed Agerer (2006), Tedersoo and Smith (2013), and consulted the DEEMY database (http://deemy.de), an information system for the characterization and determination of ectomycorrhizae. In *Russula*, if no EMM information for the species of interest was available, we assumed the EMM characteristics based on the closest species with known characteristics. To determine the closest species, we followed the phylogenetic study by Miller and Buyck (2002). Similarly, for OTUs of the genus *Hebeloma*, we followed Boyle et al. (2006).

17.3 Results

17.3.1 ECM Fungal Community Diversity

Out of the 4,046,811 original sequences, 2,068,216 passed the series of quality-filtering steps. After normalizing the library size across all samples by rarefying, we retained 1,129,660 sequences with an average length of 254.9 ± 56 bp (SD). Clustering the sequences at 97% similarity originated 10,035 OTUs. Out of these, 3148 putative chimeras and 1249 singleton OTUs were removed from downstream analysis. From the remaining 5638 non-singleton OTUs, 343 were at least 90% similar to a known ECM basidiomycete sequence in the UNITE and NCBI curated database (Kõljalg et al. 2013), and therefore classified as ECM fungal OTUs. From these, 140 occurred in a single experimental plot. After excluding these OTUs (to

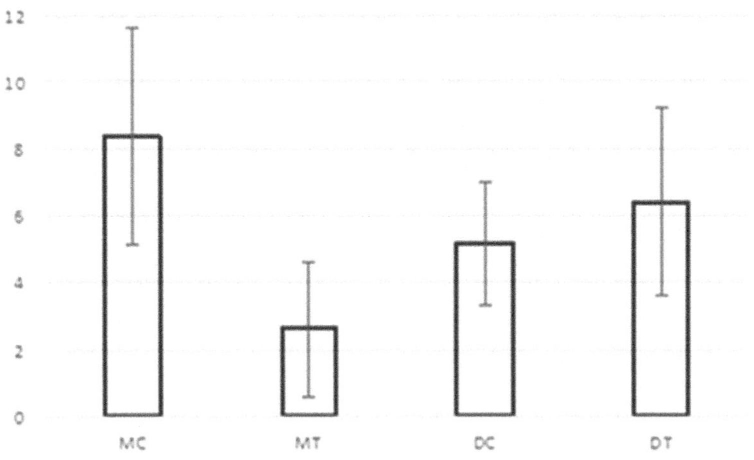

Fig. 17.2 Fisher's alpha index per treatment with standard deviation. *MC* moist tundra control plots, *MT* moist tundra warmed plots, *DC* dry tundra control plots, *DT* dry tundra warmed plots

reduce the bias in richness and to focus on the most widespread OTUs) we retained 203 OTUs.

The control plots of the dry and moist tundra had similar overall OTU richness, with 108 and 105 OTUs, respectively; while the warm plots in the dry tundra had considerably higher overall richness than the warm plots in the moist tundra, 114 and 75 OTUs, respectively. The control plots of the moist tundra had the highest average Fisher's alpha diversity and warmed plots in the dry tundra had the highest average richness. Comparing Fisher's alpha index and OTU richness between the control and the warmed plots revealed significant differences in the moist tundra ($p = 0.012$ and 0.008, respectively) but not in the dry tundra (Fig. 17.2 and Table 17.1).

The 203 ECM basidiomycete OTUs were spread in 20 genera. However, four of these accounted for ca. 80% of all OTU richness: *Tomentella* (62 OTUs, 31%), *Cortinarius* (47, 23%), *Inocybe*, (28, 14%) and *Russula* (25, 12%). Comparing average OTU richness of these genera between control and warmed plots revealed that *Tomentella*, *Inocybe* and *Russula* were significantly different in the moist tundra; while in the dry tundra there were no significant differences in OTU richness (Table 17.1).

17.3.2 The Effect of Warming on Fungal Community Composition

For the moist and dry tundra types, the GNMDS analyses with lowest stress had values of 0.0191 and 0.0938, respectively (Fig. 17.3). The ordinations showed that in the moist tundra there is a clear and distinct difference in ECM fungal community

Table 17.1 Average, plus and minus standard deviation of OTU richness per treatment

	Moist			Dry		
	MC	MT	*p*	DC	DT	*p*
Alnicola	0.8 ± 0.45	1 ± 1.00	0.621	–	–	–
Boletus	–	–	–	0.6 ± 0.55	0.6 ± 0.55	1
Ceratobasidium	–	–	–	0.4 ± 0.55	0 ± 0.00	0.178
Clavicorona	–	–	–	0.2 ± 0.45	0.4 ± 0.55	0.374
Clavulina	0.2 ± 0.45	0.4 ± 0.55	0.374	0.8 ± 0.84	0.6 ± 0.55	0.704
Cortinarius	14.8 ± 7.43	6 ± 7.38	0.132	5.8 ± 3.49	7.6 ± 7.44	0.618
Hebeloma	2 ± 0.00	1.2 ± 0.45	**0.016**	0.4 ± 0.89	0.6 ± 0.89	0.778
Hymenogaster	0.4 ± 0.55	0 ± 0.00	0.178	–	–	–
Inocybe	6.2 ± 2.39	1 ± 1.00	**0.002**	4 ± 2.83	4 ± 5.70	1
Laccaria	1.4 ± 0.89	1.4 ± 0.55	1	–	0.8 ± 1.30	0.242
Leccinum	2.8 ± 1.64	0.4 ± 0.89	**0.042**	0.4 ± 0.55	0.6 ± 0.89	0.704
Pseudotomentella	–	–	–	0.4 ± 0.55	0.8 ± 0.45	0.178
Rhizoctonia	–	–	–	0.2 ± 0.45	0.2 ± 0.45	–
Russula	6.4 ± 4.16	1 ± 1.00	**0.05**	2.8 ± 3.03	8.2 ± 6.72	0.059
Sebacina	0.8 ± 1.10	–	0.178	–	–	–
Sistotrema	2.2 ± 3.03	–	0.18	0.4 ± 0.55	0.6 ± 0.89	0.621
Tomentella	13.2 ± 5.26	2.4 ± 4.83	**0.014**	16 ± 11.11	17.2 ± 5.40	0.775
Tomentellopsis	–	–	–	–	0.4 ± 0.55	0.178
Tremellodendron	0.8 ± 1.10	–	0.178	–	–	–
Tulasnella	–	–	–	1.2 ± 1.10	0.8 ± 1.10	0.374
C/S/MDS	30.6 ± 11.17	8.4 ± 7.27	**0.004**	26.2 ± 11.17	33.8 ± 8.44	0.271
L/MDF	19.8 ± 8.35	6.4 ± 7.57	0.066	7.2 ± 3.70	9.4 ± 7.30	0.479
All OTU richness	52 ± 18.88	14.8 ± 12.32	**0.008**	33.6 ± 14.01	43.4 ± 15.47	0.251

Student's *t*-test comparison between control and warmed plots. Bold values indicate significant differences ($\alpha = 0.05$)
MT moist tundra warmed plots, *MC* moist tundra control plots, *DT* dry tundra warmed plots, *DC* dry tundra control plots

structure between the control and warmed plots; while in the dry tundra the structure of the community of the control and warmed plots were not so dissimilar. The MRPP analyses supported the results observed from the GNMDS, i.e., the community composition comparisons between control and warmed plots were significantly different in the moist tundra ($A = 0.12345835$, $P = 0.0066$), while in the dry tundra they were not ($A = 0.00147$, $P = 0.4288$). Kendall's Tau correlations analyses between community structure, relative sequence abundance and OTU richness of taxonomic groups and hyphal exploration types are summarized in Table 17.2 and significant correlations ($\alpha = 0.05$) are displayed in the GNMDS plots (Fig. 17.3).

Given that warming only had significant effect on the moist tundra community, we only considered informative the trends of relative sequence observed in the moist tundra ordination plots. *Alnicola*, *Laccaria* and OTUs with contact, short-distance and medium-distance smooth exploration type had considerably higher average relative sequence abundance in the warmed plots. On the other hand,

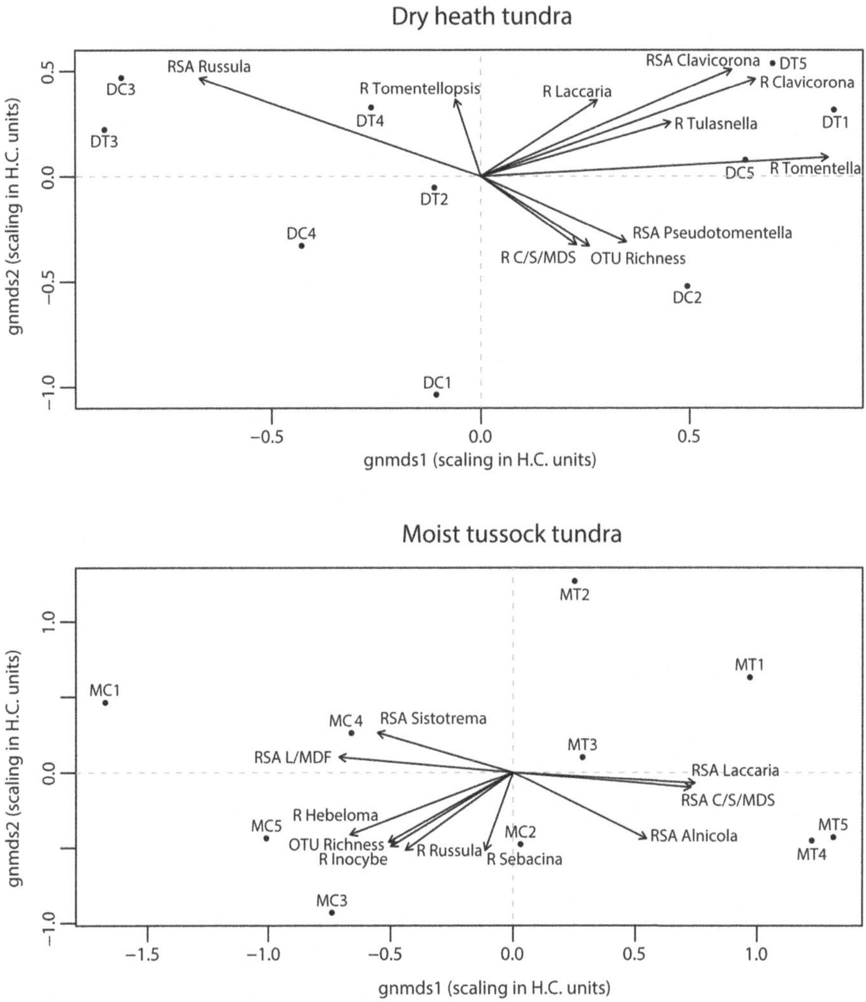

Fig. 17.3 General non-metric multidimensional scaling (GNMDS) ordination plots for fungal communities from the warmed and control plots in the dry and moist tundra types based on relative abundance of ectomycorrhizal fungal OTUs. Richness (R) and relative sequence abundance (RSA) variables of genera and hyphal exploration types that correlated significantly with any ordination axis are displayed as vectors

Sistotrema and OTUs with medium-distance fringe and long-distance exploration types had higher relative sequence abundance in the control plots (Fig. 17.3, Table 17.2).

Indicator species analyses revealed that in the moist tundra 14 OTUs were significantly correlated with the control plots and no OTU was significantly correlated with the warmed plots. On the other hand, in the dry tundra only 1 OTU was significantly correlated with control plots and no OTU was significantly correlated with

Table 17.2 Kendall's Tau correlation values between the taxonomic groups and the GNMDS ordinations

	Moist				Dry			
	Axis1		Axis 2		Axis1		Axis 2	
	p-value	Tau	*p*-value	Tau	*p*-value	tau	*P*-value	Tau
RSA *Alnicola*	0.146	0.37	**0.046**	**−0.506**	–	–	–	–
R *Alnicola*	0.377	0.241	0.062	−0.509	–	–	–	–
RSA *Boletus*	–	–	–	–	0.643	−0.119	0.781	0.072
R *Boletus*	–	–	–	–	0.522	0.183	0.670	0.122
RSA *Ceratobasidium*	–	–	–	–	0.517	−0.181	0.092	−0.470
R *Ceratobasidium*	–	–	–	–	0.602	−0.149	0.068	−0.522
RSA *Clavicorona*	–	–	–	–	**0.014**	**0.669**	0.119	0.426
R *Clavicorona*	–	–	–	–	**0.017**	**0.683**	0.210	0.358
RSA *Clavulina*	0.656	0.122	1	0	0.052	0.501	0.926	0.024
R *Clavulina*	0.909	0.033	0.909	−0.033	0.271	0.305	0.920	−0.028
RSA *Cortinarius*	0.728	−0.111	0.727	−0.111	0.369	0.225	0.209	−0.315
R *Cortinarius*	0.528	−0.159	0.087	−0.432	0.272	0.283	0.143	−0.377
RSA *Hebeloma*	0.862	−0.067	0.484	0.200	0.075	−0.487	0.373	−0.243
R *Hebeloma*	0.286	−0.304	**0.019**	**−0.669**	0.643	−0.119	0.781	0.072
RSA *Hymenogaster*	0.244	−0.325	0.897	0.036	–	–	–	–
R *Hymenogaster*	0.192	−0.373	1	0	–	–	–	–
RSA *Inocybe*	0.369	−0.225	0.719	0.090	1	−0.022	0.291	−0.289
R *Inocybe*	0.087	−0.432	**0.038**	**−0.523**	0.517	−0.181	0.092	−0.470
RSA *Laccaria*	1	0.022	**0.002**	**0.733**	0.365	0.253	0.365	0.253
R *Laccaria*	0.271	−0.305	0.764	0.083	**0.014**	**0.669**	0.119	0.426
RSA *Leccinum*	0.212	−0.328	0.291	−0.277	0.839	0.054	0.684	−0.109
R *Leccinum*	0.207	−0.337	0.065	−0.493	0.052	0.501	0.926	0.024
RSA *Pseudotomentella*	–	–	–	–	**0.020**	**0.597**	0.926	−0.024
R *Pseudotomentella*	–	–	–	–	0.369	0.225	0.209	−0.315
RSA *Rhizoctonia*	–	–	–	–	0.698	0.108	0.154	−0.398
R *Rhizoctonia*	–	–	–	–	0.075	−0.487	0.373	−0.243
RSA *Russula*	0.209	0.315	0.369	−0.225	**0.002**	**−0.764**	0.369	0.225
R *Russula*	0.365	−0.230	**0.011**	**−0.644**	1	−0.022	0.291	−0.289
RSA *Sebacina*	0.517	−0.181	0.052	−0.542	–	–	–	–
R *Sebacina*	0.146	0.368	**0.046**	**−0.506**	–	–	–	–
RSA *Sistotrema*	0.698	0.109	**0.028**	**−0.615**	0.154	−0.381	0.542	−0.163
R *Sistotrema*	0.656	0.122	1	0	0.365	0.253	0.365	0.253
RSA *Tomentella*	0.468	−0.184	0.146	−0.368	**0.009**	**0.644**	1	0.022
R *Tomentella*	0.728	−0.111	0.727	−0.111	0.839	0.054	0.684	−0.109
RSA *Tomentellopsis*	–	–	–	–	0.897	−0.036	0.154	0.398
R *Tomentellopsis*	–	–	–	–	**0.020**	**0.597**	0.926	−0.024
RSA *Tremellodendron*	0.104	−0.436	0.309	−0.272	0.698	−0.108	0.052	−0.542

Table 17.2 (continued)

	Moist				Dry			
	Axis1		Axis 2		Axis1		Axis 2	
	p-value	Tau	*p*-value	Tau	*p*-value	tau	*P*-value	Tau
R *Tremellodendron*	0.484	0.200	0.862	−0.067	–	–	–	–
RSA *Tulasnella*	–	–	–	–	**0.001**	**0.882**	0.773	0.076
R *Tulasnella*	–	–	–	–	0.698	0.108	0.154	−0.398
RSA C/S/MDS	0.862	−0.067	**0.029**	**0.556**	0.601	−0.156	0.073	0.467
R C/S/MDS	0.897	0.036	0.244	−0.325	**0.002**	**−0.764**	0.369	0.225
RSA L/MDF	0.728	0.111	**0.047**	**−0.511**	0.601	0.156	0.156	−0.378
R L/MDF	0.369	−0.225	0.719	0.090	0.154	−0.381	0.542	−0.163
Total OTU richness	1	0.022	**0.002**	**0.733**	**0.009**	**0.644**	1	0.022

RSA relative sequence abundance, *R* OTU richness, *C/S/MDS* contact, short- and medium-distance smooth hyphal exploration type, *MDF/L* medium-distance fringe and long-distance hyphal exploration type

Table 17.3 Indicator species analysis of OTUs with significant correlation (a = 0.05) with the site, their taxonomic affinity and sequence similarity with referenced species hypothesis (SH; Ver. 7.0) and known sequences from UNITE database or GenBank

OTU	Correlated site	Kõljalg et al. (2013) and UNITE classification	Similarity (%)
1281	DC	SH217480.07FU—*Tomentella coerulea* (UDB016493)	97.9
3369	MC	SH181066.07FU—*Leccinum holopus* (UDB001378)	99.6
484	MC	SH184521.07FU—*Tomentella fuscocinerea* (UDB016484)	99.6
3351	MC	SH185279.07FU—*Tomentella lateritia* (UDB016439)	97.8
181	MC	SH177933.07FU—*Tomentella coerula* (UDB018451)	98.1
4645	MC	SH177801.07FU—*Tomentella* sp. (UDB017832)	98.9
6618	MC	SH218845.07FU—*Hebeloma mesophaeum* (UDB017969)	96.2
1120	MC	SH186207.07FU—*Russula renidens* (UDB015975)	100
4313	MC	SH177782.07FU—*Tomentella badia* (UDB016188)	95.9
1124	MC	SH177782.07FU—*Tomentella badia* (UDB016188)	99.6
1625	MC	SH222334.07FU—*Cortinarius cinnamomeus* (UDB011339)	99.7
801	MC	SH180352.07FU—*Inocybe sororia* (HQ604382)	96.6
3413	MC	SH180352.07FU—*Inocybe sororia* (HQ604382)	95.9
5841	MC	SH219800.07FU—*Inocybe catalaunica* (AM882793)	96.7
219	MC	SH219800.07FU—*Inocybe catalaunica* (AM882793)	99

the warmed plots (Table 17.3). Figure 17.4 shows a close relative of *Cortinarius cinnamomeus*, a species significantly affected by warming in the moist tundra.

Fig. 17.4 *Cortinarius aurantiobasis* among arctic dwarf shrubs in moist tundra, Toolik Lake, Alaska

17.4 Discussion

There are marked shifts in fungal community composition, richness and OTU-specific abundance measures in response to long-term experimental warming in low arctic tundra in Northern Alaska. Community composition of ECM fungi responded strongly to summer warming in the moist tundra, but not in the dry tundra. Similar responses to experimental summer warming have also been observed for non-ECM fungal taxa as well, for example, in saprotrophic, lichenized, plant pathogenic and animal parasitic fungi (Geml et al. 2015; Semenova et al. 2015; Geml et al. 2016). Likely, fungi inhabiting the dry tundra are more adapted to fluctuations in environmental conditions than the moist tundra community. Moist tundra soils, being generally cool throughout the summer, tend to experience less fluctuations in temperature than dry tundra soils that are regularly exposed to higher summer temperatures and pronounced water stress in the upper layers. Because plant communities are also known to respond strongly to experimental summer warming in the moist tundra (Wahren et al. 2005; Walker et al. 2006; Pattison and Welker 2014), the greater response of ECM fungi in this tundra type may correlate with the similar trends observed for plants, although it remains unknown to what extent changes in fungal communities drive or are driven by shifts in vegetation.

ECM fungi were represented by 20 genera, but only four of these dominated the communities: *Tomentella*, *Cortinarius*, *Inocybe* and *Russula*. In the moist tundra, warming induced strong decrease in OTU richness in *Cortinarius, Inocybe* and *Russula*, while *Tomentella* OTU richness decreased only marginally. This latter is in apparent disagreement with the findings of Morgado et al. (2015); however, the differences may be explained by the fact that *Tomentella* is a species-rich genus and that several OTUs were found in a single plot, which were excluded from this chapter. Interestingly, despite the decreasing richness, none of these four major ECM fungal genera showed significant warming-induced changes in relative sequence abundance in the moist tundra, most decreased in relative abundance non-significantly. In the dry tundra, the relative abundance in *Russula* showed significant negative correlation with the treatment (GNMDS axis 1), but relative abundance of other major genera was not significantly affected. This suggests that while some species disappear from the community in response to warming, the remaining congeneric species appear to be resilient. With respect to the less diverse genera, *Laccaria* is particularly noteworthy, because of the significant response to the warming treatment both in richness (dry tundra) and relative sequence abundance (moist tundra). *Laccaria* is known to be nitrophilic and tends to respond positively to disturbance in general (Dickie and Moyersoen 2008). Indeed, *Laccaria* OTUs were virtually restricted to the warmed plots only, confirming their potential to get established more successfully in altered conditions. Overall, the general decrease in abundance in ectomycorrhizal fungi may have functional implications and the fact that several ectomycorrhizal fungi showed positive response to warming, while most were negatively affected, indicates substantial shift in the community (Geml et al. 2016).

In our dataset only a few OTUs were conspecific (>97% similarity) with edible species. These belonged to the genera *Boletus* (1 OTU) and *Leccinum* (4 OTUs each of which with more than 97% sequence similarity to distinct Species Hypothesis). *Boletus* was only detected in the dry tundra and was seemingly unaffected by warming. On the other hand, *Leccinum* was mainly detected in the moist tundra and showed significant lower richness in the warmed plots. A similar trend to that shown by *Leccinum* was recorded in *Hebeloma mesophaeum* (96.2% similarity Table 17.3), which has been reported to be widely consumed in Mexico (Carrasco-Hernandez et al. 2011). These results highlight that edible fungi responses to climate change are likely species-specific and habitat-dependent. Two examples of edible mycorrhizal fungi of the tundra environment are illustrated in Figs. 17.5 and 17.6.

Fig. 17.5 *Boletus edulis* among arctic dwarf shrubs in dry tundra, near Toolik Lake, Alaska

Fig. 17.6 *Leccinum variicolor* among arctic dwarf shrubs, Toolik Lake, Alaska

Regarding the functional types of ECM fungi, represented here by the EMM characteristics, warming appeared to favour taxa with contact, short- or medium-distance smooth mycelial (C/S/MDS) exploration types (e.g. *Laccaria*) in both tundra types, showing significant increases in richness (dry tundra) and in relative abundance (moist tundra) as opposed to the significant decrease in relative sequence abundance seen in medium-distance fringe and long-distance (MDF/L) exploration types, potentially affecting the mobilization of different nutrient pools in the soil (Morgado et al. 2015). The majority of ECM fungi with contact and short-distance exploration types with hydrophilic hyphae have been hypothesized to explore the pools of labile nutrients in the soil, since most of them showed reduced proteolytic capabilities in laboratory experiments (Lilleskov et al. 2002; Nygren et al. 2007). On the other hand, MDF/L exploration types have been linked with increased proteolytic capabilities and are expected to play important roles in nutrient translocation and colonization of new roots (Hobbie and Agerer 2010).

As fungi are a major component of arctic soil biomass, melanized hyphae may play an important role in C cycling. *Tomentella*, the most species-rich taxa with melanized hyphae in our dataset, has contrasting response to warming in dry and moist tundra. While in the dry tundra these taxa do not change in response to warming, in the moist tundra there is a sharp and significant decrease in their OTU richness. Moreover, six *Tomentella* OTUs are 'indicator species' of the moist control plots, meaning that they were present in the majority of control plots but absent from the warmed plots. One reason might be due to physiological maladaptation to the new environmental conditions, which would in turn lead to a detrimental selection as ECM fungal partners by the ECM host. Importantly, the reduction in *Tomentella* richness may lead to a reduced potential for C soil storage, since melanized hyphae are a recalcitrant component of the soil biome. However, it is necessary to understand that the fate of the C stored in any biomass is directly dependent on its turnover rate.

Here we provide evidence that long-term experimental summer warming has profound effects on community composition and abundance of arctic ECM fungi. We also underline that, while there are similarities within ECM genera, changes in occurrence and abundance in response to warming tend to be species-specific, and may be masked when communities are compared at higher taxonomic levels. Therefore, we recommend that studies of arctic ECM fungal communities and their roles in nutrient cycling take into account species-level differences.

References

Afgan E, Baker D, Beek M et al (2016) The galaxy platform for accessible, reproducible and collaborative biomedical analyses: 2016 update. Nucleic Acids Res 44(W1):W3–W10

Agerer R (2001) Exploration types of ectomycorrhizae. A proposal to classify ectomycorrhizal mycelial systems according to their patterns of differentiation and putative ecological importance. Mycol Prog 11:107–114

Agerer R (2006) Fungal relationships and structural identity of their ectomycorrhizae. Mycol Prog 5:67–107

Amend AS, Seifert KA, Bruns TD (2010) Quantifying microbial communities with 454 pyrosequencing: does read abundance count? Mol Ecol 19:5555–5565

Anderson I, Cairney JWG (2007) Ectomycorrhizal fungi: exploring the mycelial frontier. FEMS Microbiol Rev 31:388–406

Anisimov OA, Vaughan DG, Callaghan TV et al (2007) Polar regions (Arctic and Antarctic). In: Parry ML, Canziani OF, Palutikof JP et al (eds) Climate change 2007: impacts, adaptation and vulnerability. Contribution of working group II to the fourth assessment report of the intergovernmental panel on climate change. Cambridge University Press, Cambridge, England, pp 653–685

Arrigo KR, Van Dijken GL (2011) Secular trends in Arctic Ocean net primary production. J Geophys Res Oceans 116:C09011

Bjorbækmo MF, Carlsen T, Brysting A et al (2010) High diversity of root associated fungi in both alpine and arctic *Dryas octopetala*. BMC Plant Biol 10:244

Blaalid R, Carlsen T, Kumar S et al (2012) Changes in the root-associated fungal communities along a primary succession gradient analysed by 454 pyrosequencing. Mol Ecol 21:1897–1908

Blaalid R, Kumar S, Nilsson RH et al (2013) ITS1 versus ITS2 as DNA metabarcodes for fungi. Mol Ecol Resour 13:218–224

Blackwell M (2011) The fungi: 1, 2, 3 ... 5.1 million species? Am J Bot 98:426–438

Bokhorst S, Huiskes A, Aerts R et al (2013) Variable temperature effects of open top chambers at polar and alpine sites explained by irradiance and snow depth. Glob Chang Biol 19:64–74

Borner AP, Kielland K, Walker MD (2008) Effects of simulated climate change on plant phenology and nitrogen mineralization in alaskan arctic tundra. Arct Antarct Alp Res 40:27–38

Boyle H, Zimdars B, Renker C et al (2006) A molecular phylogeny of *Hebeloma* species from Europe. Mycol Res 110:369–380

Brown SP, Callaham MA, Oliver AK et al (2013) Deep ion torrent sequencing identifies soil fungal community shifts after frequent prescribed fires in a southeastern US forest ecosystem. FEMS Microbiol Ecol 86:557–566

Butler MJ, Day AW (1998) Fungal melanins: a review. Can J Microbiol 44:1115–1136

Butler MJ, Gardiner RB, Day AW (2005) Degradation of melanin or inhibition of its synthesis: are these a significant approach as a biological control of phytopathogenic fungi? Biol Control 32:326–336

Cairney JWG (2012) Extramatrical mycelia of ectomycorrhizal fungi as moderators of carbon dynamics in forest soil. Soil Biol Biochem 47:198–208

Callaghan TV, Björn LO, Chernov Y et al (2005) Tundra and polar desert ecosystems. In: Symon C, Arris L, Heal B (eds) ACIA. Arctic climate impacts assessment. Cambridge University Press, Cambridge, England, pp 243–345

Campbell BJ, Polson SW, Hanson TE et al (2010) The effect of nutrient deposition on bacterial communities in Arctic tundra soil. Environ Microbiol 12:1842–1854

Chapin FS III, Sturm M, Serreze MC et al (2005) Role of land-surface changes in Arctic summer warming. Science 310:657–660

Clemmensen KE, Michelsen A, Jonasson S et al (2006) Increased ectomycorrhizal fungal abundance after long-term fertilization and warming of two arctic tundra ecosystems. New Phytol 171:391–404

Colpaert JV, van Assche JA, Luijtens K (1992) The growth of the extramatrical mycelium of ectomycorrhizal fungi and the growth response of *Pinus sylvestris* L. New Phytol 120:127–135

Comiso JC, Hall DK (2014) Climate trends in the Arctic as observed from space. Wiley Interdiscip Rev Clim Chang 5:389–409

Davey M, Blaalid R, Vik U et al (2015) Primary succession of *Bistorta vivipara* (L.) Delabre (Polygonaceae) root associated fungi mirrors plant succession in two glacial chronosequences. Environ Microbiol 17:2777–2790

DeMarco J, Mack MC, Bret-Harte MS (2011) The effects of snow, soil microenvironment, and soil organic matter quality on N availability in three Alaskan Arctic plant communities. Ecosystems 14:804–817

Deslippe JR, Hartmann M, Mohn WW et al (2011) Long-term experimental manipulation of climate alters the ectomycorrhizal community of *Betula nana* in Arctic tundra. Glob Chang Biol 17:1625–1636

Deslippe JR, Hartmann M, Simard SW et al (2012) Long-term warming alters the composition of Arctic soil microbial communities. FEMS Microbiol Ecol 82:303–315

Díaz S, Cabido M (2001) Vive la différence: plant functional diversity matters to ecosystem processes. Trends Ecol Evol 16:646–655

Dickie IA, Moyersoen B (2008) Towards a global view of ectomycorrhizal ecology. New Phytol 180:263–265

Dickie IA, Reich PB (2005) Ectomycorrhizal fungal communities at forest edges. J Ecol 93:244–255

Dickie IA, Schnitzer SA, Reich PB et al (2005) Spatially disjunct effects of co-occurring competition and facilitation. Ecol Lett 8:1191–1200

Dufrêne M, Legendre P (1997) Species assemblages and indicator species: the need for a flexible asymmetrical approach. Ecol Monogr 67:345–366

Edgar RC (2010) Search and clustering orders of magnitude faster than BLAST. Bioinformatics 26:2460–2461

Fernández CW, Koide RT (2013) The function of melanin in the ectomycorrhizal fungus *Cenococcum geophilum* under water stress. Fungal Ecol 6:479–486

Gardes M, Dahlberg A (1996) Mycorrhizal diversity in arctic and alpine tundra: an open question. New Phytol 133:147–157

Geml J, Laursen GA, Taylor DL (2008) Molecular diversity assessment of arctic and boreal *Agaricus* taxa. Mycologia 100:577–589

Geml J, Laursen GA, Timling I et al (2009) Molecular phylogenetic biodiversity assessment of arctic and boreal *Lactarius* Pers. (Russulales; Basidiomycota) in Alaska, based on soil and sporocarp DNA. Mol Ecol 18:2213–2227

Geml J, Timling I, Robinson CH et al (2012) An arctic community of symbiotic fungi assembled by long-distance dispersers: phylogenetic diversity of ectomycorrhizal basidiomycetes in Svalbard based on soil and sporocarp DNA. J Biogeogr 39:74–88

Geml J, Gravendeel B, Neilen M et al (2014a) DNA metabarcoding reveals high fungal diversity and pH-correlated habitat partitioning in protected coastal *Salix repens* communities in the Netherlands. PLoS One 9:e99852

Geml J, Pastor N, Fernández L et al (2014b) Large-scale fungal diversity assessment in the Andean Yungas forests reveals strong community turnover among forest types along an altitudinal gradient. Mol Ecol 23:2452–2472

Geml J, Morgado LN, Semenova TA et al (2015) Long-term warming alters richness and composition of taxonomic and functional groups of arctic fungi. FEMS Microbiol Ecol 91:fiv095

Geml J, Semenova TA, Morgado LN et al (2016) Changes in composition and abundance of functional groups of arctic fungi in response to long-term summer warming. Biol Lett 12:20160503

Gihring TM, Green SJ, Schadt CW (2012) Massively parallel rRNA gene sequencing exacerbates the potential for biased community diversity comparisons due to variable library sizes. Environ Microbiol 14:285–290

Glen M, Tommerup IC, Bougher NL et al (2002) Are Sebacinaceae common and widespread ectomycorrhizal associates of *Eucalyptus* species in Australian forests? Mycorrhiza 12:243–247

Hansen J, Ruedy R, Sato M (2010) Global surface temperature change. Rev Geophys 48:RG4004

Henry GHR, Molau U (1997) Tundra plants and climate change: the international tundra experiment (ITEX). Glob Chang Biol 3:1–9

Higgins KL, Arnold AE, Miadlikowska J et al (2007) Phylogenetic relationships, host affinity, and geographic structure of boreal and arctic endophytes from three major plant lineages. Mol Phylogenet Evol 42:543–555

Hobbie EA, Agerer R (2010) Nitrogen isotopes in ectomycorrhizal sporocarps correspond to belowground exploration types. Plant Soil 327:71–83

Hobbie JE, Hobbie EA (2006) 15N in symbiotic fungi and plants estimates nitrogen and carbon flux rates in Arctic tundra. Ecology 87:816–822

Hobbie JE, Hobbie EA, Drossman H et al (2009) Mycorrhizal fungi supply nitrogen to host plants in Arctic tundra and boreal forests: 15N is the key signal. Can J Microbiol 55:84–94

Hobbie JE, Kling GW (2014) Alaska's changing Arctic. Oxford University Press, New York, p 331

Ihrmark K, Bödeker ITM, Cruz-Martínez K et al (2012) New primers to amplify the fungal ITS2 region–evaluation by 454-sequencing of artificial and natural communities. FEMS Microbiol Ecol 82:66–677

Kade A, Walker DA, Raynolds MK (2005) Plant communities and soils in cryoturbated tundra along a bioclimate gradient in the low Arctic, Alaska. Phytocoenologia 35:761–820

Karavani A, De Cáceres M, Aragón JM et al (2018) Effect of climatic and soil moisture conditions on mushroom productivity and related ecosystem services in Mediterranean pine stands facing climate change. Agric For Meteorol 248:432–440

Kauserud H, Kumar S, Brysting AK et al (2012a) High consistency between replicate 454 pyrosequencing analyses of ectomycorrhizal plant root samples. Mycorrhiza 22:309–315

Kauserud K, Heegaard E, Büntgen U et al (2012b) Warming-induced shift in European mushroom fruiting phenology. Proc Natl Acad Sci U S A 109:14488–14493

Kharuk VI, Ranson KJ, Sergey IT et al (2013) Tree-line structure and dynamics at the northern limit of the larch forest: Anabar plateau, Siberia, Russia. Arct Antarct Alp Res 45:526–537

Koide RT, Fernández C, Malcolm G (2014) Determining place and process: functional traits of ectomycorrhizal fungi that affect both community structure and ecosystem function. New Phytol 201:433–439
Kõljalg U, Nilsson RH, Abarenkov K et al (2013) Towards a unified paradigm for sequence-based identification of fungi. Mol Ecol 22:5271–5277
Lavorel S, Garnier E (2002) Predicting changes in community composition and ecosystem functioning from plant traits. Funct Ecol 16:545–556
Lilleskov EA, Fahey TJ, Horton TR et al (2002) Belowground ectomycorrhizal community change over a nitrogen deposition gradient in Alaska. Ecology 83:104–115
Lilleskov EA, Hobbie EA, Horton TR (2011) Conservation of ectomycorrhizal fungi: exploring the linkages between functional and taxonomic responses to anthropogenic N deposition. Fungal Ecol 4:174–183
Lindahl BD, Nilsson RH, Tedersoo L et al (2013) Fungal community analysis by high-throughput sequencing of amplified markers–a user's guide. New Phytol 199:288–299
Loranty MM, Goetz SJ (2012) Shrub expansion and climate feedbacks in Arctic tundra. Environ Res Lett 7:011005
Malik KA, Haider K (1982) Decomposition of 14C-labelled melanoid fungal residues in a marginally sodic soil. Soil Biol Biochem 14:457–460
Mann ME, Bradley RS, Hughes MK (1999) Northern hemisphere temperatures during the past millennium: inferences, uncertainties, and limitations. Geophys Res Lett 26:759–762
Mann ME, Jones PD (2003) Global surface temperatures over the past two millennia. Geophys Res Lett 30:15–18
Marion GM, Henry GHR, Freckman DW et al (1997) Open-top designs for manipulating field temperature in high-latitude ecosystems. Glob Chang Biol 3:20–32
Martin F, Duplessis S, Ditengou F et al (2001) Developmental cross talking in the ectomycorrhizal symbiosis: signals and communication genes. New Phytol 151:145–154
McGuire AD, Clein JC, Melillo JM (2000) Modelling carbon responses of tundra ecosystems to historical and projected climate: sensitivity of pan-Arctic carbon storage to temporal and spatial variation in climate. Glob Chang Biol 6:141–159
Mercado-Díaz J (2011) Plant community responses of the Alaskan Arctic tundra to environmental and experimental changes in climate. MSc Thesis. University of Puerto Rico. Puerto Rico
Miller SL, Buyck B (2002) Molecular phylogeny of the genus *Russula* in Europe with a comparison of modern infrageneric classifications. Mycol Res 106:259–276
Montoya JM, Raffaelli D (2010) Climate change, biotic interactions and ecosystem services. Philos T R Soc B 365:2013–2018
Morgado LN, Semenova TA, Welker JM et al (2015) Summer temperature increase has distinct effects on the ectomycorrhizal fungal communities of moist tussock and dry tundra in Arctic Alaska. Glob Chang Biol 21:959–972
Morgado LN, Semenova TA, Welker JM et al (2016) Long-term increase in snow depth leads to compositional changes in arctic ectomycorrhizal fungal communities. Glob Chang Biol 22:3080–3096
Newsham KK, Upson R, Read DJ (2009) Mycorrhizas and dark septate root endophytes in polar regions. Fungal Ecol 2:10–20
Nilsson RH, Abarenkov K, Larsson K-H et al (2011) Molecular identification of fungi: rationale, philosophical concerns, and the UNITE database. Open Appl Informat J 5:81–86
Nygren CMR, Edqvist J, Elfstrand M et al (2007) Detection of extracellular protease activity in different species and genera of ectomycorrhizal fungi. Mycorrhiza 17:241–248
O'Brien H, Parrent JL, Jackson JA et al (2005) Fungal community analysis by large-scale sequencing of environmental samples. Appl Environ Microbiol 71:5544–5550
Oksanen J, Blanchet FG, Kindt R et al (2012) Vegan: community ecology package. R package version 2.0-3. http://CRAN.R-project.org/package=vegan
Onofri S, Selbmann L, de Hoog GS et al (2007) Evolution and adaptation of fungi at boundaries of life. Adv Space Res 40:1657–1664

Pattison RR, Welker JM (2014) Differential ecophysiological response of deciduous shrubs and a graminoid to long-term experimental snow reduction and addition in moist tundra, northern Alaska. Oecologia 174:339–350

Peay KG, Kennedy PG, Bruns TD (2011) Rethinking ectomycorrhizal succession: are root density and hyphal exploration types drivers of spatial and temporal zonation? Fungal Ecol 4:233–240

Pimm SL (2009) Climate disruption and biodiversity. Curr Biol 19:R595–R601

Post E, Forchhammer MC, Bret-Harte MS et al (2009) Ecological dynamics across the Arctic associated with recent climate change. Science 325:1355–1358

Post E (2013) Ecology of climate change: the importance of biotic interactions. In: Monographs in population biology, vol 52. Princeton University Press, New Jersey, USA, p 408

Querejeta JI, Egerton-Warburton LM, Allen MF (2009) Topographic position modulates the mycorrhizal response of oak trees to interannual rainfall variability. Ecology 90:649–662

R Core Team (2017) R: a language and environment for statistical computing. R Foundation for Statistical Computing, Vienna, Austria. https://www.R-project.org/

Rast DM, Hollenstein GO (1977) Architecture of the *Agaricus bisporus* spore wall. Can J Bot 55:2251–2262

Read DJ, Leake JR, Pérez-Moreno J (2004) Mycorrhizal fungi as drivers of ecosystem processes in heathland and boreal forest biomes. Can J Bot 82:1243–1263 https://doi.org/10.1139/b04-123

Roberts DW (2016) labdsv: ordination and multivariate analysis for ecology. R package version 1.8-0. https://CRAN.R-project.org/package=labdsv

Robinson CH (2001) Cold adaptation in Arctic and Antarctic fungi. New Phytol 151:341–353

Ryberg M, Larsson E, Molau U (2009) Ectomycorrhizal diversity on *Dryas octopetala* and *Salix reticulata* in an alpine cliff ecosystem. Arct Antarct Alp Res 41:506–514

Salzer MW, Bunn AG, Graham NE et al (2014) Five millennia of paleotemperature from tree-rings in the Great Basin, USA. Clim Dynam 42:1517–1526

Schaeffer SM, Sharp E, Schimel JP et al (2013) Soil-plant processes in a high Arctic ecosystem, NW Greenland are altered by long-term warming and higher rainfall. Glob Chang Biol 19:3529–3539

Schimel JP, Bennet J (2004) Nitrogen mineralization, challenges of a changing paradigm. J Ecol 85:591–602

Schloss PD, Westcott SL, Ryabin T et al (2009) Introducing mothur: open-source, platform-independent, community-supported software for describing and comparing microbial communities. Appl Environ Microbiol 75:7537–7541

Schoch CL, Seifert KA, Huhndorf S et al (2012) Nuclear ribosomal internal transcribed spacer (ITS) region as a universal DNA barcode marker for fungi. Proc Natl Acad Sci U S A 109:6241–6246

Schöroter D, Cramer W, Leemans R et al (2005) Ecosystem service supply and vulnerability to global change in Europe. Science 310:1333–1337

Selosse M-A, Richard F, He X et al (2006) Mycorrhizal networks: des liaisons dangereuses? Trends Ecol Evol 21:621–628

Semenova TA, Morgado LN, Welker JM et al (2015) Long-term experimental warming alters community composition of ascomycetes in Alaskan moist and dry arctic tundra. Mol Ecol 24:424–437

Sharkhuu A, Plante AF, Enkhmandal O et al (2013) Effects of open-top passive warming chambers on soil respiration in the semi-arid steppe to taiga forest transition zone in northern Mongolia. Biogeochemistry 115:333–348

Sistla SA, Moore JC, Simpson RT et al (2013) Long-term warming restructures Arctic tundra without changing net soil carbon storage. Nature 497:615–618

Sterflinger K, Tesei D, Zakharova K (2012) Fungi in hot and cold deserts with particular reference to microcolonial fungi. Fungal Ecol 5:453–462

Strand AE, Pritchard SG, McCormack ML et al (2008) Irreconcilable differences: fine-root life spans and soil carbon persistence. Science 319:456–458

Sturm M, McFadden JP, Liston GE et al (2001) Snow–shrub interactions in Arctic tundra: a hypothesis with climatic implications. J Clim 14:336–345

Sturm M, Douglas T, Racine C et al (2005) Changing snow and shrub conditions affect albedo with global implications. J Geophys Res 110:G01004
Tape K, Sturm M, Racine C (2006) The evidence for shrub expansion in northern Alaska and the pan-Arctic. Glob Chang Biol 12:686–702
Tape KD, Hallinger M, Welker JM, Ruess RW (2012) Landscape Heterogeneity of Shrub Expansion in Arctic Alaska. Ecosystems 15:711–724
Tarnocai C, Canadell JG, Schuur E et al (2009) Soil organic carbon pools in the northern circumpolar permafrost region. Global Biogeochem Cy 23:GB2023
Taylor DL, Hollingsworth TN, McFarland JW et al (2014) A first comprehensive census of fungi in soil reveals both hyperdiversity and fine-scale niche partitioning. Ecol Monogr 84:3–20
Tedersoo L, May TW, Smith ME (2010) Ectomycorrhizal lifestyle in fungi: global diversity, distribution, and evolution of phylogenetic lineages. Mycorrhiza 20:217–263
Tedersoo L, Naadel T, Bahram M et al (2012) Enzymatic activities and stable isotope patterns of ectomycorrhizal fungi in relation to phylogeny and exploration types in an afrotropical rain forest. New Phytol 195:832–843
Tedersoo L, Smith ME (2013) Lineages of ectomycorrhizal fungi revisited: foraging strategies and novel lineages revealed by sequences from belowground. Fungal Biol Rev 27:83–99
Timling I, Dahlberg A, Walker DA et al (2012) Distribution and drivers of ectomycorrhizal fungal communities across the north American Arctic. Ecosphere 3:1–25
Toljander JF, Eberhardt U, Toljander YK et al (2006) Species composition of an ectomycorrhizal fungal community along a local nutrient gradient in a boreal forest. New Phytol 170:873–883
Twieg BD, Durall DM, Simard SW et al (2009) Influence of soil nutrients on ectomycorrhizal communities in a chronosequence of mixed temperate forests. Mycorrhiza 19:305–316
Urban A, Weib M, Bauer R (2003) Ectomycorrhizas involving sebacinoid mycobionts. Mycol Res 107:3–14
Väre H, Vestberg M, Eurola S (1992) Mycorrhiza and root-associated fungi in Spitsbergen. Mycorrhiza 1:93–104
Wahren CHA, Walker MD, Bret-Harte MS (2005) Vegetation responses in Alaskan arctic tundra after 8 years of a summer warming and winter snow manipulation experiment. Glob Chang Biol 11:537–552
Walker MD, Walker DA, Welker JM et al (1999) Long-term experimental manipulation of winter snow regime and summer temperature in arctic and alpine tundra. Hydrol Process 13:2315–2330
Walker DA, Raynolds MK, Daniëls FJA et al (2005) The circumpolar Arctic vegetation map. J Veg Sci 16:267–282
Walker MD, Wahren CH, Hollister RD et al (2006) Plant community responses to experimental warming across the tundra biome. Proc Natl Acad Sci U S A 103:1342–1346
Welker JM, Fahnestock JT, Jones MH (2000) Annual CO2 flux in dry and moist Arctic tundra: Field responses to increases in summer temperatures and winter snow depth. Clim Change 44:139–150
Welker JM, Molau U, Parsons AN et al (1997) Response of *Dryas octopetala* to ITEX manipulations: a synthesis with circumpolar comparisons. Glob Chang Biol 3:61–73
Welker JM, Fahnestock JT, Sullivan PF et al (2005) Leaf mineral nutrition of arctic plants in response to long-term warming and deeper snow in N. Alaska Oikos 109:167–177
White TM, Bruns T, Lee S et al (1990) Amplification and direct sequencing of fungal ribosomal RNA for phylogenetics. In: Innis MA, Gelfand DH et al (eds) PCR protocols: a guide to methods and applications. Academic Press, San Diego, USA, pp 315–321
Zhang W, Miller PA, Smith B et al (2013) Tundra shrubification and tree-line advance amplify arctic climate warming: results from an individual-based dynamic vegetation model. Environ Res Lett 8:034023
Carrasco-Hernandez V, Pérez-Moreno J, Espinosa-Hernández V et al (2011) Nutrient contents and inoculation with edible ectomycorrhizal fungi on two neotropical pines. Rev Chil Hist Nat 84: 83–96. http://dx.doi.org/10.4067/S0716-078X2011000100006

Index

J. Pérez-Moreno et al. (eds.), *Mushrooms, Humans and Nature in a Changing World*, https://doi.org/10.1007/978-3-030-37378-8

C

F

The manufacturer's authorised representative in the EU is Springer Nature Customer Service Centre GmbH, Europaplatz 3, 69115 Heidelberg, Germany. If you have any concerns regarding our products, please contact ProductSafety@springernature.com

Printed and bound by CPI Group (UK) Ltd, Croydon, CR0 4YY
05/10/2026
02228120-0017